VLSI CIRCUIT DESIGN

By:

DEEPAK GARG
B.Tech (ECE), M.Tech(VLSI), AMUACEE
Assistant Professor
Department of Electronics and Communication Engineering
IIMT Engineering College
MEERUT (U.P.)

S.K. KATARIA & SONS®
Publisher of Engineering Books
4885/109, Prakash Mahal, Dr. Subhash Bhargav Lane,
Opposite Delhi Medical Association, Daryaganj, New Delhi–110002
Phone: +91-11-43551243; Mobile+91-9871775858, 7678425209
e-mail: katariabook@gmail.com, katariabooks@yahoo.com
Website: www.skkatariaandsons.com
(e-books, subscriptions and bulk buying also available on our website)

Published by:

S.K. Kataria & Sons®

4885/109, Prakash Mahal, Dr. Subhash Bhargav Lane,
Opposite Delhi Medical Association, Daryaganj, New Delhi-110002
Phone: +91-11-43551243; Mobile: +91-9871775858, 7678425209
e-mail: katariabook@gmail.com, katariabooks@yahoo.com

First Edition: 2012; Second Edition: 2013, Reprint: 2026

Price : ₹ 495/-

Laser Type Setting by:
SARA ASSIGNMENTS, Delhi-110092.

Printed at:
A.S. OFFSET PRINTERS, Delhi-110051.

Preface to the Second Edition

I feel encouraged by the widespread response from teachers and students alike to the first edition. I am presenting this second edition, thoroughly revised and enlarged, to my readers in all humbleness. All possible efforts have been made to enhance further the usefulness of the book. The feedback received from different sources has been incorporated.

The highlights of this revised edition are:

(i) The subject contents has been developed, refined and restructured at several points.
(ii) Several new problems have also been added at the end of various chapters for the benefit of students.
(iii) Every page of the book has been read very carefully so as to improve its quality.
(iv) Some points to be remembered are also added at the end of various chapters for the benefit of students.
(v) Short questions with answers are also added in last of every chapter for the benefit of students.

I am grateful to all who have helped me directly and/ or indirectly in preparing this revised edition. I firmly believe that there is always scope for improvement and accordingly I shall look forward to receive suggestions, (which shall be thankfully acknowledged) for further enriching the quality of the text.

Deepak Garg
IIMT, Meerut

P Preface to the First Edition

I feel great pleasure in bringing out the book, 'VLSI CIRCUIT DESIGN' for students of Uttarakhand Technical University, Dehradun. This book On "VLSI CIRCUIT DESIGN" has been written in accordance with the latest syllabi, to meet the needs of the students.

This book deals with actual design applications rather than the technology of VLSI Systems. This book is written in a simple and lucid manner with the topics arranged systematically to enable the reader to get enough knowledge of the subject.

The whole text has been logically organized and spreader over 10 chapters.

Chapter-1: Discuss about the Review of MOS Transistor.

Chapter-2: Explain the different types of fabrication for MOSFET.

Chapter-3: Discuss about the design rules in VLSI.

Chapter-4: Discuss about the CMOS Inverters with different load and characteristics.

Chapter-5: Discuss about the Low-power CMOS logic circuits.

Chapter-6: Explain the Combinational MOS logic circuits. The design technique includes both logic and VLSI design in n-MOS and CMOS technologies.

Chapter-7: Explain the Dynamic logic circuits in detail.

Chapter-8: Explain the Sequential MOS logic circuits in detail.

Chapter-9: Discuss about the Semiconductor memories and Array Structure.

Chapter-10: Discuss about the concept of Testability in logic circuits.

As an author I hope that with these unique features the book will fulfill the genuine requirements of the student community. A few unsolved problems and review question have been also added at the end of each chapter for practice. In last, I have tried my best to cover all the topics in detail.

As an author I have tried to put my best in the book and the suggestions for the improvement of this book are cordially invited from my fellow Professors and students.

Deepak Garg
IIMT, Meerut

Acknowledgment

I present my humble gratitude to Prof. (Dr.) S. Sarkar (Former Professor, ECE Department, IIT Roorkee), Dr. B. K. Kaushik (Asstt. Professor, ECE Department, IIT Roorkee) without his inspiration this book was not possible.

I especially grateful to Prof. (Dr.) R. P. Agarwal (VC, Shobhit University, Meerut), Dr. D. K. Srivastava (Reader, BIET, Jhansi), Mr. Mayank Kumar Rai (Asstt. Professor, ECE Department, Thapar University, Patiala) for their time to time much needed valuable guidance.

I would like to express my sincere thanks to Mr. Yogesh Mohanji Gupta (Chairman, IIMT Group of Colleges, Meerut), Mr. Abhinav Aggarwal (Secretary General, IIMT Group of Colleges, Meerut) and Mr. Mayank Aggarwal (Managing Director, IIMT Group of Colleges, Meerut) for thier kind support and encouragement.

I would like to express my sincere thanks to Prof. (Dr.) M. K. Sharma (Director, IIMT Engg. College, Meerut) and Prof. (Dr.) Deepa Sharma (Dy. Director, IIMT Engg. College, Meerut) for thier kind support and encouragement to carry out this important task.

I would like to express my sincere thanks to Dr. J. K. Kushwaha (Asstt. Professor, IIMT, Meerut), Mr. Manoj kumar (Asstt. Professor, G.B. Pant Engg. College, Pauri Garwal), Mr. Balraj Singh (Asstt. Professor, G.B. Pant Engg. College, Pauri Garwal), Mr. Nikhil Gupta (Asstt. Professor, IIMT, Meerut), Mr. Sumit Gupta (Asstt. Professor, IIMT, Meerut), Mr. Shobhit Saxena (Asstt. Professor, COER, Roorkee), Mr. Sudheer Goswami (Asstt. Professor, COER, Roorkee), Mrs. Chanchal Garg (Lecturer, SB-BIT, Meerut), Mr. Ramesh Kumar (Sr. Lecturer, RGEC, Meerut), Mr. Sanjeev Gaba (Asstt. Professor, Graphic Era University, Dheradhun) and all others who are directly or indirectly associated with me and helped in the timely completion of this book.

I deeply express my heartful thanks to our publishers M/s S. K. Kataria & Sons for taking a lot of pains and initiative for bringing out the book in a short space of time.

Deepak Garg
IIMT, Meerut

Dedicated To My Parents
Mrs. Suman Gupta & Mr. Naveen Kumar

Contents

1

Review of MOS Transistor

Chapter

Chapter Overview

1.1 INTRODUCTION

- The development of a technology for integrated circuits (ICs) went hand in hand with the steady improvement in semiconductor materials and discrete components during the 1950s and 1960s. Consequently, silicon bipolar technology formed the basis for the IC market during the 1970s. As circuit dimensions shrink, the MOSFET (or MOS) has gradually taken over as the major technological platform for silicon ICs. The main reasons are the ease of miniaturization and high yield for MOS compared with bipolar technology. The VLSI circuits the low standby power of complementary MOS (CMOS) gates is a significant advantage compared with integrated bipolar circuits.
- The economic viability of an IOC is in large part affected by the productively that can be brought to bear on the design. This is depends on the efficiency with which the design may be converted from concept to architecture, to logic and memory, to circuit and hence a physical layout.
- The electronics industry has achieved a phenomenal growth over the last few decades, mainly due to the rapid advances in large scale integration technologies and system design applications. With the advent of very large scale integration (VLSI) designs, the number of applications of integrated circuits (ICs) in high performance computing, controls, telecommunications, image and video processing, and consumer electronics has been rising at a very fast pace. The current cutting-edge technologies such a high resolution and low bit-rate video and cellular communications provide the end-users a marvelous amount of applications, processing power and portability. This trend is expected to grow rapidly, with very important implications on VLSI design and systems design.

1.2 EVOLUTION OF VLSI

- Fig. 1.1 gives an overview of the prominent trends in information technologies over the next decade. The current leading edge technologies (such as low bit-rate video and cellular communications) already provide the end-users a certain amount of processing power and portability. This trend is expected to continue, with very important implications for VLSI and systems design. The important characteristic is that the information services tend to become more personalized, which means that the information processing devices must be more intelligent and also be portable to allow more mobility.

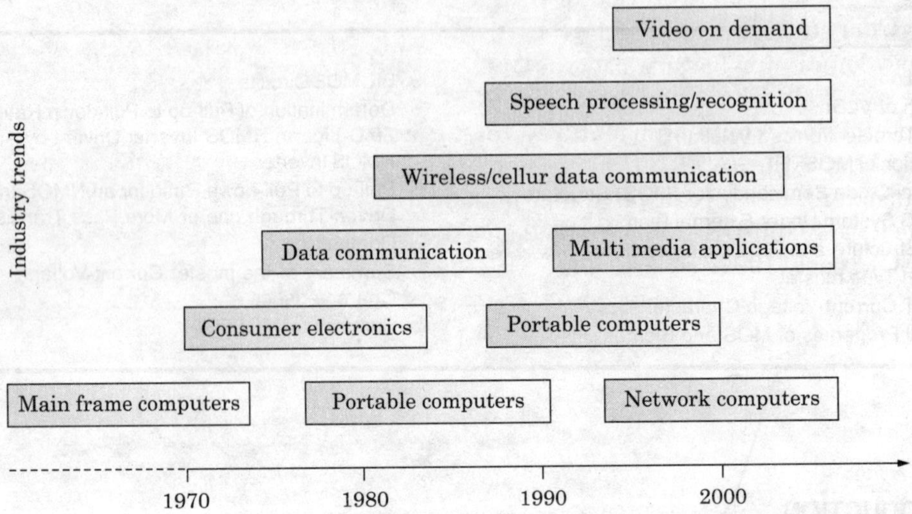

Fig. 1.1 Prominent "driving" trends in information service technologies.

- More complex functions are required in various data processing and telecommunications devices; the need to integrate these functions in a small package is also increasing. The level of integration as measured by the number of logic gates in a monolithic chip has been steadily rising for almost three decades, mainly due to the rapid progress in processing technology and interconnect technology. Table 1.1 shows the evolution of logic complexity in integrated circuits over the last three decades, and marks the *milestones* of each era.

Table 1.1 Evolution of logic complexity in integrated circuits

ERA	DATE	COMPLEXITY (# Of Logic blocks on a Chip)
Single transistor	1958	< 1
Unit logic (one gate)	1960	1
Multi-function	1962	2–4
Complex function	1964	5–20
Medium Scale Integration (MSI)	1967	20–200
Large Scale Integration (LSI)	1972	200–2,000
Very Large Scale Integration (VLSI)	1978	2,000–20,000
Ultra Large Scale Integration (ULSI)	1989	20,000-?

The monolithic integration of a large number of functions on a single chip usually provides:

- ⇒ Less area/volume and therefore, compactness
- ⇒ Less power consumption
- ⇒ Less testing requirements at system level
- ⇒ Higher reliability, mainly due to improved on-chip interconnects
- ⇒ Higher speed, due to significantly reduced interconnection length

- Fig. 1.2 shows the evolution of the minimum feature size of transistors in integrated circuits, starting from the late 1970s. In 1980, at the beginning of the VLSI era, the typical minimum feature size was 2 pm, and a feature size of 0.3 gm was expected around the year 2000. The actual development of the technology, however, has far exceeded these expectations. A minimum feature size of 0.25 gm was achieved by 1995. The first 64-Mbit DRAM and the *INTEL Pentium* microprocessor chip containing more than 3 million transistors were already available by 1994, pushing the envelope of integration density. The first 4-Gbit DRAM based on 0.15 gm manufacturing technology was announced by NEC in early 1997.

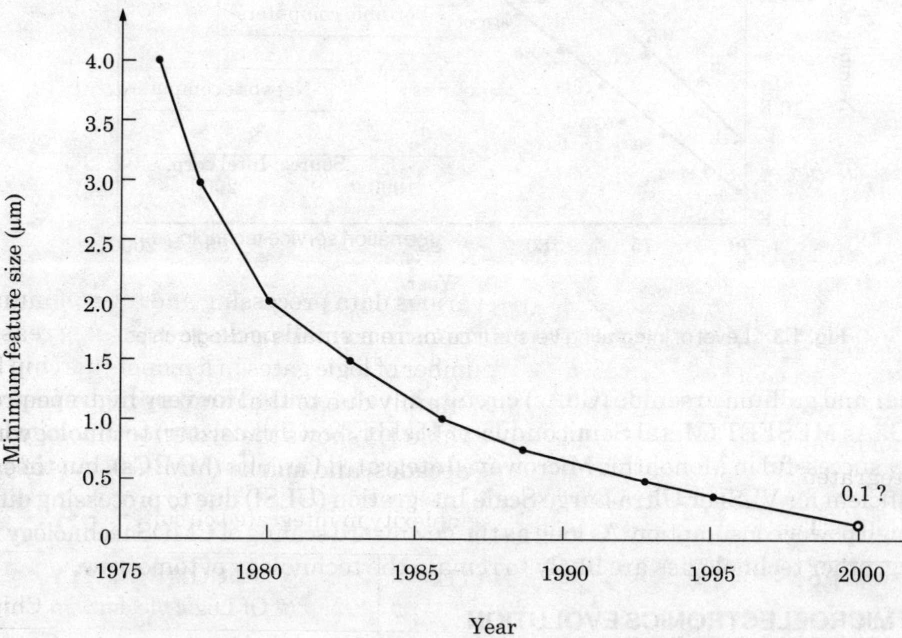

Fig. 1.2 Evolution of minimum feature size in integrated circuits over time.

- Fig. 1.3 shows the level of integration over time for memory and logic chips, starting in 1970. The number of transistors per chip has continued to increase at an exponential rate over the last three decades, effectively confirming Gordon Moore's prediction on the growth rate of chip complexity, which was made in the early 1960s (Moore's Law). Memory circuits are highly regular, and thus more cells can be integrated with much less area for interconnects. This has also been one of the main reasons why the *rate of increase* of chip complexity (transistor count per chip) is consistently higher for memory circuits.

- Digital CMOS (Complementary Metal Oxide Semiconductor) integrated circuits (ICs) have been the driving 'force behind Very Large Scale Integration (VLSI) for high-performance computing and other scientific and engineering applications. The demand for digital CMOS ICs will be continually strong due to salient features such as low power, reliable performance, circuit techniques for high speed such as using dynamic circuits, and ongoing improvements in processing technology.

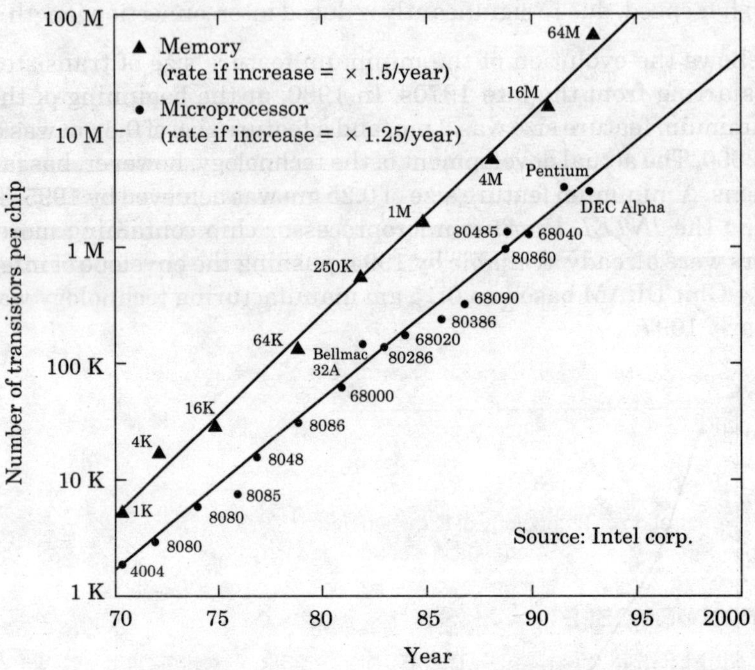

Fig. 1.3 Level of integrating versus time for memory chips and logic chips.

- Bipolar and gallium arsenide (GaAs) circuits have been used for very high speed circuits. The GaAs MESFET (Metal Semiconductor Field Effect Transistor) technology has been highly successful in Monolithic Microwave Integrated Circuits (MMICs), but they are still not efficient for VLSI or Ultra Large Scale Integration (ULSI) due to processing difficulties and high power consumption. As long as the downward scaling of CMOS technology remains strong, other technologies are likely to remain the technology of tomorrow.

1.3 ERA OF MICROELECTRONICS EVOLUTION

- The evolution of MOS technology has followed the famous Moore's law that predicts a steady decrease in gate length. Fig. 1.4 shows four periods of change in the electronics Industry in United States. Before 1860, only mechanical devices invented but in 1826 after the development of resistor and ohm's law, electro-mechanical design era started. The development of the vacuum tube in 1906 and the invention of transistor in 1947 opened the field of electronic circuit designs. The vast majority of present day electronics in the result of the invention of transistor in 1947. The invention of transistor by Willium B. Shockely, Walter H. Brattain and John Bardeem of Bell Telephone Laboratories was followed by the development of IC. The development of ICs in 1958 led to a new generation of logic families.

Since 1975, the beginning of VLSI, the frontier has moved to system organization of ICs and the associated Software designs.

- Bipolar technology has also benefited from the progress in lithography and is currently fabricated using deep UV tools with feature sizes close to 100 nm. The scaling has led to a sanification performance improvement and is further illustrated in Fig. 1.4.

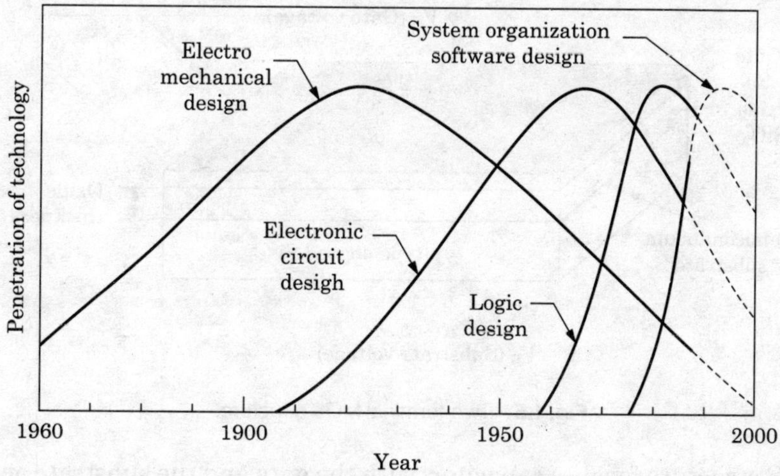

Fig. 1.4 Penetration of Technology of to the industrial output versus year for four periods of change in electronic industry.

1.4 INTRODUCTION OF MOSFET

- The MOS Field Effect Transistor (MOSFET) is the most basic device of VLSI integrated circuits. The FET devices is referred as unipolar transistor as current or *p*-cahnnel FET.
- As compared to the bipolar junction transistor (BJT), the MOS transistor occupies a relatively smaller silicon area, and its fabrication involves fewer processing steps.
- These technological advantages, together with the relative simplicity of MOSFET operation, have helped make the MOS transistor the most widely used switching device in LSI and VLSI circuits.
- In this chapter, we will examine the basic structure and the electrical behavior of nMOS (*n*-channel MOS) and pMOS (*p*-channel MOS) devices. We will consider the effects of external bias conditions on charge distribution in the MOS system and on the conductance of free carriers. It will be shown that, in field effect devices, the current flow are controlled by externally applied electric fields, and that the operation depends only on the majority carrier flow between two device terminals. The current-voltage characteristics of MOS transistors will be examined in detail, including physical limitations imposed by small device geometries and various second-order effects observed in MOSFETs.
- The nMOS transistor is used as the primary switching device in virtually all digital circuit applications, whereas the pMOS transistor is used mostly in conjunction with the nMOS device in CMOS circuits. However, the basic operation principles of both nMOS and pMOS transistors are very similar to each other.

1.5 THE METAL OXIDE SEMICONDUCTOR (MOS) STRUCTURE

- The simple two- terminal MOS structure shown in Fig. 1.5. The structure consists of three layers: The metal gate electrode, the insulating oxide (SiO_2) layer, and the *p*-type bulk semiconductor (Si) called the substrate. Here, V_G is the applied gate voltage and V_B is the substrate voltage.

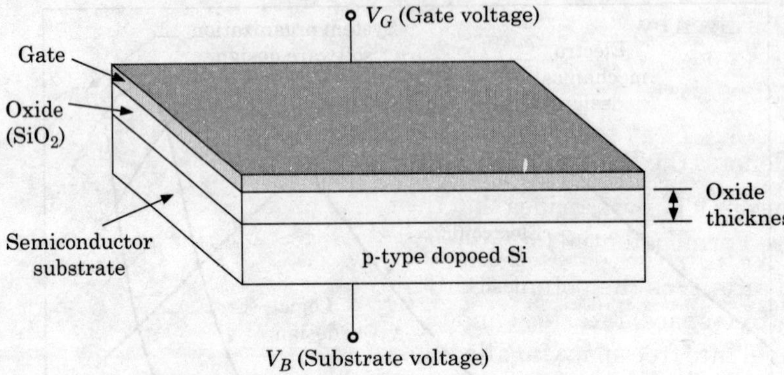

Fig. 1.5 Two-terminal MOS structure.

- The MOS structure forms a capacitor, with the gate and the substrate acting as the two terminals (plates) and the oxide layer as the dielectric. The carrier concentration and its local distribution within the semiconductor substrate can now be manipulated by the external voltages applied to the gate and substrate terminals.
- The energy band diagram of the *p*-type substrate is shown in Fig. 1.6. The band-gap between the conduction band and the valence band for silicon is approximately 1.1 eV. The location of the equilibrium Fermi level E_F within the band-gap is determined by the doping type and the doping concentration in the silicon substrate.

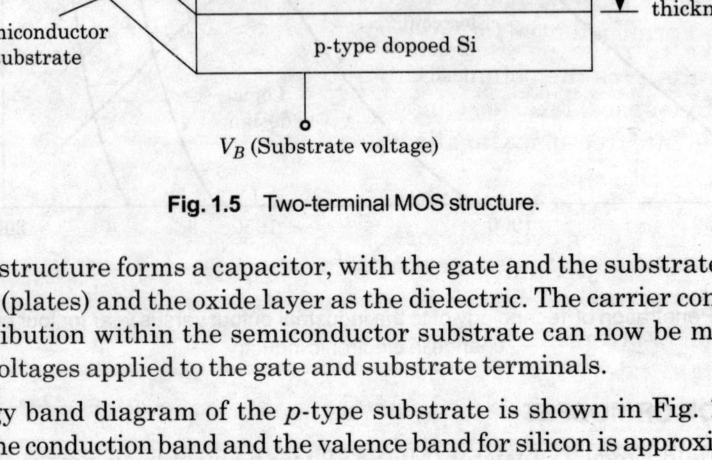

Fig. 1.6 Energy band diagram of a p-type silicon substrate.

- The Fermi potential ϕ_F, which is a function of temperature and doping, denotes the difference between the intrinsic Fermi level E_i, and the Fermi level E_F.

$$\phi_F = \frac{E_F - E_i}{q}$$

For a p-type semiconductor, the Fermi potential can be approximated by

$$\phi_{Fp} = \frac{kT}{q} \ln \frac{n_i}{N_A} \qquad ...(1.2)$$

Where as for an n-type semiconductor (doped with a donor concentration N_D), the Fermi potential is given by

$$\phi_{Fn} = \frac{kT}{q} \ln \frac{N_D}{n_i} \qquad ...(1.3)$$

Here, k denotes the Boltzmann constant and q denotes the unit (electron) charge.

The equation (1.1) gives an idea that the Fermi potential for n-type material is positive and a negative Fermi potential for p-type material.

- Electron affinity is the potential difference between the conduction band level and the vacuum (free-space) level ($q\chi$) and the energy required for an electron to move from the Fermi level into free space is called the work function $q\,\Phi_S$, and is given by

$$q\,\Phi_S = q\chi + (E_c - E_F) \qquad ...(1.4)$$

- The insulating silicon dioxide layer between the silicon substrate and the gate has a large band-gap of about 8 eV and an electron affinity of about 0.95 eV. The work function $q\,\Phi_M$ of an aluminum gate is about 1.1 eV. The energy band diagrams of metal, oxide, and semiconductor layers in a MOS system as three separate components is shown in Fig. 1.7.

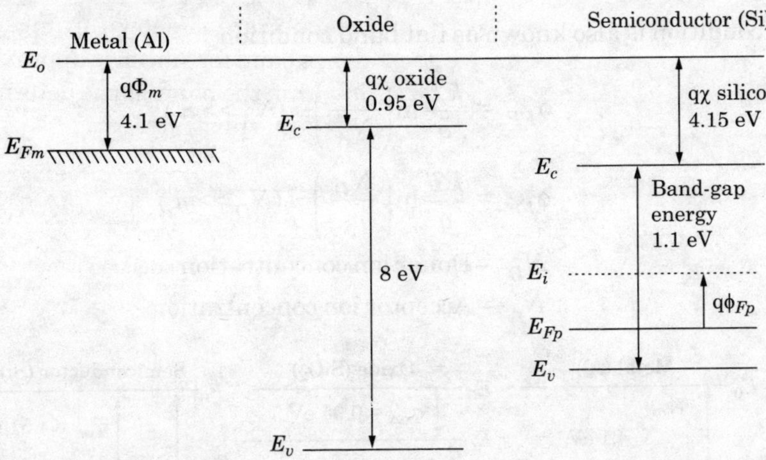

Fig. 1.7 Energy band diagrams of the components that make up the MOS system.

- The Fermi levels of all three materials must line up, as they form the MOS capacitor shown in Fig. 1.7. Also because of the work-function difference between the metal and the semiconductor, a voltage drop occurs across the MOS system. This built-in voltage drop occurs across the oxide layer and the silicon-oxide interface. The rest of the voltage drop (potential difference) occurs at the silicon surface next to the silicon-oxide interface, forcing the energy bands of silicon to bend in this region. The resulting combined energy band diagram of the MOS system is shown in Fig. 1.8. From the Fig. 1.8 we can easily observe that the surface Fermi level moves closer to the intrinsic Fermi (mid-gap) level. The Fermi potential at the surface is also called surface potential Φ_S and is smaller in magnitude than the bulk Fermi potential Φ_F.

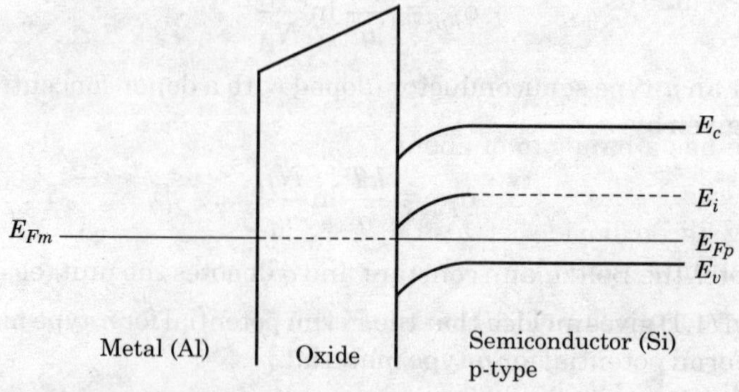

Fig. 1.8 Energy band diagram of the combined MOS system.

1.5.1 Flat Band Condition

1. The energy band is flat (Flat band condition) when there is no applied voltage.
2. The only charges that exist in the diode under any biasing conditions are those in the semiconductor and these with equal but opposite sign on the metal surface adjacent to the oxide.
3. There is no carrier transport through the oxide under direct current biasing conditions or resistivity of the oxide is infinite.

The above condition is also known as flat band condition

where, $$\phi_{FP} = \frac{kT}{q}\ln\left(\frac{n_i}{N_A}\right) \quad (N_A >> n_i)$$

and $$\phi_{Fn} = \frac{kT}{q}\ln\left(\frac{N_D}{n_i}\right) \quad (N_D >> n_i)$$

where, N_D = Donor ion concentration

N_A = Acceptor ion concentration

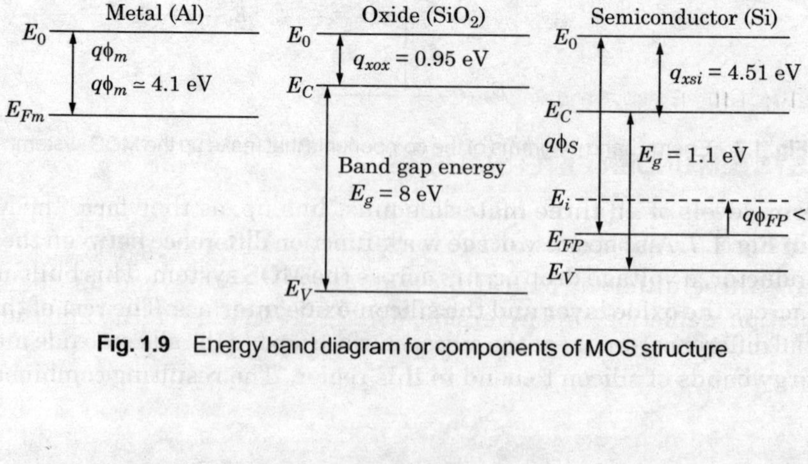

Fig. 1.9 Energy band diagram for components of MOS structure

In the semiconductor, the bottom of the conduction band E_C is located at a fixed energy separation from E_0. This energy difference between E_0 and E_C in the semiconductor is labeled as electron affinity and denoted by greek letter χ (chi). The affinity is therefore the energy needed to move an electron from the bottom of the conduction band and place it at rest outside the solid. This work function can be represented as

$$q\phi_s = q\chi + [E_C - E_F]$$

1. The substrate has a bandgap of about 1.1 eV (for silicon) and electron affinity of about 4.15 eV.
2. The insulating silicon dioxide between silicon and metal gate as large bandgap of about 8 eV and electron affinity of 0.95 V.
3. The work function of a metal (*Al*) is about 4.1 eV.

The flat band voltage V_{FB} is obtained when the applied gate voltage equals the work function difference between the gate metal and the semiconductor. If there is also a fixed charge in the oxide and or at the oxide-silicon interface, the expression for the flat band voltage must be modified accordingly.

Flat band voltage can be given as,

$$V_{FB} = \phi_m - \phi_s \text{ Volts} \quad \text{or} \quad V_{FB} = \phi_{ms} - \frac{Q_{OX}}{C_{OX}}$$

$\therefore$ Built in potential V_{bi} can be as,

$$V_{bi} = qV_{FB} = q\phi_m - q\phi_s \text{ Volts}$$

$\phi_s > \phi_m$ or $V_{bi} < 0$, For *p*-type silicon

$\phi_s < \phi_m$ or $V_{bi} > 0$, For *n*-type silicon

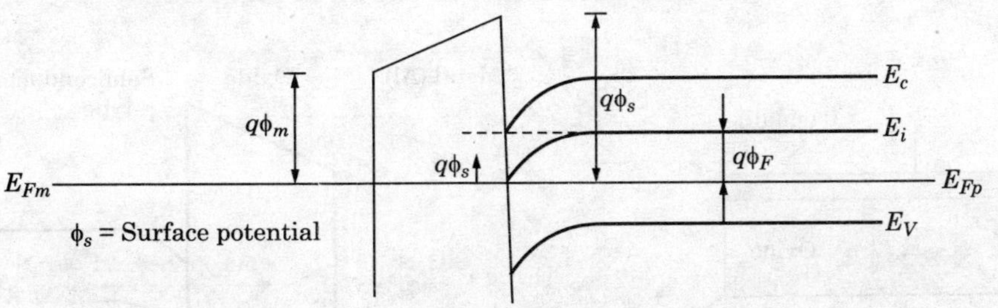

Fig. 1.10 Energy band diagram for MOS structure showing the band bending

1.6 THE MOS SYSTEM UNDER EXTERNAL BIAS (UTU-2012)

- The electrical behavior of the MOS structure under externally applied bias voltages is changed. Here, the substrate voltage is assumed to be zero ($V_B = 0$).
- There are three different operating regions can be observed for the MOS system: accumulation, depletion, and inversion, depending on the polarity and the magnitude of V_G.

Case-I

- If a negative voltage V_G is applied than the holes in the p-type substrate are attracted to the semiconductor-oxide interface.
- The majority carriers are attached towards the gate and the minority carriers are pushed into the substrate due to the negative voltage. This condition is called carrier **accumulation** on the surface (Fig. 1.11). Note that in this case, the oxide electric field is directed towards the gate electrode. The negative surface potential also causes the energy bands to bend upward near the surface.

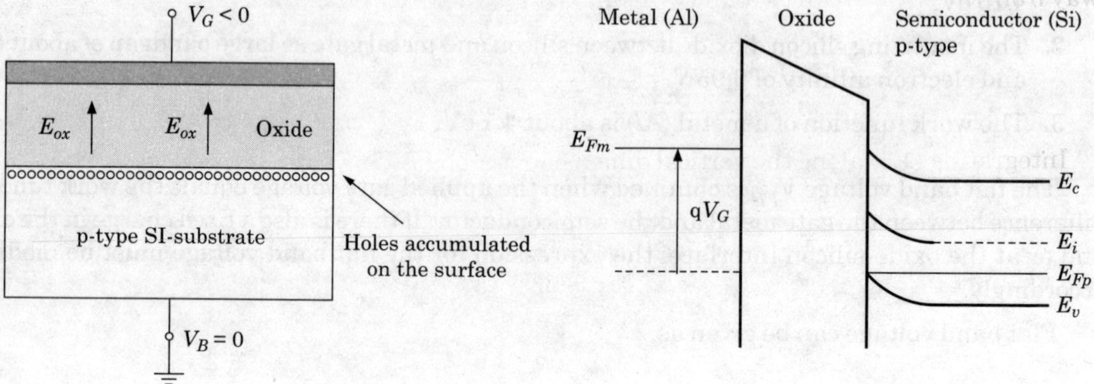

Fig. 1.11 The cross-sectional view and the energy band diagram of the MOS structure operating in accumulation region.

Case-II

- When we apply a small positive gate bias V_G to the gate electrode. The oxide electric field will be directed towards the substrate and the positive surface potential causes the energy bands to bend downward near the surface as shown in Fig. 1.12.

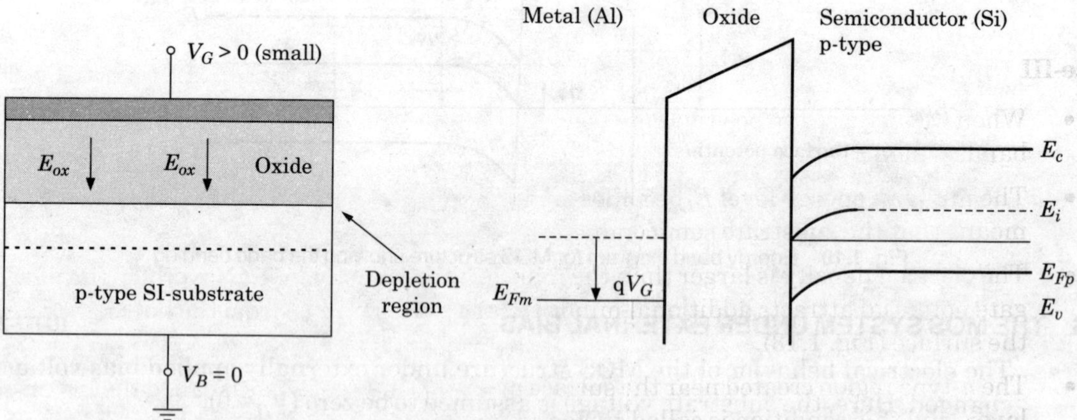

Fig. 1.12 The cross-sectional view and the energy band diagram of the MOS structure operating in depletion mode, under small gate bias.

- The majority carriers (the holes in the substrate) will be repelled back into the substrate as a result of the positive gate bias, and these holes will leave negatively charged fixed acceptor ions behind. Thus, a *depletion* region is created near the surface. The thickness x_d of this depletion region depends on the hole concentration and the surface potential Φ_S.

Assume that the mobile hole charge in a thin horizontal layer parallel to the surface is

$$dQ = -q \cdot N_A \cdot dx \qquad ...(1.5)$$

The *change* in surface potential required to displace this charge sheet dQ by a distance X_d away from the surface can be found by using the Poisson equation.

$$d\phi_s = -x \cdot \frac{dQ}{\varepsilon_{Si}} = \frac{q \cdot N_A \cdot x}{\varepsilon_{Si}} dx \qquad ...(1.6)$$

Integrating (1.5) along the vertical dimension (perpendicular to the surface) yields

$$\int_{\phi_F}^{\phi_s} d\phi_s = \int_0^{x_d} \frac{q \cdot N_A \cdot x}{\varepsilon_{Si}} \qquad ...(1.7)$$

$$\phi_s - \phi_F = \frac{q \cdot N_A \cdot x_d^2}{2\,\varepsilon_{Si}} \qquad ...(1.8)$$

Thus, the depth of the depletion region is

$$x_d = \sqrt{\frac{2\,\varepsilon_{Si} \cdot |\,\phi_s - \phi_F\,|}{q \cdot N_A}} \qquad ...(1.9)$$

The depletion region charge density is obtained by

$$Q = -q \cdot N_A \cdot x_d = -\sqrt{2q \cdot N_A \cdot \varepsilon_{Si}\,|\,\phi_s - \phi_F\,|} \qquad ...(1.10)$$

Case-III **(UTU-2012)**

- When we increase the positive gate bias. At this state, the downward bending of the energy bands will increase.
- The mid-gap energy level E_i becomes smaller than the Fermi level E_{FP} on the surface. It means that the substrate semiconductor in this region is *n*-type.
- The electron density is larger than the majority hole density in thin layer. So, the positive gate potential attracts additional minority carriers (electrons) from the bulk substrate to the surface (Fig. 1.13).
- The *n*-type region created near the surface by the positive gate bias is called the **inversion** layer, and this condition is called surface inversion.

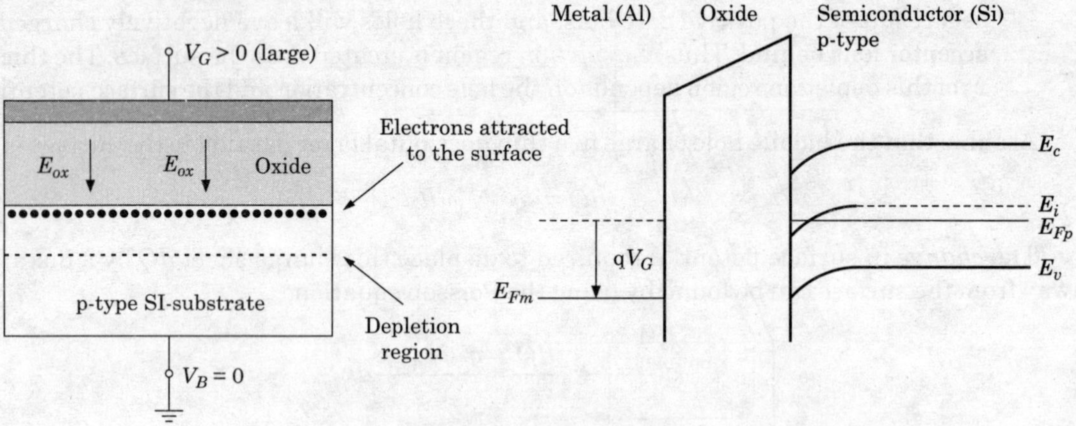

Fig. 1.13 The cross-sectional view and the energy band diagram of the MOS structure in surface inversion, under larger gate bias voltage.

- It will be seen that the thin inversion layer on the surface with a large mobile electron concentration can be utilized for conducting current between two terminals of the MOS transistor.
- When the density of mobile electrons on the surface becomes equal to the density of holes in the bulk (p-type) substrate than the surface is known as inverter.
- If we further increase in the gate voltage leads to an increase of mobile electron concentration on the surface, but not to an increase of the depletion depth.
- The depletion region depth achieved at the onset of surface inversion is also equal to the maximum depletion depth, X_{dm}, which remains constant for higher gate voltages. Using the inversion condition $\phi_s = -\phi_F$, the maximum depletion region depth at the onset of surface inversion can be found from (1.9) as follows:

$$x_{dm} = \sqrt{\frac{2 \cdot \varepsilon_{Si} \cdot |2\phi_F|}{q \cdot N_A}} \qquad ...(1.11)$$

- The creation of a conducting surface inversion layer through externally applied gate bias is an essential phenomenon for current conduction in MOS transistors.

1.7 DEVICE STRUCTURE AND PHYSICAL OPERATION

1.7.1 Device Structure

The basic structure of an n-channel MOSFET is shown in Fig. 1.14. This four terminal device consists of a p-type substrate, in which two n^+ diffusion regions, the drain and the source are formed. The surface of the substrate region between the drain and the source is covered with a thin oxide layer, and the metal gate is deposited on top of this gate dielectric.

The two n^+ regions will be the current-conducting terminals of this device. The device structure is completely symmetrical with respect to the drain and source regions; these two regions will be deferent defined only in conjunction with the applied terminal voltages and the current flow direction.

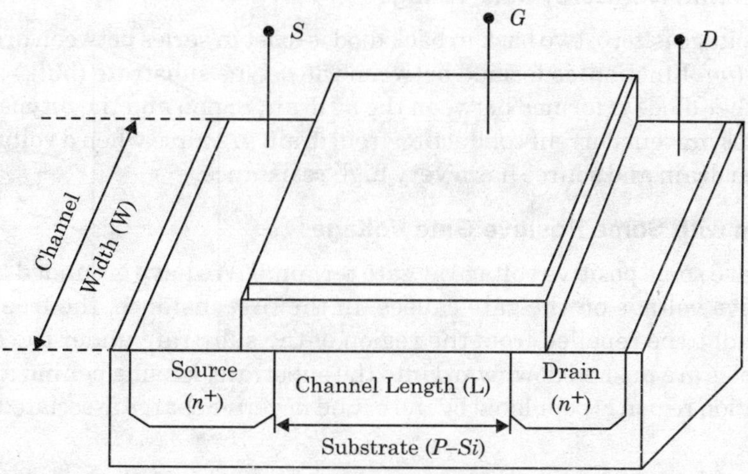

Fig. 1.14 The physical structure of an *n*-channel enhancement-type MOSFET.

A conducting channel will eventually be formed through applied gate voltage in the section of the device between the drain and source diffusion regions. The distance between the drain and source diffusion regions is the channel length *L*, and the lateral extent of the channel (perpendicular to the length dimension) is the channel width *W*.

A MOS transistor which has no conducting channel region at zero gate bias is called *enhancement type MOSFET*. If a conducting channel already exist at zero bias gate, on the other hand, the device is called depletion type MOSFET. In a MOSFET with *P*-type substrate and with n^+ source and drain regions, the channel region to be found on the surface is *n*-type. Thus, such a device with *p*-type substrate is called an *n-channel MOSFET*.

In a MOSFET with *n*-type substrate and with P^+ source and drain regions, On the other hand, the channel is *p*-type and the device is called a *p-channel MOSFET*. Thus four terminals are brought out. The gate terminal (*G*), the source terminal (*S*), the drain (*D*) terminal and the substre or body terminal (*B*). The MOSFET is also Known as **Insulated-gate FET** or IGFET. This name also arises from the physical structure of the device, emphasizing the fact that the gate electrode is electrically insulated from the device body. It is this insulator that causes the current in the gate terminals to be extremely small (of a order of 10^{-15} A).

The circuit symbol's for both *n*-channel and *p*-channel enhancement type MOSFET's are shown in Fig. 1.15.

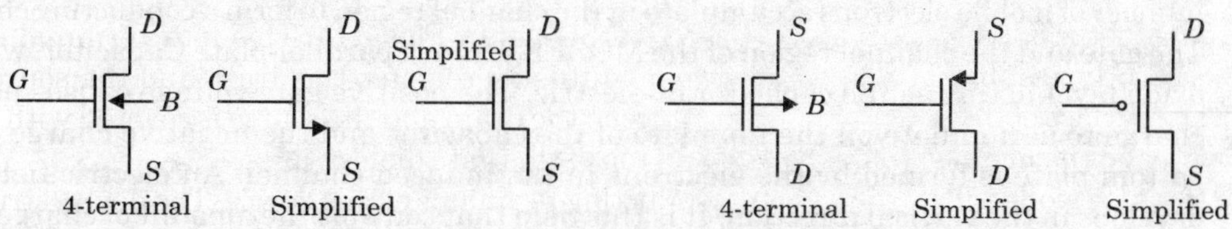

Fig. 1.15 Circuits symbols for *n*-channel and *p*-channel enhancement type MOSFET.

1.7.2 Operation with NO (Zero) Gate Voltage

When the Gate voltage is zero, two back to back diodes exist in series between drain and source. In which one diode (*pn*-Junction) is formed between the *p*-type substrate (bulk) and the n^+ source region and the other diode is formed between the n^+ drain region and the *p*-type substrate. These back to back diodes prevent current conduction from drain to source when a voltage V_{DS} is applied. The path between drain and source has a very high resistance.

1.7.3 Operation with Some Positive Gate Voltage

In this topic we give some positive voltage at gate terminal. We have grounded the source and the drain. The positive voltage on the gate causes, in the first instance, the free holes (which are positively charged) to be repelled from the region of the substrate under the gate (the channel region). These holes are pushed downward into the substrate, leaving behind a carrier-depletion region. The depletion region is populated by the bound negative charge associated with the acceptor atoms (Fig. 1.16).

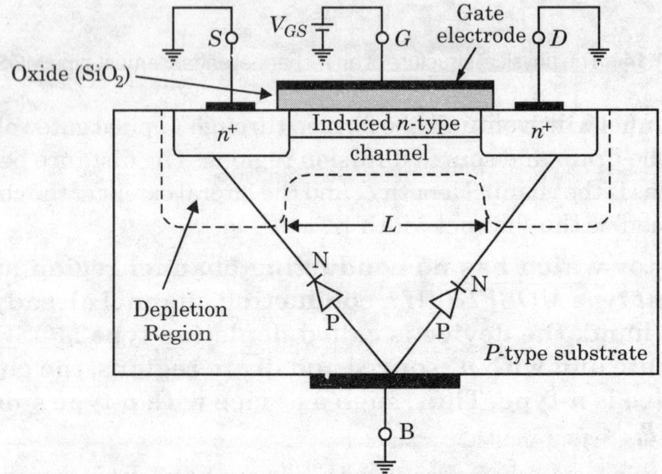

Fig. 1.16 The enhancement-type NMOS transistor with a positive voltage applied to the gate.

- Here, the positive gate voltage attracts electrons from the n^+ source and drain regions into the channel region. Due to the sufficient number of electrons accumulate on the substrate under the gate, an *n*-region is in effect created, connecting the source and drain regions (Fig. 1.16.). When we apply a voltage between drain and source, current flows through this induced *n* region, carried by the mobile electrons. So the MOSFET is called *n*-channel MOSFET. Here, the channel is created by inverting the substrate surface from *p*-type to *n*-type. Hence the induced channel is also called an **inversion layer**.
- The **threshold voltage** (V_T) of a MOSFET is the value of V_{GS} at which a sufficient number of mobile electrons accumulate in the channel region to form a conducting channel.
- The gate and the channel region of the MOSFET from a parallel-plate Capacitor, with the oxide layer acting as the capacitor di-electric. The positive gate voltage causes positive charge to accumulate on the top plate of the capacitor and the negative charge on the bottom plate is formed by the electrons in the induced channel. An electric field thus develops in the vertical direction. It is this field that controls the amount of charge in the channel, and it uses for determining the channel conductivity.

1.7.4 Operation with a Small Drain to Source Voltage V_{DS}

- When having induced a channel, we apply a positive voltage V_{DS} between drain and source as Fig. 1.17. The voltage V_{DS} causes a current I_D to flow through induced *n*-channel at small V_{DS}. Current is carried by free electrons traveling from source to drain.
- The direction of current flow is opposite to that of the flow of negative charge. The magnitude of I_D depends on the density of electrons in the channel, which in turn depends on the magnitude of V_{GS}. For $V_{GS} = V_T$ the channel is just induced and the current conducted is still negligibly small. As V_{GS} exceeds V_T, the increase in the charge carriers in the channel as in increase in the channel depth. The result is a channel of increased conductance or equivalently reduced resistance. The conductance of the channel is proportional to excess gate voltage ($V_{GS} = V_T$), Known as the **overdrive voltage** or **effective voltage**.

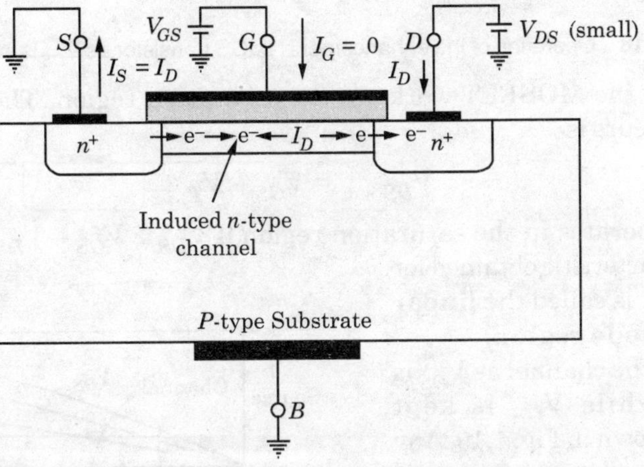

Fig. 1.17 An NMOS transistor with $V_{GS} > V_T$ and with a small V_{DS}.

- The current I_D will be proportional to $V_{GS} = V_T$. The MOSFET is operating as a linear resistance whose value is controlled by V_{GS}.
- According to the above description, The MOSFET to conduct, a channel has to be induced, Then increasing V_{GS} above the threshold voltage V_T a enhances the channel, names enhancement mode operation. The current that leaves the source terminal (I_S) is equal to the current that enters the drain terminal (I_D) and the gate current $I_G = 0$.

1.7.5 Operation as V_{DS} is Increased

Here, we consider a value of V_{GS} which is greater than V_T (threshold voltage). The V_{DS} is a voltage drop across the length of the channel. When we travel along the channel from source to drain, the voltage increase 0 to V_{DS}. So the voltage between the gate and points along the channel decreases from V_{GS} at the source voltage, we find that the channel is no longer of uniform depth as Fig. 1.18. As V_{DS} is increased, the channel becomes more tapered and its resistance increases correspondingly.

- When V_{DS} is increased to the value that reduces the voltage between gate and channel at the drain end to V_T (*i.e.* $V_{GD} = V_T$ or $V_{GS} - V_{DS} = V_T$ or $V_{DS} = V_{GS} - V_T$), the channel depth at the drain end decreases to almost zero, and the channel is said to be **pinched off**. The shape of channel and the current through the channel remains constant at the value reached for $V_{DS} = V_{GS} - V_T$.

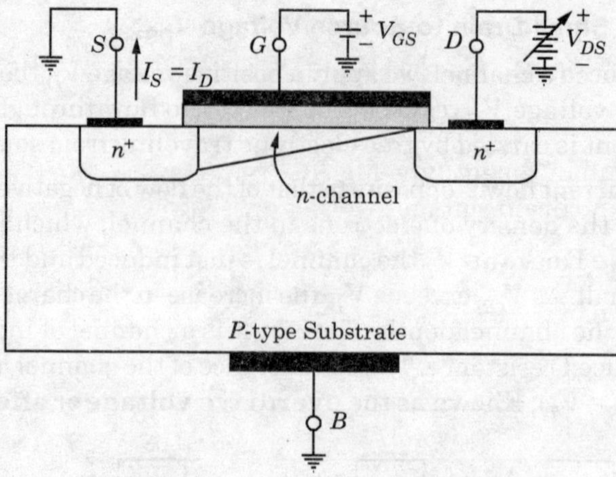

Fig. 1.18 Operation of the enhancement NMOS transistor as V_{DS} is increased.

- At this stage the MOSFET works in the saturation region. The voltage V_{DS} at which saturation occurs is

$$V_{DS\ sat.} = V_{GS} - V_T \quad \text{...(1.12)}$$

- The device operates in the saturation region if $V_{DS} \geq V_{GS} - V_{DS}$ sat. The region of the $I_D - V_{DS}$ characteristic obtained for $V_{DS} < V_{DS}$ sat is called the **linear region or triode region.**
- The graph of the channel as V_{DS} is increased while V_{GS} is kept constant is shown in Fig. 1.19. Any and simply appears across the depletion region surrounding the channel and the n^+ drain region.

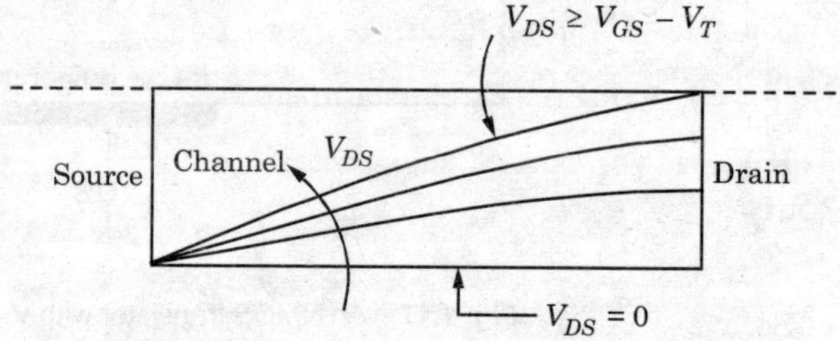

Fig. 1.19 Graph of channel when V_{DS} is increased.

1.7.6 $I_D - V_{DS}$ Characteristics and Current Equation

- Figure 1.20 shows the drain characteristics of MOSFET indicating that there are three distinct regions of operation cut off region, triode region and the saturation region. The saturation region is used if the device is to operate as an amplifier. For operation as a switch, the cut off triode regions are utilized for the operation of switch.

The device is cut off when $V_{DS} < V_T$. To operate the MOSFET in the triode region we must first induce a channel.

$$V_{DS} > V_T \quad \text{... (1.13)}$$

For the channel remains continuous, we can keep V_{DS} small. This is achieved by ensuring that gate to drain voltage is

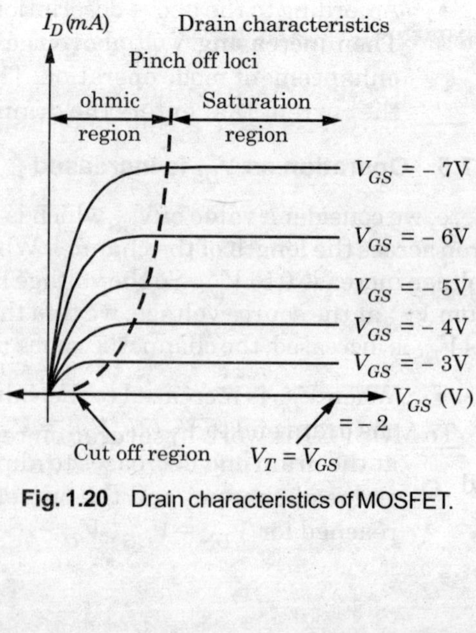

Fig. 1.20 Drain characteristics of MOSFET.

$$V_{DS} > V_T$$

so $$V_{DS} = V_{GS} - V_{DS}$$

$\therefore$ $$V_{GS} - V_{DS} > V_T$$

- The *N*-channel enhancement type MOSFET operates in the triode region when V_{GS} is greater than V_T and drain voltage is lower than the gate voltage by at least V_T volts.

For the linear or Triode Region:

The Drain current is given by

$$\boxed{I_D = \frac{\mu_n C_{ox} W}{L}\left[(V_{GS} - V_T)V_{DS} - \frac{V_{DS}^2}{2}\right]} \quad \text{...(1.14)}$$

$$I_D = \frac{K'_n W}{L}\left[(V_{GS} - V_T) - \frac{V^2{}_{DS}}{2}\right] \quad \text{...(1.15)}$$

where $K'_n = \mu_n C_{ox}$ is the process of transconductance parameter

W = Channel width

L = Channel length

C_{ox} = Gate oxide capacitance

If V_{DS} is very small, then

$$V_{DS}^2 \approx 0$$

So, $$I_D \approx \frac{K'_n W}{L}(V_{GS} - V_T)V_{DS} \quad \text{...(1.16)}$$

$$I_D \approx \frac{K'_n W}{L} V_{OV}\, V_{DS}$$

where V_{OV} is the overdrive voltage, it means that voltage which is required to overdrive the device.

- The linear relationship represents the operation of the MOS transistor as a linear resistance r_{DS} whose value is controlled by V_{GS}

$$r_{DS} = \left.\frac{V_{DS}}{I_D}\right|_{V_{DS}\approx 0} = \frac{V_{DS}}{\frac{K'_n W}{L}(V_{GS} - V_T)V_{DS}}$$

$$r_{DS} = \frac{1}{\frac{K'_n W}{L}(V_{GS} - V_T)} \quad \text{...(1.17)}$$

To MOSFET is work in saturation region, when $V_{GS} \geq V_T$

and $$V_T \geq V_{GS} - V_{DS} \quad \text{...(1.18)}$$

- The N-channel enhancement type MOSFET operates in the saturation region when V_{GS} is greater than and the drain voltage does not fall below the gate voltage by more than threshold voltage V_T volts.

So, $$V_{DS} \geq V_{GS} - V_T$$

Hence, the drain current is

$$I_D = \frac{K'_n W}{2L}(V_{GS} - V_T)^2 \quad ...(1.19)$$

1.7.7 Characteristics of N-channel Enhancement MOSFET

- The transfer characteristics and transfer characteristics of N-channel enhancement MOSFET (EMOSFET) is shown in Fig. 1.21 (*a*). Form the drain characteristics, we known that, for values of V_{GS} less than the threshold voltage level. When $V_{DS(\text{sat})} = V_{GS} - V_T$, the MOSFET have entered into the saturation region of operation.

- From the drain characteristics as the level of V_{as} increased from V_T, the resulting saturation level for I_D also increased. For levels of $V_{GS} > V_T$, the drain current is related to the applied gate to source voltage by the following non-linear relationship.

$$I_D = k_n (V_{GS} - V_T)^2 \quad ...(1.20)$$

or $$k_n = \frac{I_{D(\text{on})}}{(V_{GS(\text{on})} - V_T)^2}$$

where K_n is a constant, $I_{D(\text{on})}$ and $V_{GS(\text{on})}$ are the values for each point (after V_T) on the characteristics of the device.

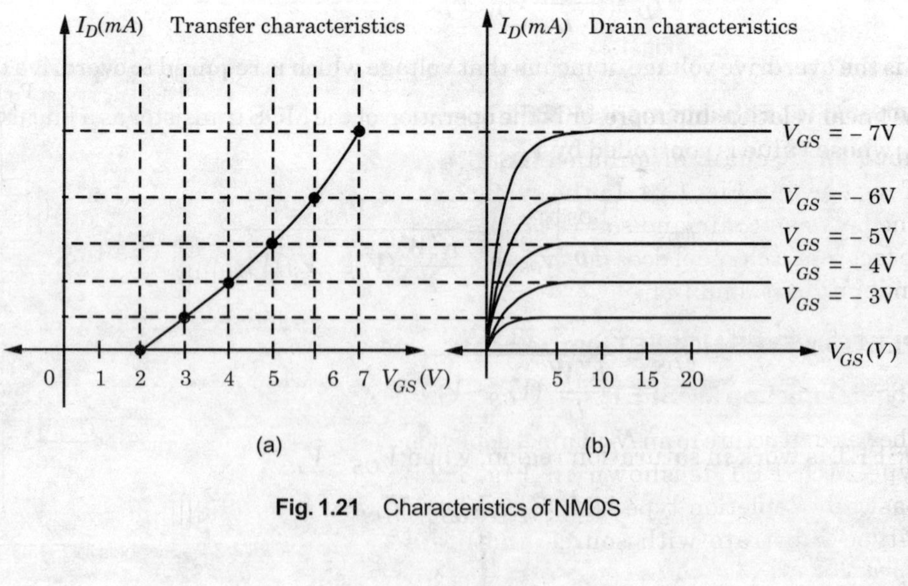

Fig. 1.21 Characteristics of NMOS

1.7.8 P-Channel Enhancement MOSFET (PMOS)

- The construction of *P*-channel enhancement type MOSFET is shown in Fig. 1.22. The construction of *P*-channel enhancement type MOSFET is exactly reverse of the *N*-channel enhancement type MOSFET. In this construction, *N*-type substrate and *P*-doped regions under the source and drain terminals. The terminals remain the same, but all the voltage polarities and the current directions are reversed.

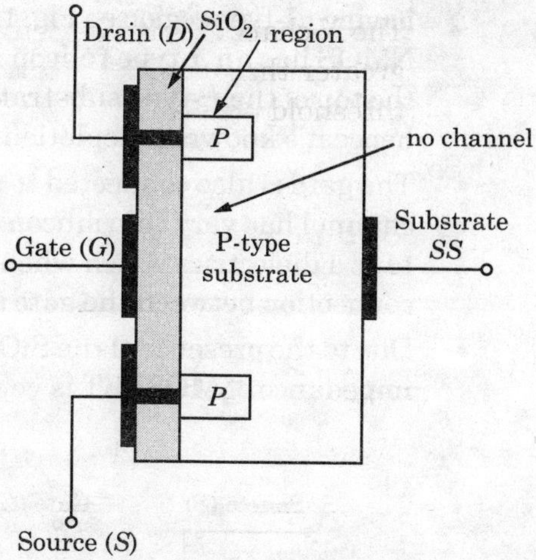

Fig. 1.22 Construction of *p*-channel EMOSFET

- The drain current characteristics of P-type EMOSFET is shown in Fig. 1.23 (*b*). Here, the drain current increases with increase in the negative gate to source voltages (V_{GS}). The transfer characteristics of this type EMOSFET is exactly a mirror image of the transfer characteristics of *n*-channel enhancement MOSFET. The drain current remains zero upto the voltage $V_{GS} < V_T$ and increases as negative V_{GS} becomes than V_T. The transfer characteristics of *P*-channel EMOSFET is shown in Fig. 1.23 (*a*).

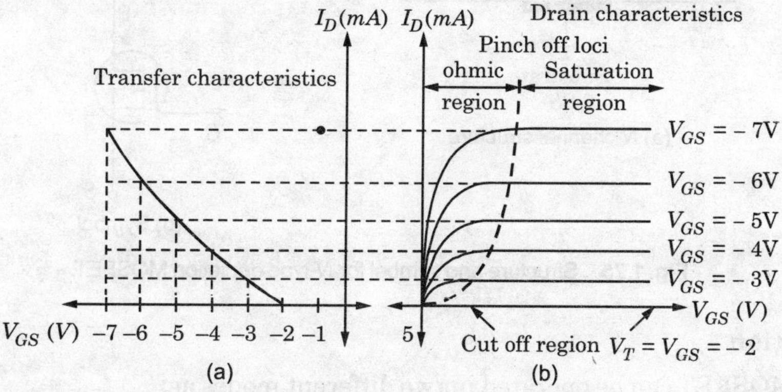

Fig. 1.23 Construction of *p*-channel E-MOSFET

1.7.9 Symbols for Enhancement MOSFET

The symbol for *N*-channel enhancement type MOSFET in shown in Fig. 1.24. In the symbol, the dashed line between drain and source is chosen to reflect the fact that a channel does not exist between the two under no bias condition.

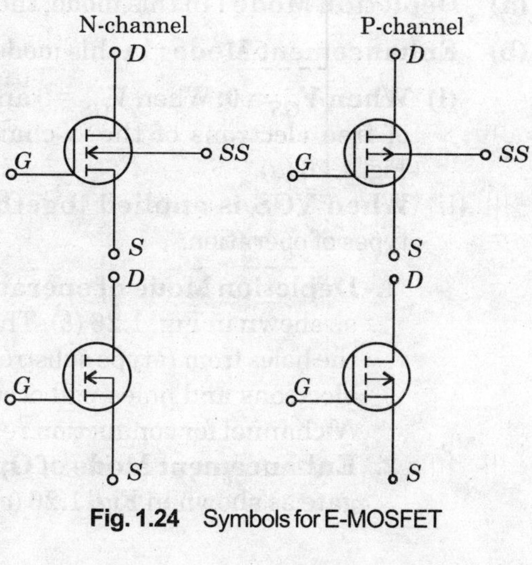

Fig. 1.24 Symbols for E-MOSFET

1.8 DEPLETION TYPE MOSFET (UPTU 2005-06)

1.8.1 *N*-type Depletion MOSFET

- The basic structure to an *N*-channel depletion type MOSFET is shown in Fig. 1.24. Basically depletion type NMOS consist of *P*-type substrate with source and drain having n^+-type region as Fig. 1.25 (*a*). It has a physically implanted channel. Thus depletion-NMOS has an *n*-type relicon region connecting the n^+ source and the n^+ drain regions at the top of the *P*-type substrate. Since the MOSFET has a physically implanted channel, hence it's known as depletion type MOSFET.

- The gate is also connected to a metal contact surface but remains inculcated from the *N*-channel has very thin silicon dioxide layer. SiO_2 is a particular type of insulator referred to as a dielectrics which when exposed to an externally applied field. There is no electrical connection between the gate terminal and the channel of MOSFET.
- Due to the presence of the SiO_2 layer between gate ferininal and *N*-type channel, the input impedance of MOSFET is very high.

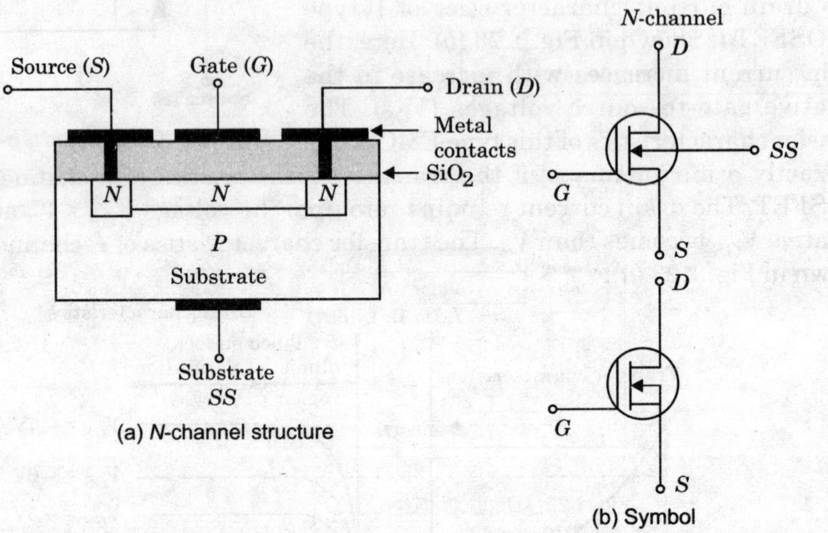

Fig. 1.25 Structure and symbol for *N*-type depletion MOSFET.

1.8.1.1. Operation

The depletion MOSFET can be operated in two different modes as:

(a) Depletion Mode : In this mode, the device operates when V_{GS} is negative.

(b) Enhancement Mode : In this mode, the device operates when V_{GS} is positive.

(i) When $V_{GS} = 0$: When $V_{GS} = 0$ and positive V_{DS} is applied, it will result in an alternation of free electrons of the *N*-channel at the drain. Hence drain current will flow as Fig. 1.26 (*a*).

(ii) When VGS is applied together with the applied V_{DS} : This give results in two types of operation:

1. **Depletion Mode of operation ($V_{GS} < 0$):** When negative V_{GS} is applied at the gate as shown in Fig. 1.26 (*b*). Then electrons towards the *P*-type substrate and attracts the holes from *P*-type substrate. Depending on negative V_{GS}, a recombination between electrons and holes will occur that will reduce the number of free electrons in the *N*-channel for conduction resulting in reduced drain current.
2. **Enhancement Mode of Operation ($V_{GS} > 0$) :** When positive V_{GS} is applied at the gate as shown in Fig. 1.26 (*c*), it will create a positive potential at the gate resulting

in drawing minority carriers (e^-) form the p-type substrate due to reverse leakage and establish new carriers through three collisions resulting between accelerating particles. As the number of carriers are increased, the drain current will increase at a rapid rate.

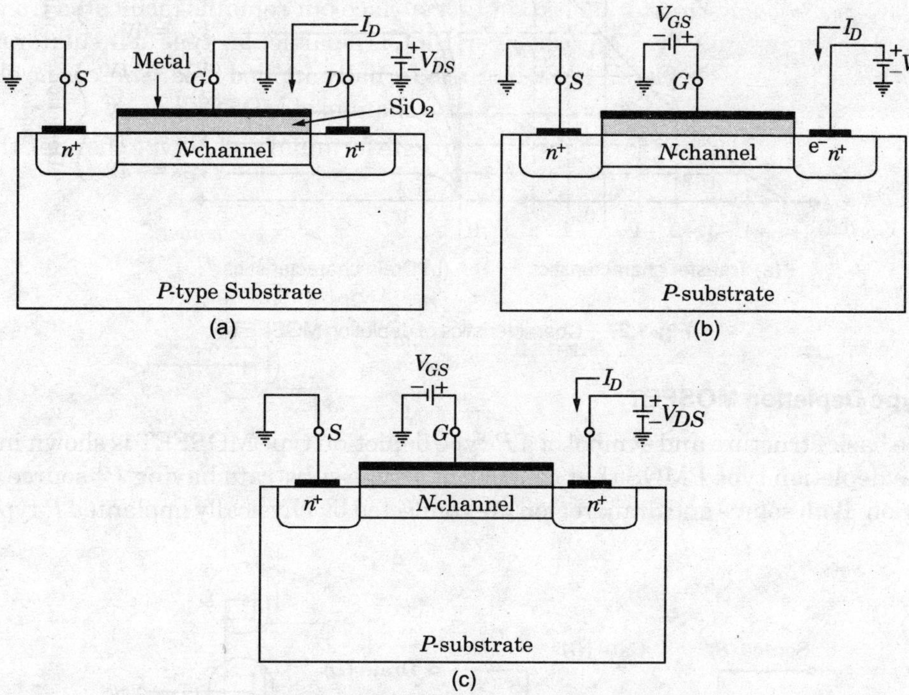

Fig. 1.26 Depection type NMOS operation.

1.8.1.2 Characteristics of Depletion NMOS

1. The drain and transfer characteristics for an N-channel depletion MOSFET is shown in Fig. 1.27. It may be noted that the region AB of the characteristics is similar to that of JFET. But the curve extends for the positive values of gate to source voltage V_{GS} also. The value of I_{DSS} represents the current from drain to source with $V_{as} = 0$, it is given by

$$I_D = I_{DSS}\left(1 - \frac{V_{as}}{V_{GS\,(off)}}\right)^2 \qquad \text{...(1.21)}$$

2. The drain characteristics having plotted for both negative and positive values of V_{as}. When $V_{GS} \geq 0$, the MOSFET operates in the depletion mode. The only difference between the JFET and DMOSFET is that JFET does not operate for positive values of V_{GS}.
3. The $V_{GS\,(off)}$ is the gate source voltage where the channel is completely cutoff and the drain current becomes zero. As the reverse gate source voltage is increased, the cross sectional area of channel decrease. Due to this I_D is decreases. At some V_{GS}, the depetion layer extended completely across the channel. At this position I_D becomes zero. The V_G at which the channel is cutoff is called cutoff voltage. It is interesting $V_{GS(off)}$ will always have the same magnitude values as V_P.

So

$$V_{GS(off)} = -V_P \qquad \text{... (1.22)}$$

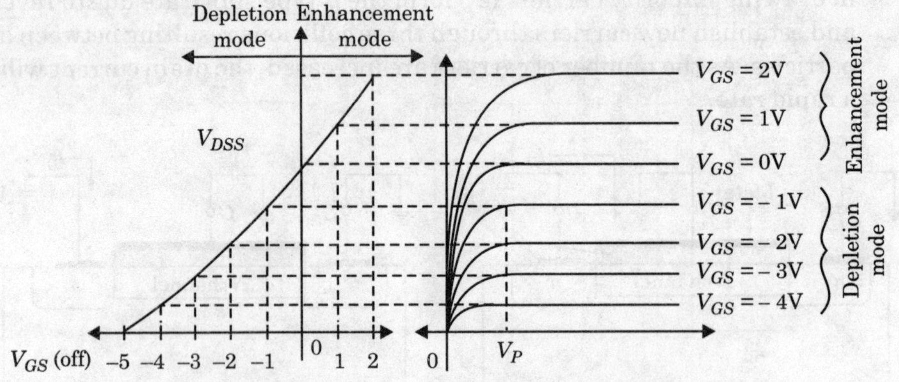

(a) Transfer characteristics (b) Drain characteristics

Fig. 1.27 Characteristics of depletion MOSFET

1.8.2 P-type Depletion MOSFET

- The basic structure and symbol of a *P*-type depletion type MOSFET is shown in Fig. 1.28. The depletion type PMOS that consists of *n*-type substrate having P^+ source and drain region. Both source and drain region are connected by physically implanted *P*-type channel.

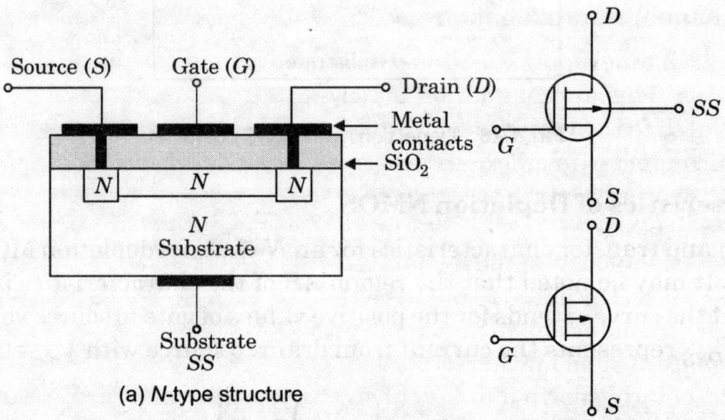

(a) *N*-type structure (b) Symbol

Fig. 1.28 Structure and symbol for *P*-type depletion MOSFET.

1.8.2.1 Characteristics of Depletion PMOS

1. The drain and transfer characteristics for a *P*-type depletion MOSFET is shown in Fig. 1.29. The characteristics of PMOS would be same as NMOS but with V_{DS} having negative values, I_D having positive values and V_{GS} having opposite polarities as Fig.1.29.
2. Due to the reversal of V_{GS} or the drain current will increase from cutoff $V_{as\,(off)}$, $V_{GS} = V_P$ in positive V_{GS} region to I_{DSS} and then continue to increase for increasing negative values of V_{GS}.

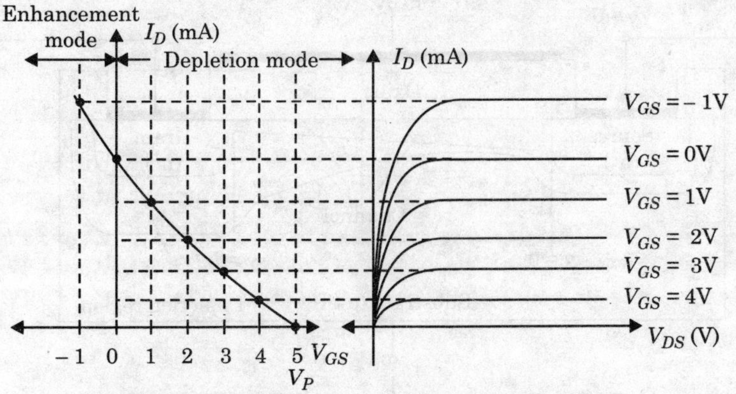

Fig. 1.29 Characteristics of depletion P-MOSFET.

1.9 MOSFET CURRENT-VOLTAGE CHARACTERISTICS

Analysis of the MOSFET current-voltage characteristics becomes a very complex matter. Here, we use the gradual channel approximation (GCA) for establishing the MOSFET current-voltage relationships. This will allow us to device relatively simple current equations that agree well with experimental results. The GCA has its limitations, especially for small-geometry MOSFETs.

1.9.1 Gradual Channel Approximation

- The gradual channel approximation (GCA) used to calculate the MOSFET current-voltage relationships. This approximation is based on the assumption that the electric charge density related to the variation of the electric field in the channel in the direction parallel to the semiconductor-insulator interface is much smaller than the electric charge density related to the variation of the electric field in the direction perpendicular to the semiconductor insulator interface.
- The cross-sectional view of the n-channel MOSFET operating in the linear mode is shown in Fig. 1.30. Here, the gate-to-source voltage is set to be larger than the threshold voltage V_{T0} to create a conducting inversion layer between the source and the drain.
- The coordinate system for this structure is defined as, the x-direction is perpendicular to the surface, pointing down into the substrate, and the y-direction is parallel to the surface. The y-coordinate origin ($y = 0$) is at the source end of the channel. The channel voltage with respect to the source will be denoted by $V_C(y)$ and the threshold voltage V_{T0} is constant along the entire channel region, between $y = 0$ and $y = L$. The electric field component E_y along the y-coordinate is dominant compared to the electric field component E_x along the x-coordinate. This assumption will allow us to reduce the current-flow problem in the channel to the y-dimension only. Note that the boundary conditions for the channel voltage V_C are:

$$V_C(y = 0) = V_S = 0$$
$$V_C(y = L) = V_{DS} \quad \text{...(1.23)}$$

and

$$V_{GS} \geq V_{T0}$$
$$V_{GD} = V_{GS} - V_{DS} \geq V_{T0} \quad \text{...(1.24)}$$

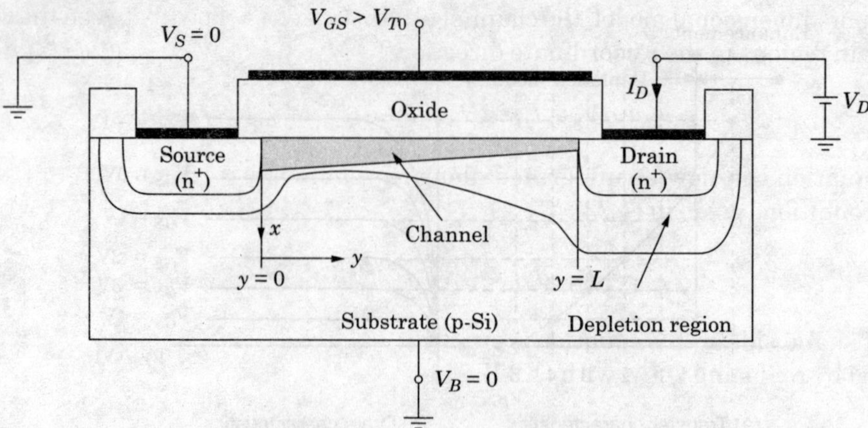

Fig. 1.30 Cross-sectional view of an *n*-channel transistor, operating in linear region.

- The channel current (drain current) I_D is due to the electrons in the channel region traveling from the source to the drain under the influence of the lateral electric field component E_y. Since the current flow in the channel is primarily governed by the lateral drift of the mobile electron charge in the surface inversion layer.
- Let $Q_I(y)$ be the total mobile electron charge in the surface inversion layer. This charge can be expressed as a function of the V_{GS} and of the channel voltage $V_C(y)$ as follows:

$$Q_I(y) = -C_{ox} \cdot [V_{GS} - V_C(y) - V_{T0}] \quad \text{...(1.25)}$$

- The spatial geometry of the surface inversion layer that indicates its significant dimensions is shown in Fig. 1.31. The thickness of the inversion layer tapers off from the source to the drain, since the gate-to-channel voltage causing surface inversion is smaller at the drain end.

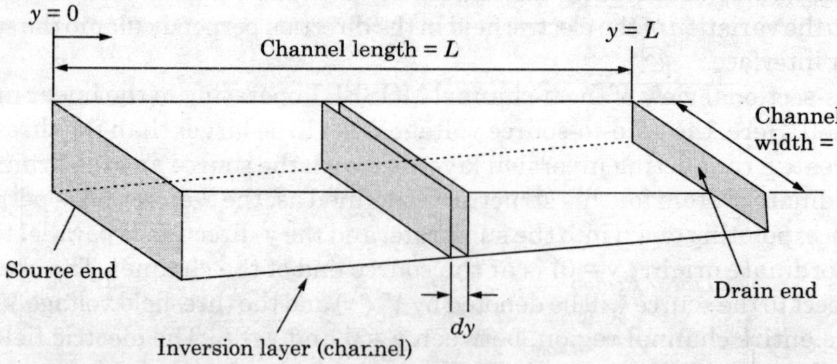

Fig. 1.31 Simplified geometry of the surface inversion layer (channel region).

- Now consider the incremental resistance dR of the differential channel segment shown in Fig. 1.31. Assuming that all mobile electrons in the inversion layer have a constant surface mobility μ_n, the incremental resistance can be expressed as follows.

$$dR = \frac{dy}{W \cdot \mu_n \cdot Q_1(y)} \quad \text{...(1.26)}$$

- The minus sign is due to the negative polarity of the inversion layer charge Q_I.
- The electron surface mobility μ_n depends on the doping concentration of the channel region and its magnitude is typically about one-half of that of the bulk electron mobility.

- In one-dimensional model, the channel (drain) current I_D flows between the source and the drain regions in the y-coordinate direction.

$$dV_c = I_D \cdot dR = \frac{I_D}{W \cdot \mu_n \cdot Q_I(y)} \cdot dy \qquad ...(1.27)$$

This equation can now be integrated along the channel, i.e., from $y = 0$ to $y = L$, using the boundary conditions given in (1.23).

$$\int_0^L I_D \cdot dy = -W \cdot \mu_n \int_0^{V_{DS}} Q_1(y) \cdot dV_c \qquad ...(1.28)$$

The left-hand side of this equation is simply equal to LI_D. The integral on the right-hand side is evaluated by replacing $Q_I(y)$ with (1.25). Thus,

$$I_D \cdot L = W \cdot \mu_n \cdot C_{ox} \int_0^{V_{DS}} (V_{GS} - V_C - V_{T0}) \cdot dV_c \qquad ...(1.29)$$

Assuming that the channel voltage V_C, is the only variable in (1.29) that depends on the position y, the drain current is found as follows.

$$I_D = \frac{\mu_n \cdot C_{ox}}{2} \cdot \frac{w}{L} \cdot \left[2 \cdot (V_{GS} - V_{T0})V_{DS} - V_{DS}^2\right] \qquad ...(1.30)$$

Equation (1.30) represents the drain current I_D as a simple second-order function of the two external voltages, V_{GS} and V_{DS}. This current equation can also be rewritten as

$$I_D = \frac{k'}{2} \cdot \frac{W}{L} \cdot \left[2 \cdot (V_{GS} - V_{T0})V_{DS} - V_{DS}^2\right] \qquad ...(1.31)$$

or $$I_D = \frac{k}{2} \cdot \left[2 \cdot (V_{GS} - V_{T0})V_{DS} - V_{DS}^2\right] \qquad ...(1.32)$$

where $$k' = \mu_n \times C_{ox} \qquad ...(1.33)$$

and $$k = k' \cdot \frac{W}{L} \qquad ...(1.34)$$

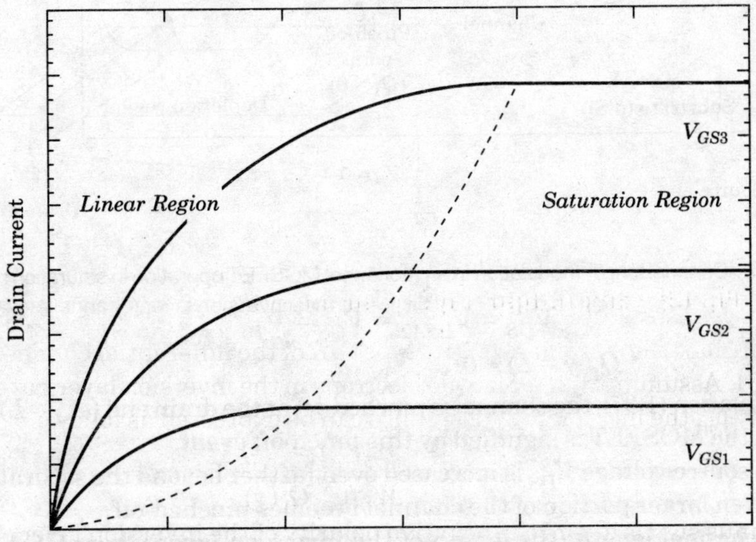

Fig. 1.32 Basic current-voltage characteristics of an n-channel MOS transistor.

- The drain current equation given in (1.31) is the simplest analytical approximation for the MOSFET current-voltage relationship. Note that, in addition to the process dependent constants k' and V_{T0}, the current-voltage relationship is also affected by the device dimensions, W and L. We will see that the ratio of W/L is one of the most important design parameters in MOS digital circuit design.
- The typical drain current versus drain voltage characteristics of an n-channel MOSFET is shown in Fig. 1.32. The parabolic boundary between the linear and the saturation regions is indicated here by the dashed line.

1.9.2 Channel Length Modulation

In this topic, we will examine the mechanisms of channel pinch-off and current flow in saturation mode in more detail. Consider the inversion layer charge Q_I that represents the total mobile electron charge on the surface, given by (1.25). The inversion layer charge at the source end of the channel is

$$Q_I(y=0) = -C_{ox} \cdot (V_{GS} - V_{T0}) \qquad ...(1.35)$$

and the inversion layer charge at the drain end of the channel is

$$Q_I(y=L) = -C_{ox} \cdot (V_{GS} - V_{T0} - V_{DS}) \qquad ...(1.36)$$

At the edge of saturation (when the drain-to-source voltage reaches V_{DSAT}), the inversion layer charge at the drain end becomes zero, according to (1.36). In reality, Q_I the channel charge does not become exactly equal to zero (remember that the GCA is just a simple approximation of the actual conditions in the channel), but it indeed becomes very small.

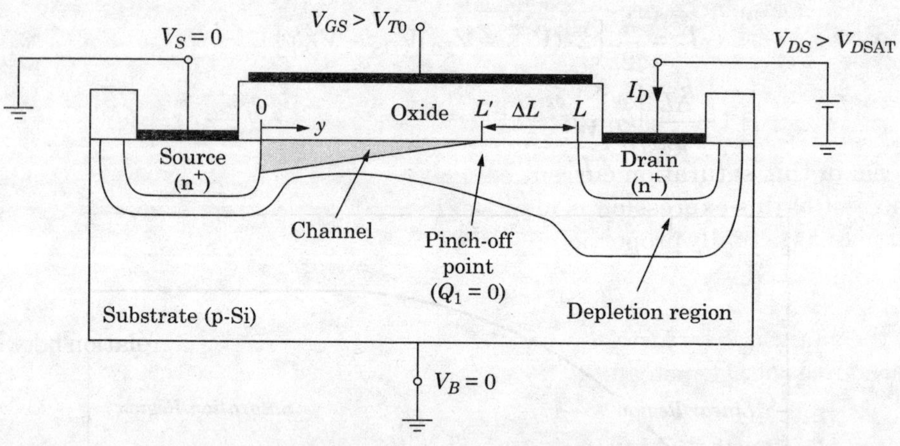

Fig. 1.33 Channel length modulation in an n-channel MOSFET operation in saturation mode.

$$V_{DS} = V_{DSAT} = V_{GS} - V_{T0} \qquad ...(1.37)$$

$$Q_I(y=L) \approx 0 \qquad ...(1.38)$$

According to equation (1.37), the channel is pinched-off at the drain end (at $y = L$). The saturation mode operation in the MOSFET is signified by this pinch-off event.

If the drain-to-source voltage V_{DS} is increased even further beyond the saturation edge so that $V_{DS} > V_{DSAT}$, an even larger portion of the channel becomes pinched-off.

The effective channel length (the length of the inversion layer where GCA is still valid) is reduced to

$$L' = L - \Delta L \quad ...(1.39)$$

where ΔL is the length of the channel segment with $Q_I = 0$ (as Fig. 1.31). Hence, the pinch-off point moves from the drain end of the channel toward the source with increasing drain-to-source voltages. The remaining portion of the channel between the pinch-off point and the drain will be in depletion mode. Since $Q_I(y) = 0$ for $L' < y < L$, the channel voltage at the pinch-off point remains equal to V_{DSAT}, i.e.

$$V_c(y = L') = V_{DSAT} \quad ...(1.40)$$

The electrons traveling from the source toward the drain traverse the inverted channel segment of length L', and then they are injected into the depletion region of length ΔL that separates the pinch-off point from the drain edge. The gradual channel approximation is valid in this region; thus, the channel current can be found.

$$I_D(\text{sat}) = \frac{\mu_n \cdot C_{ox}}{2} \cdot \frac{W}{L} \cdot (V_{GS} - V_{T0})^2 \quad ...(1.41)$$

This current equation corresponds to a MOSFET with effective channel length L', operating in saturation. Equation (1.41) accounts for the actual shortening of the channel, also called channel length modulation. Since $L' < L$, the saturation current calculated by using (1.41) will be larger than that found by using (1.35) under the same bias conditions.

As L' decreases with increasing V_{DS}, the saturation mode current I_D(sat) will also increase with V_{DS}. By approximating the effective channel length $L' = L - \Delta L$ as a function of the drain bias voltage,than we can modify (1.41) to reflect this drain voltage dependence.

First, rewrite the saturation current as follows:

$$I_D(\text{sat}) = \left(\frac{1}{1 - \frac{\Delta L}{L}} \right) \cdot \frac{\mu_n \cdot C_{ox}}{2} \cdot \frac{W}{L} \cdot (V_{GS} - V_{T0})^2 \quad ...(1.42)$$

$$\left(1 - \frac{\Delta L}{L}\right) I_D(\text{sat}) = \frac{\mu_n C_{ox}}{2} \frac{W}{L} (V_{GS} - V_{T0})^2$$

The first term of this saturation current expression accounts for the channel modulation effect, while the rest of this expression is identical to (1.35). It can be shown that the channel length shortening ΔL is actually proportional to the square root of $(V_{DS} - V_{DSAT})$.

$$\Delta L \propto \sqrt{V_{DS} - V_{DSAT}} \quad ...(1.43)$$

To simplify the analysis even further, we will use the following empirical relation between ΔL and the drain-to-source voltage instead:

$$1 - \frac{\Delta L}{L} \approx 1 - \lambda \cdot V_{DS} \quad ...(1.44)$$

Here, λ is an empirical model parameter, and is called the channel length modulation coefficient. Assuming that $\lambda V_{DS} << 1$, the saturation current given is written as:

$$I_D(\text{sat}) = \frac{\mu_n \cdot C_{ox}}{2} \cdot \frac{W}{L} \cdot (V_{GS} - V_{T0})^2 \cdot (1 + \lambda \cdot V_{DS}) \quad ....(1.45)$$

This simple current equation prescribes linear drain-bias dependence for the saturation current in MOS transistors, determined by the empirical parameter λ. The drain current versus drain-to-source voltage characteristics of an n-channel MOSFET, obtained by using (1.30) for the linear region and (1.41) for the saturation region, are shown in Fig. 1.34.

The saturation mode current increases linearly with V_{DS}. The slope of the current voltage curve in the saturation region is determined by the channel length modulation coefficient λ.

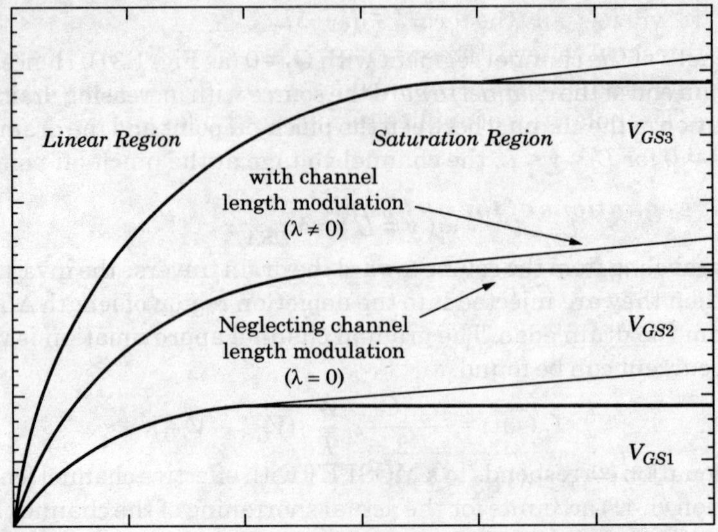

Fig. 1.34 Current-voltage characteristics of an n-channel MOS transistor, including the channel length modulation effect.

1.9.3 Substrate Bias Effect

- In many digital circuit applications, the source potential of an nMOS transistor can be larger than the substrate potential, which results in a positive source-to-substrate voltage $V_{SB} > 0$.
- In this case, the influence of the nonzero V_{SB} upon the current characteristics must be accounted for recall that the general expression for the threshold voltage V_T already includes the substrate bias term and, hence, it reflects the influence of the nonzero source-to-substrate voltage upon the device characteristics.

$$V_T(V_{SB}) = V_{T0} + \gamma \cdot \left(\sqrt{|2\phi_F| + V_{SB}} - \sqrt{|2\phi_F|}\right) \quad \text{...(1.46)}$$

We can simply replace the threshold voltage terms in linear-mode and saturation-mode current equations with the more general $V_T(V_{SB})$ term

$$I_D(\text{lin}) = \frac{\mu_n \cdot C_{ox}}{2} \cdot \frac{W}{L} \cdot \left[2 \cdot (V_{GS} - V_T(V_{SB}))V_{DS} - V_{DS}^2\right] \quad \text{...(1.47)}$$

$$I_D(\text{sat}) = \frac{\mu_n \cdot C_{ox}}{2} \cdot \frac{W}{L} .(V_{GS} - V_T(V_{SB}))^2 \cdot (1 + \lambda \cdot V_{DS}) \quad \text{...(1.48)}$$

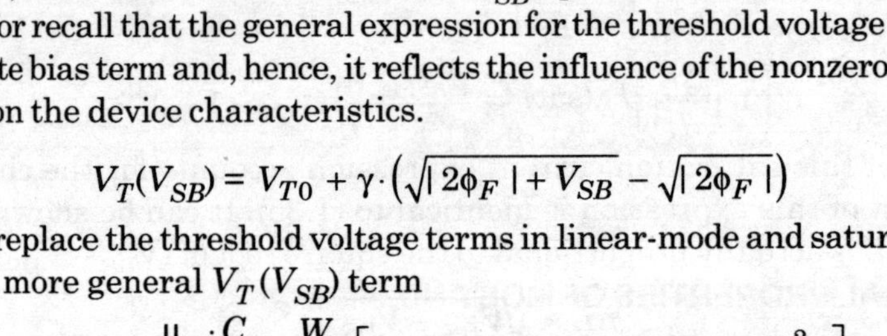

Fig. 1.35 Terminal voltages and currents of the nMOS and the pMOS transistor

In general, we will use only the term V_T instead of $V_T(V_{SB})$ to express the general (substrate-bias dependent) threshold voltage. Hence, here we will repeat the current-voltage equations derived under the first order gradual channel approximation (GCA), both for *n*-channel and for *p*-channel MOS transistors. The polarities of applied terminal voltages and the drain current directions is shown in Fig. 1.35.

Current-voltage equations of the n-channel MOSFET:

$$I_D = 0, \quad \text{for} \quad V_{GS} < V_T \qquad ...(1.49)$$

$$I_D(\text{lin}) = \frac{\mu_n \cdot C_{ox}}{2} \cdot \frac{W}{L} . \left[2 \cdot (V_{GS} - V_T)V_{DS} - V_{DS}^2\right] \quad \text{for} \quad V_{GS} \geq V_T$$

$$\text{and} \quad V_{DS} < V_{GS} - V_T \qquad ...(1.50)$$

$$I_D(\text{sat}) = \frac{\mu_n \cdot C_{ox}}{2} \cdot \frac{W}{L} . (V_{GS} - V_T)^2 \cdot (1 + \lambda \cdot V_{DS}) \quad \text{for} \quad V_{GS} \geq V_T$$

$$\text{and} \quad V_{DS} \geq V_{GS} - V_T \qquad ...(1.51)$$

Current-voltage equations of the p-channel MOSFET:

$$I_D = 0, \quad \text{for} \quad V_{GS} > V_T \qquad ...(1.52)$$

$$I_D(\text{lin}) = \frac{\mu_p \cdot C_{ox}}{2} \cdot \frac{W}{L} . \left[2 \cdot (V_{GS} - V_T)V_{DS} - V_{DS}^2\right] \quad \text{for} \quad V_{GS} \leq V_T$$

$$\text{and} \quad V_{DS} > V_{GS} - V_T \qquad ...(1.53)$$

$$I_D(\text{sat}) = \frac{\mu_p \cdot C_{ox}}{2} \cdot \frac{W}{L} . (V_{GS} - V_T)^2 \cdot (1 + \lambda \cdot V_{DS}) \quad \text{for} \quad V_{GS} \leq V_T$$

$$\text{and} \quad V_{DS} \leq V_{GS} - V_T \qquad ...(1.54)$$

1.10 ELECTRICAL PROPERTIES OF MOS AND BICMOS (UTU-2012)

In order to give better understanding of MOS and BiCMOS their electrical properties will be discussed, before starting discussions of fabrication schemes. Among these electrical properties first of all drain to source current I_{ds} as a function of V_{DS} will be given.

As it is known to everyone that

$$I_{ds} = -I_{sd} = \frac{\text{Charge induced in channel } (Q_c)}{\text{Electron transit time } (\tau)} \qquad ...(1.55)$$

The transit time, $\tau_{sd} = \dfrac{\text{Length of channel } (L)}{\text{Velocity } (v)}$, the L and W are shown in Fig. 1.36.

If E_{ds} is electric field between drain and source, μ the electron or hole mobility, the τ_{ds} is given as

$$\tau_{ds} = \frac{L^2}{\mu V_{ds}} \qquad ...(1.56)$$

typical values of μ at room temperature for electron and hole is given as

$$\mu_n = 650 \text{ cm}^2/\text{V-sec} \qquad \text{(surface)}$$

$$\mu_p = 240 \text{ cm}^2/\text{V-sec} \qquad \text{(surface)}$$

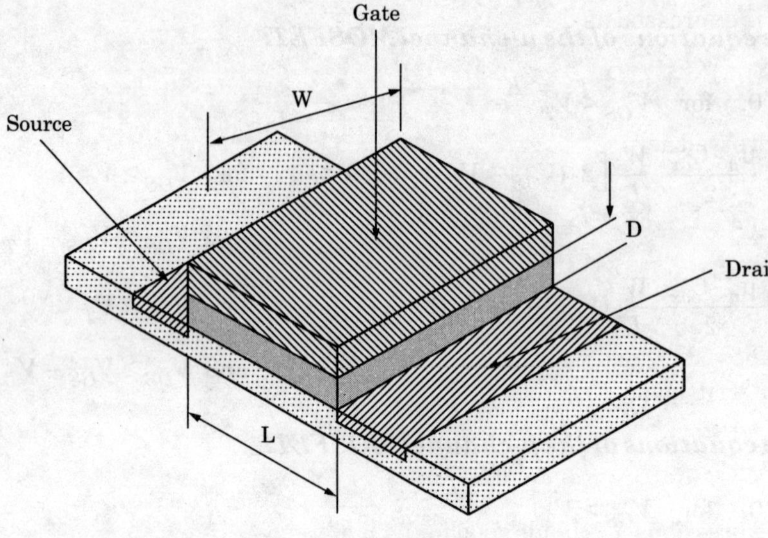

Fig. 1.36 nMOS transistor structure

The drain to source current I_{ds} is now given as

$$I_{ds} = \frac{C_g \mu}{L^2}\left[(V_{gs} - V_{th})\, V_{ds} - \frac{V_{ds}^2}{2}\right] \qquad ...(1.57)$$

or

$$I_{ds} = K\,\frac{W}{L}\left[(V_{gs} - V_{th})\, V_{ds} - \frac{V_{ds}^2}{2}\right] \qquad ...(1.58)$$

where

$$K = \frac{\varepsilon_{ins}\varepsilon_0 \mu}{D} \qquad ...(1.59)$$

in equations (1.58) and (1.59), ε_{ins} is insulator permittivity, D the oxide thickness C_g the gate capacitance and given as $C_g = \dfrac{\varepsilon_{ins}\varepsilon_0 WL}{D}$.

Some times the gate capacitance per unit area, i.e., C_0 is used in the expression of C_g and written as

$$C_g = C_0\, WL \qquad ...(1.60)$$

So, the equation of i_{ds} is also expressed as

$$I_{ds} = C_0 \mu\, \frac{W}{L}\left[(V_{gs} - V_{th})\, V_{ds} - \frac{V_{ds}^2}{2}\right] \qquad ...(1.61)$$

The above said expression is for non-saturation region. Now, the expression of I_{ds} for saturation region will be given.

The saturation starts when $V_{ds} = V_{gs} - V_{th}$; since at this point the IR drop in channel equals the effective gate to channel voltage at the drain and one can assume that the current remains fairly constant as V_{ds} increases.

Thus, the I_{ds} is expressed as

$$I_{ds} = K \cdot \frac{W}{L} \frac{(V_{gs} - V_{th})}{2} \quad ...(1.62)$$

it may also be expressed as

$$I_{ds} = \frac{\beta}{2} (V_{gs} - V_{th})^2 \quad ...(1.63)$$

or

$$I_{ds} = \frac{C_g \mu}{2L^2} (V_{gs} - V_{th})^2 \quad ...(1.64)$$

It may also be written as

$$I_{ds} = C_0 \mu \frac{W}{2L} (V_{gs} - V_{th})^2 \quad ...(1.65)$$

In the above expressions, I_{ds}, holds good for both enhancement and depletion mode devices, but the threshold voltage for nMOS depletion mode device (denoted as V_{th}) is negative.

Now, transistor's transconductance g_m and output transconductance g_{ds} expressions will be presented.

Transconductance expresses the relationship between output current I_{ds} and input voltage V_{gs} and is defined as

$$g_m = \left. \frac{\delta I_{ds}}{\delta V_{gs}} \right|_{V_{ds}} = \text{constant}$$

So, the g_m is expressed as

$$g_m = \frac{C_g \mu}{L^2} (V_{gs} - V_{th}) \quad ...(1.66)$$

or it is expressed as

$$g_m = \frac{\mu \varepsilon_{ins} \varepsilon_0}{D} \frac{W}{L} (V_{gs} - V_{th}) \quad ...(1.67)$$

Alternatively, it may be expressed as

$$g_m = \beta (V_{gs} - V_{th}) \quad ...(1.68)$$

The output transconductance, g_{ds} is expressed as

$$g_{ds} = \frac{\delta I_{ds}}{\delta V_{gs}} = \lambda \cdot I_{ds} \ \alpha \left(\frac{1}{L}\right)^2 \quad ...(1.69)$$

Here, the strong dependence on the channel length is demonstrated as

$$\left(\lambda \propto \frac{1}{L}\right) \text{ and } I_{ds} \propto \frac{1}{L} \quad \text{for MOS devices.}$$

Another, parameters of MOS transistor is the frequency response. It may be obtained from ω_0, where

$$\omega_0 = \frac{g_m}{C_g} = \frac{\mu}{L^2}(V_{gs} - V_{th}) \quad [\text{also } \omega_0 = 1/\tau_{ds}] \quad ...(1.70)$$

This expression shows the switching speed which depends on gate voltage above threshold and on carrier mobility and inversely as the square of channel length. A fast circuit requires that g_m be as high as possible. It has been also observed that the electron mobility on a (100) oriented n type inversion layer surface (μ_m) is larger than that on a (111) oriented surface, and is in fact three times as large as hole mobility on a (iii) oriented p-type inversion layer. Surface mobility is also dependent on the effective gate voltage ($V_{gs} - V_{th}$).

1.10.1 BiCMOS Process

An *npn* vertical bipolar transistor can be integrated into n-well CMOS process with the addition of a p-base diffusion region. The characteristics of this device depend on the base width and emitter area. The base width is determined by the difference in function depth between the n^+ and the p-base diffusion. The emitter area is determined by the functional area of the n^+ diffusion at the emitter lead. The n-well serves as the collector of the *npn* transistor has a β in the range of 50 to 100 and a cut-off frequency of greater than 10 GHz.

Normally, an n^+ buried layer is used to reduce the series resistance of the collector since the n-well has a very high resistivity. However, this would further complicate the process with the introduction of p-type epitaxy and one more marking step. Other variation on the bipolar transistor include poly emitter an self-aligned base contact to minimize parasitic effects.

However, the electrical properties of BiCMOS process is same as CMOS and MOS Technology. The BiCMOS is shown in Fig. 1.37.

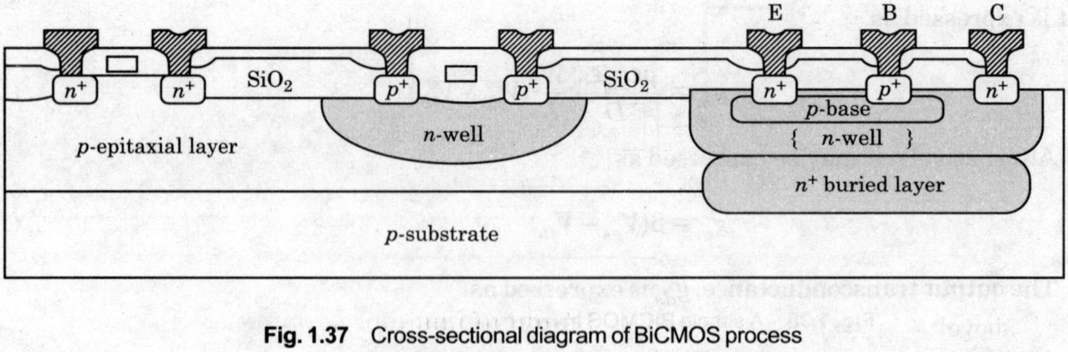

Fig. 1.37 Cross-sectional diagram of BiCMOS process

1.11 BiCMOS CIRCUITS

- When bipolar and MOS technology are merged, the resulting circuits are referred to as BiCMOS circuits.
- High gain vertical *npn* transistors with their collectors tied to the positive rail and medium-gain lateral *npn* transistors are both compatible with conventional CMOS processing.
- BiCMOS gates can be used to improve the performance of line drivers and sense amplifiers. They combine the traditional low power dissipation of CMOS with the output drive capability and low propagation delay of bipolar technology.

1.11.1 BiCMOS Inverters

- The circuit arrangement of BiCMOS inverter by using a pair of push-pull bipolar devices to provide driving capability is shown in Fig. 1.38. The MOS devices provide a high input impedance and BJT devices provide current drive and low output impedance.
- When the input signal is high, the pull-down FET short-circuits the base of Q_2 to its collector, converting it to a diode. The pull-up FET is OFF and Q_1 has no base drive. The output sees a very high resistance to the positive rail and a low resistance to ground rail, resulting in a low output value. Transistor Q_2 can sink large amounts of current.
- The reverse happens when the input signal does low, causing Q_1 to source current and pull the output high, while Q_2 is cut off as shown in Fig. 1.38.
- The circuit dissipates essentially no standby power and the low power consumption of the CMOS circuit is preserved.
- The output logic levels will be good and will be close to the rail voltages since $V_{CE\,sat}$ is quite small and V_{BE} is approximately + 0.7 volts.
- The inverter has a high input impedance and low output impedance.
- The inverter has high noise margins.

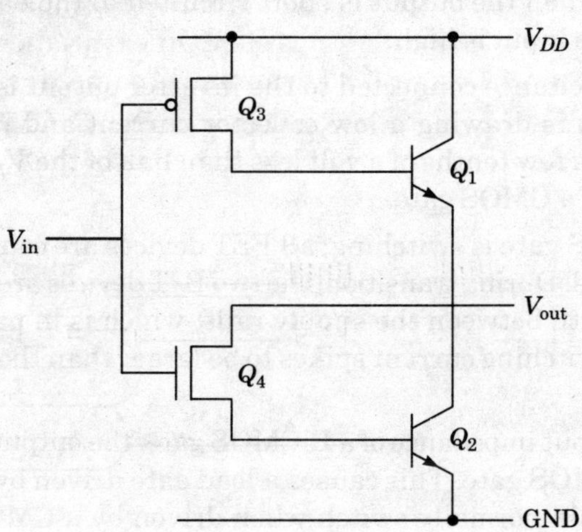

Fig. 1.38 A simple BiCMOS Inverter with zero static current flow.

- An improved BiCMOS inverter is shown in Fig. 1.39. In this configuration, transistor Q_4 turns ON when the input signal goes high, pulling node a down and discharge Q_1 quickly. As the potential at node a drops, Q_6 is turned OFF, allowing Q_5 to drive Q_2 on hard and pull the output low. When the input goes low, Q_3 turns ON while Q_4 turns OFF. This allows Q_1 to turn ON fast. Q_5 is OFF and Q_6 turns ON when the voltage at node a rises, allowing Q_2 to discharge quickly.

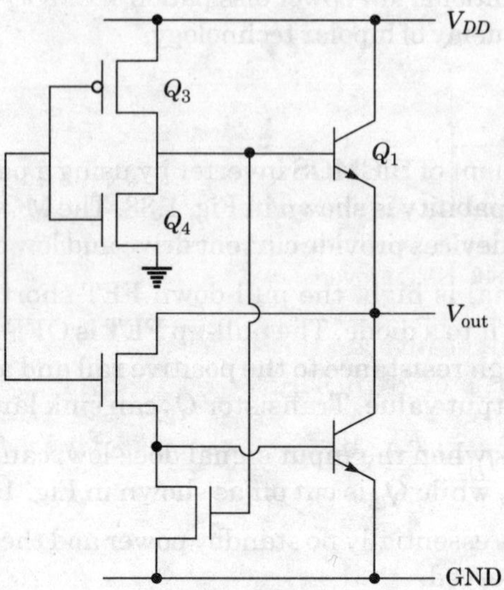

Fig. 1.39 A faster circuit (improved) BiCMOS inverters.

- The output high level voltage is given by $V_{OH} = V_{DD} - V_{BE}$, where V_{BE} is the voltage drop across the base-emitter junction of Q_1, while the output low level voltage is given by $V_{OL} = V_{BE}$ of Q_2 when the output is short circuited to the base of Q_2 through Q_5 which conducts when the input is high.
- When a load capacitance connected to the inverter output is fully charged (discharged transistor Q_1 (Q_2) is drawing a low collector current and its V_{BE} is small. The noise margins are then a few tenths of a volt less than half of the V_{DD} supply voltage or almost as good as those of a CMOS gate.
- When the BiCMOS gate is switching, all FET devices are conducting and there is a path from V_{DD} to ground. During transition, the two BJT devices are also ON, providing another low impedance path between the supply rails, which is in parallel with the MOS path. This causes the switching current spikes to be larger than those of a conventional CMOS gate.
- Due to the low output impedance of a BiCMOS gate, the output switches much faster than the output of a CMOS gate. This causes a load gate driven by a bipolar output to switch faster then it would normally switch when driven by a CMOS gate and the switching current spikes to be narrower for the BiCMOS. The overall result is lowered power dissipation per gate per frequency.

1.12 DETERMINATION OF PULL-UP TO PULL-DOWN RATIO ($Z_{P.U.}/Z_{P.D.}$) FOR AN NMOS INVERTER DRIVEN BY ANOTHER NMOS INVERTER

Consider the arrangement shown in Fig. 1.40 in which an inverter is driven from the output of another similar inverter. Consider the depletion mode transistor for which $V_{GS} = 0$ under all conditions, and further assume that in order to cascade inverter without degradation of levels, we aim to meet the requirement:

$$V_{in} = V_{out} = V_{in}$$

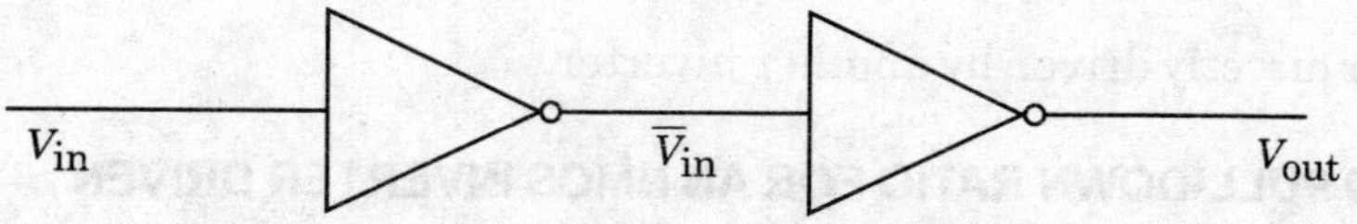

Fig. 1.40 NMOS inverter driven directly by another inverter.

For equal margins around the inverter threshold, we set $V_{GS} = V_{in} = 0.5\, V_{DD}$ for Enhancement type. At this point, both transistor are in saturation and

$$I_D = \frac{K'_n}{2} \cdot \frac{W}{L} (V_{GS} - V_T)^2$$

In depletion mode,

$$I_D = \frac{K'_n}{2} \cdot \frac{W_{P.U.}}{L_{P.U.}} (0 - V_{Td}^2)$$

where V_{Td} is the threshold voltage in depletion mode.

In Enhancement mode,

$$I_D = \frac{K'_n}{2} \cdot \frac{W_{p.d}}{L_{p.d}} (V_{inv} - V_T)^2$$

After equating the above equation

$$\frac{W_{P.d}}{L_{P.d}} (V_{inv} - V_T)^2 = \frac{W_{P.d}}{L_{P.d}} (0 - V_{Td}^2)$$

where, $W_{p.d}$, $L_{p.d.}$, $W_{P.U.}$ and $L_{P.U.}$ are the width and lengths of dull-down and pull-up transistors respectively.

Now, $$Z_{P.d.} = \frac{L_{P.d}}{W_{P.d}} \quad \text{and} \quad Z_{P.U.} = \frac{L_{P.U.}}{W_{P.U.}}$$

so, $$\frac{(V_{inv} - V_T)^2}{Z_{P.d.}} = -\frac{V_{Td}^2}{Z_{P.U.}}$$

Simplifying $$V_{inv} = V_T - \frac{V_{Td}}{\sqrt{Z_{P.U.} / Z_{p.d.}}} \qquad ...(1.71)$$

Here, we consider some typical values are follows:

$$V_T = 0.2V_{DD},\quad V_{Td} = 0.6V_{DD} \quad \text{and} \quad V_{inv} = \frac{V_{DD}}{2}$$

so

$$0.5 = 0.2 + \frac{0.6}{\sqrt{Z_{P.U.}/Z_{P.d.}}}$$

$$\Rightarrow \quad \sqrt{\frac{Z_{P.U.}}{Z_{P.d.}}} = 2 \quad \Rightarrow \quad \boxed{Z_{P.U.} = 4Z_{P.d.}}$$

For an inverter directly driven by another inverter.

1.13 PULL-UP TO PULL-DOWN RATIO FOR AN NMOS INVERTER DRIVEN THROUGH ONE OR MORE PASS TRANSISTOR

We consider the arrangement of Fig. 1.41 in which the input to the inverter 2 comes from the output inverter 1 but passes through one or more NMOS pass transistors. The connection of pass transistors in series will degrade the logic 1 level into inverter 2 so that the output will not be a proper logic 0 level. If the point A is at 0 V, and B point is thus at V_{DD}, but the voltage into inverter 2 at point C is now reduced from V_{DD} by the threshold voltage of the series pass transistor. However, there can be no voltage drop in the channels, since no static current flows through the pass transistors. With all pass transistors connected to V_{DD}, there is a loss of V_{Tn}.

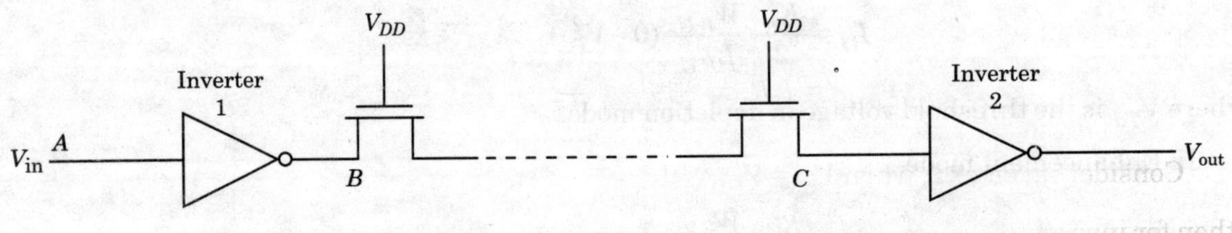

Fig. 1.41 Pull-up to pull-down ratio for inverting logic coupled bypass transistors

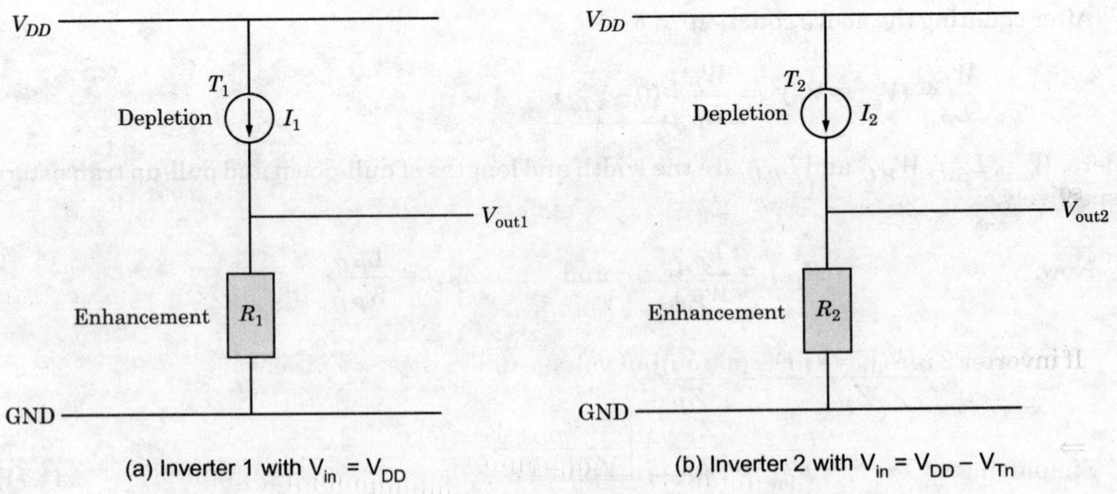

Fig. 1.42 Equivalent circuit of inverters 1 and 2.

Consider inverter 1 as shown in Fig. 1.42(*a*). If the input is at V_{DD}, then the pull-down transistor T_2 is conducting but with a low voltage across it, therefore, it is in its resistive region represented by R_1. Meanwhile, the pull-up transistor T_1 is in saturation and is represented as a current source.

For the pull-down transistor,

$$I_{DS} = K_n' \frac{W_{P.d.1}}{L_{P.d.1}} \left[(V_{DD} - V_T)\, V_{DS,1} - \frac{V_{DS1}^2}{2} \right]$$

$$\therefore \qquad R_1 = \frac{V_{DS,1}}{I_{DS,1}} = \frac{1}{K_n'} \cdot \frac{L_{P.d.1}}{W_{P.d.1}} \left[\frac{V_{DS,1}}{(V_{DD} - V_T)V_{DS,1} - \frac{V_{DS,1}^2}{2}} \right]$$

if V_{DS1} is very small, so $\frac{V_{DS}}{2} \approx 0$

$$\text{Thus} \qquad R_1 = \frac{1}{K_n} Z_{P.d.1} \left(\frac{1}{V_{DD} - V_T} \right)$$

Now, for the depletion mode pull-up transistor in saturation, with $V_{GS} = 0$

$$I_1 = I_{DS} = K_n' \frac{W_{P.U.1}}{L_{P.U.1}} \left(\frac{-V_{Td}^2}{2} \right)$$

$$V_{\text{out }1} = I_1 R_1 = \frac{Z_{P.d.1}}{Z_{P.U.1}} \left(\frac{1}{V_{DD} - V_T} \right) \frac{V_{Td}^2}{2}$$

Consider inverter 2, when $V_{\text{in}} = V_{DD} - V_{Tn}$

then for inverter 2,

$$R_2 = \frac{1}{K_n'} Z_{P.d.2} \frac{1}{[(V_{DD} - V_{Tn}) - V_T]}$$

$$I_2 = K_n' \frac{1}{Z_{P.U.2}} \frac{(-V_{Td})^2}{2}$$

$$\text{so} \qquad V_{\text{out }2} = I_2 R_2$$

$$= \frac{Z_{P.d.2}}{Z_{P.U.2}} \left(\frac{1}{V_{DD} - V_{Tn} - V_T} \right) \left(\frac{-V_{Td}^2}{2} \right)$$

If inverter 2 is to have the same output voltage under these conditions.

$$\Rightarrow \qquad \frac{Z_{P.U.2}}{Z_{P.d.2}} = \frac{Z_{P.U.1}}{Z_{P.d.1}} \left(\frac{V_{DD} - V_T}{V_{DD} - V_{Tn} - V_T} \right) \qquad \text{...(1.72)}$$

Here, we consider some typical values are follows:

$$V_T = 0.2\ V_{DD},\quad V_{Tn} = 0.3\ V_{DD}$$

$$\frac{Z_{P.U.2}}{Z_{P.d.2}} = \frac{Z_{P.U.1}}{Z_{P.d.1}}\left(\frac{0.8}{0.5}\right)$$

$$\therefore \quad \frac{Z_{P.U.2}}{Z_{P.d.2}} = 2\left(\frac{Z_{P.U.1}}{Z_{P.d.1}}\right) = \frac{8}{1}$$

Summarizing for an NMOS transistor:

- An inverter driven directly from the output of another should have a

$$\frac{Z_{P.U.}}{Z_{P.d.}} \text{ ratio} \geq \frac{4}{1}$$

- An driven through one or more pass transistors should have a inverter

$$\frac{Z_{P.U.}}{Z_{P.d.}} \text{ ratio} \geq \frac{8}{1}$$

1.14 COMPARISON

1.14.1 Comparison of N-with P-channel MOSFET's

1. P-channel MOSFET is much easier and cheaper to produce than the N-channel device.
2. The drain resistance of P-channel MOSFET is three times higher than that for an identical N-channel MOSFET.
3. The hole mobility is nearly 2.5 times lower than the electron mobility. Thus, a P-channel MOSFET occupies a larger area than an N-channel MOSFET having the same I_D rating.
4. The N-channel MOSFET is smaller for the same complexity the P-channel device.
5. The N-channel MOSFET has the higher packing density, which makes it faster in switching applications due to the smaller junction areas and lower in herent capacitances.

1.14.2 Comparison of DMOSFET's with EMOSFET's

S. No.	DMOSFET's	EMOSFET'S
1.	Drain Current (I_D) becomes zero when $V_{GS} = 0$	Drain current becomes zero when $V_{GS} < V_T$
2.	Channel between source and drain is present	Channel is absent between source and drain.
3.	For an N-channel DMOSFET, I_D is decreases when V_{GS} becomes more and more negative.	For *N*-channel EMOSFET, I_D increases as V_{as} becomes more and more positive mode.
4.	An insulating layer (SiO_2) is present between gate and channel.	The insulating oxide layer is present between Gate and substrate.
5.	It is operated in two modes: (*a*) Depletion mode (V_{GS} is negative) (*b*) Enhancement mode (V_{GS} is positive)	It is operated in enhancement mode only.

1.15 SUMMARY OF THE MOSFET CURRENT-VOLTAGE CHARACTERISTICS

NMOS-Transistor	*PMOS-Transistor*
Symbol 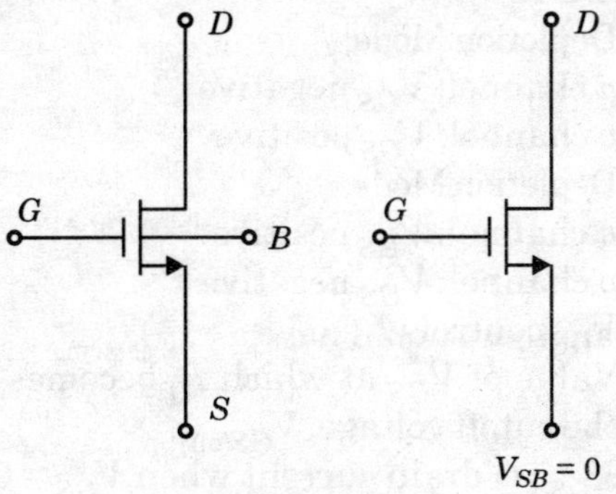 Fig. 1.43	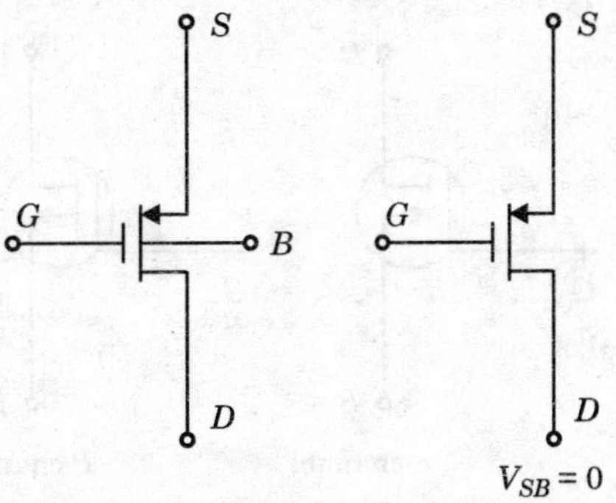 Fig. 1.44
Overdrive Voltage	***Overdrive Voltage***
$V_{OV} = V_{GS} - V_T$ $V_{GS} = V_T + V_{OV}$	$V_{OV} = V_{GS} - V_T$ $V_{GS} = \mid V_T \mid + \mid V_{OV} \mid$
Operation in the linear region:	***Operation in the linear region:***
$\Rightarrow$ Constants (a) $V_{GS} \geq V_T \Leftrightarrow V_{OV} \geq 0$ (b) $V_{DG} \geq V_T \Leftrightarrow V_{DS} \geq V_{GS} - V_T$ $\Rightarrow$ I – V Characteristics: $I_D = \mu_n C_{ox} \left(\frac{W}{L}\right)\left[(V_{GS} - V_T)V_{DS} - \frac{V_{DS}^2}{2}\right]$	$\Rightarrow$ Conditions (a) $V_{GS} \leq V_T \Leftrightarrow V_{OV} \leq 0 \Leftrightarrow V_{GS} \geq \mid V_T \mid$ (b) $V_{GS} \leq \mid V_T \mid \Leftrightarrow V_{DS} \leq V_{GS} \geq V_T$
Operation in the Saturation region:	***Operation in the Saturation region:***
$\Rightarrow$ Conditions (a) $V_{GS} \geq V_T \Leftrightarrow V_{OV} \geq 0$ (b) $V_{DG} \geq V_T \Leftrightarrow V_{DS} \geq V_{GS} - V_T$ $\Rightarrow$ I – V Characteristics: $I_D = \frac{\mu_n C_{ox}}{2}\left(\frac{W}{L}\right)(V_{GS} - V_T)^2 (1 + \lambda V_{DS})$ $\Rightarrow$ Conditions $\epsilon_0 = 8.854 \times 10^{-12}\ F/m$ $\epsilon_{ox} = 3.9\epsilon_0 = 3.45 \times 10^{-11}\ F/m$ $\epsilon_S = 11.7\epsilon_0 = 1.04 \times 10^{-10}\ F/m$ $q = 1.602 \times 10^{-19} C$	$\Rightarrow$ Constants (a) $V_{GS} \leq V_T \Leftrightarrow V_{OV} \leq 0 \Leftrightarrow V_{GS} \geq \mid V_T \mid$ (b) $V_{GS} \leq \mid V_T \mid \Leftrightarrow V_{DS} \leq V_{GS} \geq V_T$

1.16 SUMMARY OF MOSFET

D-MOSFET's 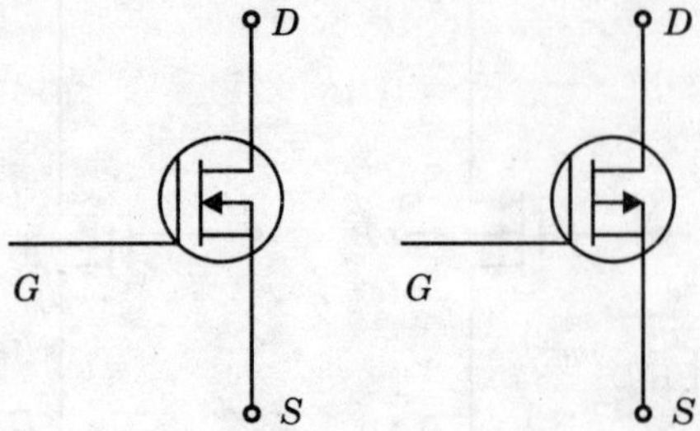 *n*-channel *P*-channel **Fig. 1.45**	Expect that it can be operated in enhancement mode, the D-MOSFET characteristics are same as JFET. ⇒ Depletion Mode: *n* channel: V_{GS} negative *p* channel: V_{GS} positive ⇒ Depletion Mode: *n* channel: V_{GS} positive *p* channel: V_{GS} negative ⇒ V_{GS} controls I_D ⇒ Value of V_{GS} at which I_D becomes zero is the cutoff voltage, $V_{GS(\text{off})}$. ⇒ V_{DSS} is drain current when $V_{GS} = 0$ ⇒ Transfer characteris: $$I_D = I_{DSS}\left(1 - \frac{V_{GS}}{V_{GS(\text{off})}}\right)^2$$
E-MOSFET's 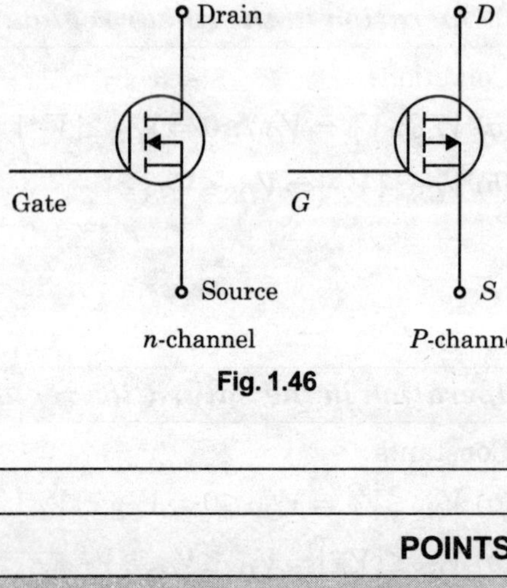 *n*-channel *P*-channel **Fig. 1.46**	There is no depletion mode and characteristics differ from *D*-MOSFET. ⇒ enhancement mode: ⇒ *n* channel: V_{GS} positive *p* channel: V_{GS} negative V_{GS} control I_D: ⇒ Value of V_{GS} at which I_D becomes zero is the cutoff voltage, $V_{GS(\text{th})}$. ⇒ Transfer characteristics: $$I_D = K(V_{BS} - V_{GS(\text{th})})^2$$ ⇒ *K* in formula can be calculated by substituting data sheet value $I_{D(\text{on})}$ for I_D and V_{GS} at which $I_{D(\text{on})}$ is specified for V_{GS}.

POINTS TO BE REMEMBER

1. Accumulation mode is when $V_{GS} < V_T$.
2. The mask is often known as Thinox.
3. High absolute power consumption and electrical asymmetry are the two major drawbacks of NMOS process.
4. The fermi potential value for typical *p*-type silicon substrate is –0.3 V.
5. The threshold voltages in short-channel devices also have the tendency to drift overtime.
6. Due to silicon dioxide, no carriers can move either from gate electrode to channel or channel to gate electrode and hence there is effectively node gate current, i.e. gate current is zero ($I_G = 0$) and the channel is capacitively coupled to the gate via the electric field in the oxide. This capacitor is called gate oxide capacitor (C_{ox}, in which one plate is metal and the other plate is semiconductor material and these two plates are separated by SiO_2.

7. For current to flow in the MOS device, it is required to apply both terminal voltages gate-source and drain-source. If any one of the terminal voltage is zero, no current will flow. Therefore, drain current is a function of these two terminal voltages.
8. Threshold voltage decreases as length of MOS (L) decreases, and increasing drain-source voltage also causes fall in threshold voltage.
9. If the channel length is equal to twice of the vertical dimensions, the device behaves as long channel MOSFET. In long channel MOSFET, the threshold voltage is insensitive to channel length and drain voltage. In short channel device, 2-D effects become significant and the minimum potential, which determines the threshold voltage, is basically controlled by drain voltage rather than gate voltage.

SOLVED EXAMPLES

Example 1.1: An enhancement type nMOS transistor $V_T = 0.7V$ with has its source terminal grounded and a 1.5V applied to the gate. In what region does the device operate for *(a)* $V_D = 0.56V$ (b) $0.9V = V_D$.

Solution: We know that

The conditions for various regions for enhancement type NMOS:

Cutoff region $V_{GS} < V_T$

Linear region $V_{DS} \leq V_{GS} - V_T$

Saturation region $V_{DS} \geq V_{GS} - V_T$

Here, $V_S = 0, V_G = 1.5V, V_T = 0.7V$

(a)

$$V_D = 0/7V$$

$$V_{DS} = V_D - V_S = 0.5 - 0 = 0.5V$$

$$V_{GS} = V_G - V_S = 1.5 - 0 = 1.5V$$

(i) V_{GS} is not less than V_T.

(ii)

$$V_{DS} \leq V_{GS} - V_T$$

$$\Rightarrow \quad 0.5 \leq 1.5 - 0.7$$

$$\Rightarrow \quad 0.5 \leq 0.8$$

At $V_D = 0.5$, the device is in linear region.

(b)

$$V_D = 0.9V$$

$$V_{DS} = 0.9 - 0 = 9V$$

$$V_{GS} = V_G - V_S = 1.5 - 0 = 1.5V$$

(i) V_{GS} is not less than V_T. Hence cutoff region condition is not satisfied.

(ii)

$$V_{DS} \leq V_{GS} - V_T$$

$$\Rightarrow \quad 0.9 \leq 1.5 - 0.7$$

$$\Rightarrow \quad 0.9 \leq 0.8$$

$\Rightarrow$ Linear region condition is also not satisfied.

(iii)

$$V_{DS} \geq V_{GS} - V_T$$

$$0.9 \geq 1.5 - 0.7$$

$$0.9 \geq 0.8$$

At the device is in saturation region.

Example 1.2: An enhancement type nMOS transistor with $V_T - 0.7V$ conducts a current $I_D = 100$ μA when $V_{GS} = V_{DS} = 1.2V$. Find the value of I_D for $V_{GS} = 1.5V$ and $V_{DS} = 3V$. Also calculate the value of drain to source V_{DS} for small V_{DS} and $V_{GS} = 3.2V$.

Solution: Given that

$$V_T = 0.7V, I_D = 100\mu A \text{ when } V_{GS} = V_{DS} = 1.2V$$

Check for this condition:

For linear region $\quad V_{DS} \le V_{GS} - V_T$

$\Rightarrow \quad 1.2 \le 1.2 - 0.7$

∵ This condition not satisfied.

For saturation region

$$V_{DS} \ge V_{GS} - V_T$$

$\Rightarrow \quad 1.2 \ge 1.2 - 0.7$

$\Rightarrow \quad 1.2 \ge 0.7$

∵ This condition is satisfied.

At this $V_{GS} = V_{DS} = 12V$, the device is in saturation region.

The current at drain for saturation

$$I_D = \frac{\mu_n C_{OX} W}{2L}(1.2-0.7)^2$$

$$\Rightarrow \quad 100 \times 10^{-6} = \frac{\mu_n C_{OX} W}{2L}(1.2-0.7)^2$$

$$\Rightarrow \quad \frac{\mu_n C_{OX} W}{2L} = \frac{10^{-4}}{(0.5)} = 4\times10^{-4}$$

For $V_{DS} = 3V$ and $V_{GS} = 1.5V$ the device is in saturation region.

So, that $\quad I_D = \frac{\mu_n C_{OX} W}{2L}(V_{GS} - V_T)^2$

$$\Rightarrow \quad I_D = 4\times10^{-4}(3-1.5)^2$$

$$\Rightarrow \quad I_D = 256 \times 10^{-6} \text{ Amp.}$$

Drain to source resistance r_{DS} for $V_{GS} = 3.2V$

$$r_{DS} = \frac{1}{\frac{\mu_n C_{OX} W}{L}(V_{GS} - V_T)}$$

$$= \frac{1}{8\times10^{-6}(3.2-0.7)} = 500\Omega$$

Example 1.3: Design the circuit shown in Fig. 1.47. So that the transistor operates at $I_D = 0.5$mA and $V_D = +5$V. The nMOS transistor has

$$V_T = 0.7\text{V}, \mu_n C_{OX} = 100\frac{\mu A}{v^2}\ L = 1\ \mu m \text{ and } W = 32\ \mu m.$$

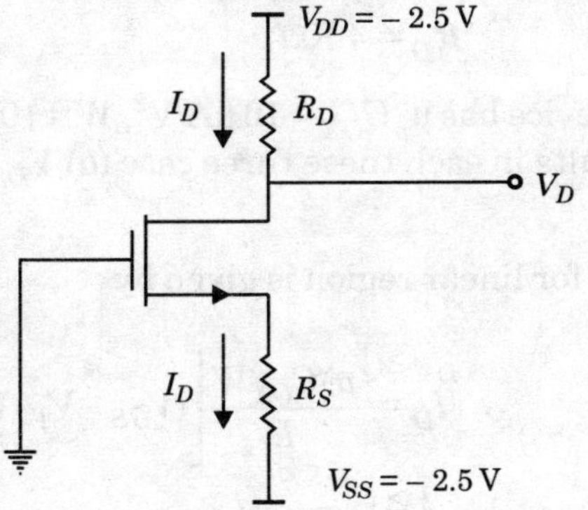

Fig. 1.47

Solution: Here $V_D = 0.5$V is greater than V_G. So that the nMOS transistor is operating in the saturation region, $V_{DG} \leq V_T$ $\because V_D \leq V_T$

For saturation region, the drain current

$$I_D = \frac{\mu_n C_{OX} W}{2L}(V_{GS} - V_T)^2$$

$$\Rightarrow \quad 5 \times 10^{-4} = \frac{100\times10^{-6}}{2\times10^{-6}}.32\times10^{-6}\,(V_{GS} - 0.7)^2$$

$$\Rightarrow \quad (V_{GS} - 0.7)^2 = \frac{5\times10^{-4}}{16\times10^{-4}} = 0.3125$$

$$\Rightarrow \quad V_{GS} = 1.259 \text{ Volt}$$

From Fig. 1.47, $V_G = 0$ so the source voltage

$$\Rightarrow \quad V_{GS} = V_G - V_S$$

$$\Rightarrow \quad 1.259 = 0 - V_S$$

$$\Rightarrow \quad V_S = -1.259 \text{ Volt}$$

The required value of R_S is determined by

$$R_S = \frac{V_S - V_{SS}}{I_D} = \frac{-1.259 - (-2.5)}{5\times10^{-4}}$$

$$\Rightarrow \quad R_S = 2.482\ K\Omega$$

For establish a dc voltage of + 0.5V at the drain, we must select R_D as follows:

$$R_D = \frac{V_{DD} - V_D}{I_D}$$

$$\Rightarrow \quad R_D = \frac{2.5 - 0.5}{0.5 \times 10^{-3}}$$

$$\Rightarrow \quad R_D = 4\ \text{K}\Omega$$

Example 1.4: If an NMOS device has $\mu_n C_{OX} = 10\ \mu\text{A/V}^2$, $W = 10\mu\text{m}$ and $L = 1\mu\text{m}$. Find the value of drain current that the results in each these three case (*a*) $V_{DS} = 0.6\text{V}$ (*b*) $V_{DS} = 0.9\text{V}$. For $V_{GS} = 1.5\text{V}$ and $V_T = 0.7\text{V}$.

Solution: The drain current for linear region is given by

and $$I_D = \frac{\mu_n C_{OX} W}{L}\left[(V_{GS} - V_T)V_{DS} - \frac{V_{DS}{}^2}{2}\right]$$

The drain current for saturation region

$$I_D = \frac{\mu_n C_{OX} W}{2L}[V_{GS} - V_T]^2$$

(*a*) When $V_{DS} = 0.6\text{V}$

For linear region $\quad V_{DS} \le V_{GS} - V_T$

$$\Rightarrow \quad 0.5 \le 1.5 - 0.7$$

$$\Rightarrow \quad 0.5 \le 0.8$$

This device is works in linear region.

Hence, $$I_D = \frac{\mu_n C_{OX} W}{L}\left[(V_{GS} - V_T)V_{DS} - \frac{V_{DS}^2}{2}\right]$$

$$\Rightarrow \quad I_D = \frac{100 \times 10^{-6} \times 10 \times 10^{-6}}{1 \times 10^{-6}}\left[(1.5 - 0.7) \times 0.6 - \frac{(0.5)^2}{2}\right]$$

$$\Rightarrow \quad I_D = 300\ \mu\text{A}$$

(*b*) When $V_{DS} = 0.6\text{V}$

For linear region $\quad V_{DS} \le V_{GS} - V_T$

$$\Rightarrow \quad 0.9 \le 1.5 - 0.7$$

$$\Rightarrow \quad 0.9 \le 0.8$$

So, this device can not work is linear region.

For saturation region, $\quad V_{DS} \ge V_{GS} - V_T$

$$\Rightarrow \qquad 0.9 \geq 1.5 - 0.7$$

$$\Rightarrow \qquad 0.9 \geq 0.8$$

At this voltage, device works in saturation region.

Hence, the drain current

$$I_D = \frac{\mu_n C_o \times W}{2L}(V_{GS} - V_T)^2$$

$$\Rightarrow \qquad I_D = \frac{100\times10^{-6}\times10\times10^{-6}}{2\times10^{-6}}(1.5-0.7)^2$$

$$I_D = 320\ \mu A$$

Example 1.5: The drain current $I_D = 0.4$ mA for the circuit as Fig. 1.48. Find the value of R and V_D. The nMOS transistor has $V_T = 2V$, $\mu_n\ C_{ox} = 20\ \mu A/V^2$, $L = 10\ \mu m$ and $W = 100\ \mu m$.

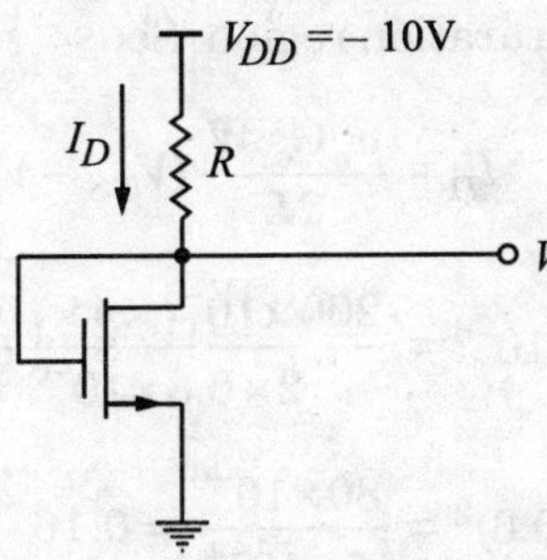

Fig. 1.48

Solution: According to the Fig. 1.48.

$$V_{GS} = V_{DS}$$

$\therefore$ But $V_s = 0$ because the source is connected to ground.

$$V_a - V_s = V_D - V_S$$

$$\Rightarrow \qquad V_G = V_D$$

Here, the MOSFET is operates in saturation region.

$$I_D = \frac{\mu_n C_{ox} W}{2L}(V_{GS} - V_T)^2$$

$$\Rightarrow \qquad 4 \times 10^{-4} = \frac{20\times10^{-6}\times100\times10^{-6}}{20\times10^{-6}}(V_{GS}-2)^2$$

$$\Rightarrow \qquad (V_{GS}-2)^2 = \frac{4\times10^{-4}}{10^{-4}} = 4$$

$$\Rightarrow \qquad V_{GS} = 2 + 2 = 4\text{Volt}$$

So that, the drain voltage will be

$$V_D = V_G = 4 \text{ Volt}$$

The required value for R

$$R = \frac{V_{DD} - V_D}{I_D}$$

$$\Rightarrow \quad R = \frac{10-4}{4\times10^{-4}}$$

$$\Rightarrow \quad R = 15\ \text{K}\Omega$$

Example 1.6: Design the circuit in Fig. 1.49. To obtain a current I_D of 80 μA. Find the value required for R and find the value of the dc voltage V_D. Let the nMOS transistor have $V_T = 0.6$V, μ*n Cox* = 200 μA/V^2, λ = 0.8 *μm* and *w* = 4 *μm*.

Solution: According to the Fig. 1.49.

$$V_G = V_D$$

and the MOSFET is operating in saturation region. So

$$I_D = \frac{\mu_n C_{ox} W}{2L}(V_{GS} - V_T)^2$$

$$\Rightarrow \quad 80 \times 10^{-6} = \frac{200\times10^{-6} \times 4\times10^{-6}}{2\times0.8\times10^{-6}}(V_{GS} - 0.6)^2$$

$$\Rightarrow \quad (V_{GS} - 0.6)^2 = \frac{80\times10^{-6}}{5\times10^{-4}} = 0.16$$

$$\Rightarrow \quad V_{GS} = 1\text{V}.$$

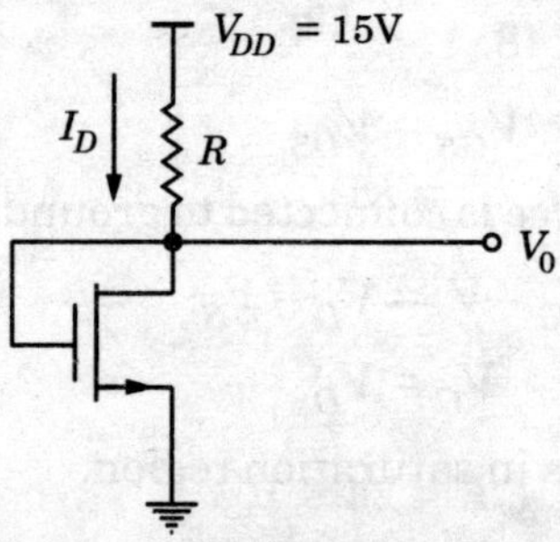

Fig. 1.49

Hence, the drain voltage will be $V_D = V_G = 1$V

The required value for R can be found as follows:

$$R = \frac{V_{DD} - V_D}{I_D}$$

$$\Rightarrow \quad R = \frac{15-1}{80\times10^{-6}}$$

$$\Rightarrow \quad R = 175\ \text{k}\Omega$$

Example 1.7: An *N*-channel E-MOSFET has the following parameters:

$$I_{D\,(ON)} = 4\ mA \text{ at } V_{GS} = 10\text{V and } V_{GS\,(th)} = 5\text{V}.$$

Calculate its drain current for $V_{as} = 8$V.

Solution: We know that, for saturation region.

$$K = \frac{I_{D\,(ON)}}{[V_{GS} - V_{GS\,(th)}]^2} = \frac{4\times10^{-3}}{(10-5)^2}$$

$$= 160 \text{ mA/V}^2$$

Now, using the value of K, I_D can be found thus,

$$I_D = K\,(V_{GS} - V_{GS\,(th)})^2$$

$$\Rightarrow \quad I_D = 160\times10^{-5}\,[8-5]^2$$

$$\Rightarrow \quad I_D = 16 \text{ mA}$$

Example 1.8: For the given circuit in Fig. 1.50. Calculate V_{GS}, I_D and V_{DS}.

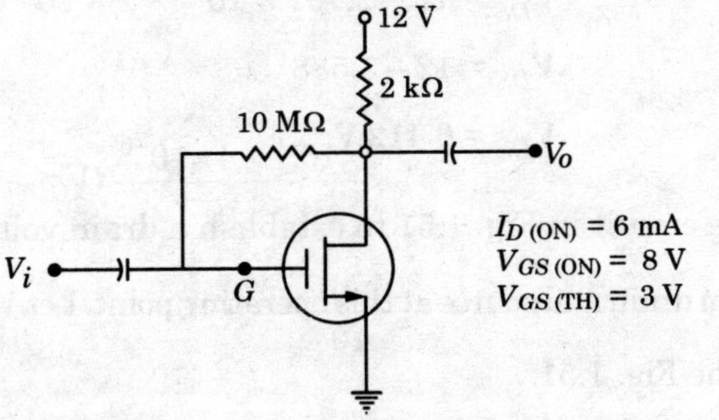

Fig. 1.50

Solution: For saturation region

$$I_{D(ON)} = K[V_{GS(ON)} - V_T]^2$$

$$\Rightarrow \quad K = \frac{I_{D\,(ON)}}{[V_{GS\,(ON)} - V_T]^2} = \frac{6\times10^{-3}}{(8-3)^2}$$

$$\Rightarrow \quad K = 24 \times 10^{-5} \text{ A/V}^2$$

$$V_{GS} = V_{DD} - I_D R_D$$

$$\Rightarrow \quad V_{GS} = 12 - I_D\,R_D$$

We have

$$I_D = K(V_{GS} - V_T)^2$$

$$\Rightarrow \quad I_D = K(12 - I_D\,R_D - V_T)^2$$

$$\Rightarrow \quad I_D = 24 \times 10^{-5}\,(12 - 2 \times 10^3 . I_D - 3)^2$$

$$\Rightarrow \quad I_D = 24 \times 10^{-5}\,(81 - 10^{-3} \times 36\,I_D + 4 \times 10^6\,I_D^2)$$

$$I_D = 0.01944 - 8.64\,I_D + 96 I_D^2$$

With the help of quadratic equation using formula $\dfrac{-b \pm \sqrt{b^2 - 4ac}}{2a}$

$$\Rightarrow \qquad I_D = \frac{+9.64 \pm 4.2755}{1920}$$

$$I_D = 2.794 \text{ mA} \quad \text{or} \quad 7.2477 \text{ mA}$$

If we calculate the value of V_{DS} taking $I_D = 7.24$ mA, we get

$$V_{DS} = V_{DD} - I_D R_D$$

$$V_{DS} = 12 - 7.2477 \times 10^{-3} \times 2 \times 10^3$$

$$V_{DS} = -2.495 \text{ V}$$

Practically, the value of V_{DS} must be positive, hence $I_D = 7.2477$ mA is invalid. Now, calculating values of V_{DS} taking $I_D = 2.794$ mA.

$$V_{DS} = 12 - 2.797 \times 10^{-3} \times 2 \times 10^3$$

$$V_{DS} = 12 - 5.588$$

$$V_{DS} = 6.412 \text{ V}$$

Example 1.9: Design the circuit in Fig. 1.51 to establish a drain voltage of 0.1 V. What is the effective resistance between drain and source at this operating point. Let $V_T = 1$V and $K'_n \dfrac{W}{L} = 1\ mA$.

Solution: According to the Fig. 1.51.

The Gate voltage is greater than the drain voltage of the MOSFET. So the MOSFET is operating in Linear region.

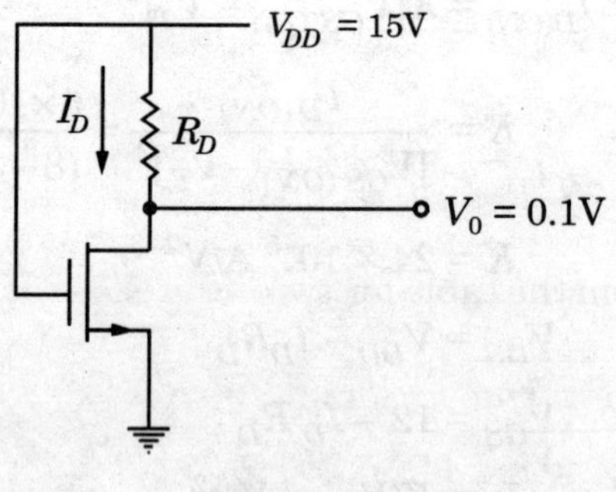

Fig. 1.51

Hence, the drain current I_D is given by

$$I_D = K'_n \frac{W}{L}\left[(V_{GS} - V_T)\, V_{DS} - \frac{V_{DS}^2}{2}\right]$$

$$\Rightarrow \qquad I_D = 1\times10^{-3}\left[(5-1)\times0.1-\frac{(0.1)^2}{2}\right]$$

$$\Rightarrow \qquad I_D = 395\ \mu\text{A}$$

The required value of R_D can be found as follows:

$$R_D = \frac{V_{DD}-V_0}{I_D}$$

$$\Rightarrow \qquad R_D = \frac{5-0.1}{395\times10^{-6}}$$

$$R_D = 12.4\ \text{K}\Omega.$$

Example 1.10. Observe the table given below. Determine the type of the MOS transistor and calculate the parameters K_n, V_{T0} and γ. Assume $\phi_F = -0.3$ V.

$V_{GS}(V)$	$V_{DS}(V)$	$V_{SB}(V)$	$I_D(\mu A)$
3	3	0	97
4	4	0	235
5	5	0	433
3	3	3	59
4	4	3	173
5	5	3	347

Solution: MOS transistor is ON. When $V_{GS} > 0$ and $V_{DS} > 0$.

Thus the transistor must be an *n*-channel MOSFET. Assuming that it is an enhancement type MOSFET, which operates in saturation region $V_{DS} = V_{GS}$, neglecting body effect.

$$I_D = \frac{K_n}{2}(V_{GS} - V_T)^2$$

$$\Rightarrow \qquad \sqrt{I_D} = \sqrt{\frac{K_n}{2}}(V_{GS} - V_T)$$

Taking two observations from the table for two difference I_{D_1} and I_{D_2}, we found that

$$\sqrt{\frac{K_n}{2}} = \frac{\sqrt{I_{D_1}} - \sqrt{I_{D_2}}}{V_{GS_1} - V_{GS_2}} = \frac{\sqrt{433} - \sqrt{97}}{5-3} = 5.48 \times 10^{-3}\ \text{A/V}^2$$

$$\Rightarrow \qquad K_n = 2 \times (5.48 \times 10^{-3})^2 = 60\ \mu\text{A/V}^2.$$

Now, we can calculate threshold voltage V_{T0} zero biasing as

$$V_{T0} = V_{GS} - \sqrt{\frac{2I_D}{K_n}} = 1.2\ \text{Volt}$$

To calculate γ, first we have to calculate threshold V_T, considering body effect so.

$$V_T(V_{SB} = 3V) = V_{GS} - \sqrt{\frac{2I_D}{K_n}} = 4 - \sqrt{\frac{2.173}{60}} = 1.6 \text{ Volts}$$

Now

$$\gamma = \frac{V_T(V_{SB} = 3) - V_{T0}}{\sqrt{(2\phi_F) + V_{SB}} - \sqrt{|2\phi_F|}} = \frac{1.6 - 1.2}{\sqrt{3 + 0.6} - \sqrt{0.6}}$$

$$\gamma = 0.36 \text{ Volt}$$

Example 1.11. Calculate the drain current in an NMOS transistor for $V_{GS} = 0$ V, 1 V and 2 V with $V_{DS} = 0.1$ V. The device parameter are $W = 16$ μm, $L = 2$ mm, $V_{Th} = 1$ V and $\mu_n G_{ox} = 50$ μA/V^2.

Solution: The drain current

$$I_D = \mu_n C_{ox}\left(\frac{W}{L}\right)\left[(V_{GS} - V_{th})V_{DS} - \frac{V_{DS}^2}{2}\right]$$

when $V_{th} \geq V_{GS}$, the channel is not present and carrier conduction does not take place,

so $\quad I_D = 0$

for $\quad V_{GS} = 0$ V and 1 V

for $\quad V_{GS} = 2$ V

$$I_D = 50 \times 10^{-6} \times \frac{16 \times 10^{-6}}{2 \times 10^{-6}}\left[(2-1)0.1 - \frac{(0.1)^2}{2}\right] = 38\ \mu\text{A}$$

Example 1.12. For the arrangement in Fig. 1.52(*a*) plot the on-resistance of M_1 as a function of V_G. Assume $\mu_n C_{ox} = 50$ μA/V^2, $W_L = 10$ and $V_{th} = 0.7$ V.

Note that the drain terminal is open

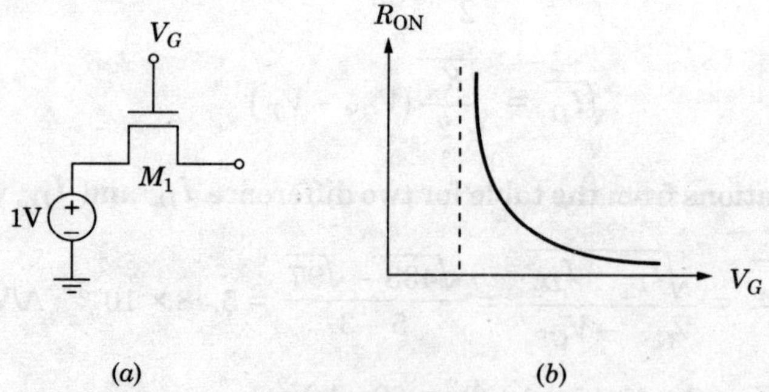

(*a*) (*b*)

Fig. 1.52

Solution: Since the drain terminal is open $I_D = 0$ and $V_{DS} = 0$. Thus the device is on, it operates in the deep triode region. For $V_G < 1\text{ V} + V_{th}$, M1 of and $R_{on} = \infty$.

For $V_G > 1\text{ V} + V_{th}$, then

$$R_{on} = \frac{1}{\mu_n C_{ox}\left(\frac{W}{L}\right)(V_{GS} - V_{th})}$$

$$= \frac{1}{50 \times 10^{-6} \times 10(V_G - 1 - 0.7)}$$

The result is plotted in Fig. 4.30(*b*).

Example 1.13. Consider an ideal n-channel MOSFET with parameters L = 1.25 μm, μ_n = 650 Cm²/V-s, $C_{ox} = 6.9 \times 10^{-8}$ F/Cm², and V_{TH} = 0.65 V. Design the channel width W such that I_D(sat) = 4 mA for V_G = 5 V.

Solution:

$$I_{D(\text{sat})} = \frac{W}{L} \cdot \frac{\mu_n C_{ox}}{2} (V_G - V_{TH})^2$$

$$4 \times 10^{-3} = \frac{W(650)\,(6.9 \times 10^{-8})}{2(1.25 \times 10^{-4})} \cdot (5 - 0.65)^2$$

$\therefore$ $$4 \times 10^{-3} = 3.39\ W$$

$\therefore$ $$W = 11.8\ \mu\text{m}$$

Example 1.14. Consider an n-channel MOSFET with W = 15 μm, L = 2 μm and $C_{ox} = 6.9 \times 10^{-8}$F/cm². Assume that the drain current in the non saturation region for V_{DS} = 0.10 V is I_D = 35 μA at V_{GS} = 1.5 V and I_D = 75 μA at V_{GS} = 2.5 V.

Solution: The inversion carrier mobility is find out by these analysis:

$\therefore$ The equation we can write,

$$I_{D2} - I_{D1} = \mu_n C_{ox} \cdot \frac{W}{L}(V_{GS2} - V_{GS1})V_{DS}$$

Here $V_{GS1} = V_{\text{th}}$ assumed,

$\therefore$ $$75 \times 10^{-6} - 35 \times 10^{-6} = \left(\frac{15}{2}\right)\mu_n (6.9 \times 10^{-8})(2.5 - 1.5)(0.10)$$

$\therefore$ $$\mu_n = 773\ \text{cm}^2/\text{V-s}$$

and V_{th} is given by

$\therefore$ $$75 \times 10^{-6} = \left(\frac{15}{2}\right)(773)\,(6.9 \times 10^{-8})\,(2.5 - V_{TH})(0.10)$$

$\therefore$ $$V_{TH} = 0.625\ \text{V}.$$

Example 1.15. Calculate the threshold V_T with no substrate bias for NMOS silicon gate device that has $N_A = 10^{15}/cm^3$ the thickness is 700° A. The positive ions per unit area at the Si–SiO$_2$ interface is 2 × 10^{10}/cm^2, $\phi_{Fsub} = 0.29$ V, $n_i = 1.4 \times 10^{10}/cm^3$ at 300°K, $\varepsilon_{ox} = 3.5 \times 10^{-13}$ F/cm, the relative permittivity of silicon is 11.8.

Solution: The surface electrostatic potential ϕ_F is given by

$$\therefore \quad \phi_{F(sub)} = \frac{kT}{q}\ln\left(\frac{n_i}{N_A}\right)$$

$$= \frac{300 \times 1.38 \times 10^{-23}}{1.6 \times 10^{-19}}\ln\left(\frac{1.4 \times 10^{10}}{10^{15}}\right)$$

$$\therefore \quad \phi_{F(sub)} = -0.347 \text{ V}$$

Here we assume, that $\phi_{F(gate)}$ is equal to conduction band potential:

$$\phi_{F(gate)} = 0.55 \text{ V}$$

and

$$\phi_{ms} = \phi_{F(sub)} - \phi_{F(gate)}$$

Here

$$= -0.347 - 0.55 = -0.897 \text{ V}$$

$$\therefore \quad Q_B = -\sqrt{2qN_A\varepsilon_{si}\,|-2\phi_{F(sub)}\,|}$$

$$= -\sqrt{2 \times 1.6 \times 10^{-19} \times 1.03 \times 10^{-12}\,|-2 \times (-0.347)}$$

$$\therefore \quad Q_B = -4.782 \times 10^{-8} \text{ C/cm}^2$$

Hence,

$$\therefore \quad Q_{ox} = q\,N_{ox}$$

$$= 1.6 \times 10^{-19} \times 2 \times 10^{10}$$

$$\therefore \quad Q_{ox} = 3.2 \times 10^{-19} \text{ C/cm}^2$$

The oxide capacitance is given by

$$C_{ox} = \frac{\varepsilon_{ox}}{t_{ox}}$$

$$\therefore \quad C_{ox} = \frac{3.5 \times 10^{-13}}{700 \times 10^{-8}} = 5 \times 10^{-8} \text{ F/cm}^2$$

If there is no substrate bias, the threshold voltage can be calculated as:

$$\therefore \quad V_{T0} = \phi_m - 2\phi_{F(sub)} - \frac{Q_{BO}}{C_{ox}} - \frac{Q_{ox}}{C_{ox}}$$

$$= -0.897 + 0.694 + 0.956 - 0.064$$

$$\therefore \quad V_{T0} = 0.689 \text{ V}$$

Example 1.16. Calculate the capacitance developed across a reverse biased *p-n* junction with $N_D = 2.1 \times 10^{20}/\text{cm}^3$ and $N_A = 10^{17}/\text{cm}^3$. The junction area is 22 μm × 16 μm and intrinsic carrier concentration is $1.45 \times 10^{10}/\text{cm}^3$.

Solution: For find the junction capacitance first find out the zero bias junction capacitance. Include this in the expression for large signal junction capacitance.

The potential ϕ_0 is given

$$\phi_0 = \frac{kT}{q}\ln\left(\frac{N_A N_D}{n^2}\right) = 0.026\ln\left(\frac{10^{17} \times 2.1\times10^{20}}{(1.45\times10^{10})^2}\right) = 1.0177 \text{ V}$$

and the zero bias junction capacitance is given by

$$C_{j0} = \sqrt{\frac{\varepsilon_{si} \times q}{2}\left(\frac{N_A N_D}{N_A + N_D}\right)\frac{1}{\phi_b}}$$

$$= \sqrt{\frac{11.7\times 8.85\times10^{-14} \times 1.6\times10^{-19}}{2}\left(\frac{10^{17} \times 2.1\times10^{20}}{10^{17} + 2.1\times10^{20}}\right)\frac{1}{1.0177}}$$

$$= 2.852 \times 10^{-6} \text{ F/cm}^2.$$

The large signal junction capacitance in reverse bias of – 5 V with initial voltage of 0 V (zero bias) is obtained by finding the voltage equivalence factor as:

$$K_{eq} = \frac{-2\sqrt{\phi_0}}{V_2 - V_1}\left(\sqrt{\phi_0 - V_2} - \sqrt{\phi_b - V_1}\right)$$

$$= \frac{-2\sqrt{1.017}}{-5-0}\left(\sqrt{1.0177-(-5)} - \sqrt{1.0177-0}\right) = 0.582$$

Hence the average junction capacitance is given as:

$$C_{\text{eq}} = A \cdot C_{j0}\, k_{eq}$$

Here A is junction area,

i.e., $A = 22 \text{ μm} \times 16 \text{ μm} = 352 \text{ μm}^2$

∴ $C_{eq} = 352 \times 2.852 \times 10^{-6} \times 0.582$

=58.42 × 10–15 Farads.

Example 1.17. An *n*-channel MOSFET is described by following parameters:

$N_A = 10^{17}/\text{cm}^3$

Source/drain doping $N_D = 1.2 \times 10^{20}/\text{cm}^3$

Side wall doping $N_A(\text{SW}) = 4.8 \times 10^{16}/\text{cm}^3$

Oxide layer thickness $t_{ox} = 400°$ A and

Junction depth $xj = 1$ mm

Find the zero bias capacitance in both the cases.

Solution: The built in potential is given by

$$\phi_0 = \frac{kT}{q}\ln\left[\frac{N_A N_D}{n_i}\right] = 0.026\ln\left[\frac{10^{17}\times 1.2\times 10^{20}}{(1.45\times 10^{10})^2}\right] = 1.0031$$

$$\phi_{OSW} = \frac{kT}{q}\ln\left[\frac{N_{A(SW)} N_D}{n_i}\right] = 0.026\ln\left[\frac{4.8\times 10^{6}\times 1.2\times 10^{20}}{(1.45\times 10^{10})^2}\right] = 0.984$$

and the zero bias junction capacitance per unit area is given by

$$C_{j0} = \sqrt{\frac{\varepsilon_{si}\, q}{2}\left[\frac{N_A N_D}{N_A + N_D}\right]\frac{1}{\phi_0}}$$

$$= \sqrt{\frac{11.7\times 8.85\times 10^{-14}\times 1.6\times 10^{-19}}{2}\left[\frac{10^{17}\times 1.2\times 10^{20}}{10^{17}+1.2\times 10^{20}}\right]\cdot\frac{1}{1.0031}}$$

$$= \sqrt{8.251\times 10^{-12}} = 2.872\times 10^{-6}\ \text{F/cm}^2$$

$$C_{j0\,sw} = \sqrt{\frac{\varepsilon_{si}\, q}{2}\left[\frac{N_{A(SW)} N_D}{N_{A(SW)} + N_D}\right]\cdot\frac{1}{\phi_{OSW}}}$$

$$= \sqrt{\frac{11.7\times 8.85\times 10^{-14}\times 1.6\times 10^{-19}}{2}\left[\frac{4.8\times 10^{16}\times 1.2\times 10^{20}}{4.8\times 10^{16}+1.2\times 10^{20}}\right]\cdot\frac{1}{0.984}}$$

$$= 2.009\times 10^{-6}\ \text{F/cm}^2.$$

Example 1.18. Bandgap of silicon is 1.1 eV. If the equilibrium fermi potential of *p*-type doped silicon substrate is 0.42 eV. Calculate the built in potential difference across the MOS system. Also find the flat band voltage.

Solution: Given that

$$q\,\phi_m = 4.1\ \text{eV} \quad \text{for Aluminium}$$

The silicon work function is given by,

$$q\,\phi_s = q + [E_C - E_{FP}]$$

$$= 4.15 + [E_C - E_i + E_i - E_p]$$

$$= 4.15 + 0.55 + 0.42 = 5.12\ \text{eV}$$

and the built in potential is given by

$$V_{bi} = q\,\phi_m - q\,\phi_s$$

$$= 4.1\ \text{eV} - 5.12\ \text{eV} = -1.02\ \text{eV}$$

The built in potential is –1.02 eV.

Example 1.19. In an NMOS transistor operating at 300°K following measurements are done:

$$V_G = 4 \text{ V}$$
$$V_{DS} = 5 \text{ V}$$
$$V_{SB} = 2.6 \text{ V}$$
$$I_D = 148 \text{ mA}$$

The process parameters are,

$$W/L = 1, \quad t_{ox} = 500°\text{A}, \quad |2\phi_F| = 0.68 \text{ V} \quad \text{and} \quad N_A = 1.2 \times 10^{16}/\text{cm}^3$$

Find the threshold voltage, electron mobility and body effect coefficient γ. Assume oxide charge $Q_{ox} = 0$.

Solution:

$$\therefore \qquad \phi_{F\,(\text{sub})} = -0.34 \text{ V (Given)}$$
$$\phi_{F\,(\text{gate})} = 0.55 \text{ V}$$
$$\therefore \qquad \phi_{ms} = \phi_{F\,(\text{sub})} - \phi_{F\,(\text{gate})}$$
$$= -0.34 - 0.55 = -0.89 \text{ V}$$

The threshold voltage V_{T0} is given by

$$V_{T0} = \phi_{ms} - 2\phi_{F(\text{sub})} - \frac{Q_{BO}}{C_{ox}} - \frac{Q_{ox}}{C_{ox}}$$

$$= -0.89 - 0.68 + \frac{\sqrt{2 \times 16 \times 10^{-19} \times 1.2 \times 10^{16} \times 1.04 \times 10^{-12} \times 0.68}}{\dfrac{3.97 \times 8.854 \times 10^{-14}}{500 \times 10^{-8}}}$$

$$= -0.89 - 0.68 + 0.742$$

$$V_{TO} = -0.82 \text{ V}$$

and the body effect coefficient γ is calculated as

$$\gamma = \frac{\sqrt{2q\,N_A \varepsilon_{si}}}{C_{ox}}$$

$$= \frac{\sqrt{2 \times 1.6 \times 10^{-9} \times 1.2 \times 10^{16} \times 1.04 \times 10^{-12}}}{7.02 \times 10^{-8}}$$

$$\gamma = 0.90$$

The V_{TH} for NMOS:

$$V_{Tn} = V_{TO} + \gamma\left(\sqrt{|-2\phi_{F(\text{sub})} + V_{SB}|} - \sqrt{|2\phi_F|}\right)$$

$$= -0.82 + 0.92\left(\sqrt{|-0.68 + 2.6|} + \sqrt{|0.68|}\right)$$

$$V_{Tn} = -0.41 \text{ V}$$

so, $$V_G - V_{Tn} = 4.0 + 0.41 < V_{DS} = 5 \text{ V}$$

∴ The device is saturation region.

$$I_D = \mu_n C_{ox} \frac{W}{L}(V_G - V_{TH})^2$$

$$148 \times 10^{-6} = \mu_n \times 7.02 \times 10^{-8} \times 1(4 - (-0.41))^2$$

∴ $$\mu_n = 108.4 \text{ cm}^2/\text{s}.$$

Example 1.20. For a MOSFT the following parameters are given $V_{T0} = 0.7$ V, $r = 0.2$, $|2\phi_F| = 0.84$ V, $K' = 20 \times 10^{-6}$ A/V^2, $V_G = 2.9$ V, $V_{DS} = 5$ V and $I_D = 193.6$ μA. Find the width-length ratio. Assume the substrate bias voltage is zero.

Solution: Here, substrate bias voltage is zero:

$$V_{TH} = V_{T0} = 0.7 \text{ V}$$

Now $V_G - V_{TH} < V_{DS}$. So the transistor is in saturation region. There current relation can be given as:

$$I_D = K' \frac{W}{L}(V_G - V_{TH})^2$$

$$193.6 \times 10^{-6} = 20 \times 10^{-6} \frac{W}{L}(2.9 - 0.7)^2$$

∴ $$\frac{W}{L} \cong 2$$

Hence if Length $L = 1$ μm then width $W = 2$ μm.

Example 1.21. Calculate the threshold voltage of nmos transistor with substrate grounded. The substrate doping density $N_A = 2.1 \times 10^{15}/\text{cm}^3$, polysilicon gate doping density of $N_D = 1 \times 10^{20}/\text{cm}^3$, gage oxide thickness ($t_{ox}$) is 900°A and oxide interface fixed change density N_{ox} is $1.2 \times 10^{10}/\text{cm}^2$. Assume no threshold adjust implantation impurities are added.

Solution:

∴ $$\phi_{F(sub)} = \frac{kT}{q} \ln\left(\frac{n_i}{N_A}\right) = 0.026 \ln\left(\frac{1.45 \times 10^{20}}{2.1 \times 10^{15}}\right)$$

∴ $$\phi_{F(\text{sub})} = -0.308 \text{ V}$$

Since N_D of polysilicon gate is very high, it is assumed that $\phi_{F(\text{gate})}$ is equal to the conduction band potential, so

$$\phi_{F(\text{gate})} = 0.55 \text{ V}$$

The metal semiconductor work function for polysilicon gate is given by

$$\phi_{ms} = \phi_{F(\text{sub})} - \phi_{F(\text{gate})}$$
$$= -0.308\text{ V} - 0.55\text{ V} = -0.858\text{ V}$$

The depletion region charge density is calculated as

$$Q_B = -\sqrt{2q\,N_A\,\varepsilon_{si}\,|-2\phi_{F(\text{sub})}|} \quad \text{negative sign = nmos}$$
$$= -\sqrt{2\times1.6\times10^{-19}\times2.1\times10^{15}\times11.7\times8.85\times10^{-14}\,|-2\times(-0.308)|}$$
$$= -\sqrt{4.263\times10^{-16}}$$

$\therefore$ $$Q_B = -2.064\times10^{-8}\text{ C/cm.}$$

and $$Q_{ox} = q\,N_{ox}$$
$$= 1.6\times10^{-19}\times1.2\times10^{10}$$
$$= 1.92\times10^{-19}\text{ C/cm}^2$$

The gate oxide capacitance is given by

$$C_{OX} = \frac{\varepsilon_{ox}}{t_{ox}} = \frac{3.97\times8.85\times10^{-14}}{900\times10^{-8}}$$

$\therefore$ $$C_{OX} = 3.9\times10^{-8}\text{ F/cm}^2$$

When the substrate is grounded, the threshold voltage can be calculated as:

$$V_{TO} = \phi_{ms} - 2\phi - \frac{Q_{BO}}{C_{OX}} - \frac{Q_{ox}}{C_{OX}}$$
$$= -0.858 + 0.616 + 0.529 - 0.049$$
$$V_{T0} = +0.238\text{ V}$$

Example 1.22. A MOS transistor has threshold voltage of 0.85 V. The body effect coefficient $\gamma = 0.64$. Calculate the threshold voltage for $V_{SB} = 4$ V and $2\phi_F = -0.58$ V. **(UTU-2012)**

Solution: As the $V_{SB} \neq 0$, threshold voltage is calculated as,

$$V_T = V_{TO} + \phi\left(\sqrt{|-2\phi_F| + V_{SB}} - \sqrt{|-2\phi_F|}\right)$$
$$= 0.85 + 0.64\left(\sqrt{0.58+4} - \sqrt{0.58}\right)$$

$\therefore$ $$V_T = 1.732\text{ V}$$

Example 1.23. Find the depletion- layer with x_d, the depletion region charge Q_{BO}, the threshold voltage with no source to-body voltage V_{THO}, and body factor γ of a device with the following physical parameters $t_{ox} = 400°\text{A}$, $N_A = 1.5\times10^{16}/\text{cm}^3$, $N_D = 10^{18}/\text{cm}^3$, $N_{ox} = 5\times10^{10}\text{ cm}^2$.

Solution:

$$\phi_F = \frac{kT}{q}\ln\left(\frac{n_i}{N_A}\right) = 0.0259\ln\left(\frac{1.45\times10^{10}}{1.5\times10^{16}}\right) = -0.3587\text{ V}$$

The depletion region width is calculated as follows:

$$x_d = \sqrt{\frac{2\varepsilon_{si}\,|\phi_s - \phi_F|}{qN_A}}$$

$$x_d = \sqrt{\frac{2\times 1.03\times 10^{-12}\times 0.717}{1.6\times 10^{-19}\times 1.5\times 10^{16}}}$$

$$x_d = 0.518\ \mu\text{m}.$$

and the charge in the channel at the onset of strong inversion is given by:

$$Q_B = -\sqrt{2n\,qN_A\varepsilon_{si}\,|-2\phi_F|}$$

$$= -\sqrt{2\times 1.6\times 10^{-19}\times 1.5\times 10^{16}\times 1.03\times 10^{-12}\times 0.717}$$

$$= -59.7\times 10^{-9}\ \text{C/cm}^2$$

and

$$\therefore \qquad \phi_{F(\text{gate})} = \frac{KT}{q}\ln\left(\frac{N_D}{n_i}\right) = 0.0259\ln\left(\frac{10^{18}}{1.45\times 10^{10}}\right) = 0.4675\ \text{V}$$

The work difference between metal work function and semiconductor work function is give by

$$\therefore \qquad \phi_{ms} = \phi_{F(\text{sub})} - \phi_{F(\text{gate})} = -0.3587 - 0.4675 = -0.826\ \text{V}$$

$$C_{ox} = \frac{\varepsilon_{ox}}{t_{ox}} = \frac{0.345\times 10^{-12}}{4\times 10^{-6}} = 86\ \text{nF/cm}^2$$

$$Q_{ox} = q\,N_{ox} = 1.6\times 10^{-19}\,(5\times 10^{10}) = 8\ \text{nC/cm}^2$$

So, the threshold voltage V_{THO} is :

$$V_{\text{THO}} = \phi_{ms} - 2\phi_F - \frac{Q_{ox}}{C_{ox}} - \frac{Q_{BO}}{C_{ox}}$$

$$= -0.826 + 0.717 - 0.093 + 0.692 = 0.490\ \text{V}$$

Hence, the body factor:

$$\gamma = \frac{\sqrt{2q\,\varepsilon_{si}\,N_A}}{C_{ox}}$$

$$\gamma = \frac{\sqrt{3.2\times 10^{-19}(1.035\times 10^{-12})(1.5\times 10^{16})}}{86\times 10^{-19}} = 0.817\ V^{1/2}$$

Example 1.24. Calculate the ion implant does necessary to change the threshold voltage V_{THO} of the device in Ex. 1.23 to +1 V or to – 4 V. Assume 100 percent ionisation of implanted material.

Solution:

For $\qquad V_{THE} = +1.0\ \text{V}$

$$N_I = \frac{Q_I}{q} = \frac{C_{ox}}{q}(V_{\text{THE}} - V_{\text{THO}}) = \left(\frac{86\times10^{-9}}{1.6\times10^{-19}}\right)(1-0.49)$$

$$= 0.275 \times 10^{12} \text{ ions/cm}^2$$

For $\quad V_{THD} = -1.0 \text{ V}$

$$N_I = \frac{Q_1}{q} = \frac{C_{ox}}{-q}(V_{\text{THD}} - V_{\text{THO}})$$

$$= \left(\frac{86\times10^{-9}}{1.6\times10^{-19}}\right)(1-0.49) = 2.275 \times 10^{12} \text{ ions/cm}^2$$

For $\quad V_{THD} = -4.0 \text{ V}$

$$N_I = \frac{Q_1}{q} = \frac{C_{ox}}{-q}(V_{\text{THD}} - V_{\text{THO}})$$

$$= \left(\frac{86\times10^{-9}}{-1.6\times10^{-19}}\right)(-4-0.49) = 2.425 \times 10^{12} \text{ ions/cm}^2$$

Example 1.25. Calculate the drain current in an NMOS transistor for $V_G = 0$ V, 1 V and 2 V with $V_{DS} = 0.2$ V. The device parameter are $W = 8$ μm, $L = 1.5$ μm, $V_{TH} = 1$ V and $\mu_n C_{ox} = 26$ μA/V².

Solution: The drain current is given by:

$$I_D = \frac{\mu_n C_{ox}}{2}\frac{W}{L}[2(V_G - V_{TH})V_{DS} - V_{DS}^2]$$

For $\quad V_G = 0$ V

∵ $V_G < V_{TH}$ the channel is not present and carrier conduction does not take place.

For $\quad V_G = 1$ V

∵ $V_G \le V_{TH}$ conduction does not take place and $I_D = 0$

For $\quad V_G = 2$ V

∵ VG > VTH transistor is in linear region

Hence, $$I_D = 26\times10^{-6} \times \frac{8\times10^{-6}}{1.5\times10^{-6}}[2(2-1)0.2-(0.2)^2]$$

$$= 49.92 \quad \mu\text{A}$$

Example 1.26. Consider, an n-channel silicon MOSFET at $T = 300$°K. Assume the substrate is doped to $N_A = 3 \times 10^{16}/\text{cm}^3$ and assume $t_{ox} = 500$°A and $V_{SB} = 1$ V. Calculate the change in the threshold voltage due to an applied source to body voltage.

Solution:

$$\therefore \quad \phi_F = \frac{kT}{q}\ln\left(\frac{N_A}{n_i}\right) = 0.0259 \ln\left(\frac{3\times10^{16}}{1.45\times10^{10}}\right) = 0.376 \text{ V}$$

and

$$\therefore \quad C_{ox} = \frac{\varepsilon_{ox}}{t_{ox}} = \frac{3.9\times 8.854\times 10^{-14}}{500\times 10^{-8}} = 6.9\times 10^{-8}\ \text{F/cm}^2$$

Hence, change in threshold voltage is calculated as:

$$\Delta V_{TH} = \frac{\sqrt{2q\,\varepsilon_{si}\,N_A}}{C_{ox}}\left[\sqrt{2\,\phi_F + V_{SB}} - \sqrt{2\phi_F}\right]$$

$$= \frac{\sqrt{2\times 1.6\times 10^{-19}\times 11.7\times 8.854\times 10^{-14}\times 3\times 10^{16}}}{6.9\times 10^{-18}}$$

$$\times\left[\sqrt{(2\times 0.376)+1} - \sqrt{2\times 0.376}\right]$$

$$\therefore \quad \Delta V_{TH} = 1.445(1.324 - 0.867) = 0.66\ \text{V}$$

Example 1.27. Find the average value of C_{db} of the NMOS transistor junction with 5 V reverse bias voltage having the following parameters:

1. $N_A = 2\times 10^{15}/\text{cm}^3$
2. N_A (side wall) $= 2.1\times 10^{16}/\text{cm}^3$
3. $N_D = 10^{19}/\text{cm}^3$
4. $X_j = 0.5\ \mu\text{m}$
5. $t_{ox} = 200°\text{A}$
6. $L_D = 0.2\ \mu\text{m}$
7. $A = 55\ \mu\text{m}^2$
8. $P = 30\ \mu\text{m}$.

Solution: The value of C_{ox} is given by

$$\therefore \quad C_{ox} = \frac{\varepsilon_{ox}}{t_{ox}} = \frac{3.9\times 8.85\times 10^{-14}}{200\times 10^{-8}}$$

$$\therefore \quad C_{ox} = 1.73\times 10^{-7}\ \text{F/cm}^2$$

and the value of body effect coefficient (γ):

$$\therefore \quad \gamma = \frac{\sqrt{2qN_A\varepsilon_{si}}}{C_{ox}} = \frac{\sqrt{2\times 1.6\times 10^{-19}\times 2\times 10^{15}\times 11.7\times 8.85\times 10^{-14}}}{1.73\times 10^{-7}}$$

$$\gamma = 0.15\ \text{V}^{1/2}$$

So,

$$\therefore \quad \phi_0 = \frac{kT}{q}\ln\left(\frac{N_A\,N_D}{ni^2}\right) = 0.0259\ln\left[\frac{2\times 10^{15}\times 10^{19}}{(1.45\times 10^{10})^2}\right]$$

$$\therefore \quad f_0 = 0.837\ \text{V}$$

and

$$\phi_{0(sw)} = \frac{kT}{q}\ln\frac{N_{A(sw)}\cdot N_D}{ni^2} = 0.0259\ln\left[\frac{2.1\times 10^{16}\times 10^{19}}{(1.45\times 10^{10})^2}\right]$$

$$\phi_{0(sw)} = 0.898\ \text{V}$$

Hence, the rain-substrate junction capacitance can be calculated as:

$$C_{db} = A \times C_{j0} \times k_{eq} + P \times C_{j(sw)} \times k_{eq(sw)}$$
$$= (55 \times 10^{-8} \times 1.41 \times 10^{-4} \times 0.54) + (30 \times 10^{-8} \times 2.2 \times 10^{-10} \times 0.56)$$
$$= 4.18 \times 10^{-11} + 3.696 \times 10^{-17}$$

Example 1.28. For an NMOS transistor $K_n = 20$ mA/V^2 and $V_{TH} = 1.5$ V is operated at $V_{GS} = 5$ V and $I_D = 100$ µA find V_{DS}.

Solution: The Drain current is give by

$$\therefore \quad I_{D(sat)} = \frac{K_n}{2}(5 - 1.5)^2 = 122.5 \text{ mA}$$

Thus transistor is in linear region,

$$\therefore \quad I_{D(LIN)} = 100 \text{ mA}$$

$$I_{D(liner)} = \frac{K_n}{2}\left[2(V_{GS} - V_{TH})V_{DS} - V_{DS}^2\right]$$

$$100 \times 10^{-6} = \frac{20 \times 10^{-6}}{2}[2(5 - 1.5)V_{DS} - V_{DS}^2]$$

$$\Rightarrow \quad 7V_{DS} - V_{DS}^2 - 10 = 0$$

$$\Rightarrow \quad V_{DS} = 2 \text{ V} \quad \text{or} \quad VDS = 5 \text{ V}$$

Since in linear region, $V_{DS} < V_{GS} - V_{th}$

$$\therefore \quad V_{DS} = 2\text{V}$$

Example 1.29. Find the flatband voltage V_{FB} and the voltage V_{TH2} of the device of Ex. 4.14. Find the threshold voltage of this device with $V_{SB} = 0$ V.

Solution:

$$Q_{ox} = 8 \times 10^{-9} \text{ C/cm}^2$$
$$\phi_{ms} = -0.826 \text{ V}$$
$$C_{ox} = 860 \times 10\text{–}8 \text{ F/cm}^2$$

$$\therefore \quad V_{FB} = \phi_{ms} - \frac{Q_{ox}}{C_{ox}} = -0.826 - \frac{8 \times 10^{-19}}{860 \times 10^{-8}} = -0.919 \text{ V}$$

$$V_{TH2} = -2\phi_F - \frac{Q_{B0}}{C_{ox}} = 0.717 + \frac{59.7 \times 10^{-9}}{860 \times 10^{-8}} = 1.409 \text{ V}$$

Example 1.30. Find the threshold voltage and body factor *r* for an *n*-channel transistor with an n^+ silicon gate if $t_{ox} = 200°$ A $N_A = 3 \times 10^{16}$/cm^3, gate doping, $N_D = 4 \times 10^{19}$/cm and of the positively charged ions at the oxide-silicon interface per area is 10^{10}/cm^2.

Solution:

Assuming temperature $T = 300$°K

$$\therefore \quad \phi_{F(sub)} = \frac{kT}{q}\ln\left(\frac{n_i}{N_A}\right) = 0.026 \ln\left(\frac{1.45 \times 10^{10}}{3 \times 10^{16}}\right)$$

$$\therefore \qquad \phi_{F(\text{sub})} = -0.378 \text{ V}$$

The equilibrium electrostatic potential for the n^+ polysilicon gate is calculated by

$$\therefore \qquad \phi_{F(\text{gate})} = \frac{kT}{q} \ln\left(\frac{N_D}{n_i}\right) = 0.026 \ln\left(\frac{4\times10^{19}}{1.45\times10^{10}}\right)$$

$$\therefore \qquad \phi_{F(\text{gate})} = 0.563 \text{ V}$$

the ϕ_{ms} can be found as:

$$\phi_{ms} = \phi_{F(\text{sub})} - \phi_{F(\text{gate})}$$

$$= -0.378 - 0.563 = -0.941 \text{ V}$$

and $$C_{ox} = \frac{\varepsilon_{ox}}{t_{ox}} = \frac{3.9\times8.854\times10^{-14}}{200\times10^{-8}} = 1.727\times10^{-7} \text{ F/cm}^2$$

The fixed charge in the depletion region Q_{BO} is given as:

$$Q_B = -\sqrt{2qN_A\varepsilon_{si}\,|2\phi_{F(\text{sub})}|}$$

$$= -\sqrt{2\times1.6\times10^{-19}\times3\times10^{9}\times11.7\times8.85\times10^{-14}\,|-2(-0.378)|}$$

$$\therefore \qquad Q_B = -8.66\times10^{-8} \text{ C/cm}^2$$

$$\therefore \qquad Q_{OX} = qN_{OX} = 1.6\times10^{-19}\times10^{10} = 1.6\times10^{-8} \text{ C/cm}^2$$

The threshold voltage is given by

$$\therefore \qquad V_{TO} = \phi_{ms} - 2\phi_{F(\text{sub})} - \frac{Q_{BO}}{C_{ox}} - \frac{Q_{OX}}{C_{OX}}$$

$$= -0.941 + 0.756 + \frac{-8.66\times10^{-8}}{1.727\times10^{-7}} - \frac{1.6\times10^{-8}}{1.727\times10^{-7}}$$

$$= -0.941 + 0.756 + 0.501 - 0.0926 = 0.2234 \text{ V}$$

Hence, the body factor is found as:

$$\gamma = \frac{\sqrt{2qN_A\varepsilon_{si}}}{C_{ox}} = \frac{\sqrt{2\times1.6\times10^{-19}\times3\times10^{16}\times11.7\times8.854\times10^{-14}}}{1.727\times10^{-7}}$$

$$\therefore \qquad \gamma = 0.5774$$

Example 1.31. Consider a n-channel MOSFET with $N_A = 3\times10^{16}/\text{cm}^3$ and $t_{ox} = 450°\text{A}$. Assume that we want to limit the threshold shift to $\Delta V_{TH} = 0.2$ V. Calculate the narrow channel width.

Solution: We Know that, the oxide capacitance is given by

$$\therefore \qquad C_{OX} = \frac{\varepsilon_{OX}}{t_{OX}} = \frac{3.9\times8.85\times10^{-14}}{450\times10^{-8}} = 7.67\times10^{-8} \text{ F/cm}$$

Calculation for X_{dT}:

$$\therefore \quad X_{dT} = \sqrt{\frac{4\varepsilon_{si}\,\phi_F}{qN_a}}$$

$$= \sqrt{\frac{4(11.7)\,(8.85\times10^{-14})\,(0.376)}{1.6\times10^{-19}\times3\times10}} = 0.18\ \mu\text{m}$$

The width of channel is given by

$$W = \frac{q.N_A\pi\cdot x_{dt}}{C_{ox}\ \Delta V_{th}}$$

$$= \frac{1.6\times10^{-9}\times3\times10^{16}\times(0.18\times10^{-4})^2\,\pi}{0.2\times7.67\times10^{-8}} = 1.59\ \mu\text{m}.$$

Example 1.32. Consider PMOS that has gate oxide thickness of t_{ox} = 60 Å, the hole mobility 220 cm^2/V-sec, aspect ratio W/L = 12/1. Assume that V_{DD} = 3.3 V, V_{TP} = –0.7 V, ε_{ox} = 0.345 × 10^{-12}

(i) Calculate the process transconductance k'_p

(ii) Find the device transconductance k_p and resistance R_{eqp}.

Solution: (i) We know that

$$k'_p = \mu_p \cdot C_{ox}$$

$$k'_p = \mu_p \cdot \frac{\varepsilon_{ox}}{t_{ox}}$$

$$k'_p = 220 \times \frac{0.345\times10^{-12}}{60\times10^{-8}}$$

$$k'_p = 126.5\ \mu\text{A/V}^2 \quad \textbf{Ans}$$

(ii) The device trans conductance k_p is given by

$$k_p = k'_p \cdot \left(\frac{W}{L}\right)$$

$$k_p = 126.5\times10^{-6}\times12$$

$$k_p = 15.8\ \text{mA/V}^2 \quad \textbf{Ans}$$

The channel resistance

$$R_{eqp} = \frac{1}{k_p(V_{DD} - |V_{TP}|)}$$

$$= \frac{1}{15.8\times10^{-6}(3.30.7)}$$

$$= 253.3\ \text{kW} \quad \textbf{Ans}$$

Example 1.33. PMOS transistor has substrate doping $N_D = 10^{16}/cm^3$ and gate doping of $10^{20}/cm^3$. The oxide charge density is $Q_{ox} = 4 \times 10^{10}$, $q = 6.4$ nc/cm^2 and the thickness of oxides 1000 Å. Calculate the threshold voltage with zero substrate bias V_{THO} and the body effect coefficient γ drine the relation used.

Solution: The built in potential for substrate is give by

$$\phi_F(\text{sub.}) = \frac{kT}{q}\ln\left(\frac{N_D}{n_i}\right)$$

$$= 0.0259 \ln\left(\frac{10^{16}}{1.45\times 10^{10}}\right)$$

$$= 0.407 \text{ volt}$$

and the built in potential for gate

$$\phi_F(\text{gate}) = \frac{kT}{q}\ln\left(\frac{N_D}{n_i}\right)$$

$$= 0.0259 \ln\left(\frac{1.45\times 10^{10}}{10^{20}}\right) = -0.648 \text{ volt}$$

The metal semiconductor work function is given by

$$\phi_{ms} = \phi_F(\text{sub.}) - \phi_F(\text{gate})$$

$$\phi_{ms} = 0.407 - (-0.648)$$

$$\phi_{ms} = 1.055 \text{ volt}$$

and

$$Q_B = \pm\sqrt{2q\, N_D\, \varepsilon_{si}\, |2\phi_F|}$$

$$= \sqrt{2\times 1.6\times 10^{-19}\times 10^{16}\times 1.03\times 10^{-12} 10.8141}$$

$$= 1.637 \times 10^{-7} \text{ c/cm}^2$$

$$C_{ox} = \frac{\varepsilon_{ox}}{t_{ox}}$$

$$= \frac{3.5\times 10^{-13}}{1000\times 10^{-8}} = 3.5\times 10^{-8} \text{ F/cm}^2$$

$$V_{To} = \phi_{ms} - 2\phi_F - \frac{Q_B}{\varepsilon_{ox}} - \frac{Q_{ox}}{C_{ox}}$$

$$= 1.055 - 0.814 - 4.67 - 0.011$$

$$= -4.44 \text{ volt}$$

The body effect coefficient

$$\gamma = \sqrt{\frac{2q\, N_D\, \varepsilon_{si}}{C_{ox}}}$$

$$\gamma = \sqrt{\frac{2\times 1.6\times 10^{-19}\times 10^{16}\times 1.03\times 10^{-12}}{3.5\times 10^{-8}}}$$

$$\gamma = 5.187$$

SHORT ANSWERS TYPE QUESTIONS

Q.1. What are enhancement mode devices.

Ans. The *n*-channel devices that are cut off with zero gate bias.

Q.2. Why drain current is zero in the absence of gate voltage even if drain voltage is applied in nMOS enhancement?

Ans. Since the drain and source regions are isolated by the substrate, any drain-to-source voltage in the absence of gate voltage is extremely low because the structure is analogous to two diodes connected back to back.

Q.3. Why does drain current reduces due to the body effect?

Ans. The body effect will cause the population of electrons in the channel to decrease, which will be manifested externally as a decrease in drain current. In other words, we can say that body acts as another gate, referred as back gate. Applying the negative potential on the body will reduce the population of electrons.

Q.4. Drain saturation current, in the presence of the velocity saturation effect, does not depend upon the channel length. Explain physically.

Ans. The transit time of the carriers in the channel is proportional to L (assuming that they travel with constant maximum velocity). The total mobile charge in the channel is also proportional to L. Hence, the current, which is the ratio of this charge to the transit time, is independent of L.

Q.5. It has been found experimentally that lower gate-source voltage is required to produce the same drain current in the short channel device as in the long channel device. Explain the reason.

Ans. One reason for this is barrier lowering. The effect of decreasing L is to deplete more of the region under the inversion layer. This deeper depletion region is accompanied by a larger surface potential which makes the channel more attractive for electrons. Thus, the device conductors more current than what would be predicated from the long channel theory for a given gate-source voltage.

Q.6. Discuss the mobility varication with substrate bias.

Ans. The effect of substrate bias on the mobility is simply that the surface field is increased at the onset of inversion because of the increase in substrate impurity concentration, except that there is no variation in the impurity concentration inside the inversion layer. It was experimentally found that except near the threshold voltage where there is a small difference due to substrate biasing, the mobility curve of $V_{BS} = 0$ is well represented for various substrate biases. As very high effective transverse electric field, the effect of substrate bias on the mobility is almost negligible. The mobility is affected by substrate bias only.

Q.7. What is the charge sharing?

Ans. When two or more capacitors at different potentials are tied together, charge sharing will take place.

REVIEW QUESTIONS

1. Explain the constructional details of a MOSFET. Draw the schematic symbol for an *N*-channel MOSFET.
2. With the help of neat sketches and characteristic curves explain the operation of the MOSFET (*N* and *P*-type both).
3. Explain why a MOSFET can be operated with positive or negative gate voltage.
4. Draw two biasing circuits for an enhancement type MOSFET.
5. Explain the operation of *n* MOSFET with NO Gate voltage.
6. Explain the Enhancement type *n* MOSFET with diagram. **(UPTU 2005-05)**

7. What do you understand by strong inversion in MOSFET? **(UTU-2012)**

8. Draw and Explain the *V–I* characteristics of *n* type depletion-MOSFET.

9. Draw and Explain the *V–I* characteristics of *n* type Enhancement-MOSFET.

10. What is the significant difference between construction of an enhancement type MOSFET and depletion type MOSFET? Explain with suitable diagram. **(UPTU 2005-06)**

11. Describe briefly, the construction of a MOSFET in enhancement mode. Draw its characteristics. Draw the circuit diagram of a MOSFET operating in enhancement mode. **(UPTU 2006-07)**

12. Explain the Body effect of *n*-MOSFET.

13. Explain the difference between the following.

(*a*) DMOSFET and EMOSFET.

(*b*) *N*-channel MOSFET and *P*-channel MOSFET.

14. Compare the MOSFET and JFET. **(UPTU 2003-04)**

15. Can a depletion type MOSFET operate in enhancement mode? If yes, why and how and of no, why? **(UPTU 2003-04)**

16. Explain the MOSFET structure and operation with neat sketch. Also show that it can be used as voltage controlled linear resistor. **(UPTU 2007-08)**

17. Explain the electrical properties of the MOSFET, CMOS and BiCMOS. **(UTU-2012)**

❑❑❑

2

Chapter

MOSFET Fabrication

Chapter Overview

- Introduction
- VLSI Fabrication
- Silicon Wafer Preparation
- Electronic Grade Silicon
- Crystal Growth Techniques
- Steps of Wafer Processing
- Fabrication Process Flow: Basic Steps
- CMOS Fabrication
- Thermal Aspects of Processing
- Latch Up

2.1 INTRODUCTION

The fundamentals of MOS chip fabrication and the major steps of the process flow will be discussed in this chapter. Here our main target on the general outline of the process flow and on the interaction of various processing steps, which ultimately determine the device and the circuit performance characteristics. The links between the fabrication process, the circuit design process and the performance of the chip are very strong.

The circuit designer must have a working knowledge of chip fabrication to create effective designs and to optimize the circuits with respect to various manufacturing parameters.

We requires the both *n*-channel (nMOS) and *p*-channel (pMOS) transistors on the same chip substrate for the fabrication of CMOS. To accommodate both nMOS and pMOS devices, special regions must be created in opposite type substrate of semiconductor. These regions are known as *wells* or *tubs*.

The fabrication process sequence for the *n*-well CMOS integrated circuit is shown in Fig. 2.1. This fabrication process steps have not detailed fabrication step. The processing steps are given below:

1. Firstly, the process starts with the creation of the *n*-well regions for *p*MOS transistors, by impurity implantation into the substrate.
2. After that, a thick oxide is grown in the regions surrounding the *n*MOS and *p*MOS active regions. The thin gate oxide is subsequently grown on the surface through thermal oxidation.
3. After this step, the creation of n^+ and p^+ regions (source, drain and channel-stop implants) are take place.
4. At the last create the contact windows and deposit the pattern metal layer by metallization.

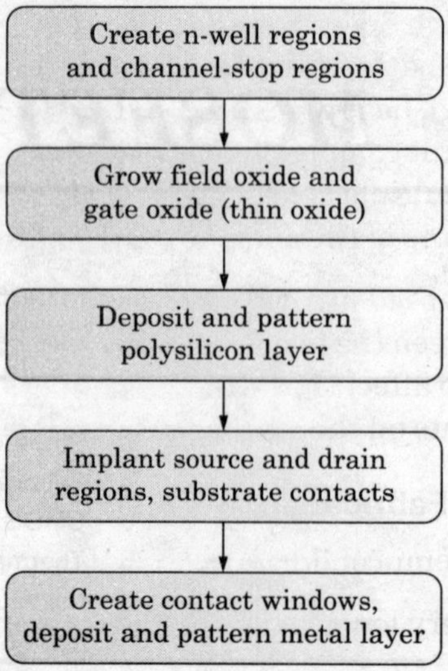

Fig. 2.1 Simplified process sequence for fabrication of *n*-well CMOS IC.

2.2 VLSI FABRICATION

VLSI stands for "Very large scale integration". This is the field which involves packing more and more logic devices into smaller and smaller areas. Thanks to VLSI, circuits that would have taken boardfuls of space can now be put into a small space few millimetres across. This has opened up a big opportunity to do things that were not possible before.

VLSI fabrication involves producing the product (chip or IC) from the raw materials. In other words, it is defined as the various steps or processes employed to realise semi-conductor devices and electronic circuits on a silicon chip. Fabrication is a multiple step sequence of photographic and chemical processing steps during which electronic circuits are gradually created on a wafer made of pure semi-conductor material. From the above discussion we can say that fabrication of an integrated circuit resembles the preparation of a sandwich in which on a good slice of bread as a substrate, the sauces diffuse and several layers are harmoniously placed one on top of the other to satisfy the customer's taste. All is sealed with the final slice of bread that keeps the whole product together if the underlying layers are planar.

Idea of Fabrication

Is to selectively dope certain regions of the semi-conductor to form *N* junctions and inter-connect the doped region to form devices and circuits.

2.2.1 Dealing with VLSI Circuits

Digital VLSI circuits are pre-dominately CMOS based. The need of dealing VLSI ckts is due to certain factors :

(i) Circuit delays: Large complicated circuits running at very high frequencies have one big problem to tackle-the problem of delays in propagation of signals through gates and wires even for areas a few micrometers across. The operation speed is so large that the delays add up, they can actually becomes comparable to the clock speeds.

(ii) Power: Another effect of high operation frequencies is increased consumption of power. Due to this devices consume batteries faster and heat dissipation increases and hence the surface areas have decreased, heat poses a major threat is the stability of the circuit itself.

(iii) Layout: Laying all the circuit components is task common to all branches of electronics. In this VLSI, there can be multiple layers of different materials on the same silicon, these can he different arrangements of the smaller parts for the same component and so on.

Here the power dissipation and speed in a circuit present a trade-off, if we try to optimism one, the other is affected. The choice between the two is determined by the way we choose the layout the circuit components. Layout-can also affect the fabrication of VLSI chips making it either easy or difficult to implement the components on the silicon.

2.2.2 Need of Silicon for VLSI Fabrication

Silicon is the most commonly used semi-conductor material because of the following reasons:

- Abundance in nature and very low cost
- Relatively easy process
- Reliable high volume fabrication
- Silicon can be used for much higher temperature than any other semi-conductor
- Silicon has high intrinsic resistivity of 230,000 Ωcm as compared to Ge (47 Ωcm) which is used for high voltage rectifying devices and infrared sensing devices.

2.2.3 Steps Involved in VLSI Fabrication

The VLSI fabrication involves the following steps i.e.,

- Crystal growth – means formation of single crystal
- Thermal oxidation – i.e.., formation of field oxide and gate oxide in MOSFET
- Diffusion, ion implantation for doping impurities
- Photolithographic process for defining pattern
- Etching of insulator, metal or silicon by chemical solution or plasma
- Deposition of polycrystalline silicon, silicon oxide, silicon nitride by CVD
- Deposition of metallisation layer by physical vapour deposition for Al and electroplating for copper.

2.3 SILICON WAFER PREPARATION

- The processing of silicon wafers to produce integrated circuits involves a good deal of chemistry and physics. In order to alter the surface conditions and properties, it is necessary to use both inert and toxic chemicals, specific and unusual conditions, and to manipulate those conditions with both plasma-state elements and with RF (Radio Frequency) energies.
- Starting with thin, round wafers of silicon crystal, in diameters of 150, 200, and 300 mm, the processes described here build up a succession of layers of materials and geometries to produce thousands of electronic devices at tiny sizes, which together function as Integrated Circuits ICs). In this chapter we study silicon wafer processing and silicon oxidation techniques which plays very important role in the IC fabrication.

2.3.1 Why Silicon?

- The fabrication of semiconductor devices has been based on the use of silicon as the premier semiconductor. Two other semiconductors, germanium (Ge) and gallium arsenide (GaAs), present special problems while silicon has certain specific advantages not available with the others.
- At 300°K silicon has a band gap of 1.12 eV, while germanium's band gap is 0.66 eV. Because of this small band gap, the intrinsic carrier density of germanium at $T = 300°K$ is about $2{,}5 \times 10^{13}\ cm^{-3}$. At temperatures of about 400°K. this density becomes $10^{15} cm^{-3}$, which is comparable to the lower range of doping densities used. This property limits its use to low temperature applications at less than 350°K.
- The other semiconductor of major interest is gallium arsenide. In spite of its attractive electrical properties, gallium arsenide crystals have a high density of crystal defects, which limit the performance of devices made from it.
- Silicon is an abundant clement and occurs naturally in the form of sand. It can be refined using simple purification and crystal growth techniques. It also exhibits suitable physical properties for fabricating active devices with good electrical characteristics. In addition silicon can be easily oxidized to form an excellent insulator, (SiO_2) or glass.
- This native oxide is useful, for constructing capacitors and M'OSFETs. It also serves as a diffusion barrier that masks against unwanted impurities from diffusing into the high purity silicon material. This masking property allows selective alternation of electrical properties in the silicon.
- Thus active and passive elements can be built on the same piece of material. The components can be interconnected to form a monolithic IC.

2.4 ELECTRONIC GRADE SILICON

- Electronic Grade Silicon (EOS) is the raw material for the preparation of single-crystal silicon. It is the poly crystalline material of high purity and one of the purest materials available, pure electronic grade silicon requires doping elements be in the parts per billion (ppb) range. The major impurities are boron, phosphorous carbon etc.

2.4.1 Production Process of EGS

- The production of electronic grade silicon is the multistep process. First step is the Metallurgical Grade Silicon (MGS) production in submerged-electrode arc furnace. The furnace is charged with quartzite and carbon.

  In the furnace a number of reactions takes place, the overall reaction being,

$$SiC_{(solid)} + SiO_{2(solid)} \rightarrow Si_{(liquid)} + SiO_{(gas)} + CO_{(gas)}$$

- MGS available in above process solified at purity of 98%. The next step is to pulverize the silicon mechanically and react it with anhydrous hydrogen chloride to form trichlorosilane ($SiHCl_3$).

$$Si_{(solid)} + 3HCl_{(gas)} \rightarrow (SiHCl_{3(gas)} + H_{2(gas)} + \text{heat}$$

  $SiHCl_3$ contains many unwanted chlorides, which are removed by fractional distillation,

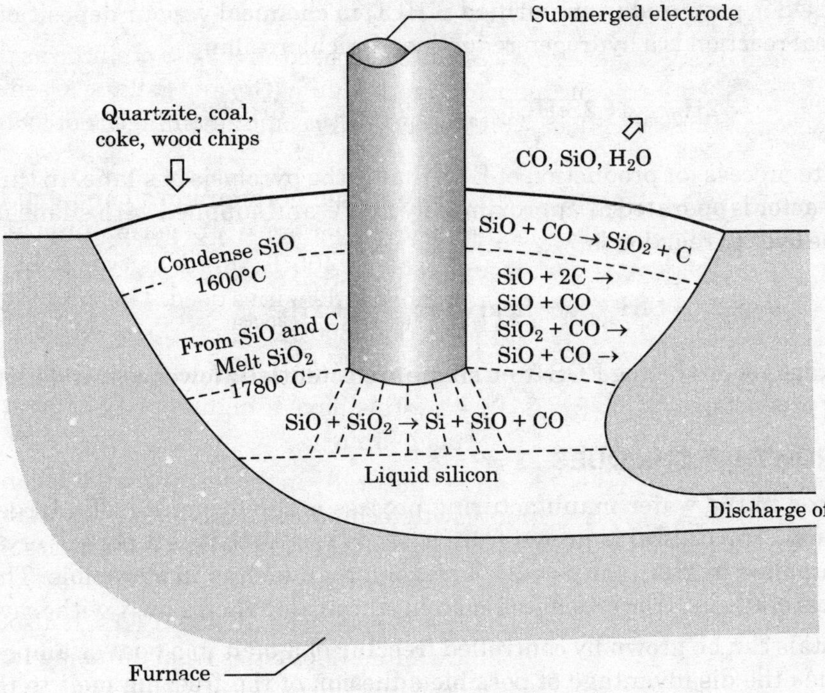

Fig. 2.2 Submerged-electrode arc furnace for production of MGS

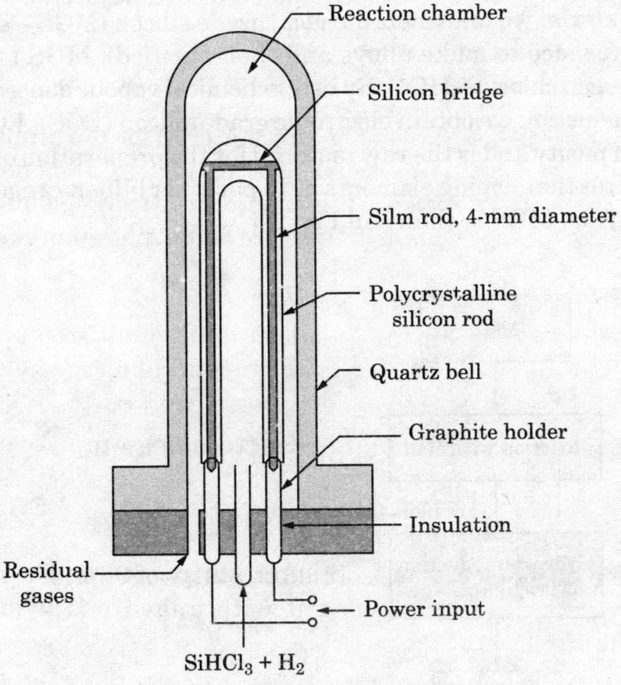

Fig. 2.3 CVD reactor for ECS production

- The final EOS is prepared from purified $SiCl_3$ in chemical vapour deposition process The chemical reaction is a hydrogen reduction of trichlorosilane

$$2H_{2(gas)} + 2SiHCl_{3(gas)} \rightarrow 2Si_{(solid)} + 6HCl_{(gas)}$$

- An alternate process for production of ECS that is the pyrolysis of silane. In this process the CVD reactor is operated at approximately 900°C and supplied with silane instead of $SiHCl_3$. The overall reaction is

$$SiH_{4(gas)} + heat \rightarrow Si_{(solid)} + 2H_{2(gas)}$$

- The advantages of producing EGS from silane are potentially lower cost and less harmful reaction by products.

2.5 CRYSTAL GROWTH TECHNIQUES

- The first step in the wafer manufacturing process is the formation of a large, perfect silicon crystal. The crystal is grown from a 'seed crystal' that is a perfect crystal. The silicon is supplied in granular powder form, and then melted in a crucible. The seed is immersed carefully into the crucible of molten silicon, and then slowly withdrawn.
- Single crystals can be grown by controlled freezing of a melt in a boat or ampoule. This approach has the disadvantage of possible adhesion of the freezing melt to the walls. Contact of-this type can initiate the formation of dislocations, and special care must be taken to minimize this effect.
- As a starting material we take metallurgical grade silicon (MGS – silicon in original form is treated in a furnace to make alloys and then purified). MGS is further treated with anhydrous hydrogen chloride (HC1). By using chemical vapour deposition techniques (CVD) and hydrogen reduction we obtain electronic grade silicon (EGS), EGS is a polycrystalline material of high purity and is the raw material for the preparation of single crystal silicon. Pure EGS requires that doping elements be in parts per billion range. The whole process of obtaining EGS from MGS is shown in Fig. 2.4

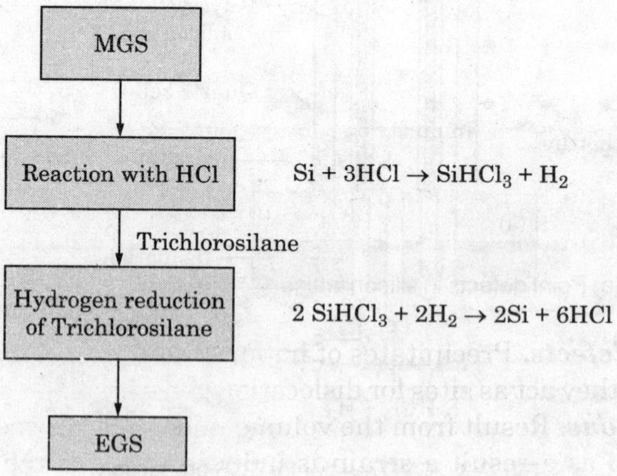

Fig. 2.4 Steps in formation of EQS from MQS

- EGS has very small impurity levels measured in parts per million (ppm) to parts per billion (ppb). EGS is polycrystalline material and is the raw material for the preparation of single crystal silicon. In crystal growth the polycrystalline silicon rods (EGS) are converted to even purer and defect free single crystals. The single crystals are defined as:
- Crystals are solids which are characterized by long range order. The atoms, ions or molecules are in a regular array of identical units over long range. In single crystals this regularity is maintained throughout the entire volume of crystals. So the crystal have well defined planes as crystal faces. In polycrystalline solid, the regularity is maintained over limited portion of the crystal called grain. The processing characteristics and some material properties of the wafers depend on the orientation of the crystals from which wafers were obtained. For example the <111> planes have the highest density of atoms on the surface so they are suitable for single crystal growth.
- In real crystals, due to thermal agitation, the atoms are displaced from their location and give rise to the defects described as
  - ➢ ***Point defects.*** Any non silicon atom incorporated in the lattice by replacing a host silicon atom as shown in Fig. 2.5(a)
  - ➢ ***Dislocation.*** There are two types of dislocations, screw and line. A line dislocation is an extra plane of atoms *XY* inserted into the lattice as shown in Fig. 2.5(b)
  - ➢ ***Area defects.*** The two common are defects are twins and grain boundaries. Twining is a change in the crystal orientation across a twin plane. Grain boundary is a transition between crystals having no particular orientation relationship with me another. The crystals showing area defects are "not suitable for IC fabrication as they represent large area discontinuity in the lattice.

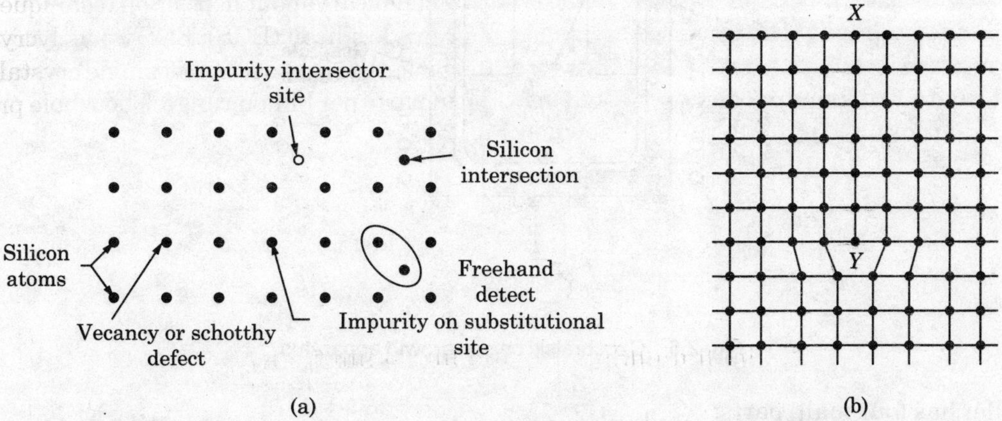

Fig. 2.5 (a) Point defects in silicon lattice. (b) Dislocation in a cubic silicon lattice.

  - ➢ ***Volume defects.*** Precipitates of impurity or dopant atom constitutes the volume defects as they act as sites for dislocation generation.
  - ➢ ***Dislocations.*** Result from the volume mismatch between the precipitate and the lattice and as a result a strain is induced which is relieved by the formation of dislocation.

- When crystals are grown, a phase change from solid, liquid or gas phase to crystalline solid phase occurs. There are mainly two crystal growth techniques widely used in industry – Czochralski technique and Bridgeman technique where former is widely used for growing silicon crystals and latter is used for growing Gallium arsenide crystals.

2.5.1 The Czochralski (CZ) Technique

As the name suggests, this technique was developed by Czochralski and this process is mainly used to grow silicon crystals from which silicon wafers are produced. The process is characterized by a liquid-solid mono-component growth system. This technique involves the solidification of atoms from a liquid phase at an interface. The apparatus used for CZ crystal growth is also known as "puller". This is shown in Fig. 2.6. It is a very bulky apparatus weighing about 17,600 kg and 6.5 m in height. The speed of growth is determined by the number of .sites on the face of the crystal and the specifics of heat transfer at the interface.

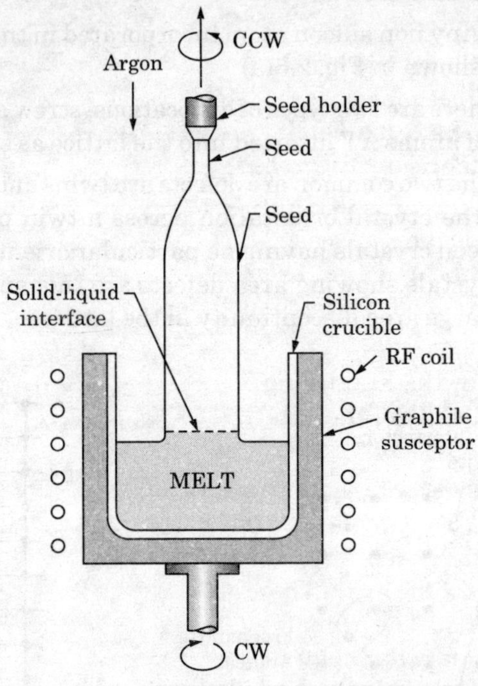

Fig. 2.6 Czochralski crystal growth apparatus

The puller has four main parts:

1. ***Furnace.*** It includes a fused silica crucible, a graphite susceptor, a rotation mechanism, a heating element and a power supply. The crucible contains the melt (molten silicon). It should be unreactive to silicon and have thermal stability. The graphite susceptor is used to support the silica crucible. The susceptor is placed on a pedestal whose shaft is connected to a motor that provides rotation. To melt the charge, radio frequency heating or resistance heating is used and finally is connected to a dc power supply.
2. ***Crystal pulling mechanism.*** This mechanism should control the pull rate and crystal rotation. That is the seed crystal is simultaneously pulled up and rotated to meet proper growth.

3. ***Ambient control.*** *This* includes an inert gas source (*e.g.*. Argon) a flow control and an exhaust system. The gas source must meet the purity requirements.
4. ***Control system.*** The control system includes control of process parameters like temperature, crystal diameter, pull rate, rotation speed. Programmed process steps are provided through microprocessor based control and infrared sensors are used to judge any change.

Working. Polycrystalline silicon is placed in the crucible and the furnace is heated to melt silicon. A suitably oriented seed crystal is placed over the crucible in the seed holder. The seed is dipped into the melt. Some part the seed melts and some touch the liquid surface. The seed is then slowly pulled up. So freezing at the solid-liquid interface occurs and single crystal is formed. The seed crystal in this form is known as ingot.

- In the growth process, a known amount of dopant is added to the melt to obtain desired doping concentration in the grown crystal. Boron is usually added for p-type material and phosphorus for n-type material.
- The ingot so obtained is hard and brittle. To give a shape to the ingot to form wafers, it is treated with a series of mechanical and chemical processes. The material used for cutting and shaping silicon is usually diamond. After cutting and polishing the silicon wafers are packaged to be used to fabricate integrated circuits.
- The two ends of ingot is known as seed and tang. The first step in silicon shaping operations is the removal of seed and tang ends using a circular saw. In the left over crystal resistivity and perfection analysis is carried on. The portions that fail to show acceptable resistances and perfections are also cut off. These cut off portions can be recycled and can also be used for other purposes,
- At the time of crystal growth, the growing process is not uniform so that there is a variation in diameter along the length of the ingot, "This is because of the inability to control the growth rate properly.
- After the crystal is grown, the seed is removed and the surface is grinded so that the diameter of the material is defined. A rotating cutting tool makes multiple passes down a rotating ingot until the chosen diameter is attained. After that one or more flats are ground along the length of the ingot, There are usually two flats known as primary (major) and secondary (minor) flat. The primary flat helps to form an array of square die (chip) on the wafer and the secondary flat helps to identify the orientation and conductivity type of the material.
- After grinding the ingot is sliced into wafers by diamond saws. The surface orientation, thickness, taper (variation in thickness) and bow (curvature) are determined from slicing. The wafers are usually 10 to 30 cm in diameter and 400 to 600 Jim thick. The surface of the wafer is then polished to a mirror finish using chemical and mechanical polishing to obtain a smooth, defect free surface. Wafer polishing can be done on a single wafer in a batch processing manner. The polishing pad is made up of artificial fabric. The wafers are mounted on a fixture, pressed against this pad under high pressure and are rotated. A mixture of polishing slurry and water, dripped on to the pad does the polishing. The wafers are now ready for the fabrication of integrated circuits. The schematic for the water shaping operations are shown in Fig. 2.7.

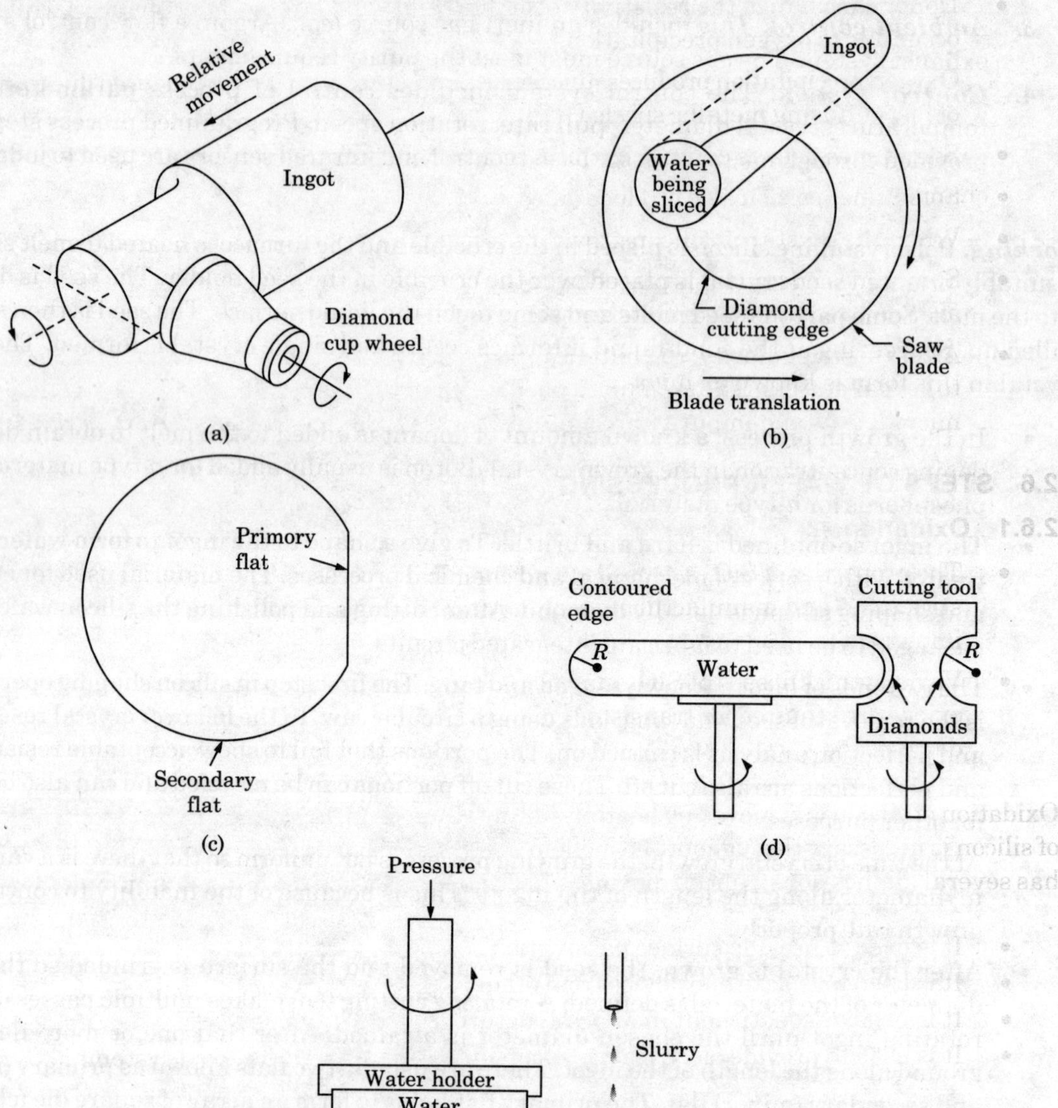

Fig. 2.7 Wafer shaping operations: (a) Wafer grinding, (b) Wafer slicing, (c) Flats, (d) Edge contouring, (e) Wafer pollshing.

2.5.2 Effects of Undesirable Impurities on Silicon Crystal Properties

- There are two types of undesirable impurities i.e., carbon and oxygen.
- Oxygen in silicon arises from the dissolution of the crucible during growth and carbon occurs due to its transportation from the graphite parts in the furnace to the melt.
- Due to oxygen impurity following affects are produced i.e.,
  - ⇒ Donor formation
  - ⇒ Improvement of yield strength
  - ⇒ Generation of defects

- Donor can change the resistivity of crystal and yield strength in beneficial while defects occur due to oxygen precipitation.

  Oxygen precipitation produces junction leakage currents which is produced due to attraction of fast-diffusing metallic species.
- Sometimes defects capture harmful impurities known as gettering.
- Sometimes carbon also produce defects.
- When the crystal of .silicon grows it is known as ingots or boules.
- Sometimes irregular shapes of ingot are cut and twining (gross crystalline imperfections) are removed by cutting. During this process there is a loss of 50%.
- After this irregularity, the butt (tang) of the ingot (a slice cut from that position) is etched to reveal defects. The etchant is used for the polished wafers to delineate other types of microdefects and impurity precipitates.

2.6 STEPS OF WAFER PROCESSING

2.6.1 Oxidation

- The properties of SiO_2 are used to make many silicon integrated circuits rely with different structures and manufacturing techniques. So, reliable manufacture of SiO_2 is extremely important.
- Various thicknesses of SiO_2 may be required, depending on the particular process. Thin oxides are required for transistor gates; thicker oxides might be required for higher voltage devices.

Oxidation of silicon is achieved by heating silicon wafers in an oxidizing atmosphere. The oxidation of silicon is necessary throughout the modem integrated circuit fabrication process. Silicon dioxide has several applications: Silicon dioxide is used as a mask against diffusion or implant of dopant.

- It is used to provide surface passivation.
- It is used to isolate one device from other.
- It is used as constituent in MOS structure.
- It is used to provide electrical isolation in multilevel metallization system.

Several techniques for growing the oxide layers have been developed, such as:

Thermal Oxidation: Thermal oxidation of silicon is performed in an open-ended quartz tube kept in a furnace in the temperature range of 1000°C to 1200°C. The oxidizing ambient can be either dry oxygen, or steam.

Wet Oxidation: In the case of wet oxidation, as shown in Fig. 2.8, high purity oxygen enters the inlet side of a bubbler that contains high purity water maintained at a constant temperature below its boiling point. The temperature is usually between 9000 C and 1000° C. The O_2 becomes saturated with water vapor and exits through the outlet of the bubbler in to the furnace. This gas mixture oxidizes the Si surface with the following chemical reaction.

$$Si + 2H_20 \rightarrow SiO_2 + 2H_2\uparrow$$

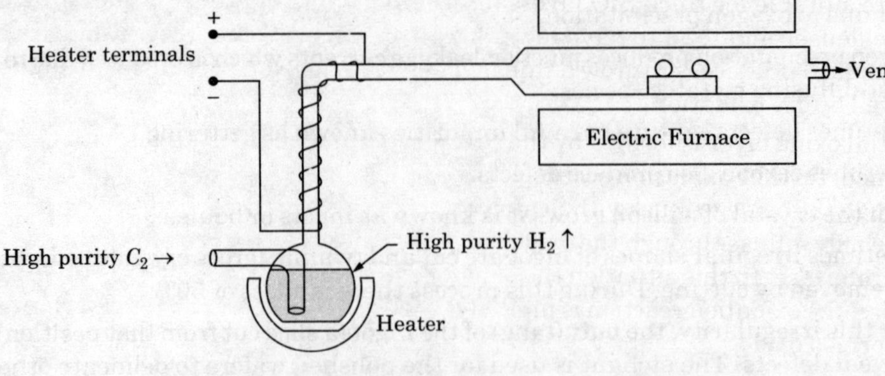

Fig. 2.8 Schematic diagram of wet oxidation system

Dry Oxidation: The schematic diagram of a dry oxidation system is shown in Fig. 2.9. When the oxidizing atmosphere is pure oxygen. Temperatures are in the region of 1200° C to achieve an acceptable growth rate. Si wafers are cleaned to remove all contaminants. The cleaned wafers are dried and loaded in to a quartz boat, which is then placed inside the furnace heated to the desired temperature and the O_2 gas from the source is passed directly through the quartz tube. Following reaction takes place inside the furnace. Dry oxidation forms a better quality oxide than wet oxidation. It is used to form thin, highly controlled gate oxides, while wet oxidation may be used to form thick field oxides.

$$Si + O_2 \rightarrow SiO_2$$

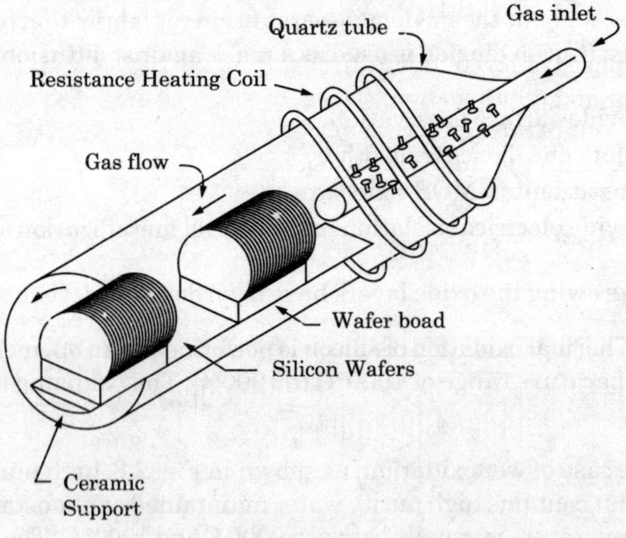

Fig. 2.9 Schematic diagram of Dry oxidation furnace system

Steam Oxidation: For steam oxidation, the O_2 supply at the inlet of the bubbler is blocked, and the water in the bubbler is heated to its boiling point. The resulting steam flows through the furnace and oxidizes the Si wafers.

The wet oxidation rate is much higher than the dry oxidation rate, making the wet oxidation method more suitable for thick SiO layers (exceeding 1 μm). However, oxides grown in the dry ambient are denser and tend to have fewer structural defects and a lower density of interface states. A standard practice is to grow a thin layer of dry oxide, followed by wet oxidation, which is again followed by dry oxidation.

The initial oxide layer is formed by the reaction of the surface Si atoms with O_2. Once a SiO_2 layer is formed, further oxidation occurs at the Si-SiO_2 interface such that the oxidizing species diffuse through the oxide to react with Si atoms at the interface. For thin SiO_2 layers, the oxidizing species can easily diffuse through the oxide, and the oxide growth is limited by the reaction rate at the Si-SiO_2 interface. In this situation the oxide thickness varies linearly with time. In the case of thicker oxides, the oxidation reaction is limited by the diffusion of the oxidizing species through the previously formed oxide.

2.6.2 Deposition

Deposition involves placing dopant material onto the silicon surface and then driving it into the bulk using a thermal diffusion step. This can be used to build deep junctions.

A process of deposition is **Vapor-Phase Technique (Chemical Vapor Deposition-CVD)** can be used for the deposition. The CVD occurs when heated gases react in the vicinity of the wafer and produce a product that is deposited on the silicon surfaced CVD also used to lay down thin films of material later in the CMOS process.

The CVD reactor is shown in Fig. 2.10. A gas mixture containing either N_2 or Ar with 2 and about 1 percent SiH_4 is used. The SiO_2 growth rate depends on the substrate temperature and the rate of gas flow. The chemical reaction, which takes place in the reactor, the temperatures between 250°C and 500°C, is given as:

$$SiH_4 + 2O_2 \rightarrow SiO_2 + 2H_2O\uparrow$$

Layer of SiO_2 prepared at low temperatures are not very dense but can be identified by a heat treatment at 900°C for about half an hour. Denser oxide layers can be prepared with a reaction of SiH_4 and Co or CO_2 at temperatures between 900 and 1100°C. The gas passed through the reactor is a mixture of H_2 or N_2 with CO_2 (or CO) + 1 percent SiH_4. The relevant chemical reaction in case of CO_2 is

$$SiH_4 + 4CO_2 \rightarrow SiO_2 + 2H_2O\uparrow + 4CO\uparrow$$

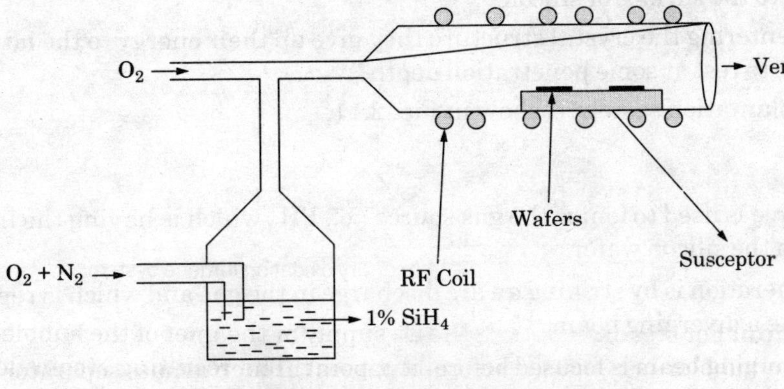

Fig. 2.10 Schematic of CVD System

2.6.3 Ion Implantation

It is defined as :

"Process by which energetic impurity atoms can be introduced into a single crystal substrate in order to change its electronic properties i.e. (conduction property)."

2.6.3.1 Requirement for Ion Implantation

- Ion sources which is to be introduced into crystal substrate.
- Ion implantation is carried out with ion energies in 50 to 500 KeV range.
- After ion sources, beam deflection is carried out.
- Then at last scanning prior to impingement on the substrate.

2.6.3.2 Features of Ion Implantation

- It is an alternative to diffusion as a means for junction fabrication (may be Gate, drain or source) in semi-conductor technology.
- Impurity atoms are free from contamination due to mass separation techniques.
- It is carried out at high vacuum conditions.
- It provides more uniform surface coverage than diffusion because dopant incorporation during diffusion is sensitive to variation in the electronic character of the surface.
- on implantation is carried out at room temperature and hence variety of masks are used (i.e., silica, aluminium, photoresist) for selective doping. So many self-aligned mask techniques are used for device fabrication which is not with diffusion.
- It is non-equilibrium process and hence many dopants can be introduced at concentration in excess of their equilibrium solid solubility.
- In this method we can add controlled amount of species specific region of wafer and hence it is important application in the threshold control of MOS devices.

2.6.3.3 Demerits of Ion Implantation

- The equipment used is very expensive.
- It can cause damage to the semi-conductor means it damage the target, and displaces many atoms for each implanted ion.

2.6.3.4 Ion Implantation System

Principle "A beam of impurity ions is accelerated to kinetic energies in the range of ions of kV and then is directed to the surface of silicon.

Then after entering the crystal structure they give up their energy to the lattice in collisions and finally come to rest at some penetration depth."

The ion implantation system is show in Fig. 2.11.

Operation

- Ion source is used to ionise the gas source i.e., PH_3 which is having the impurity 'P' to be input on the silicon wafer.
- Ions generation is by striking an arc discharge in the gas and which is repelled from their source as a diverging beam.
- This diverging beam is focused before- it passes through a mass separator which is used to direct only the ions by dopant species through the narrow aperture.

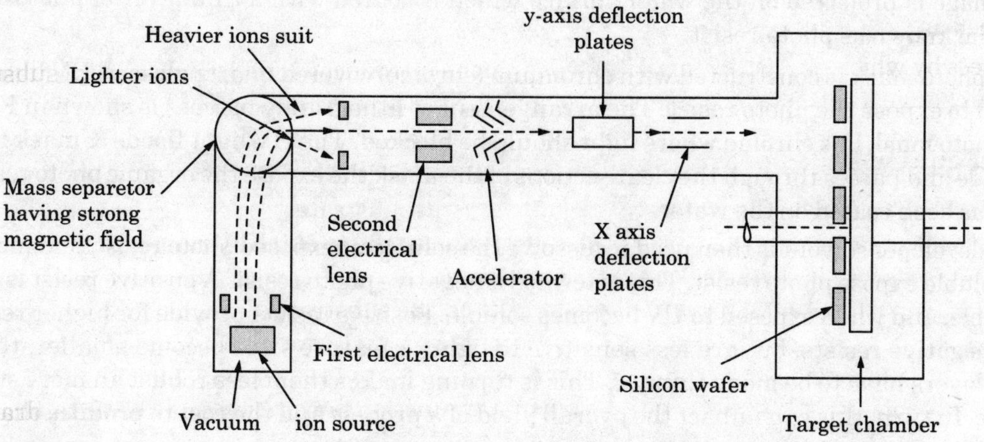

Fig. 2.11

- Second lens focus their beam and then pass it to through the accelerator which is used to increase the kinetic energy of ions before they strike the target and then it is implanted on the silicon wafer.
- Range of accelerating voltages are from 20 kV o 250 kV.
- Sometime some implanters have mass separation after the ions are accelerated to high energy.
- Dopant concentration to be deposited on the wafer is determined by monitoring the current in the ion beam.
- Scanning is used for uniformly distributed ions because the ion he am is small.
- Target chamber is used for producing uniformity by scanning electrostatically over the surface of wafer.

2.6.4 Photolithography

Photolithography is the most complicated, expensive and critical process in main stream microelectronic fabrication. Recounts for about 1/3rd of the total fabrication cost, a percentage that is rising day by day. Recall that regions of dopants polysillicon, metal, and contacts are defined using masks. Lithography can be classified:

(a) Optical lithography
(b) Electron lithography
(c) X-ray lithography
(d) Ion lithography

For instance, in places covered by the mask, ion implantation might not occur or the dielectric or metal layer might be left intact. In areas where the mask is absent, the implantation can occur, or dielectric or metal could be etched away. The patterning is achieved by a process called photolithography, from the Greek photo (light), lithos (stone), and graphe (picture), which literally means "carving pictures in stone using light."

Optical Lithography an optical source on the tip is used to shine light through the mask. The image is projected on the wafer surface which is coated with a thin layer of photosensitive material known as photo-resist.

A photo mask is constructed with chromium (chrome) covered quartz glass. A UV light source is used to expose the photo resist. The arrangement of lithography process is shown in Fig. 2.12. The photo mask has chrome where light should be blocked. The UV light floods & mask from the backside and passes through the clear sections of the mask the expose the organic photo resist (PR) that has been coated on the wafer.

A developer solvent is then used to dissolve the soluble unexposed photo resist, leaving islands of insoluble expose photo resist. This is termed a negative photo resist. A positive resist is initially insoluble, and when exposed to UV becomes soluble. Positive resists provide for higher resolution than negative resists, but are less sensitive to light. A feature sizes become smaller, the photo resist layers have to be made thinner. This is turning makes them less robust an more subject to failure. In turn, this can impact the overall yield of a process and the cost to produce the chip.

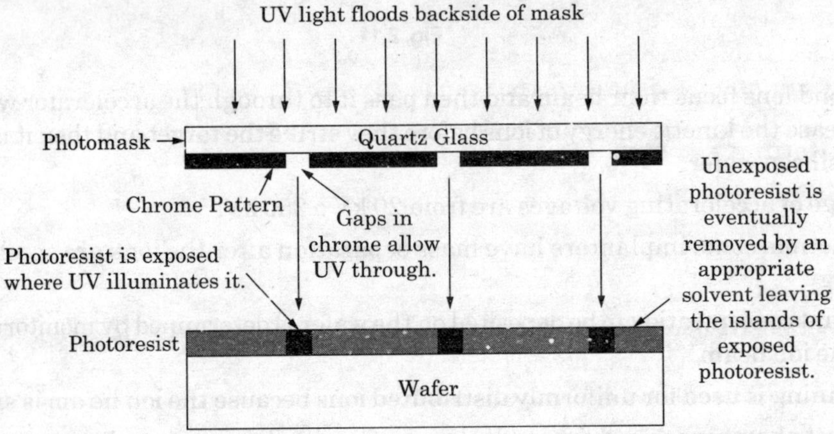

Fig. 2.12 Photo-masking with a negative resist

2.6.4.1 Optical Resist

Photo resists are of two types positive and negative.

- Negative resist on exposure to light become less soluble in developer solution, while a positive resist becomes more soluble. Commercial negative resist such as Kodak Microneg-747 consist of two parts a chemically inert poly-isoprene rubber which is the film forming component and a photoactive agent. The photoactive agent on exposure to light reacts with the rubber to form cross links between rubber molecules making rubber less soluble in an organic developer solvent.
- The reactive species from during the exposure can react with oxygen and be rendered ineffective for cross linking. Therefore, the resist is usually exposed in a N_2 atmosphere.
- The developer solvent dissolves the unexposed resist. The exposed resist swells as the uncross linked molecules are dissolved. The swelling distorted the pattern features and limits resolutions a-3 times the initial film thickness.

- Positive resist have two components – a resin and a photoactive compound dissolved in a solvent. The photoactive compound is a dissolution inhibitor when it is destroyed by exposure to light the resin more soluble in an aqueous developer solution. The unexposed regions do not swell much in the developer solution. So, higher resolution is possible with positive resist. The sensitivity i of most standard resist peaks in (300–400) nm spectral range eg. of commercially available positive resist are MP2400 and HPR-206.
- Photo resist are being developed for exposure at shorter wave length where higher resolution is possible few such deep ultra violet resist are poly methyl methacylate (PMMA) sensitive for $\lambda < 250$ nm. and poly-butain sulfide sensitive for $\lambda < 200$ nm. and MP–2400 sensitive for $\lambda = 250$ nm.
- Projection printing is normally used, in which lenses between the reticle and wafer focus the pattern on the wafer surface. Older techniques include contact printing, where the mask and wafer are in contact, and proximity printing, where the mask and wafer are close but not touching. The original form of both techniques can cause damage to the mask and are limited in feature definition in the range of 1-2 μm. However, newer approaches to proximity printing are being proposed for future processes. The reticle can be the same size as the area to be patterned (1X) or larger. For instance, 2.5X and 5X steppers with optical reduction have been used the industry. The wavelength of the light source influences the minimum feature size that can be printed.

2.6.4.2 History/Reviews for lithography

- In the 1980s, mercury lamps with 436 nm or 365 run wavelengths were used. At the 0.25 mm process generation, excimer laser with 248 nm (deep ultraviolet) were adopted and have been used down to the 180 nm node. Currently 193 nm argon-fluorid lasers are used for the critical layers down to the 90 nm node. The critical layers are those that define the device behavior. A example would be the gate (polysillicon), source/drain (diffusion), first metal and contact masks. "The ASML TWINSCAN 1250 step and scan lithography tool uses a 193 nm laser to expose up to 150 wafers per hour. It has an alignment tolerance of nm and can process features down to 70 nm. 157 nm fluorine lasers are in development for processes below 90 nm. In the future, 13.4 nm extreme ultraviolet (EUV) light sources may be used, but at present these sources are costly and require prohibitively expensive reflective optics. Each successive UV stepper is more expensive and the throughput of the stepper decrease. This is just another contributory issue to the spiraling cost of chip manufacturing.
- Wavelengths comparable to or greater than the feature size cause distortion in the patterns exposed on the photo resist Resolution enhancement techniques precompensate for this distortion so the desired patterns are obtained. These technique involve modifying the amplitude, phase, or direction of the incoming light. The ends of a line in a layout receive less light that the center, causing no uniform exposure.
- Optical proximity correction (OPC) makes small changes to the patterns on the masks to compensate for these local distortion Multiple lines running in parallel on a mask behave as a diffraction grating. Phase shift masks (PSM) vary the thickness of the mask to change the phase such that light from adjacent lines are out of phase and cancel where no light is desired. Off—ax illumination can also improve contrast for certain types of dense, repetitive patterns. Using these techniques, the resolution can be extended to one-eighth the wavelength of the light.

2.6.5 Well and Channel Formation

Varying proportions of donor and acceptor impurities can be achieved using epitaxy, deposition, or implantation. Epitaxy involve growing a single-crystal film on the silicon surface (which is already a single crystal) by subjecting the silicon wafer surface to a elevated temperature and a source of dopant material.

2.6.5.1 Epitaxy

- The term epitaxy literally means *"arranged upon"*. Epitaxy is a combination of two Greek words epi–"upon" and taxie "odered" is a term applied to processes used to grow a thin crystalline layer (typically a few new nanometers to a few micrometers) on an already existing crystalline substrate such that the film has the same lattice structure as the substrate.
- Epitaxy can be used to produce a layer of silicon with fewer defects than the native wafer surface and also can help prevent latch up. Foundries usually provide a choice of epi (with epitaxial layer) or non-epi wafers. Microprocessor designers usuall prefer to use epi wafers for device uniformity of performance. The cross sectional view of a selective Epitaxial process shown in Fig. 2.13.
- Expitaxy can be further classified into
  - → Vapor phase epitaxy (VPE).
  - → Liquid phase epitaxy (LPE).
  - → Molecular beam epitaxy (MBE).

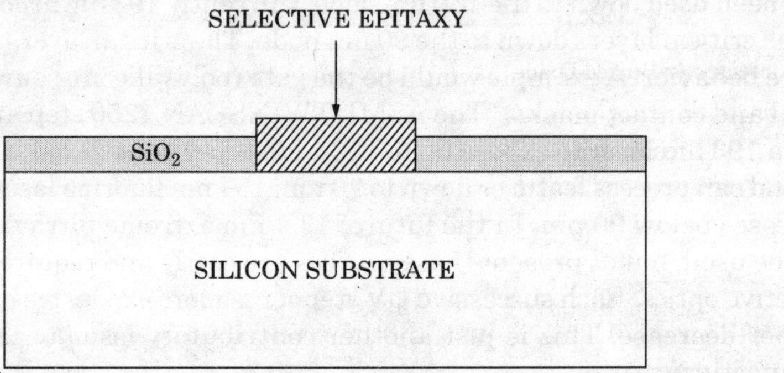

Fig. 2.13 Cross sectional view of selective epitaxial process.

2.6.6 Isolation

In a CMOS process, individual devices are need to be isolated from one to another so that they do not have unexpected interactions. The source and drain of transistors form reverse-biased *p-n* junctions with the substrate or well, isolating them from their neighbors.

- Next, the formation of any parasitic MOS channels must be prevented. This is commonly achieved using a thin gate oxide for transistors and a much thicker field oxide elsewhere. The thicker oxide increases the threshold voltage to a value above the supply voltage and so prevents a channel from forming in the substrate unless there is an over voltage condition (actually, these field devices are used for I/O protection).

- In addition to using the thick oxide, the substrate in areas where transistors are not required can be further implanted with dopants to create channel-stop diffusion. The implant increases the impurity concentration in the substrate, which in turn raises the threshold voltage and prevents inversion of an unwanted channel. In early metal gate processes that had a uniform (thin) oxide layer, the channel-stop diffusion surrounded each transistor and was the only method isolating transistors.
- Local Oxidation of Silicon (LOCOS) was used to produce varying oxide thicknesses. A common problem with LOCOS-based processes was the transition between thick and thin oxide, which extended some distance laterally because of the way the oxide was grown; this in turn limited the packing density of transistors.
- The isolation step that is used to achieve isolation between devices in processes at and below the 180 nm node is to form insulating trenches of SiO_2 that surround active areas. This is called shallow trench isolation (STI).
- The diagram of shallow trench isolation is shown in Fig. 2.14. STI starts with a pad oxide and a silicon nitride layer which act as the masking layers. Openings in the pad oxide are then used to etch into the well or substrate region. A linear oxide is then grown to cover the exposed silicon. The pad oxide and nitrides are removed and a Chemical Mechanical Polishing (CMP) step is used to panelize the structure.

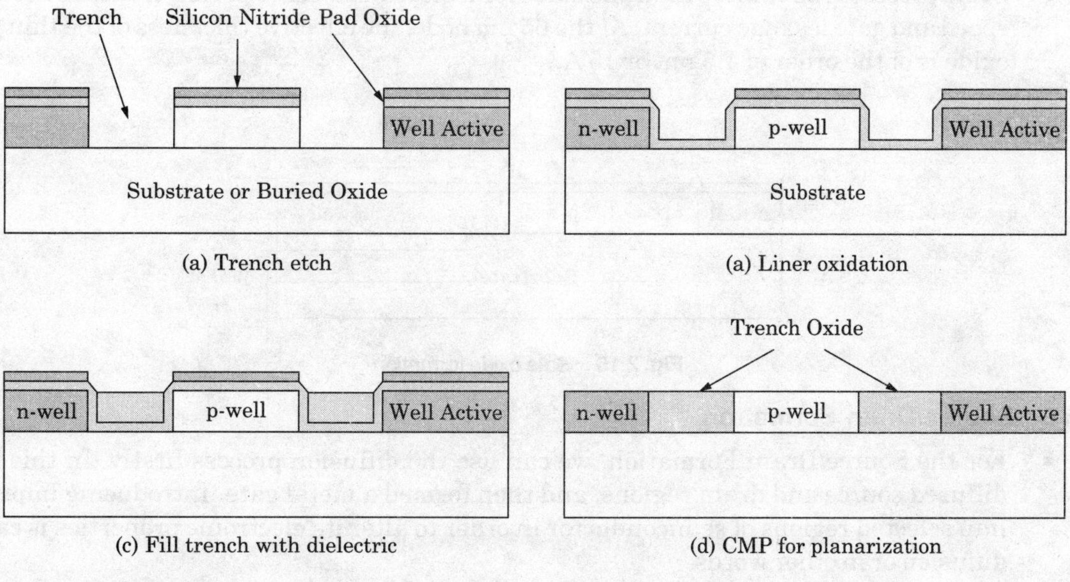

Fig. 2.14 Diagram for Shallow Trench Isolation.

- CMP is the combination of mechnical and chemical techniques which combines a mechanical grinding action in which the rotating wafer is contacted by a stationary polishing head while an abrasive mixture is applied. The mixture also reacts chemically with the surface to aid in the polishing action. CMP is used to achieve fiat surfaces, which are of central importance in modern processes with many layers. The presence of a deep n-well and/or trench isolation makes it easier to isolate noise-sensitive (analog or memory) portions of a chip from digital sections.

- Trench isolation also permits nMOS and pMOS transistors to be placed closer together because the isolation provides a higher source/drain breakdown voltage that a source or drain diodes start to conduct in the reverse-biased condition. The breakdown voltage must exceed the supply voltage and is determined by the junction dimensions and doping levels of the junction formed.
- In the case of a twin-well process, only one mask need be defined because the other well is by definition its complement. Triple-well processes have to define at least two masks, one for the deep well and the other for either nwell or p-well.

2.6.7 Gate Oxide

- The formation of the gate oxide (commonly silicon dioxide (SiO_2)) is the next step in the process for the transistors. In the case of STI-defined source/drain regions, the gate oxide is grown on top of the panelized structure mat occurs at the stage shown in Fig. 2.15.
- The oxide structure is called the gate stack. This term arises because current processes seldom use a pure SiO_2 gate oxide, but prefer to produce a stack that consists of a few atomic layers, each 3-4 A thick, of SiO_2 for reliability, overlaid with a few layers of an oxynitrided oxide. The presence of the nitrogen increases the dielectric constant, which decreases the effective oxide thickness. Being able to use a thicker oxide improves the robustness of the process.
- Some processes offer more than one oxide for logic transistors to permit tradeoffs between speed and gate leakage current. At the 65 nm node, the effective thickness of the thin gate oxide is of the order of 1.5 nm or 15 A.

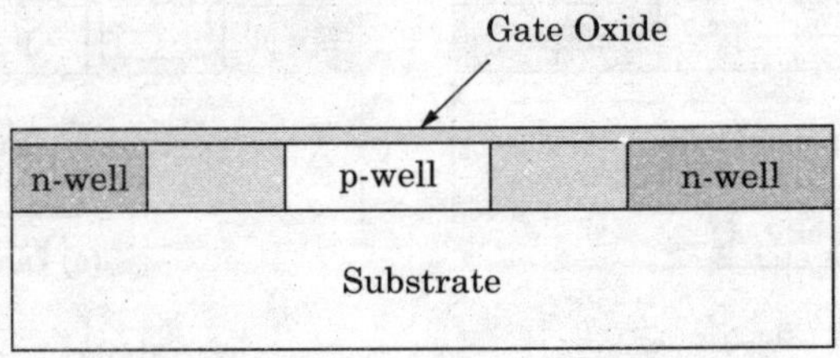

Fig. 2.15 Gate oxide formation

2.6.8 Source/Drain Formation

- For the Source/Drain Formation, we can use the diffusion process firstly. In this first diffused source and drain regions, and then formed a metal gate. Introducing impurity into selected regions of semiconductor in order to alter its electronic properties is called diffusion or in other words

"Diffusion is a process of introducing controlled amount of doponts into semiconductor."

- If the gate was misaligned, it could fail to cover the entire channel and lead to a transistor that never turned ON. To prevent this, the metal gate had to overhang the source and drain by more than the alignment tolerance of the process. This created large parasitic gate-to-source and gate-to- drain overlap capacitances that degraded switching speeds. When silicon is deposited on SiO_2 or other surfaces without crystal orientation, it forms polycrystalline silicon, called poly-silicon or simply poly.

- An annealing process is used to control the size of the single crystal domains and to improve the quality of the polysillicon. The poly-silicon gate serves as a mask to allow precise alignment of the source and drain with the gate. This process is called a self-aligned poly-silicon gate process. Aluminum could not be used because it would melt during formation of the source and drain.

The steps to define the gate, source, and drain in a self-aligned polysillicon gate are as follows:

⇒ Grow gate oxide wherever transistors are required (area = source + drain + gate) - elsewhere there will be thick oxide Fig. 2.16(a).
⇒ Deposit polysillicon on chip Fig. 2.16(b).
⇒ Pattern polysillicon (both gates and interconnect) Fig. 2.16(c).
⇒ Etch exposed gate oxide—that is, the area of gate oxide that was not covered by polysillicon; at this stage, the chip has windows down to the well or substrate wherever a source/drain diffusion is required Fig. 2.16(d).
⇒ Implant pMOS and nMOS source/drain regions Fig. 2.16(e).

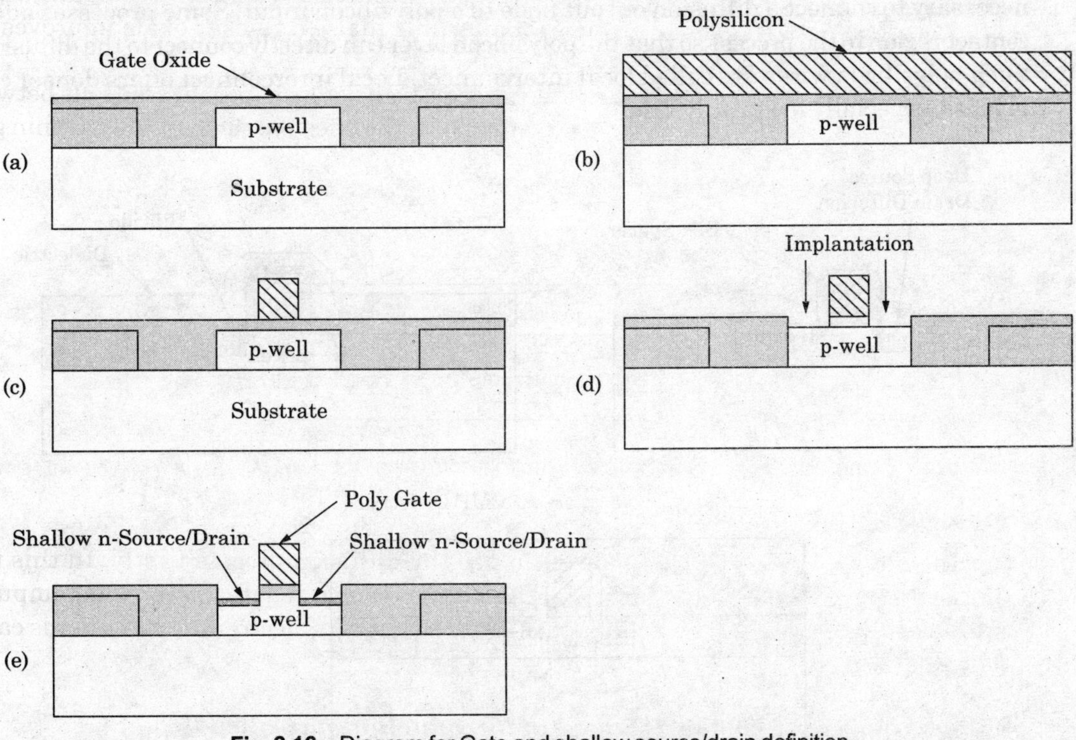

Fig. 2.16 Diagram for Gate and shallow source/drain definition.

- The source/drain implant is relatively low, typically in the range $10^{18} - 10^{20}$ cm^{-3} of impurity atoms. Such a lightly doped drain structure reduces the electric field at the drain junction (the junction with the highest voltage), which improves the immunity of the device to hot electron damage. The lightly doped drain implants exhibit low capacitance but high resistance.
- A silicon nitride (Si_3N_4) spacer along the edge of the gate serves as a mask to define the location of this deeper diffusion, as shown in Fig. 2.17(a).

- The polysillicon gate and source/drain diffusion have high resistance due to the resistively of silicon. Modern processes form a surface layer of a refractory metal (Tantalum, molybdenum, titanium, cobalt) on the silicon to reduce the resistance.
- In a Polycide process, only the gate polysilicon is silicate. In a salicide process (usually implemented as a self- aligned silicidization from whence comes the synonymous term salicide) both gate poly-silicon and source/drain regions are solicited. This process lowers the resistance of the poly-silicon interconnect and/or the source and drain.
- The resultant structure with gate and source/drain regions silicided is shown in Fig. 2.17(b). In addition, SiO_2 or an alternative dielectric has been used to cover all areas prior to the next processing steps.
- A structure where CMP has been employed is shown in Fig. 2.17(c). Achieving a very flat finish allows layers to be stacked vertically without incurring the problems of metal having to traverse rapid transitions in surface height as shown in Fig. 2.17(b), which can lead to breaks and a plethora of design rules that relate to metal edges.
- Polysilicon over diffusion normally forms a transistor gate, so a short metal 1 wire is necessary to connect a diffusion output node to a polysilicon input. Some processes add a contact region to the process so that the polysilicon layer can directly connect to the diffusion. Such polysilicon wires are called local interconnect. Local interconnect offers denser cell layouts, especially in static RAMs.

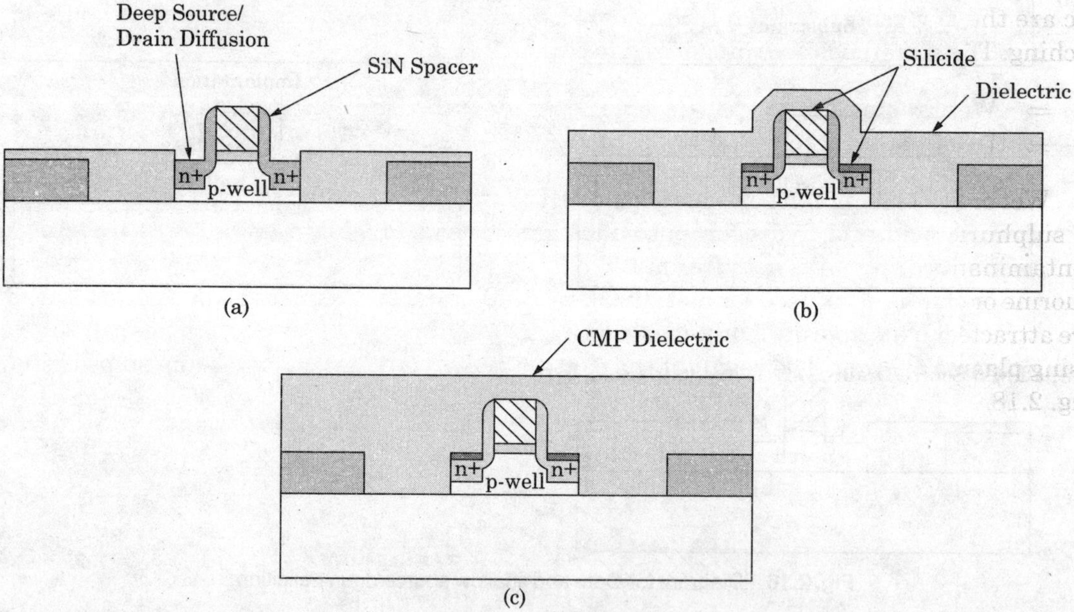

Fig. 2.17 Structure of various Lightly Doped Drain.

2.6.9 Contacts and Metallization

- Contacts cuts are made to source, drain, and gate according to the contact mask. These are holes etched in the dielectric at the end of the source/drain step covered in the previous section. Aluminum (Al) is commonly used for wires but tungsten (W) can be used as a plug to fill the contact holes. In some processes, the tungsten can also be used as a local interconnect layer.

- Metallization is a process by which metal layers can be deposited for making ohmic contacts, interconnects gate etc. The metal that is used to provide ohmic contacts for integrated circuits has to meet several requirements. For example, the metal must have high electrical conductivity and should be able to make good ohmic contacts to the semiconductor. Furthermore, it should have good adherence to the underlying surface and must not corrode under normal operating conditions. No single metal is known to perfectly meet all these requirements for a given semiconductor. However, in the case of silicon, Al meets most of the requirements to a sufficient degree and is widely used for device interconnections, and in many cases, for making ohmic contacts. In some instances, where Al does not meet all the requirements of metallization, multilayered structures are used. A typical example of such a structure that is used for Si is the Au-based system PtSi-Ti-Pt-Au. Here Au is the main conductor, and each of the other layers meets some of the requirements mentioned above.
- Most common methods used for metal deposition are: Vacuum Deposition using Resistive Heating or Electron Beam, Sputtering, and Plating. Plating may involve electroplating or electro less plating.

ETCHING: The etching in IC fabrication is used for the selective removal of undesired metallic and dielectric layers, which may be typically a semiconductor film of silicon dioxide, aluminum, silicon nitride or poly silicon. Precise controls of dimensions, operator safety, ecology cost efficiency etc are the key aspects of etch technology. A number of techniques are developed to carry out etching. These methods are categorized in two main categories:

⇒ Wet Etching
⇒ Dry Etching

Wet or dry etching can be used to remove unwanted metal. Piranha solution is a 3:1 to 5:1 mix of sulphuric acid and hydrogen peroxide that is used to clean wafers of organic and metal contaminants or photo resist after metal patterning. Plasma etching is a dry etch process with fluorine or chlorine gas used for metallization steps. The plasma charges the etch gas ions, which are attracted to the appropriately charged silicon surface. Very sharp etch profiles can be achieved using plasma etching. The result of the contact and metallization patterning steps is shown in Fig. 2.18.

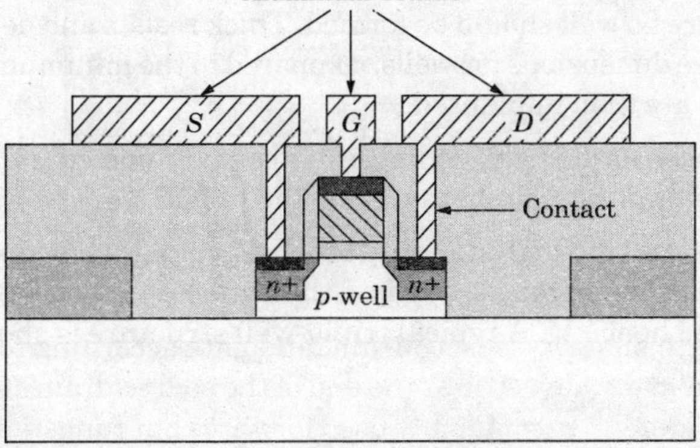

Fig. 2.18 Aluminum Metallization

2.6.10 Passivation

The final processing step is to add a protective glass layer called passivation or over-glass that prevents the ingress of contaminants. Openings in the passivation layer allow connection to I/O pads and test probe points if needed. After passivation, farther steps can be performed such as bumping, which allows the chip to be directly connected to a circuit board using plated solder bumps in the pad openings.

2.6.11 Metrology

- Metrology is the science of measuring. Everything that is built in a semiconductor process has to be measured to give feedback to the manufacturing process. This ranges from simple optical measurements of line widths to advanced techniques to measure thicknesses of thin films and defects such as voids in copper interconnect.
- A natural requirement exists for in real-time measurements so that the manufacturing process can be controlled in a direct feedback manner.
- Optical microscopes are used to observe large structures and defects, but are no longer adequate for structures smaller than the wavelength of visible light (0.5μm).
- Scanning electron microscopy (SEM) is used to observe very small features. An SEM raster scans a structure under observation and observes secondary electron emission to produce an image of the surface of the structure.
- Energy Dispersive Spectroscopy (EDX) bombards a circuit with electrons causing X-ray emission. This can be used for imaging as well.

2.6.12 Steps of CMOS Processes

Define the well regions is the basic step in CMOS processes. In a triple-well process.

Firstly a deep n-well is driven into the p-type substrate using high-energy (MeV) ion implantation as opposed to a thermal diffused operation. This avoids the improvement in temperature, which improves throughput and reliability.

A 2-3 MeV implantation can yield a 2.5-3.5um deep n-well. Such a well has a peak dopant concentration just under the surface and for this reason is called a retrograde well. This can enhance device performance by providing improved latch-up characteristics and reduced susceptibility to vertical punch-through. A thick (3.5-5.5μm) resist has to be used to block the high energy implantation where no well should be formed. Thick resists and deep implants necessarily lead to fairly coarse feature dimensions for wells, compared to the minimum feature size. Shallower n-well and p-well regions are then implanted.

After the wells have been formed, the doping levels can be adjusted (called a threshold implant) to set the desired threshold voltages for both nMOS and pMOS transistors.

For a given gate and substrate material, the threshold voltage (V_t) depends on the doping level in the substrate (N_A), the oxide thickness (t_{ox}) and the surface state charge (Q_{fc}). The implant can affect both N_A and Q_{fc} and hence V_t. A typical triple-well structure is shown in Fig. 2.19.

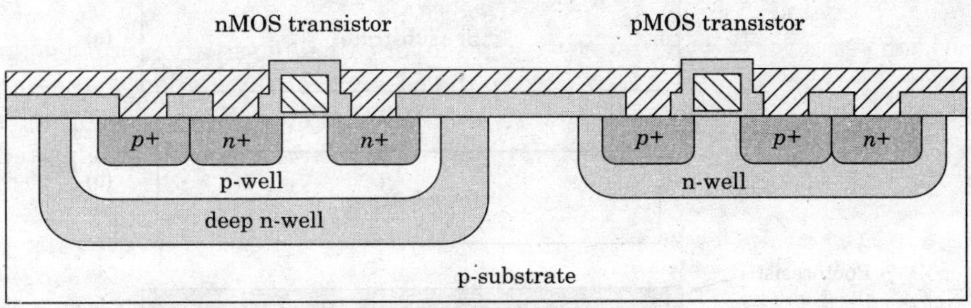

Fig. 2.19 Well structure in triple-well process.

2.7 FABRICATION PROCESS FLOW: BASIC STEPS

For the fabrication of any device such as nMOS, CMOS etc., we can use the following technological process step by step:

Oxidation: Oxidation refers to chemical process of reduction of silicon with oxygen to form silicon dioxide, SiO2.

Diffusion: Diffusion is the process of introducing controlled amount of dopants into semiconductors.

Ion-Implantation: Ion-implantation is the process of introduction of high energy charged particles into the substrate. The dopant atoms are vaporized, accelerated and targeted at the substrate.

Epitaxy: The word epitaxy is made from two Greek words epi + taxis i.e., epi means upon and taxis mean ordered. Thus growth of thin ordered crystalline layer on a crystalline substrate is known as epitaxy. During epitaxy the substrate acts as the seed crystal and the crystal can be grown below the melting point.

Epitaxy is generally used to enhance the performance of bipolar-transistor and other CMOS IC's also.

Lithography: Lithography is the process of transferring the patterns of geometric shapes on a mask. The pattern define the regions of the integrated circuit that have to be fabricated.

Etching: The process of removing material selectively is known as etching.

Metallization: Metallization is the formation of metal films used for inter-connections, ohmic contacts and rectifying metal-semiconductor contacts.

The integrated circuit is a set of patterned layers of doped silicon, polysilicon, and insulating silicon dioxide. A layer must be patterned before the next layer of material is applied on the chip. The process used to transfer a pattern to a layer on a chip is known as lithography. Since each layer has its own distinct patterning requirements.

The required process flow steps for patterning of silicon dioxide is shown in Fig. 2.20. These steps are discussed below:

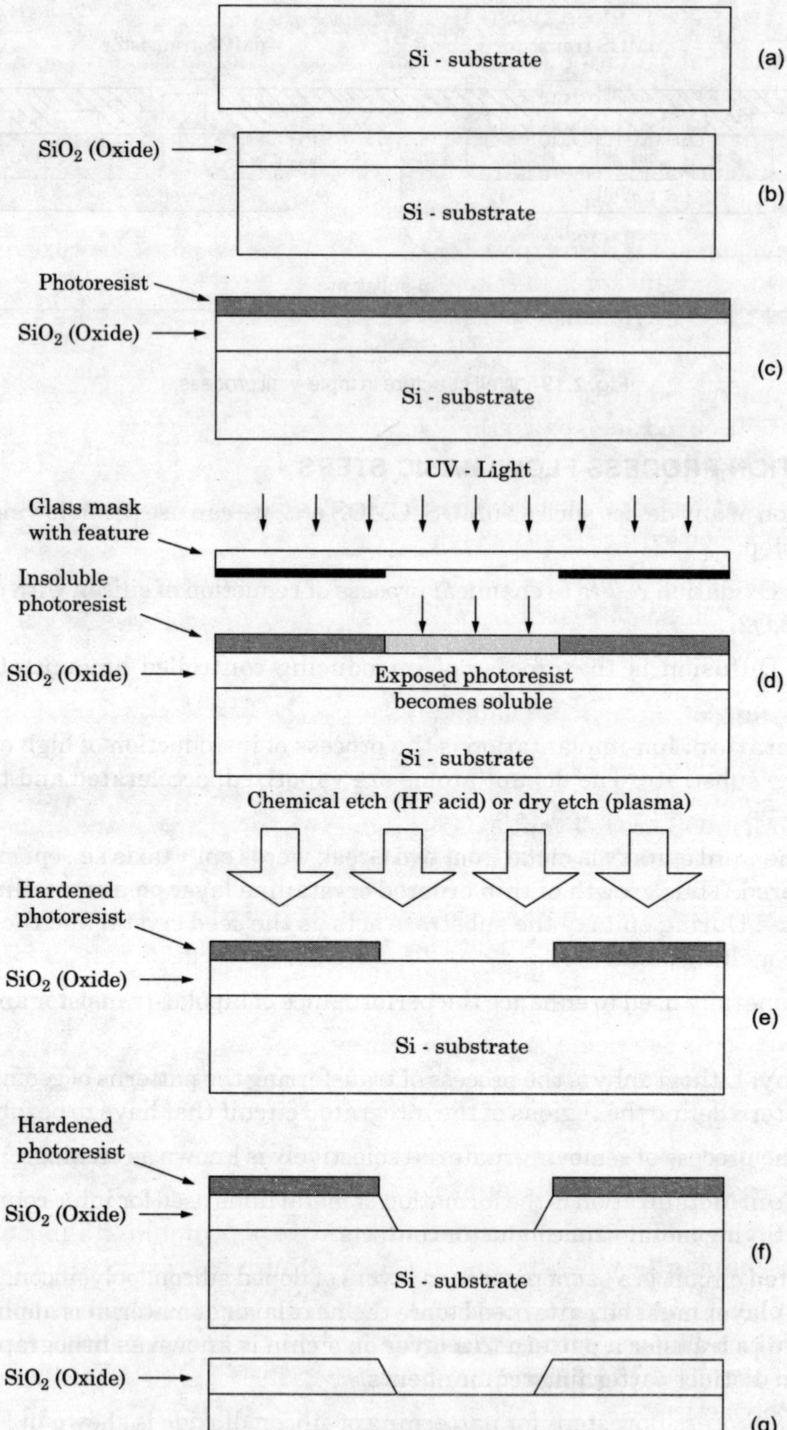

Fig. 2.20 Process steps required for patterning of silicon dioxide.

Step 1: First, we take a silicon substrate as Fig. 2.20(a).

Step 2: The thermal oxidation is used to created oxide layer of about 1 mm thickness on the substrate [Fig. 2.20(b)].

Step 3: After that, the entire oxide surface is covered with the a layer of photoresist [Fig. 2.20(c)]. This photoresist is essentially a light-sensitive, acid-resistant organic polymer, initially insoluble in the developing solution.

Step 4: The photoresist is then exposed to UV light through a mask, which defines those regions into which diffusion is to take place together with transistor channels. In the areas where the UV light can be pass through the photoresist is exposed and becomes soluble [Fig. 2.20(d)].

Step 5: The unexposed portions are subsequently etched away together with the underlying SiO_2 (silicon dioxide) so that the wafer surface is exposed in the window defined by mask [Fig. 2.20(e)].

Step 6: After the Step-5 (or) at the end of Step-5, we obtain an oxide window that reaches down to the silicon surface [Fig. 2.20(f)].

Step 7: The remaining photoresist is removed from the silicon dioxide surface by using another solvent. The leaving patterned silicon dioxide facture on the surface as shown in [Fig. 2.20(g)].

2.7.1 Fabrication of the nMOS transistor

The fabrication of NMOS can be considered a 'standard' process. The brief introduction to the general aspects of the polysilicon gate NMOS fabrication process will now be given. The fabrication process used for NMOS are relevant to CMOS and BiCMOS which may be viewed as involving additional fabrication steps. The NMOS fabrication sequence is shown in Fig. 2.21.

Step 1: Firstly, we take a thin wafer cut from a single crystal of silicon of high purity. In this wafer *p*-impurities are introduced as the crystal for grown. Such wafers are typically 75 to 150 mm in diameter and 0.4 mm thick and are doped with, e.g. boron to impurity concentrations of $10^{15}/cm^3$ to $10^{16}/cm^3$. At this doping concentrations, the wafer has resistivity in the range of 25 Ω-cm to 2 Ω-cm.

Step 2: A layer of silicon dioxide (SiO_2) of about 1 μm thick is grown all over the surface of the wafer by using the oxidation technology. The layer used as protective layer, acts as a barrier to dopants during processing and to provide a common insulating substrate onto which other layers may be grown.

Step 3: The entire oxide surface is covered with a layer of photoresist. This photoresist layer is deposited onto the wafer and spun to achieve an even distribution of the required thickness.

Step 4: The photoresist is then exposed to UV light through a mask, which defines those regions into which diffusion is to take place together with transistor channels. Here, the unexposed areas are left unaffected and may be dissolved in proper etchant, whereas exposed areas are polymerized.

Step 5: The unexposed portions are subsequently etched away together with the underlying silicon dioxide (SiO_2) so that the wafer surface is exposed in the window defined by the mask.

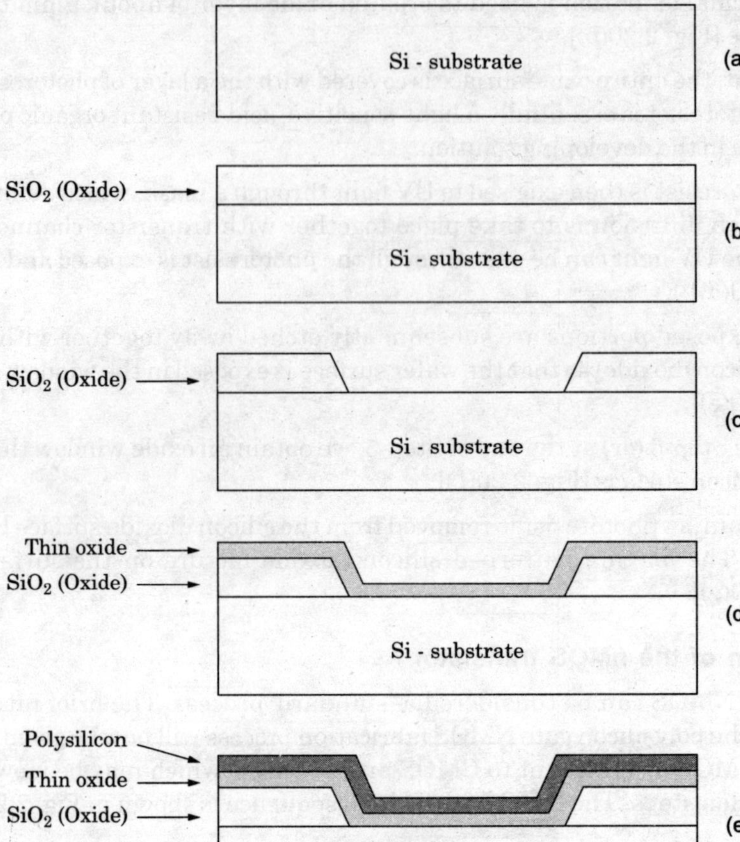

Si - substrate
(a)
SiO$_2$ (Oxide)
(b)
Si - substrate
SiO$_2$ (Oxide)
(c)
Si - substrate
Thin oxide
SiO$_2$ (Oxide)
(d)
Si - substrate
Polysilicon
Thin oxide
SiO$_2$ (Oxide)
(e)
Si - substrate
Polysilicon
Thin oxide
SiO$_2$ (Oxide)
(f)
Si - substrate
Polysilicon
SiO$_2$ (Oxide)
(g)
Si - substrate

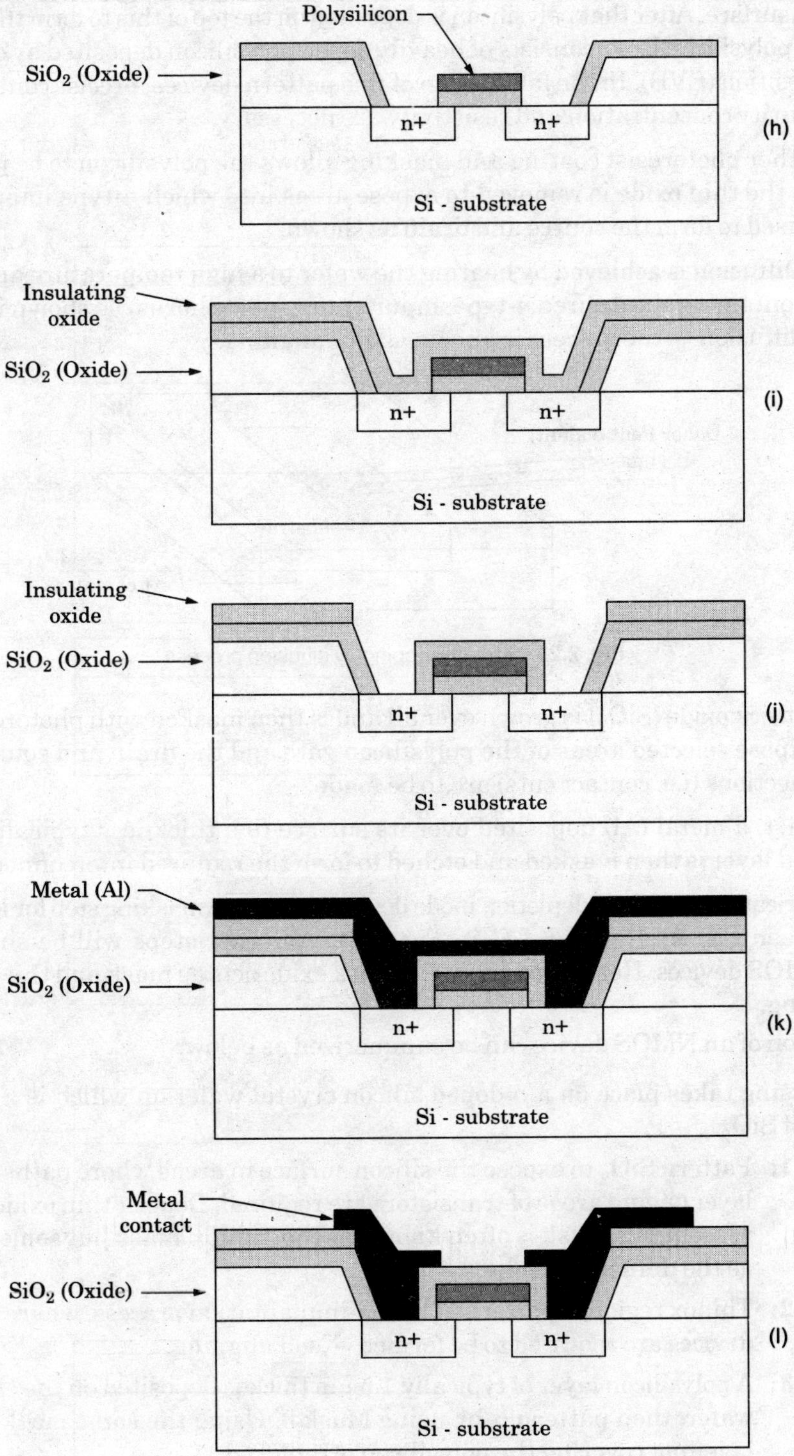

Fig. 2.21 Process flow for the fabrication of *n*-type MOSFET on *p*-type silicon.

Step 6: The remaining photoresist is removed and a thin layer of SiO_2 is grown over the entire chip surface. After that polysilicon is deposited on the top of this to form the gate structure. The polysilicon layer consists of heavily doped polysilicon deposited by chemical vapour deposition (CVD). In the fabrication of fine pattern devices, precise control of thickness, impurity concentration and resistivity are necessary.

Step 7: Further photoresist coating and masking allows the polysilicon to be patterned. After that the thin oxide is removed to expose areas into which n-type impurities are to be diffused to form the source and drain as shown.

- Diffusion is achieved by heating the wafer to a high temperature and passing a gas containing the desired n-type impurity (e.g. phosphorus) as shown in Fig. 2.22. The diffusion — the process is known as self-aligning.

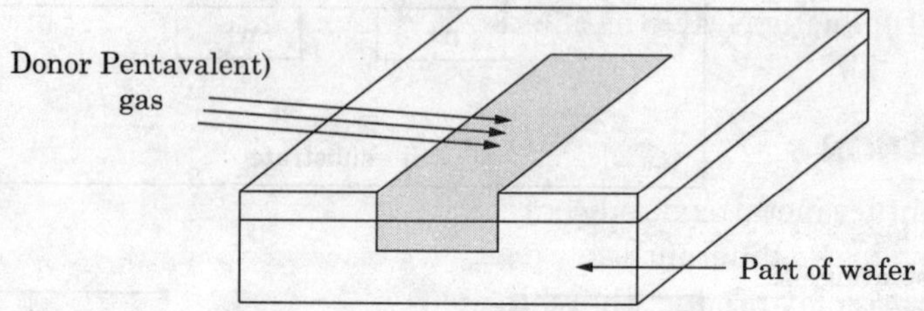

Fig. 2.22 Impurity doping by diffusion process.

Step 8: The thick oxide (SiO_2) is grown over all and is then masked with photoresist and etched to expose selected areas of the polysilicon gate and the drain and source areas where connections (i.e. contact cuts) are to be made.

Step 9: At last, a metal (Al) deposited over its surface to a thickness typically of 1 μm. This metal layer is then masked and etched to form the required interconnection pattern.

For the fabrication process of depletion mode devices, one more masking step for ion-implantation is added between Fig. 2.21(e) and (f) in Fig. 2.21. All other steps will be similar to that of fabricating NMOS devices. Here, once again the thick oxide acts as mask and this process stage is also self-aligning.

The function of an NMOS device can be summerized as below:

- Processing takes place on a p-doped silicon crystal wafer on which is a grown a 'thick' layer of SiO_2.
- **Mask 1:** Pattern SiO_2 to expose the silicon surface in areas where paths in the diffusion layer or gate areas of transistors are required. Deposit thin oxide overall for this reason, this mask is often known as the 'thinox' mask but some texts refer to it as the diffusion masks.
- **Mask 2:** Thinox region is patterned by ion-implantation process, where depletion mode devices are required to be formed — self aligning.
- **Mask 3:** A polysilicon layer of typically 1.5 μm thick is deposited on the entire area of the wafer, then pattern it by using Mask 3. Using the same mask, the thin oxide layer not covering the polysilicon is removed.
- **Mask 4:** Diffuse n^+ regions into areas where thinox has been removed. Thus transistor drains and sources are self-aligned with respect to the gate structure.

- **Mask 5:** Growing thick oxide layer over the entire substrate and then etching for contact cuts.
- **Mask 6:** The metal is deposited and then patterned with Mask 5.
- **Mask 7:** This will be required for over glassing processing step.

2.7.1.1 Advantages of nMOS Fabrication

1. This process require less photolithographic steps.
2. It has high functional density, good speed/power performance.
3. It has tremendous future potential due to its scalability.

2.2.1.2 Disadvantages of nMOS Fabrication

The major drawbacks of *n*MOS process are its high absolute power consumption and its electrical asymmetry. CMOS is replacing NMOS as the 'standard' process because it minimizes both of the above disadvantages.

2.8 CMOS FABRICATION

The CMOS (complementary metal oxide silicon) technology is recognized as the leader VLSI systems technology. CMOS provides an inherently low power static circuit technology that has the capability of providing lower power-delay product than bipolar, *n*MOS or GaAs technologies.

The four main CMOS technologies are:

1. p-well process
2. n-well process
3. Twin-tub process
4. Silicon on insulator process.

2.8.1 The p-Well Process

In this process of CMOS, the structure consists of an *n*-type substrate in which *p*-type devices may be formed by suitable masking and diffusion. In order to accommodate *n*-type devices, a deep *p*-well is diffused into the *n*-type substrate as shown in Fig. 2.23.

The diffusion must be carried out with special care since the *p*-well doping concentration and depth with affect the threshold voltages as well as the breakdown voltages of the *n*-transistors. To achieve low threshold voltages (0 to 1.0 V), we need either deep well diffusion or high well resistivity. However, deep wells require larger spacing between the *n*-type and *p*-type transistors and wires because of lateral diffusion and hence a large chip area.

The *p*-wells act as substrates for the *n*-devices within the parent *n*-substrate and the two areas are electrically isolated.

In summary, typical processing steps are:

- **Mask 1:** Defines the areas in which the deep *p*-well diffusions are to take place.
- **Mask 2:** Defines the thin oxide regions, namely those areas where the thick oxide is to be stripped and thin oxide grown to accommodate *p* and *n*-transistors and wires.
- **Mask 3:** Used to pattern the polysilicon layer which is deposited after the thin oxide.
- **Mask 4:** A p^+ mask is now used (to be in effect "Anded" with Mask 2) to define all areas where *p*-diffusion is to take place.

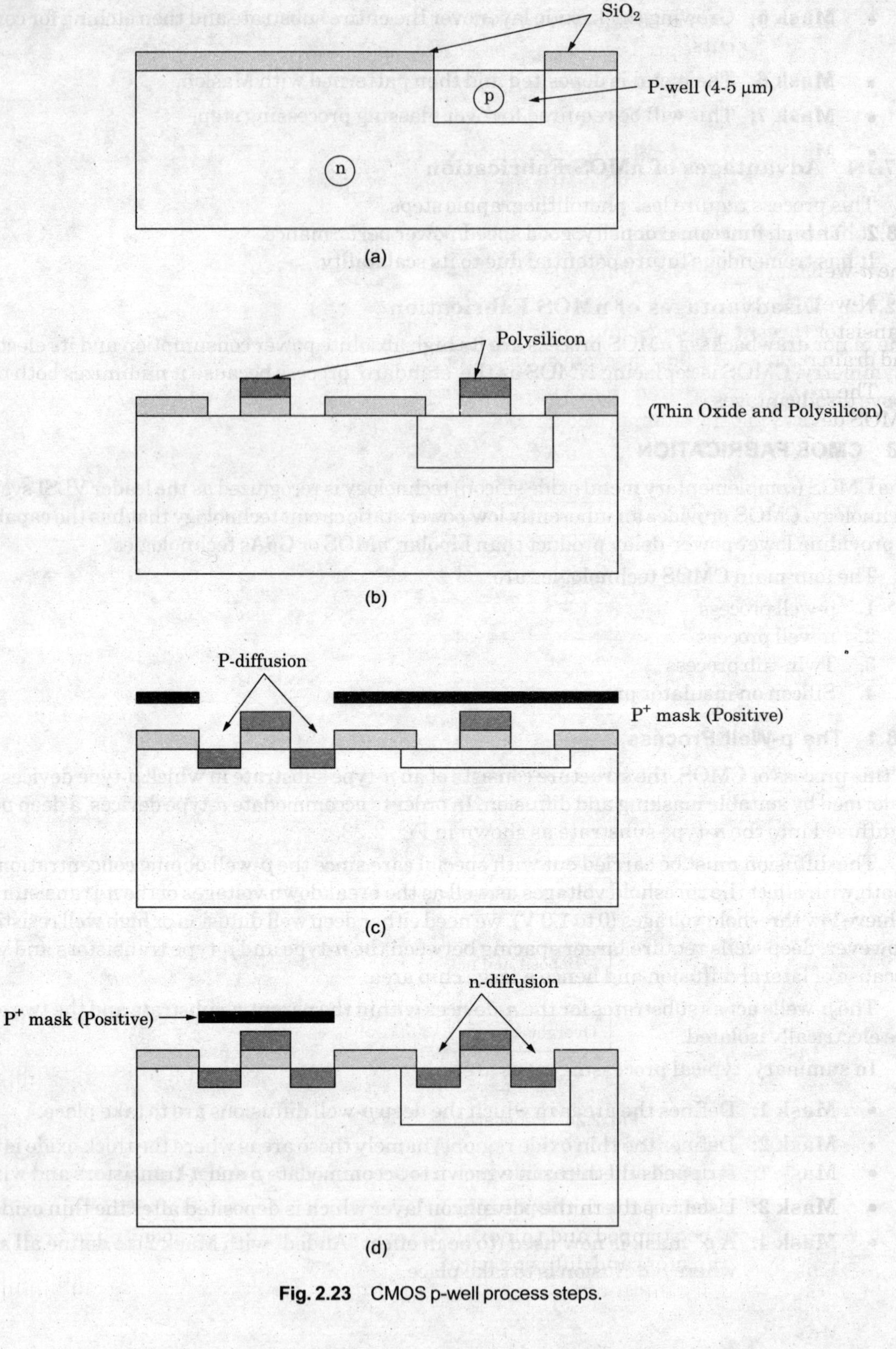

Fig. 2.23 CMOS p-well process steps.

- **Mask 5:** This is usually performed using the negative form of the p^+ mask and defines these areas where n-type diffusion is to take place.
- **Mask 6:** Contact cuts are now defined.
- **Mask 7:** The metal layer pattern is defined by this mask.
- **Mask 8:** An overall passivation (over glass) layer is now applied and this mask is needed to define the openings for access to bonding pads.

2.8.2 The n-Well Process

The n-well fabrication has gained wide acceptance.

N-well CMOS circuits are superior to p-well because of the lower substrate bias effect on transistor threshold voltage and inherently lower parasitic capacitances associated with source and drain regions. The flow diagram of the fabrication for nMOS process are illustrated in Fig. 2.24.

The n-wells are created in p-type substrate. The typical processing steps for fabrication of CMOS devices may be summerized as below:

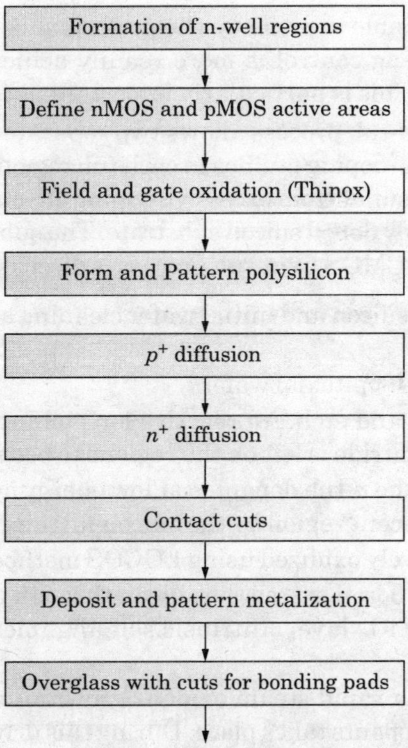

Fig. 2.24 Fabrication steps for nMOS processes.

- **Mask 1:** It defines the areas in which the deep n-well diffusions has to take place.
- **Mask 2:** It defines the thin oxide regions, *i.e.*, those areas in which the thickoxide layer is to be stripped and thinoxide is to be grown fcr accommodating n- and p-type transitors and diffusion wires.

- **Mask 3:** This mask is used to pattern the polysilicon layer, which is to be deposited after thin oxide.
- **Mask 4:** In this a n^+ mask is used to define the areas for all types of diffusion of *p*-type impurity.
- **Mask 5:** It is generally performed using negative form of n^+ mask and with Mask 2. It is used to define the areas for *p*-type diffution.
- **Mask 6:** Contact cuts are defined.
- **Mask 7:** The metal layer pattern is defined.
- **Mask 8:** At this stage, the overall passivation layer is applied and by this mask, the openings for bonding pads are defined.

2.8.3 The Twin-Tub Process

A logical extension of the *p*-well and the *n*-well approaches is the twin-tub fabrication process.

In this process, we start with a substrate of high resistivity *n*-type material and then create both *n*-well and *p*-well regions.

Through this process, it is possible to preserve the performance of the *n*-transistors without compromising *p*-transistors. Doping control is more readily achieved and some relaxation in manufacturing tolerance results. This is particularly important as far as latchup is concerned.

Twin-Tub Process: The twin-tub process allows two separate tubs to be implant into very lightly doped silicon. This allows the doping profiles in each tub region to be tailored independently so that neither type of devices will suffer from excessive doping effects. The lightly doped silicon is an epitaxially grown layer on heavily doped silicon substrate. The substrate can be either *n*-type or *p*-type. The process sequence for a CMOS twin-tub process is discussed as:

1. The initial wafer chosen n+ silicon and initial wafer cleaning and back silicon implantation process takes place.
2. Deposition of near intrinsic *V*-epitaxial wafer.
3. Grow the nitride sandwich and etch. An reactive ion etching may be done to maintain a sharp edge definition. The nitride is left on the regions to become p-tub.
4. Phosphors in implanted as the *n*-tub dopant is at low energy and enters the exposed silicon but is masked from the adjacent region by the silicon nitride.
5. The wafers are then selectively oxidized using LOCOS method over the n-tub regions.
6. The nitride is stripped and boron extras the silicon through the thin oxide but is masked from the *n*-tub by thicker SiO_2 layer. Intrinsic self alignment taken place in the entire process.
7. Oxide is now stripped and a rapid steam oxide is grown. A separate anneal step is now incorporated. Drive-in the dopants takes place. During this drive-in, lateral diffusion of both n and p dopants takes place. Due to LOCOS the silicon surface in the two tub regions is not the same and a link at the edge exists.
8. After the tub formation, the formation of field oxide and gates is same as that for nMOS process.
9. The threshold adjustment implants can be made into the channel regions of the devices to adjust the threshold voltages of the *n* and *p* channel transistors.
10. The sources and drains are self-aligned to the gate. Selective implantation of the n-channel and *p*-channel source/drains to form n^+ regions for the *n*-channels and implanted into all

the sources and drains. This is followed with a selective implant of phosphorus or arsenic into the n-channel source/drain regions sat a higher dose so that it over-compensates the bottom.

11. After the subsequent thermal cycles the phosphorus or arsenic completely covers the boron vertically and laterally.
12. A phosphorus glass layer (PSG) is taken deposited and windows are etched.
13. Aluminium metallization is defined using dry etching.
14. The final layer is a plasma deposited silicon nitride layer that seals the devices and provides mechanical scratch protection.

The twin-tub CMOS technology provides the basis for separate optimization of the p-type and n-type transistors. The arrangement of an inverter (CMOS) using the twin-tub CMOS Technology is shown in Fig. 2.25.

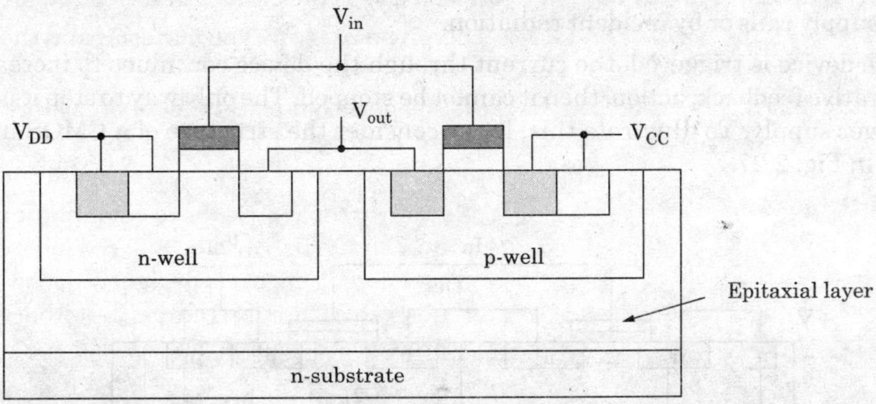

Fig. 2.25 Twin-tub structure.

2.9 THERMAL ASPECTS OF PROCESSING

The sequences of high temperature are different for making nMOS and CMOS device which are shown in Fig. 2.26.

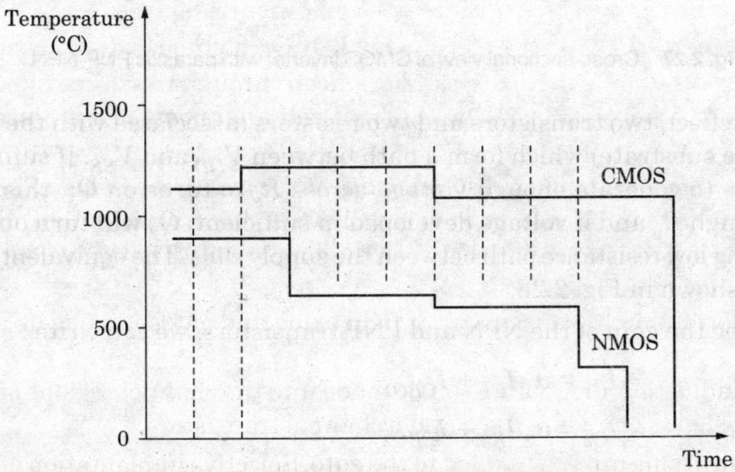

Fig. 2.26 Thermal sequence difference between nMOS and CMOS processes.

The CMOS p-well process has a high temperature p-well diffusion process (1100 to 1350°C) but n-MOS process having no such requirement. Because of simplicity, ease of fabrication and high density per unit area of nMOS circuits, many of the earlier IC designs, still in current use, have been fabricated using nMOS tehnology.

2.10 LATCH UP

- In CMOS circuits, the parasitic NPN and PNP bipolar junction transistors (BJT) forms a thyristor or PNPN structure. When the thyristor is triggered by noise or glitch, the BJTs become ON, and it leads to a formation of short circuit between the power and ground lines, known as CMOS latch-up.
- Latch-up is a condition in which parasitic components give rise to establishment low-resistance conducting path between V_{DD} and V_{SS} with disastrous results. Careful steps during fabrication are considered in order to avoid these problems. This may be induced by glitches on the supply rails or by incident radiation.
- One the device is triggered, the current through the device continues to increase due to regenerative feedback, action, then it cannot be stopped. The only way to stop it is to switch the power supply. To illustrate this, let us consider the structure of a CMOS inverter as shown in Fig. 2.27.

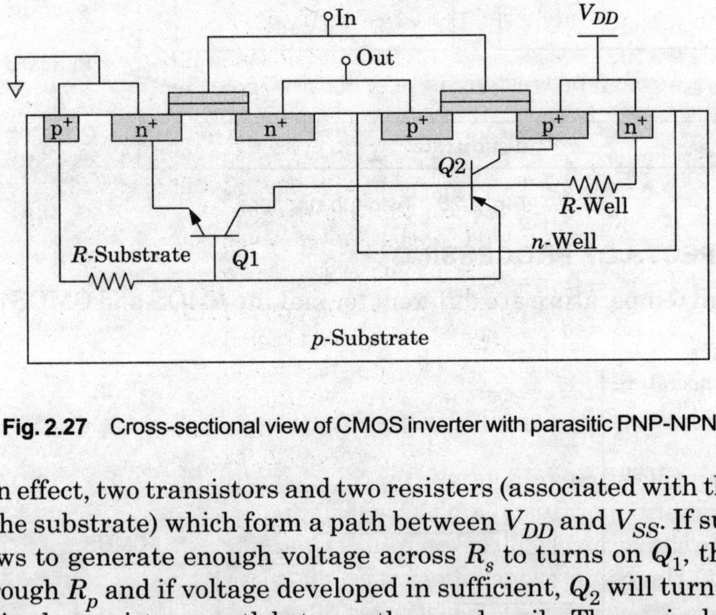

Fig. 2.27 Cross-sectional view of CMOS inverter with parasitic PNP-NPN

- These are in effect, two transistors and two resisters (associated with the p-well and with regions of the substrate) which form a path between V_{DD} and V_{SS}. If sufficient substrate current flows to generate enough voltage across R_s to turns on Q_1, this will then draw current through R_p and if voltage developed in sufficient, Q_2 will turn on, establishing a self-sustaining low resistance path between the supply rails. The equivalent circuit of PNPN structure is shown in Fig. 2.28.

If α_1 and α_2 be the gain of the NPN and PNP transistors, we can write

$$I_{C1} = \alpha_1 I_{E1} + I_{CBO1} \quad ...(2.1)$$

$$I_{C2} = \alpha_2 I_{E2} + I_{CBO2} \quad ...(2.2)$$

Lets, $$I = I_{E2} + I_W \quad ...(2.3)$$

$$I = I_{E1} + I_S \quad ...(2.4)$$

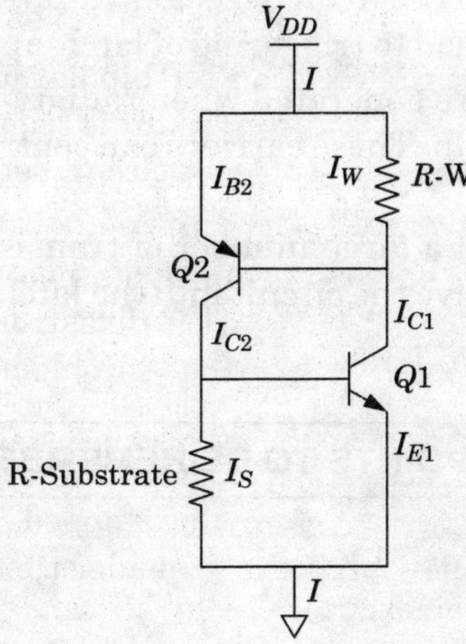

Fig. 2.28 Equivalent circuit of PNPN structure

The total current can be written as

$$I = I_{C1} + I_{C2}$$

$$I = \alpha_1 I_{E1} + I_{CBO1} + \alpha_1 I_{E2} + I_{CBO2}$$

$$I = \alpha_1(I - I_S) + \alpha_2(I - I_W) + I_{CBO1} + I_{CBO2}$$

$$I(1 - \alpha_1 - \alpha_2) = I_{CBO1} + I_{CBO2} - (\alpha_1 I_S + \alpha_2 I_W)$$

$$I = \frac{I_{CBO1} + I_{CBO2} - (\alpha_1 I_S + \alpha_2 I_W)}{1 - (\alpha_1 + \alpha_2)} \quad ...(2.5)$$

where, $$I_{CBO} = I_{CBO1} + I_{CBO2} \quad ...(2.6)$$

- When the sum of the current gains approaches unity $\{\alpha_1 + \alpha_2 = 1\}$, the current become very large. The circuit operates in a positive feedback loop. Due to the regenerative feedback action, there can be a very large current flowing through the device, damaging the device components. Hence, in order to keep the current limited, the parasitic transistors gain must be low and the parasitic resistances must be small.

2.10.1 Techniques for Latch-Up Prevention

There are some techniques are used/include for the latch-up prevention.

1. Use of p^+ and n^+ guad rings around nMOS and pMOS connected to the ground and V_{DD}, respectively.
2. Reduction of *R*-well and *R*-substrate as much as possible by placing substrate and well contacts as close as possible to the source connect.
3. Keeping sufficient spacing between the nMOS and pMOS transistors.

2.10.2 System-Level Approaches to Avoid the Latch-Up

1. Power supplies must be OFF before plugging aboard. A plus-in in power supply ON condition can cause a surge voltage to appear on the signal pins which could trigger the latch-up.
2. The electrostatic discharge (ESD) can trigger latch-up. ESD protection circuits need special care such that it does not lead to triggering of latch-up.
3. The electron-hole praise are generated when radiation, such as X-rays, cosmic or alpha-rays, penetrate into the chip. These carriers can contribute to well or substrate currents, leading to latch-up.
4. Simultaneous switching of a large number of transistors can cause noise in the power/ground lines, which can drive the circuit into the latch-up. This must be checked through simulation.

POINTS TO BE REMEMBER

1. Accumulation mode is when $V_{GS} < V_T$.
2. The mask is often known as Thinox.
3. High absolute power consumption and electrical asymmetry are the two major drawbacks of NMOS process.

SHORT ANSWERS TYPE QUESTIONS

Q. 1. What is the active area? Discuss the role of channel doping in MOS fabrication.

Ans. Active areas are those areas where the transistor are created. The desired threshold voltage of a MOS device can be adjusted by controlling the channel doping for the given gate material and oxide thickness. Channel doping shown not be either too high or too low because if channel doping is too high, then surface mobility is reduced drastically than bulk mobility and if doping is too low, then the depletion regions of source and drain start nearly touching each other. This condition is called *punch-through condition.* One method to reduce punch through for short channel MOS devices is to implant boron sufficiently deep into the channel region to raise the substrate concentration without changing the surface concentration.

A shallow implant is used to adjust the threshold voltage. Boron implant is to raise the threshold voltage whereas arsenic/phosphorous implants to reduce the threshold voltage.

Q. 2. What are the main advantages of non-self aligned process for MOSFET?

Ans. The disadvantages of the process are:

(*a*) If there is a large overlap of gate electrode and contact "tubs", it will result in a large parasitic capacitance.

(*b*) If there is no overlap of gate electrode and contact "tubs", MOSFET does not work.

Q. 3. Why silicon is preferred over germanium? Explain.

Ans. Silicon is preferred over germanium due to the following reasons.

(*i*) Si devices can operate upto 150 degree Celsius as compared to 100 degrees for Ge.

(*ii*) Ge is more costly than that of silicon.

(*iii*) Si grows a stable oxide (Silicon dioxide) which is important in the fabrication of IC as compared to Ge oxide which is unsuited for device application.

(*iv*) Undoped resistively of germanium in 47 ohm-cm as that of 23×10^4 ohm-cm for Si due to which high-voltage rectifying devices are practical with Si.

(v) Ge is difficult to dope as compared to silicon.

(vi) Silicon is easily available near the sea beach as a raw material for silicon wafer preparation.

(vii) Ge has high junction leakage current or reverse current as compared to that of silicon.

(viii) Growth of silicon oxide layer does not requires any special arrangements, whereas for growing Ge oxide, special plant has to be installed which increases the cost of the finished product.

Q. 4. What is a twin-well process?

Ans. A logical extension of the *p*-well and the *n*-well approaches is the twin-tub process.

Q. 5. What are Enhancement Mode Devices.

Ans. The *n*-channel devices that are cut off with zero gate bias.

REVIEW QUESTIONS

1. With the help of IC process flow diagram, describe briefly, the basic considerations for IC processing for an *n*-channel, polysilicon MOS.
2. Discuss and describe the basic design rules for miniaturizing VLSI circuits, in brief.
3. List different steps of IC manufacturing and explain briefly.
4. How many diffusions are needed to form a BJT? Explain the purpose of each diffusion step.
5. What are the processing steps in *p*-well process of CMOS fabrication? Summarise only.
6. Explain twin-tub process with suitable sketch.
7. Explain all the steps of nMOS fabrication.
8. Explain nMOS fabrication with suitable sketch.
9. Discuss various thermal aspects of processing.
10. What is latch-up? Discuss in detail.
11. Explain CMOS Latch-up. How it is prevented?
12. Explain Latch-up phenomenon is CMOS technology with suitable diagram. How can it be avoided?
13. What is electronic grade silicon? Discuss Czochralski crystal growth method in detail.
14. What are the steps of fabricating an enhancement mode NMOS transistor, Aid your answer with suitable diagrams.
15. Explain ion-implantation.

❑❑❑

Chapter

3 MOS Circuit Design Processes

Chapter Overview

- Introduction
- MOS Layers
- Layout Designs
- Why Design Rules
- Stick Diagrams
- Full-Custom Mask Layout Design

3.1 INTRODUCTION

Design processes are aided by simple concepts such as stick and symbolic diagram. The key element of the design process is a set of design rules. Design rules are used to produce workable mask layouts from which the various layers in silicon will be formed or patterned. Design rules specify to the designer certain geometric constraints on the layout artwork so that the patterns on the processed wafer will preserve the topology and geometry of the designs.

3.2 WHY DESIGN RULES

There are several error mechanisms that contribute to deviations in feature-shapes from the shapes defined in a designer's artwork. The purpose of design rules is to guarantee that under the worst cumulative variations of the processes, the circuit does not fail to operate, i.e., separate features do not merge and small features do not split, thereby preserving the original topology of the intended circuit. The rules must also guarantee that the electrical parameters such as the resistance and capacitance of the wires, which are determined by the physical dimensions of the features, are not altered by the process-variations to a point that may cause a serious degradation in performance.

One of the fundamental difficulties in specifying design rules is that the fabrication processes are undergoing rapid evolutionary changes. The most important parameter that characteristics any process is the feature size. As long as the processing steps are not altered drastically, a set of design rules expressed in terms of feature size will have the best chance of survivability.

With improvements in process technology, the feature size will scale down without changing the design rules.

3.2.1 Advantages of Generalized Design Rules

1. Ease of learning because they are scalable, portable and durable.
2. Increased designer efficiency due to fewer levels and fewer rule.
3. Longevity of designs that are simple, abstract and have minimum clutter.
4. Design efficiency accomplished by compaction, layout rule checking, electrical rule checking, simulation and verification.
5. Automatic translation to final layout.

3.3 MOS LAYERS

- We know that the MOS circuits are formed with four basic layers — *n*-diffusion, *p*-diffusion, polysilicon and metal. These layers are isolated from one another by thick or thin (thinox) silicon dioxide insulating layers. The thinoxide mask region having *n*-diffusion, *p*-diffusion and transistor channels. When polysilicon and thinox regions cross one another then a transistor is formed. In some processes, there may be second metal layer and also, in some processes, a second polysilicon layer. Layers may be joined together where contacts are formed.
- The basic MOS transistor properties can be modified by the use of an implant within the thinox region and this is used in NMOS circuits to produce depletion mode transistors.
- Bipolar transistors can be included in this design process by adding extra layers to be CMOS process. This is referred to as BiCMOS technology.

3.4 STICK DIAGRAMS **(UTU-2012)**

- Stick diagram are used to convey layer information through the use of a colour code. Here, the designer draws a freehand sketch of a layout using coloured lines to represent the various process layers such as diffusion, metal and poly silicon. Where polysilicon crosses the diffusion, transistors are created and where metal wires pin diffusion or polysilicon, contacts are formed.
- In NMOS design, green colour is used for *n*-diffusion, blue for metal, red for polysilicon, yellow for implant and block for contact areas.
- A stick diagram is not exact model of layout. The stick diagram represents the rectangles with lines which represents wires and component symbols.
- The colour-coding has been complemented by monochrome encoding of the lines so that black and white copies of stick diagrams do not lose the layer information.
- the mask layout information, which is also colour coded may also be hatched for monochrome encoding as shown in Fig. (3.1 to 3.2). The encoding chosen are shown and illustrated in colour as in monochrome form (as Fig. 3.1 to 3.3).
- The colour and monochrome encoding scheme used has been evolved to cover NMOS and CMOS processes. In order to use the encoding schemes, the simple set required for a single metal NMOS design is set as Fig. 3.1.
- For a double-metal CMOS *p*-well process, the required encodings are extended by those given in Fig. 3.2.
- The stick diagram of inverter circuit in NMOS and in *p*-well CMOS technology is shown in Fig. 3.3.
- Having conveyed information and topology by using stick or symbolic diagrams, these diagrams are relatively easily turned into mask layouts as, for example, the transistor stick diagrams of Fig. 3.4 stressing the ready translation into mask layout form. In order that the mask layouts produced during design will be compatible with the fabrication processes, a set of design rules are setout for layouts.

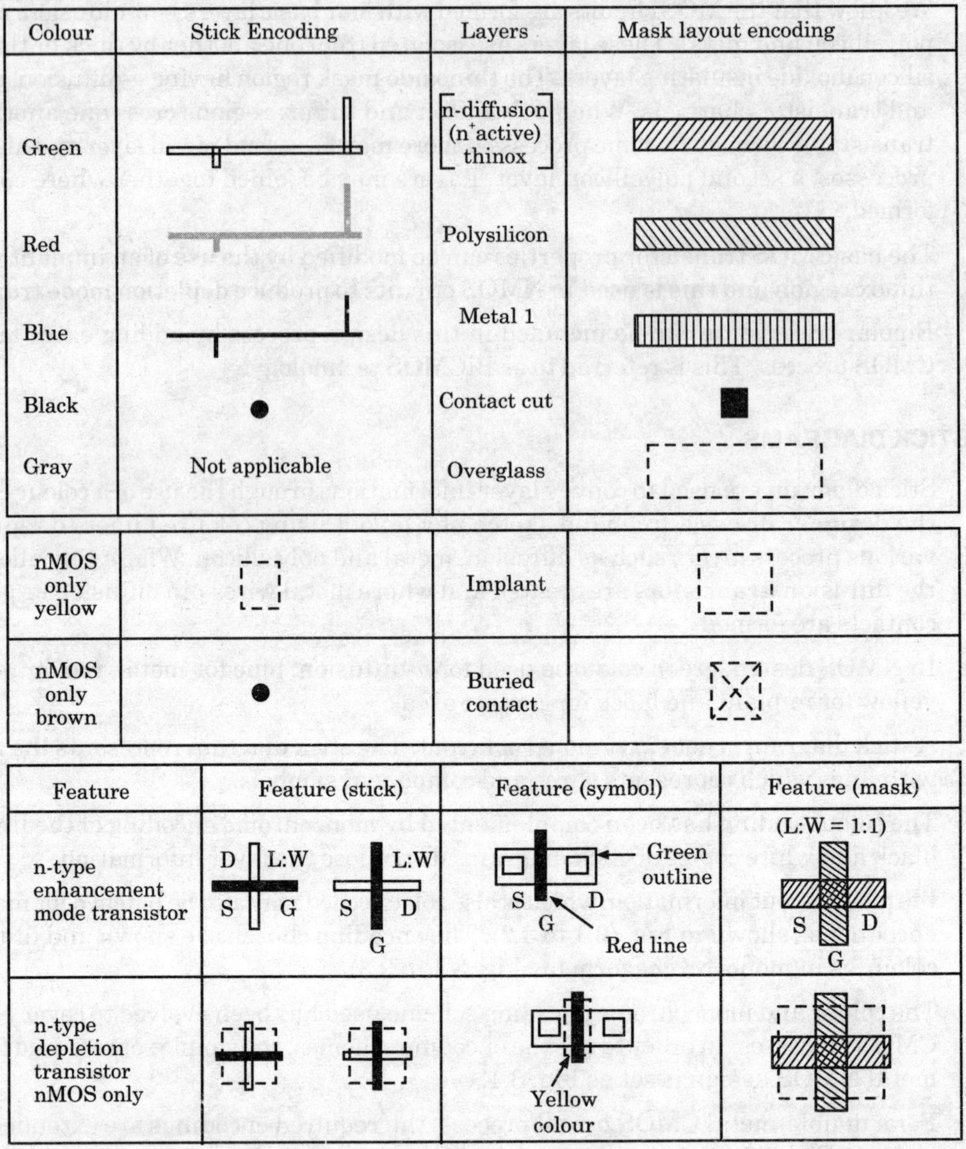

Fig 3.1 Encoding for a simple single metal NMOS process.

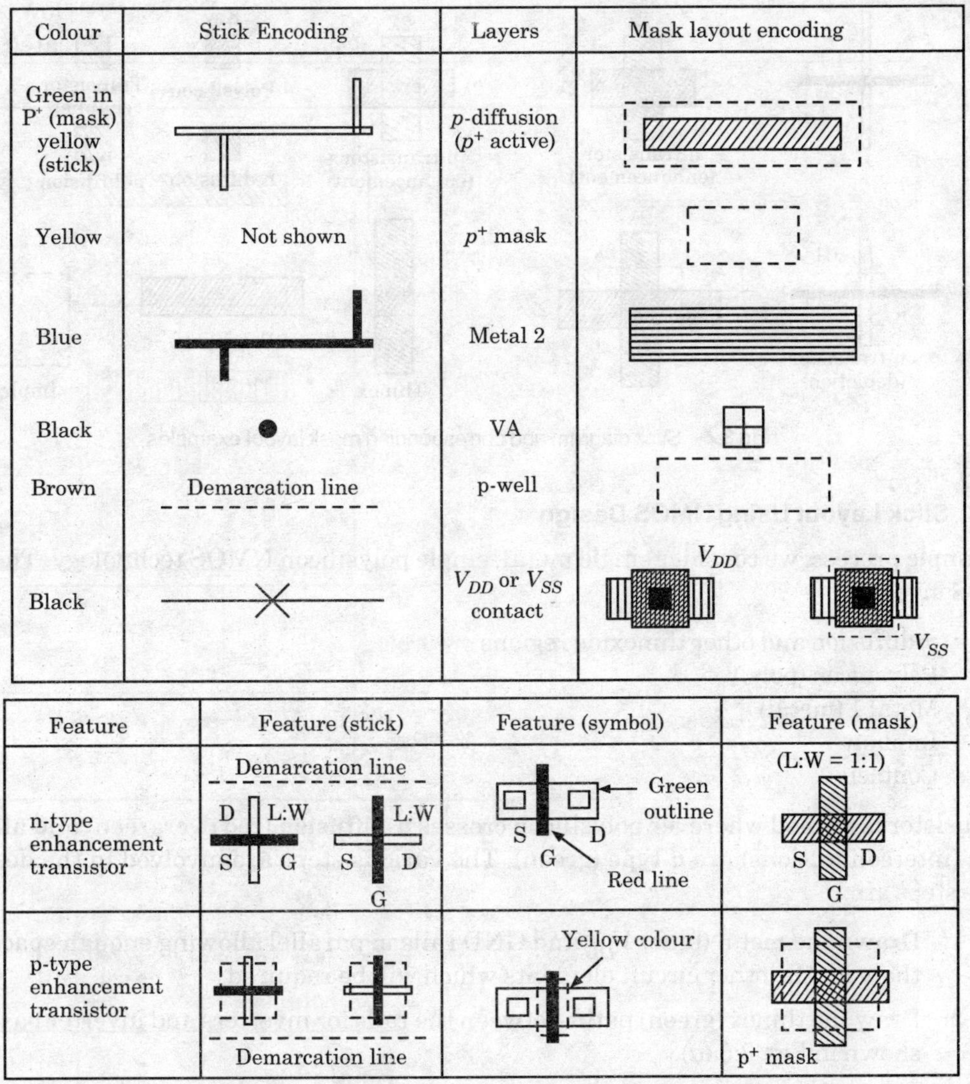

Fig 3.2 Encoding for double metal CMOS *p*-well process.

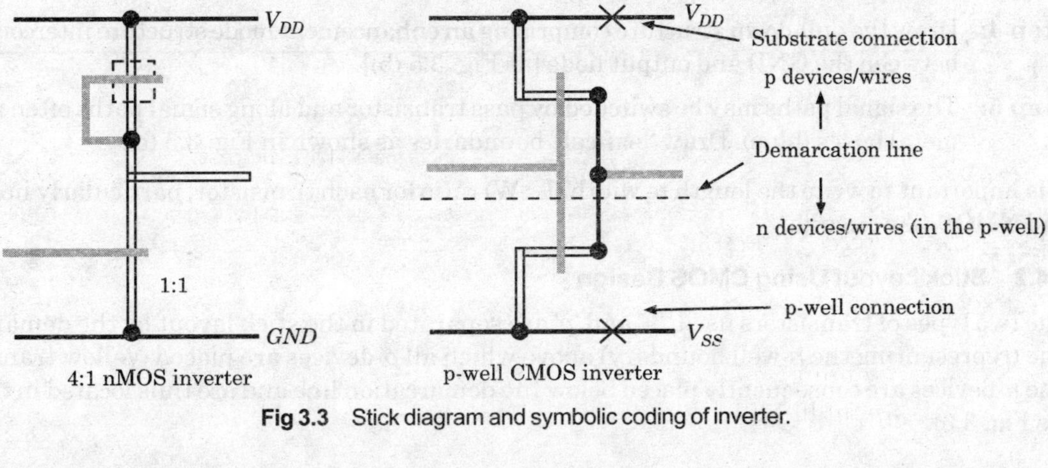

Fig 3.3 Stick diagram and symbolic coding of inverter.

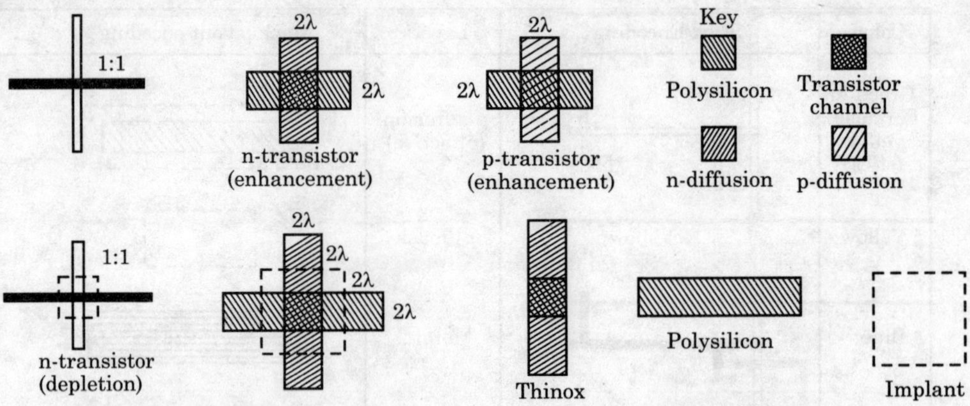

Fig 3.4 Stick diagram and corresponding mask layout examples

3.4.1 Stick Layout Using NMOS Design

For simple process, we consider single metal, single polysilicon NMOS technology. The layout of NMOS involves:

- *n*-diffusion and other thinoxide regions
- Polysilicon (poly.)
- Metal 1 (metal)
- Implant
- Contacts

A transistor is formed wherever polysilicon crosses *n*-diffusion (red overgreen) and all diffusion wires (interconnections) are *n*-type (green). The various steps are involved in the design style. These steps are:

Step 1: Drawn the metal (blue) V_{DD} and GND rails in parallel allowing enough space between them for the other circuit elements which will be required.

Step 2: Draw the thinox (green) paths between the rails for inverters and inverter based logic as shown in Fig. 3.5 (*a*).

Step 3: Draw the pull-up structure which comprise a depletion mode transistor interconnected between the output point and V_{DD}.

Step 4: Draw the pull-down structure comprising an enhancement mode structure interconnected between the GND and output node [as Fig. 3.5 (b)].

Step 5: The signal paths may be switched by pass transistor and along signal paths often require metal buses (blue). Draw 'leaf-cell' boundaries as shown in Fig. 3.5 (*c*).

It is important to write the length to width (L : W) ratio for each transistor, particularly in NMOS and NMOS-like circuits.

3.4.2 Stick Layout Using CMOS Design

The two types of transistors used, '*n*' and '*p*' are separated in the stick layout by the demarcation line (representing the *p*-well boundary) above which all *p*-devices are placed (yellow transistor). The *n*-devices are consequently placed below the demarcation line and are thus located in the well (as Fig. 3.6).

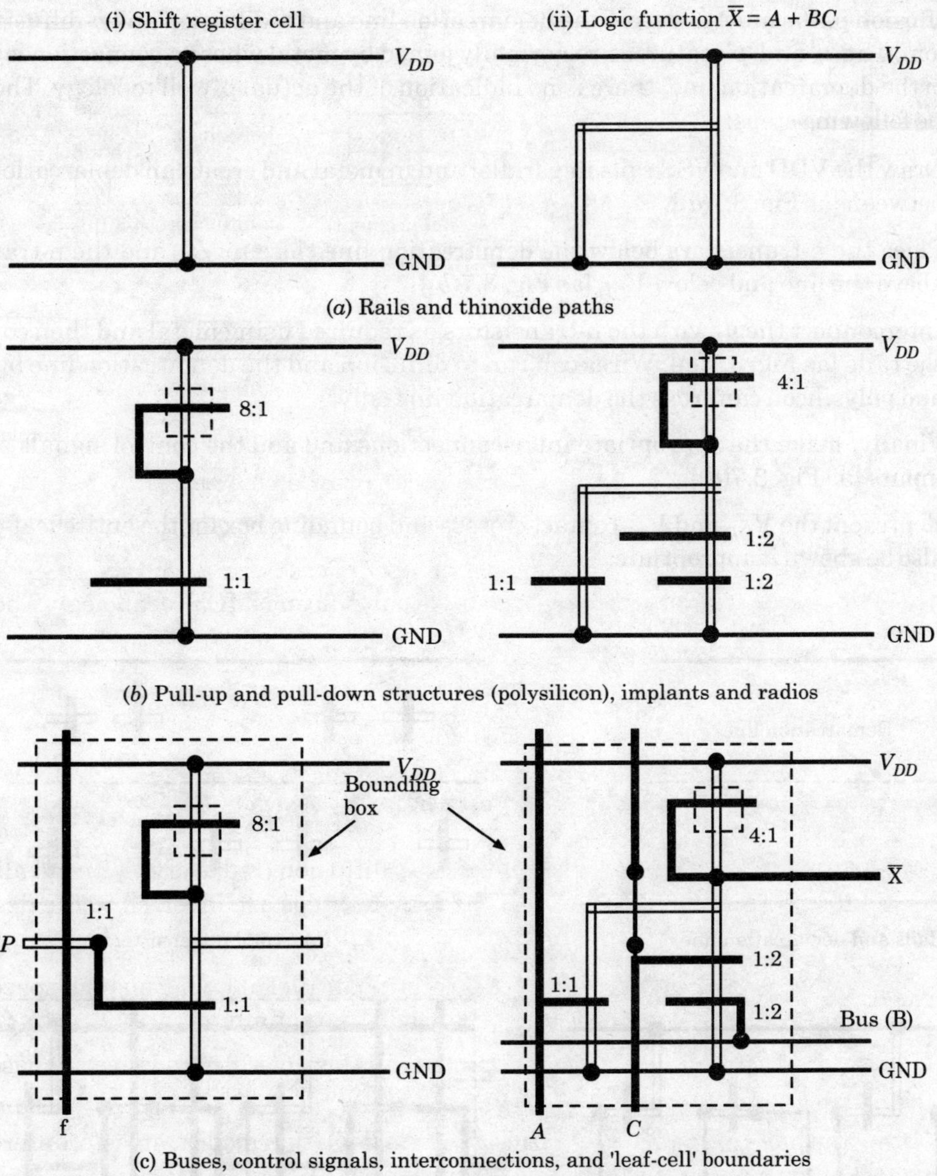

Fig. 3.5 Examples of stick diagrams using NMOS design.

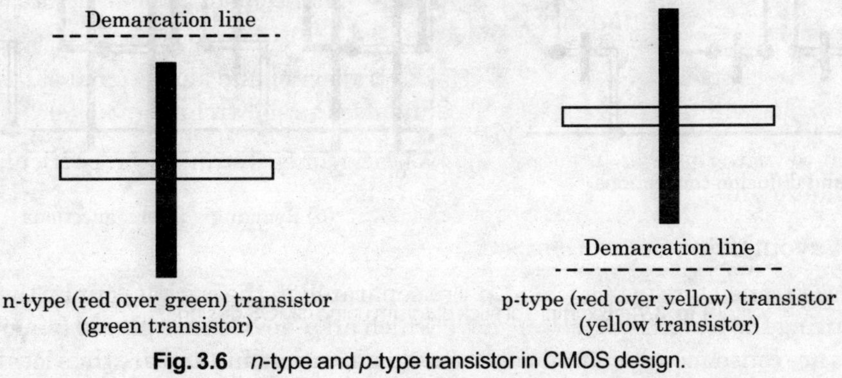

Fig. 3.6 *n*-type and *p*-type transistor in CMOS design.

The diffusion paths must not cross the demarcation line and *n*-diffusion and *p*-diffusion wires must not join. The '*n*' and '*p*' features are normally joined by metal where a connection is needed. Apart from the demarcation line, there is no indication of the actual *p*-well topology. The design includes the following steps:

Step 1: Draw the VDD and VSS rails in parallel and in metal and create an demarcation line in between [as Fig. 3.7(*a*)].

Step 2: Place the *n*-transistors below the demarcation line close to V_{SS} and the p-transistors above the line and below V_{DD} [as Fig. 3.7(*b*)].

Step 3: Interconnect the *n*- with the *p*-transistors as required using metal and then connect to the rails [as Fig. 3.7(*c*)]. Wires can run to diffusion and the demarcation line but metal and polysilicon can cross the demarcation line only.

Step 4: Finally, make the appropriate inter-connections and add the control signals and gate inputs [as Fig. 3.7(*d*)].

Step 5: Represent the V_{DD} and V_{SS} contact crosses and bounding box for the entire leaf-cell may also be shown if appropriate.

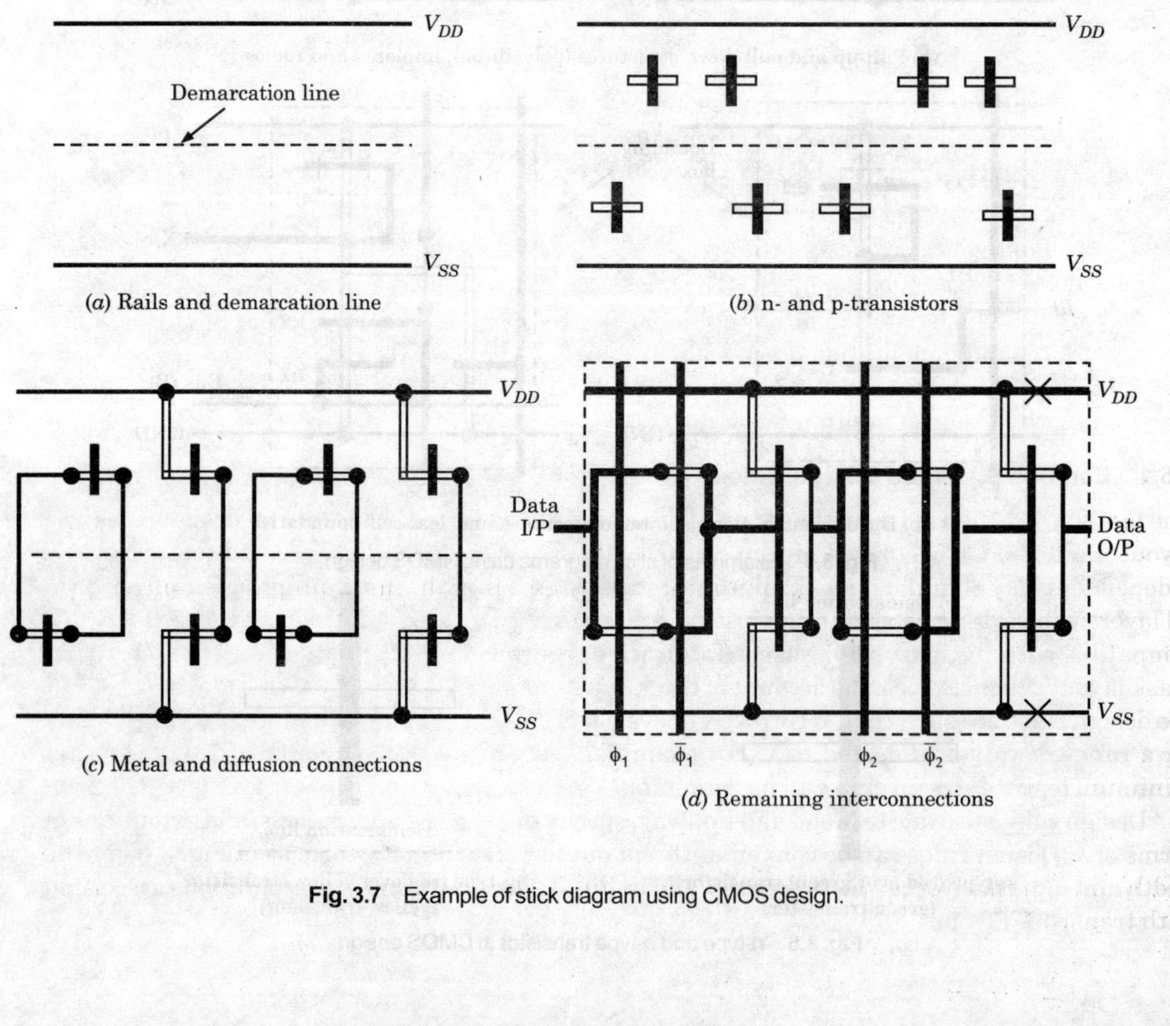

Fig. 3.7 Example of stick diagram using CMOS design.

3.5 LAYOUT DESIGNS

- The design rules are the effective interface between the circuit designer and the fabrication engineer.
- Design rules govern the layout of the individual components and the interactions-spacings and electrical connections between those components. They determine the low-level properties of chip designs: how small individual logic gates can be made; how small the wires connecting gates can be made. Hence, the parasitic resistance and capacitance which determine delay.
- The design rules having mainly two issues:
  (i) the geometrical reproduction of features that can be reproduced by the mask-making and lithographical process.
  (ii) the interactions between different layers.
- We define layout design as follows:

  The processes of creating an accurate physical representation of engineering drawing (net list) that conforms to constraints imposed by the manufacturing process, the design flow and the performance requirements shown to be feasible by simulation.
- There are several approaches that can be taken in describing the design rules. These include

  ⇒ *Micron design rules:*
  - — stated at some micron resolution
  - — usually given as a list of minimum feature sizes and spacing for all masks required in a given process.
  - — normal style for industry.

  ⇒ *Lambda-based design rules:*
  - — These rules popularized by Mead and Conway are based on a single parameter, λ, which characterizes the linear feature — the resolution of the complete wafer implementation process and permits first order scaling.
  - — They have been widely used, particularly in the educational context and in the design of multiproject chips.

3.5.1 Lambda (λ) based Design Rules

The lambda, λ design rules are based on Mead and Conway work and in general design rules and layout methodology are based on the concept of λ which provide a process and feature size—independent way of making mask dimensions to scale. In lambda based design rules, all paths in all layers will be dimensioned in λ units and subsequently λ can be allocated an appropriate value compatible with the feature size of the fabrication process. This concept means t that the actual mask layout design takes little account of the value subsequently allocated to the feature size, but the design rules are such that, if correctly obeyed, the mask layouts will produce working circuits for a range of values allocated to λ. For example, λ can be allocated a value of 1.0 μm so that minimum feature size on chip will be 2 μm (2λ).

Design rules, also due to Mead and Conway, specify line widths, separations, and extensions in terms of λ. Design rules can be conveniently set out in diagrammatic form as in Fig. 3.8, for the width and separation of conducting paths, and in Fig. 3.9 for extensions and separations associated with transistor layouts.

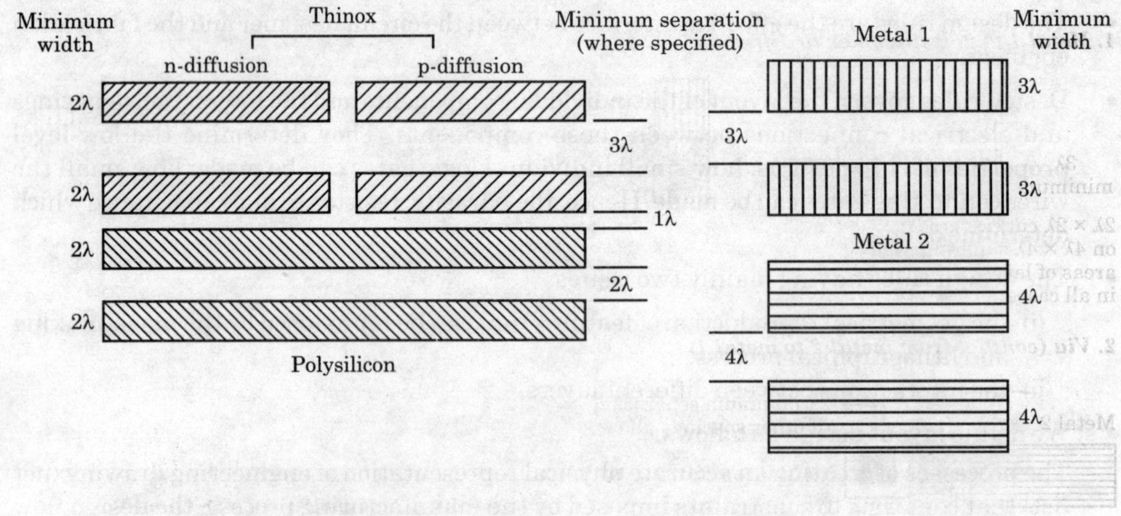

Fig. 3.8 Design rules for wires (NMOS and CMOS).

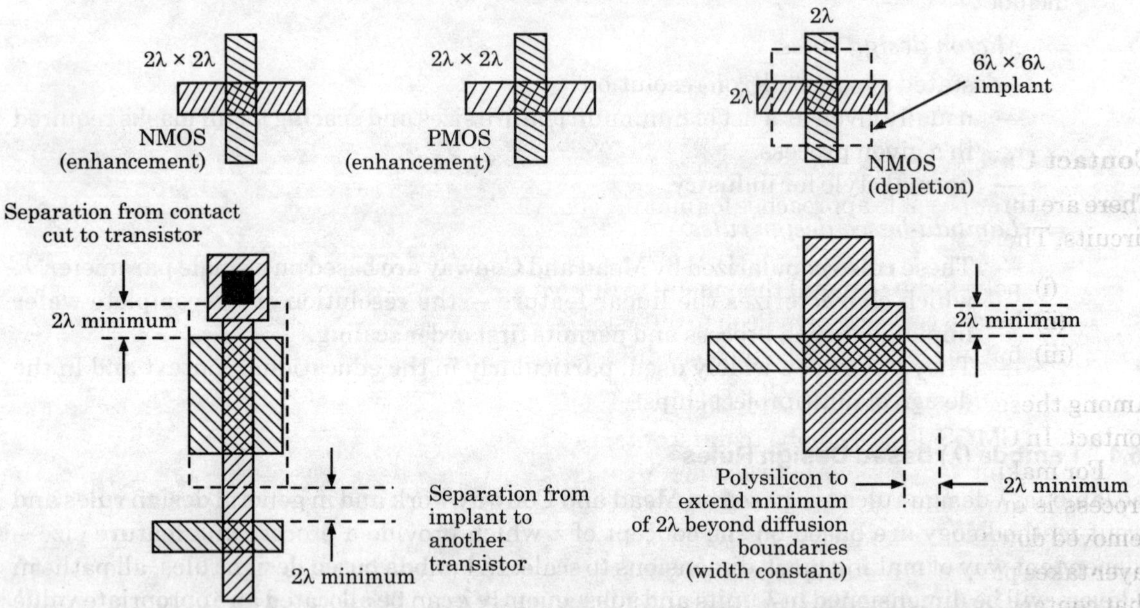

Fig. 3.9 Transistor design rules (NMOS, PMOS, and CMOS).

As per design rules, the contacts between layers are set out in Fig. 3.10 and 3.11. Here, it will be observed that connection can be made between two or, in the case of NMOS designs, three layers.

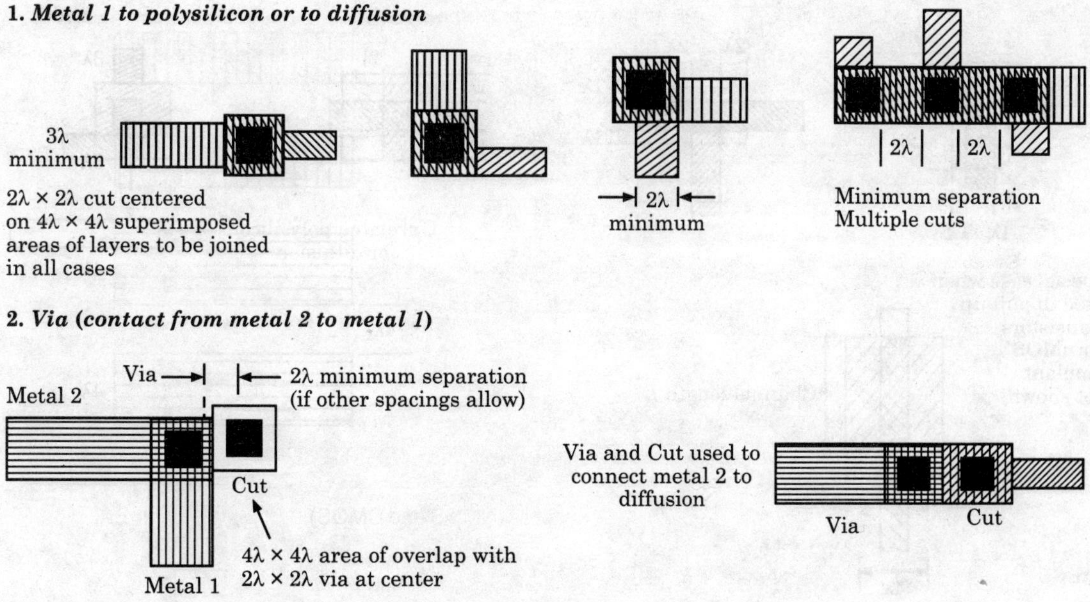

Fig. 3.10 Contacts (NMOS and CMOS).

Contact Cuts

There are three possible approaches for making contacts between polysilicon and diffusion in NMOS circuits. There are

(i) polysilicon to metal then metal to diffusion
(ii) buried contact polysilicon to diffusion
(iii) butting contact (polysilicon to diffusion using metal)

Among these, the buried contact is the most widely used giving economy in space and a reliable contact. In CMOS designs, poly. to diff. contacts are almost made via metal.

For making connections between metal and either of the other two layers (as in Fig. 3.10), the process is quite simple. The $2\lambda \times 2\lambda$ contact cut indicates an area in which the oxide is to be removed down to the underlying polysilicon or diffusion surface. When the deposition of the metal layer takes place, the metal is deposited through the contact cut areas onto the underlying areas so that contact is made between the layers.

For connecting diffusion to polysilicon using the butting contact approach as in Fig. 3.11, the process is rather more complex. In effect, a $2\lambda \times 2\lambda$ contact cut is' made down to each of the layers to be joined. The layers are butted together in such a way that these two contact cuts become contiguous. Since the polysilicon and diffusion outlines overlap and thinoxide under polysilicon acts as a mask in the diffusion process, the polysilicon and diffusion layers are also butted together. The contact between the two butting layers is then made by a metal overlay as shown in Fig. 3.11. If the butting contact is made as shown in Fig. 3.12(b), the cross sectional view of butting contact will make the contact more realistic.

1. Buried contact

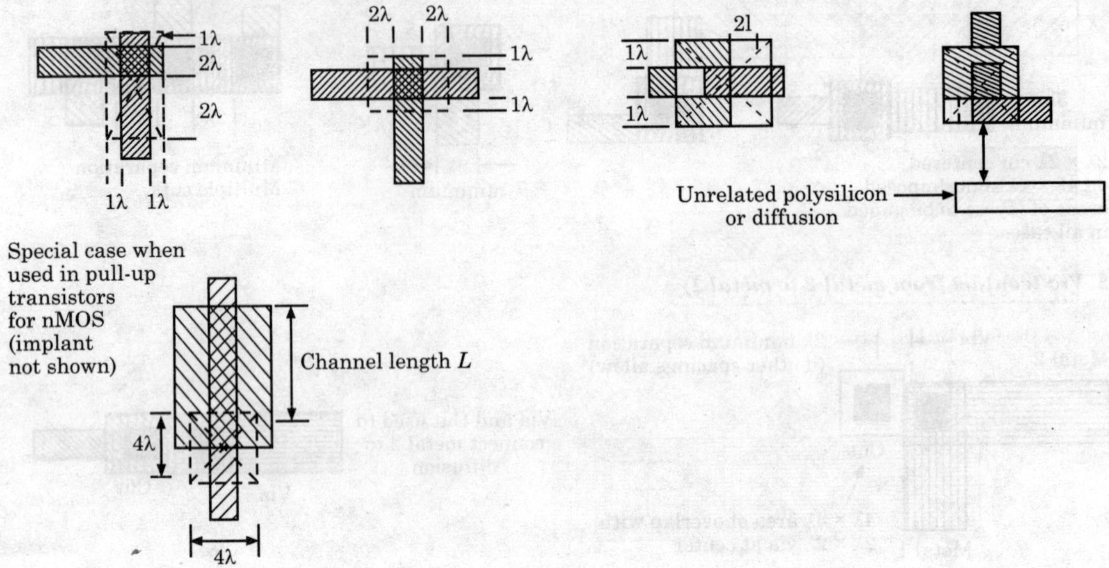

2. Butting contact

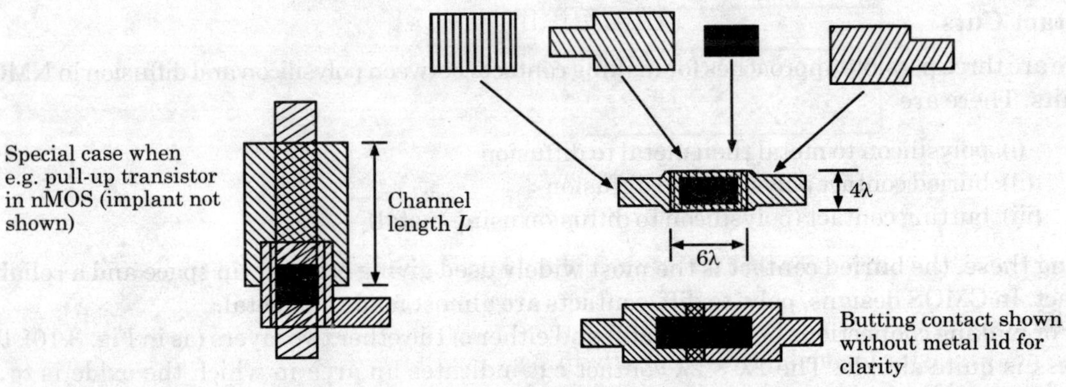

Fig. 3.11 Contacts polysilicon to diffusion (NMOS only).

The buried contact approach shown in Fig. 3.11 and 3.12(a) is simpler, since the contact cut (broken line) in this case indicating where the thinoxide is to be removed to reveal the surface of the silicon wafer before polysilicon is deposited. Thus, the polysilicon is deposited directly on the underlying crystalline wafer. When diffusion takes place, impurities will diffuse into the polysilicon as well as into the diffusion region within the contact area. Thus, a satisfactory connection between polysilicon and diffusion is ensured. Buried contacts can be smaller in area than their butting contact counterpart and, since they use no metal layer, they are subject to fewer design rule restrictions in a layout. Figure 3.12(c) shows the cross-section obtained wing a via contact and a cut.

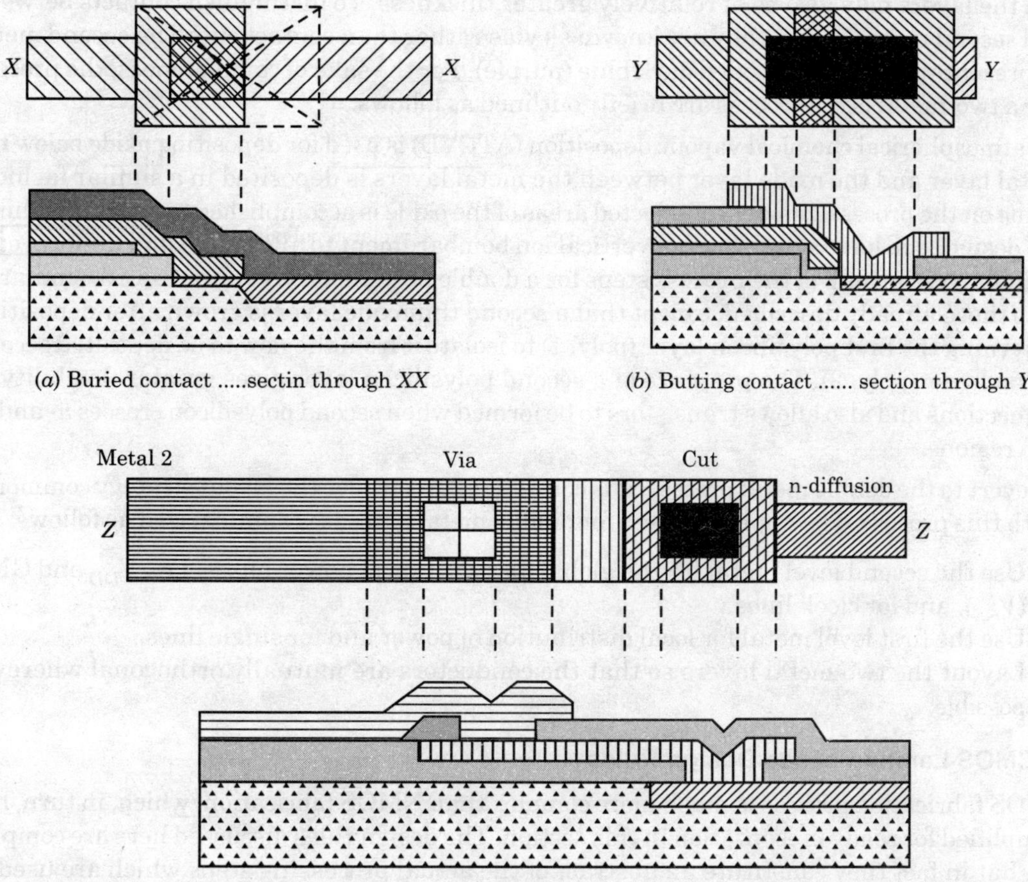

(a) Buried contact .....sectin through XX

(b) Butting contact ...... section through YY

(c) Metal 2-via-metal 1-cut-n-diffusion connection ... section through ZZ

Fig. 3.12 Cross-sections of some contact structures.

The design rules in this case ensure that a reasonable contact area is achieved and that there will be no transistor formed unintentionally in series with the contact. The rules are such that they also avoid the formation of unwanted diffusion to polysilicon contacts and protect the gate oxide of any transistor in the vicinity of the buried contact cut area.

3.5.2 Double Metal MOS Process Rules

A powerful extension to the process so far described is provided by a second metal layer. This gives a much greater degree of freedom, for example, in distributing global V_{DD} and V_{SS} (GND) rails in a system. From the overall chip interconnection aspect, the second metal layer in particular is important and, although the use of such a layer is readily envisaged, its disposition relative to (and details of) its connection to other layer using metal 1 to metal 2 contacts, are called vias. Usually, second level metal layers are coarser than the first (conventional) layer and the isolation layer between the layers may also be of relatively greater thickness. To distinguish contacts between first and second metal layers, they are known as vias rather than contact cuts The second metal layer representation is colour-coded dark blue (purple). For the sake of completeness, the process steps for a two-metal layer process are briefly outlined as follows.

The atmospherical chemical vapour deposition (ATCVD) is used for depositing oxide below the first metal layer and the oxide layer between the metal layers is deposited in a similar fashion. Depending on the process, removal of selected areas of the oxide is accomplished by plasma etching, which is designed to have a high level of vertical ion bombardment to allow high and uniform etch rates. Similarly, the bulk of the process steps for a double polysilicon layer process are similar in nature to those already described, except that a second thinoxide layer is grown after depositing and patterning the first polysilicon layer (poly. 1) to isolate it from the now to be deposited second polysilicon layer (poly. 2). The presence of a second polysilicon layer gives greater flexibility in interconnections and also allows transistors to be formed when second polysilicon crosses *n*- and *p*-diffusion regions.

To revert to the double metal process, it is convenient to consider the layout strategy commonly used with this process. The approach used for double metal process is summarized as follows:

1. Use the second level metal for the global distribution of power buses, i.e., V_{DD} and GND (V_{SS}), and for clock lines.
2. Use the first level metal for local distribution of power and for single lines.
3. Layout the two metal layers so that the conductors are mutually orthogonal wherever possible.

3.5.3 CMOS Lambda-based Design Rules

The CMOS fabrication process is much more complex than NMOS fabrication, which, in turn, has been simplified for ready presentation in this section. The design rules discussed here are complex enough, but in fact they constitute an abstract of the actual processing steps which are used to produce the chip. In a CMOS process, for example, the actual set of industrial design rules may well comprise more than 100 separate rules. The particular rules for *p*-well CMOS process is shown in Fig. 3.13.

The additional rules are concerned with those features unique to *p*-well CMOS, such as the *p*-well and p^+ mask and the special 'substrate' contacts.

Although the CMOS rules in total may seem difficult to comprehend for the new designer, once use has been made of the simpler NMOS rules, the transition to CMOS is not hard to achieve. The real key to success in VLSI design is to put it into practice and this text attempts to encourage the students to follow the above said rules.

NMOS and CMOS design rules as specified by MOSIS, the MOS implementation system of the university of Southern California Information Science Institute are given in the following Tables 3.1 and 3.2. Here λ represents a small parameter specifying the feature size of the fabrication process.

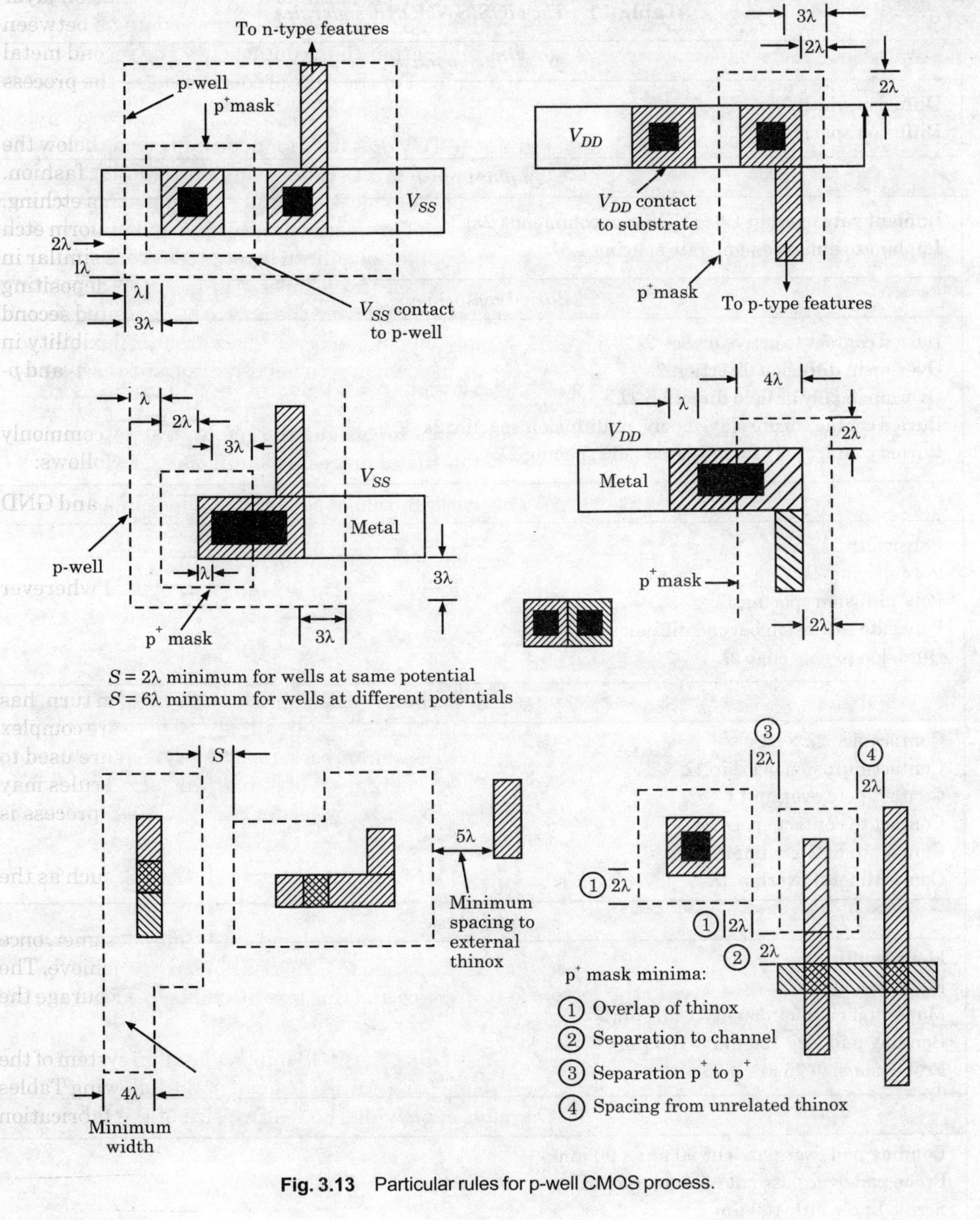

Fig. 3.13 Particular rules for p-well CMOS process.

Table 3.1 *The MOSIS NMOS design rules*

N^+ diffusion mask
Diffusion width 2λ Diffusion spacing 3λ
Implant mask
Implant gate overlap 1.5λ (MOSIS recommends 2λ) Implant to enhancement gate spacing 1.5λ
Buried contact mask
Buried contact to active device 2λ Overlap in diffusion direction 2λ Overlap in poly or field direction 1λ Buried contact to unrelayed poly, or diffusion spacing 2λ Buried contact poly to unrelated poly spacing 2λ
Poly mask
Poly width 2λ Poly spacing 2λ Poly-diffusion spacing 1λ Poly gate extension beyond diffusion 2λ Diffusion to poly edge 2λ
Contact mask
Contact size 2λ × 2λ Contact-diffusion overlap 1λ Contact-poly overlap 1λ Contact to contact space 2λ Contact to FET channel 2λ Contact-metal overlap 1λ
Metal mask
Metal width 3λ Metal spacing 3λ Maximum current density 1 mA/μm Bonding pad size 100 μm × 100 μm Probe pad size 75 μm × 75 μm
Overglass mask
Bonding pad overglass cut 90 μm × 90 μm Probe pad overglass cut 65 μm × 65 μm Scribe line width 100 μm Feature distance from scribe line 50 μm Feature distance from bonding pad 40 μm Bonding pad separation 80 μm Bonding pad pitch 180 μm

Table 3.2 *The MOSIS portable CMOS design rules*

n-well and p-well mask
n-well or *p*-well width 6λ *n*-well or *p*-well spacing 6λ
n^+ or p^+ active (diffusion or implant) mask
Active width 3λ Active to active spacing 3λ Source/drain active to well-edge 6λ Substrate/well contact, active to well-edge 3λ
Poly mask
Poly width 2λ Poly spacing 2λ Gate overlap of active 2λ Active overlap of gate 2λ Field poly to active 1λ
p-select, n-select mask
Select-space (overlap) to (of) channel 3λ Select-space (overlap) to (of) active 2λ Select-space (overlap) to (of) contact to well or substrate 1λ
Simpler contact to poly mask
Contact size 2λ × 2λ Active overlap of contact 2λ Contact-to-contact spacing 2λ Contact-to-gate spacing 2λ
Denser contact to poly mask
Contact size 2λ × 2λ Poly overlap of contact 1λ Contact spacing on the same poly 2λ Contact spacing on a different poly 5λ Contact to non-contact poly 4λ Space to active short run 2λ Space to active long run 3λ
Simpler contact to active mask
Contact size 2λ × 2λ Active overlap of contact 2λ Contact/contact spacing 2λ Contact/gate spacing 2λ

Denser contact to active mask
Contact size = 2λ × 2λ Active overlap of contact = 1λ Contact/contact spacing on same active = 2λ Contact/contact spacing on different active = 6λ Contact to different active = 5λ Contact to gate spacing = 2λ Contact to field poly, short run = 2λ Contact to field poly, long run = 2λ
Metal 1 mask
Width = 3λ Metal 1 to metal 1 = 3λ Overlap of contact to poly = 1λ Overlap of contact to active = 1λ
Via mask
Size = 2λ × 2λ Via-to-Via separation = 2λ Metal 1/via overlap = 1λ Space to poly or active edge = 2λ Via to contact spacing = 2λ
Metal 2 mask
Width = 3 λ Metal 2/Metal 2 spacing = 4 λ Metal overlap of via = 1 λ
Overglass mask
Bonding pad = 100 μm × 100 μm Probe pad = 75 μm × 75 μm Pad to glass edge = 6 μm

3.6 FULL-CUSTOM MASK LAYOUT DESIGN

- The design of physical layout is very useful for circuit performance (area, speed, and power dissipation) since the physical structure directly determines the trans-conductance's of the transistors, the parasitic capacitances and resistances.
- The mask layout of logic gates are justifiable only in special circumstances where the area and/or the performance of the circuit must be optimized under very tight constraints.
- The automated layout generation (e.g., using a standard cell library, computer aided placement-and-routing) is typically preferred for the design of most digital VLSI circuits.
- The physical (mask layout) design of CMOS logic gates is a process which starts with the circuit topology (to realize the desired logic function) and to realize the desired performance specifications.
- The designer can only estimate the total parasitic load at the output node, based on the fan-out, the number of devices, and the expected length of the interconnection lines.

- If the logic gate contains many (> 4 to 6) transistors, the topological graph representation. A simple stick diagram layout is drawn, which shows the locations of the transistors and the local interconnections between the transistors with the locations of the contacts.
- After this, the mask layers are drawn according to the layout design rules. This procedure may require several small iterations in order to accommodate all the design rules, but the basic topology should not change very significantly.
- Following the final DRC (Design Rule Check), a circuit extraction procedure is performed on the-finished layout to determine the actual transistor sizes. The result of the extraction step is usually a detailed SPICE input file, which is automatically generated by the extraction tool. Now, the actual performance of the circuit can be determined by performing a SPICE simulation, using the extracted net-list.
- If the simulated circuit performance (e.g., transient response times or power dissipation) do not match the desired specifications, the layout must be modified and the whole process

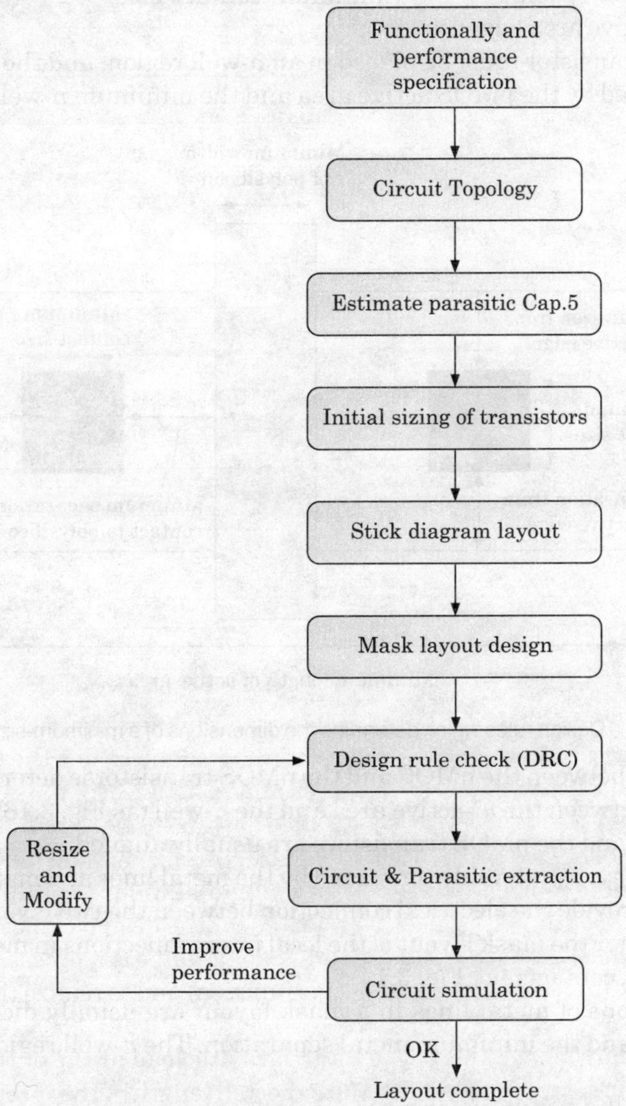

Fig. 3.14 Typical design flow for the production of a mask layout.

must be repeated. The layout modifications are usually concentrated on the (W/L) ratios of the transistors (transistor resizing), since the width-to-length ratios of the transistors determine the device transconductance and the parasitic source/drain capacitances.

- The flow diagram of this iterative process is shown in Fig. 3.14.

3.6.1 CMOS Inverter Layout Design

- In this topic, we will discuss about the mask layout design of a CMOS inverter. Here, we consider the design of inverter with minimum-size transistors.
- The width of the active area is determined by the minimum diffusion contact size and the minimum separation from diffusion contact to both active area edges. The width of the polysilicon line over the active area (which is the gate of the transistor) is typically taken as the minimum poly width (Fig. 3.15). Then, the minimum overall length of the active area is determined by the following sum: (minimum polysilicon width) + 2 × (minimum poly-to-contact spacing) + 2 × (minimum contact size) + 2 × (minimum spacing from contact to active area edge).
- The pMOS transistor must be placed in an n-well region, and the minimum size of the n-well is dictated by the pMOS active area and the minimum n-well overlap over $n+$.

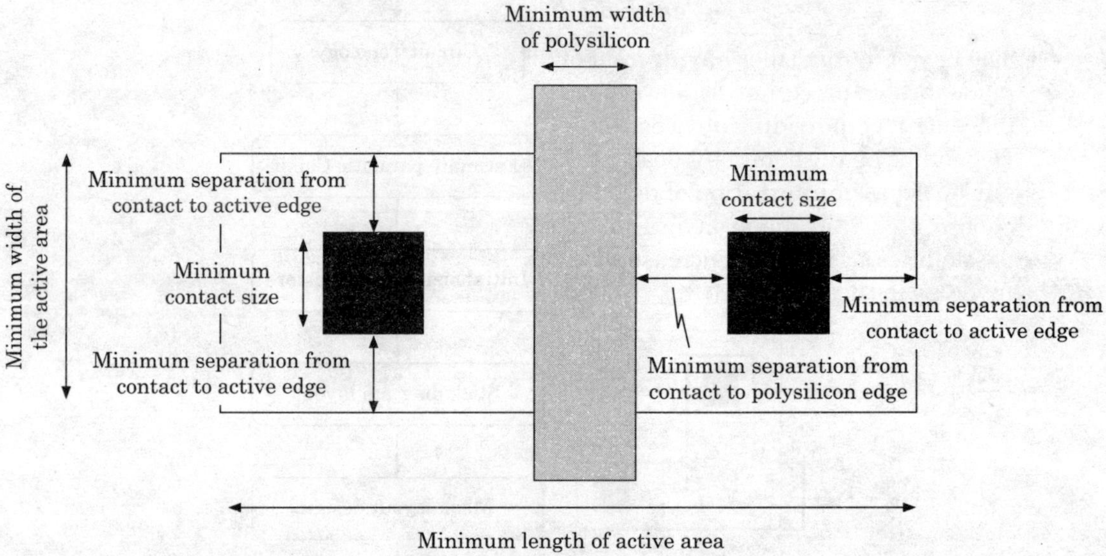

Fig. 3.15 Design rules which determine the dimensions of a minimum-size transistor.

- The distance between the nMOS and the pMOS transistor is determined by the minimum separation between the n^+ active area and the n-well (as Fig. 3.16). The poly-silicon gates of the nMOS and the pMOS transistors are usually aligned.
- All the signals are preferably connected by the metal lines and metal-poly-silicon contacts are used to provide the electrical connection between the two layers.
- The last step for the mask layout is the local interconnections in metal for the output node, V_{DD} and G_{ND} contacts (as Fig. 3.17).
- The dimensions of metal lines in a mask layout are usually dictated by the minimum metal width and the minimum metal separation. The n-well region has a V_{DD} contact for proper biasing.

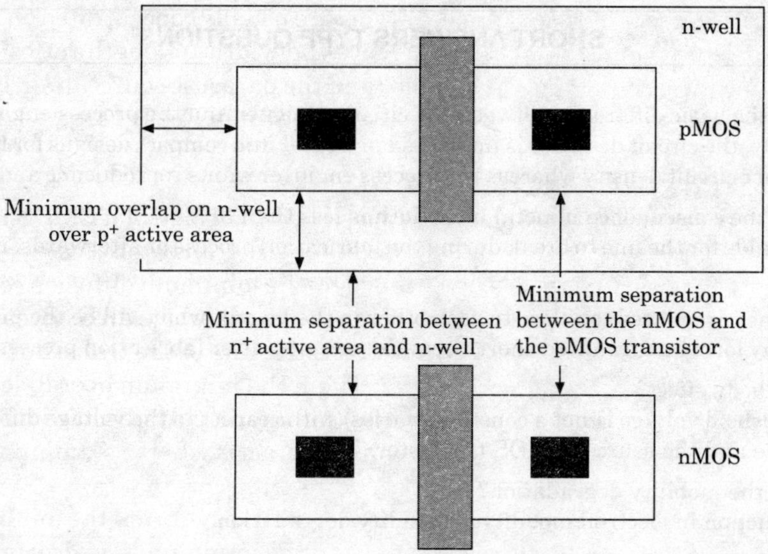

Fig. 3.16 Design rules which determine the separation between the nMOS and the pMOS transistor of the CMOS inverter.

- The layout design rules having a set of limitations for the mask geometry. The full-custom layout design process still allows a large number of variations in terms of device sizing, the placement of individual devices, and the routing of interconnections between the devices.
- Depending on the dominant design criteria and design constraints (minimization of overall silicon area, minimization of delay times, placement of input/output pins, etc.), one can choose a certain mask layout design over other alternatives. The number of layout possibilities also tends to increase with circuit complexity, i.e., with the number of transistors involved in the design.

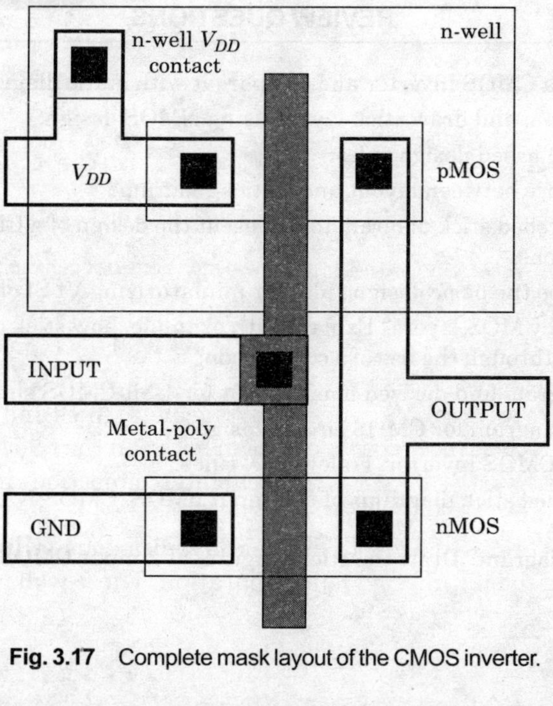

Fig. 3.17 Complete mask layout of the CMOS inverter.

SHORT ANSWERS TYPE QUESTIONS

Q. 1. What is the basic difference between the circuit designer and the process engineer?

Ans. Generally, the circuit designer is interested in tighter and compact designs for higher performance and higher circuit density whereas the process engineer wants reproducible and high-yield process.

Q. 2. What is the consequence if metal line width is less than 3λ?

Ans. It is possible for the line to break during the fabrication process or afterwards, resulting in an open circuit.

Q. 3. If two lines are placed too close to each other in the layout, what will be the problem?

Ans. They may form an unwanted short circuit by merging after fabrication process.

Q. 4. What is body effect?

Ans. The threshold voltage is not a constant (varies) with respect to the voltage difference between the substrate and the source of MOS transistor.

Q. 5. What is the mobility degradation?

Ans. The reduction in electron mobility is mobility degradation.

Q. 6. What is the purpose of design rule?

Ans. Design rule is to guarantee that under the worst cumulative variations of the processes, the circuit does not fail to operate.

Q. 7. What is the use of stick diagram?

Ans. It is used to convey layer information through the use of colour code.

Q. 8. Give three possible approaches for making contacts between polysilicon and diffusion in NMOS circuit.

Ans. (*i*) Polysilicon to metal then metal to diffusion.

(*ii*) Buried contact polysilicon to diffusion.

(*iii*) Butting contacts.

REVIEW QUESTIONS

1. Draw the layout of a CMOS inverter and compare it with static diagram. **[UPTU 2007-08]**
2. Explain stick diagram and draw stick layout using NMOS design.
3. Explain Lambda (λ) based design rules. **[UPTU 2003-04]**
4. What is the difference between layout and a stick diagram?
5. Construct a colour coded stick diagram to represent the design of a CMOS circuit that implements the following functions.
6. Discuss and describe the basic design rules for miniaturizing VLSI circuits in brief.
7. What do you mean by MOS layers? Explain with example, how stick diagrams are used to conveys layer information's through the use of a colour code?
8. Draw symbolic diagram and derived mask layout for 1-bit CMOS shift register cell.
9. Describe the stick diagram for CMOS circuit design. **[UPTU 2005-06]**
10. Draw the layout of CMOS inverter. Follow all λ-rules. **[UPTU 2004-05]**
11. Draw the colour coded stick diagrams of two input nMOS, CMOS NAND and NOR gates. **[UPTU 2007-08]**
12. What is the stick diagram? Draw the stick diagram of 2-input CMOS NAND gate. **[TUT-2012]**

❑❑❑

4 Chapter
MOS Inverters

Chapter Overview

4.1 INTRODUCTION

- The inverter is the most fundamental logic gate that performs a Boolean operation on a single input variable. The electrical behavior of these circuits (NAND gates, adders, microprocessors, etc.) can be derived completely by extrapolating the results obtained for inverters.
- The logic symbol and the truth table of the ideal inverter are shown in Fig. 4.1. In MOS inverter circuits, both the input variable A and the output variable B are represented by node voltages, referenced to the ground potential.
- The DC voltage transfer characteristic (VTC) of the ideal inverter circuit is shown in Fig. 4.2. The voltage V_{th} is called the inverter threshold voltage.

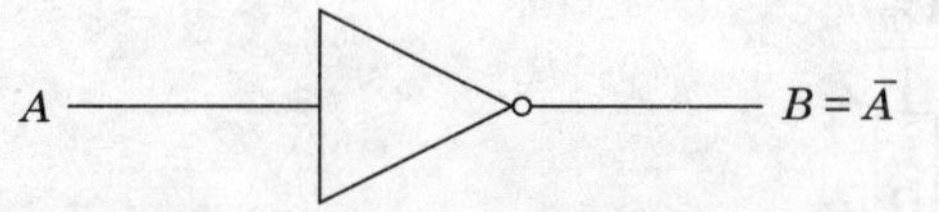

A	B
0	1
1	0

Fig. 4.1 Logic symbol and truth table of the inverter.

- For any input voltage between 0 and $V_{th} = V_{DD}/2$, the output voltage is equal to V_{DD} (logic "1"). The output switches from V_{DD} to 0 when the input is equal to V_{th}.
- For any input voltage between V_{th} and V_{DD}, the output voltage assumes a value of 0 (logic "0").
- Thus, an input voltage $0 < V_{in} < V_{th}$ is interpreted by this ideal inverter as a logic "0" while an input voltage $V_{th} < V_{in} < V_{DD}$ is interpreted as a logic "1".
- Since the transition between states is abrupt, the dynamic power dissipation is zero. The input circuit of the ideal inverter has no loading effect on the driving signal and the inverter output is capable of driving an arbitrary numbers of similar gates.

- The ideal inverter characteristics are applicable to multiple input gates without any loading effects. The DC characteristics of actual inverter circuits will obviously differ in various degrees from the ideal characteristic shown in Fig. 4.2.

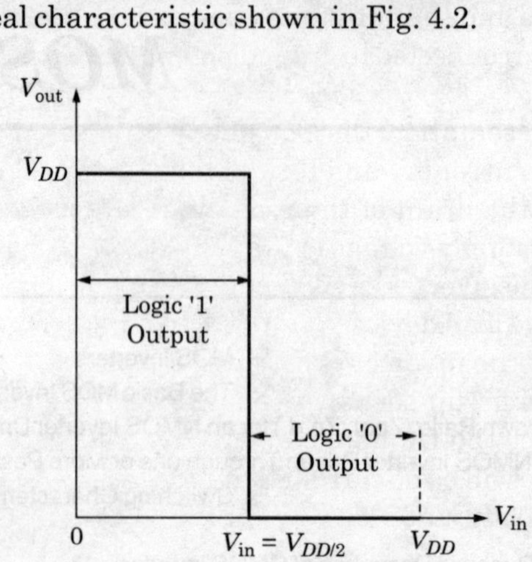

Fig. 4.2 Voltage transfer characteristic (VTC) of an ideal inverter.

- The accurate estimation and the manipulation of the shape of VTC for various inverter types are actually important parts of the design process. The performance of a gate is determined by the following characteristics.
  - ➢ The range of voltage that corresponds to logic levels 1 or 0.
  - ➢ The switching speed.
  - ➢ The static and dynamic power dissipation.
  - ➢ The range of input voltages for which the output is undefined.
  - ➢ The loading Effects.

4.2 MOS INVERTERS

- The generalized circuit structure of an nMOS inverter is shown in Fig. 4.3.

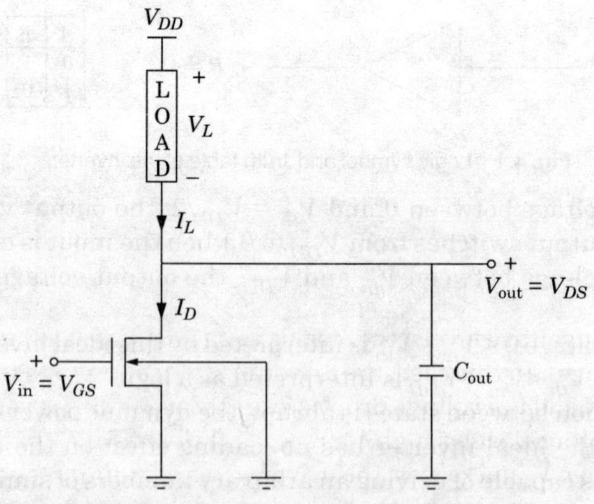

Fig. 4.3 General circuit structure of an nMOS inverter.

- For the nMOS transistor, the input voltage is equal to the gate-to-source voltage ($V_{in} = V_{GS}$), while the output voltage of the circuit is equal to the drain-to-source voltage ($V_{out} = V_{DS}$). The source and the substrate terminals of the nMOS transistor, also called the *driver* transistor, are connected to ground potential. So, the source-to-substrate voltage is $V_{SB} = 0$.
- In this generalized representation, the *load* device is represented as a two-terminal circuit element with terminal current I_L and terminal voltage $V_L(I_L)$. One terminal of the load device is connected to the drain of the n-channel MOSFET and other one terminal is connected to the power supply voltage (V_{DD}).
- The characteristics of the inverter circuit actually depend very strongly upon the type and the characteristics of the load device.
- The output terminal of the inverter is connected to the input of another MOS inverter. Consequently, the next circuit seen by the output node can be represented as a lumped capacitance, Cout. Since the DC gate current of an MOS transistor is negligible for all practical purposes, there will be no current flow into or out of the input and output terminals of the inverter in DC steady state.

4.3 LOGIC GATES CHARACTERISTICS

4.3.1 Voltage Transfer Characteristic (VTC)

- The typical voltage transfer characteristics of a realistic nMOS inverter is shown in Fig. 4.4. The voltage transfer characteristic describing V_{out} as a function of V_{in} under DC conditions. We have already seen that the drain current of MOSFET is a function of the gate-source voltage and the drain source voltage. So that

$$I_D(V_{in}, V_{out}) = I_L(V_L) \quad \text{...(4.1)}$$

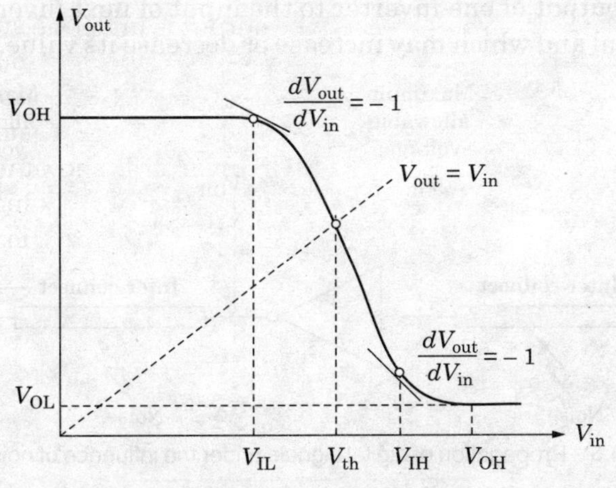

Fig. 4.4 Typical voltage transfer characteristic (VTC) of a realistic nMOS inverter.

- For very low input voltage levels, the output voltage V_{out} is equal to the high value of V_{OH} (output high voltage). In this case, the driver nMOS transistor is in cut-off. Due to this no current is conduct and the voltage drop across the load device is very small in magnitude, and the output voltage level is high.
- As the input voltage V_{in} increases, the driver transistor starts conducting a certain drain current, and the output voltage starts to decrease. Due to this, the drop in the output

voltage level does not occur abruptly, such as the vertical drop assumed for the ideal inverter VTC, but rather gradually and with a finite slope.

- There are two critical voltage points in the characteristics, when the slope $V_{out}/(V_{in})$ characteristic becomes equal to -1, i.e.,

$$\frac{dV_{out}}{dV_{in}} = -1 \qquad \text{...(4.2)}$$

- The smaller input voltage where the slope $V_{out}/(V_{in}) = -1$ is called the input low voltage V_{IL}, and the larger input voltage at which $V_{out}/(V_{in}) = -1$ is called the input high voltage V_{IH}. V_{IL} and V_{IH} play significant roles in determining the noise margins of the inverter circuit. Further increase in the input voltage, decrease the output voltage.
- At a point where the input voltage is equal to V_{OH} then the output corresponding to it is known as output low voltage V_{OL}.
- The threshold voltage of the inverter is defined as a voltage at which $V_{in} = V_{out}$. Thus, we define five voltages points on the VTC curve known as critical voltages, V_{OL}, V_{OH}, V_{IL}, V_{IH} and V_{th}. The functional definitions for the first four of these critical voltages are given below.

  V_{OH}: Maximum output voltage when the output level is logic "1".
  V_{OL}: Minimum output voltage when the output level is logic "0".
  V_{IL}: Minimum input voltage which can be interpreted as logic "1".
  V_{IH}: Maximum input voltage which can be interpreted as logic "0".
- The ability of an inverter to interpret an input signal within a voltage as either logic "0" or as logic "1" depends on the tolerance. These external noises have the characteristics of increasing or decreasing the value of their levels.
- For example, we have considered the tandem connection of inverters as shown in Fig. 4.5. The inverters are capacitive or inductive coupled through interconnection. When the signal passes from the output of one inverter to the input of next inverter stage noise may be added to the signal and which may increase or decrease its value.

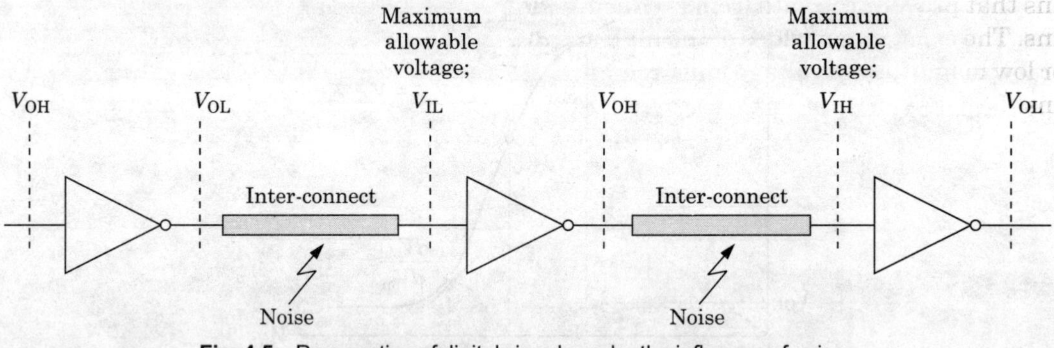

Fig. 4.5 Propagation of digital signals under the influence of noise.

4.3.2 Logic Swing and Transition Width

Two important parameters that are obtained from voltage difference of VTC critical voltages on a mutual axis are logic swing and transition width.

(a) *Logic swing:* Logic swing is define as the magnitude of voltage difference between the output high and low voltage levels. Hence, the logic swing is equal to

$$V_{LS} = V_{OH} - V_{OL}$$

(b) *Transition – Width:* The transition width is the amount of voltage change that is required to the input voltage to cause a change in the output voltage from the high to the low level (or vice versa). Hence from the VTC the transition width is given by

$$V_{TW} = V_{IH} - V_{IL}$$

4.3.3 Noise in the Logic Gates

4.3.3.1 Noise

Variation in the steady state voltage levels of digital circuits (i.e., the logic 0 and the logic 1 states) are undesirable and cause logic errors if the fluctuation from the desired or specified voltage levels is too great. This variation of steady state voltage levels in digital circuits is referred to as voltage level degradation and it termed as noise.

4.3.3.2 Noise Margins

Since variations in the high and low logic levels occur, terminology is used to describe the fluctuation, from the idealized inverter VTC. The definition of noise tolerances of digital circuits, called noise margin and denoted by *DM*. The noise immunity of the circuit increases with Noise margin.

Tow noise margin will be defined; the noise margin for low signal levels (NM_L) and the noise margin for high signal levels (NM_H).

$$NM_L = V_{IL} - V_{OL}$$
$$NM_H = V_{OH} - V_{IH}$$

The noise margin is shown in Fig. 4.6. Here the shaded areas indicate he valid regions of the input and output voltages, and the noise margins are shown is the amount of variation in the signal levels that can be allowed while the signal is transmitted from the output of one gate to the input of the next gate.

The noise margin in high state NM_H and noise margin in the low state NM_L represent safety margins that prevent the gate from producing erroneous logic recisions in the presence of noise margins. The exact magnitudes of the high and low voltage level is not important. However, the high or low magnitude of voltage must remain in the range of voltages that provide positive noise margins.

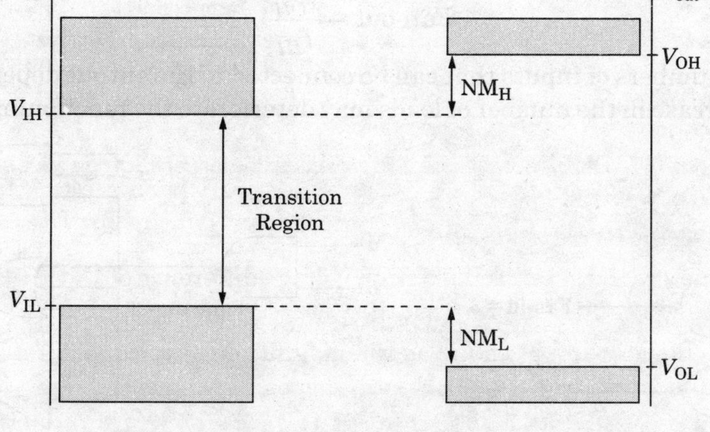

Fig. 4.6 Definition of noise margins NM_L and NM_H.

4.3.3.3 Noise Immunities

The quantity noise immunity is the ability of a gate to reject noise. The noises immunity of the circuit increases with noise margin (*NM*). The high and low noise immunities are quantitatively defined as the quotient of the noise sensitivities (*NS*) and the logic swing (VL_S) as follows:

$$NI_H = \frac{NS_H}{VL_S}$$

and

$$NI_L = \frac{NS_L}{VL_S}$$

4.3.3.4 Noise Sensitivities

The effect of input variations are quantified in terms of the noise sensitivities. The high and low noise sensitivities are defined as the difference between the input and midpoint voltage for V_{IN} at V_{OH} and V_{OL} respectively.

$$NS_H = V_{OH} - V_M$$
$$NS_L = V_M - V_{OL}$$

where V_M = Midpoint voltage.

4.3.4 Fan-In and Fan-Out

Fan-In: • The fan-in of a logic gate is the number of inputs the gate has in the logic path being exercised. The diagram of CMOS gates for may result in an undefined output due to excessive loading.

Fan-Out: • The fan-out of a logic gate is the total number of gate inputs that are driven by a gate output. This is usually expressed in terms of minimum-sized inverter as unity. The circuit diagram of CMOS gates for the fan-out is shown in Fig. 4.7(b).

- The fan-out is calculated from the amount of current available in the output of a gate and the amount of current needed to drive each gate input. For example, from Fig. 4.7(b) the output of first stage has a logic high output equal to V_{OH} and corresponding to it a current I_{OH} is supplied to the input of other gates. Input of each gate requires a current I_{IH} for its operation. Hence Fan-out of the gate is calculated from the ratio of current output to current input, i.e.,

$$\text{Fan-out} = \frac{I_{OH}}{I_{IH}}$$

- Thus the numbers of inputs that can be connected to the output depends on the current ratios. Increase in the number of loads may deteriorate the functioning of the device.

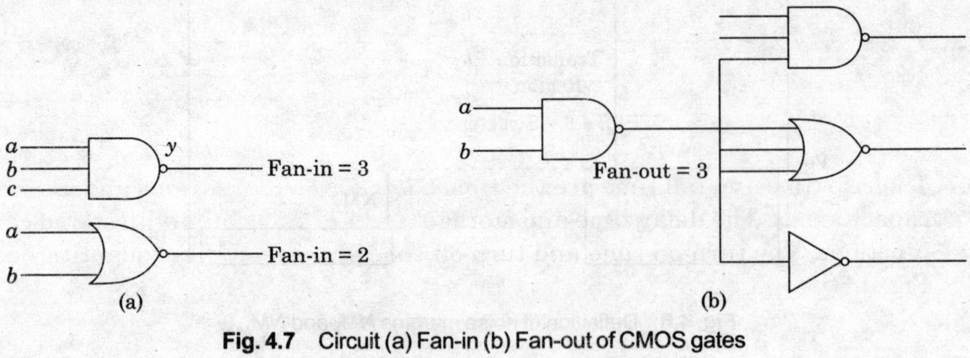

Fig. 4.7 Circuit (a) Fan-in (b) Fan-out of CMOS gates

4.3.5 Switching Speed of Logic Gate

An input pulse and output response of a logic inverting circuit is shown in Fig. 4.8. On the output response, which does not necessarily reach V_{CC} (usually $V_{OH} < V_{CC}$), the 10% V_{OH} and 90% V_{OH} points are market on both the rising and falling edges of the output voltage.

- The characteristic switching speed times are then defined in general in terms of these output values as follows:

t_d = delay time

t_r = rise time

t_s = storage time

t_f = falling time

t_{on} = turn on time = $t_r + {}_{td}$

t_{off} = turn of time = $t_f + t_s$

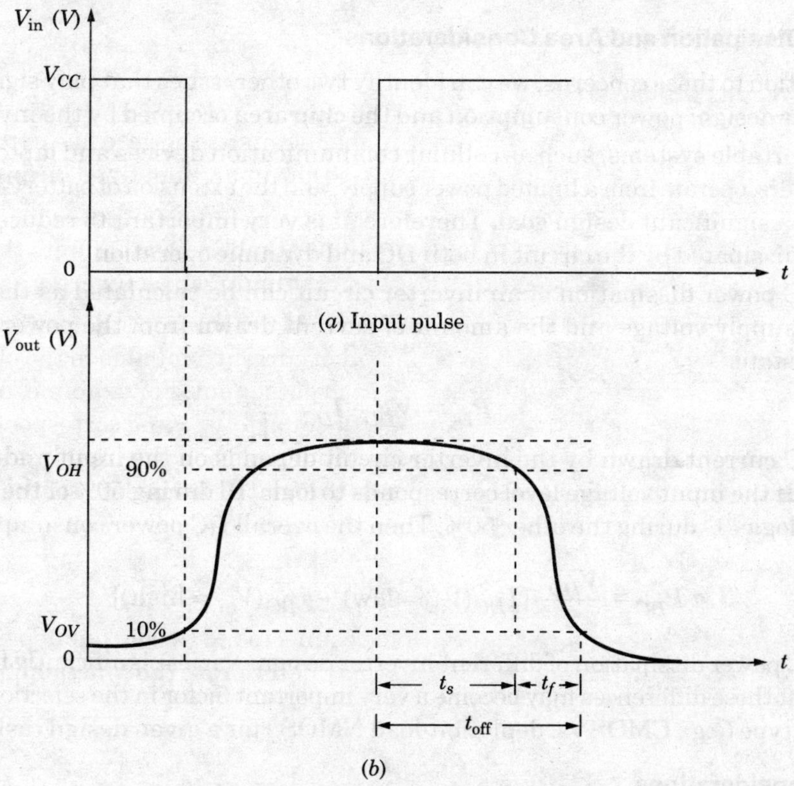

Fig. 4.8 Switching speed of logic gate

- The rise time and fall time are the time associates with charging and discharging load capacitances. The delay time and storage time are associated with stored charge of *p-n* junctions. The turn-on time and turn off time are sum of these quantities as shown in Fig. 4.8.

4.3.6 Propagation Delay

- Propagation delay is measured as a difference in time between the input and output signals reaching the 50% in their respective transitions. The 50% point in the voltage level corresponding to one half of the total transition between V_{OH} and V_{OL}.

$$V_{50\%} = \frac{V_{OH} + V_{OL}}{2}$$

- The propagation delay on the high-low output transition τ_{PHL} and that of the low or high transition is τ_{PHH}. In general case, these two delays will not be equal and the average propagation delay τ_p is defined by

$$\tau_p = \frac{\tau_{PHL} + \tau_{PLH}}{2}$$

- Average propagation delay is one figure of merit that is commonly used to compare the performance of different logic families.

4.3.7 Power Dissipation and Area Considerations

- In addition to these concerns, we can identify two other issues that play significant roles in inverter design: power consumption and the chip area occupied by the inverter circuit.
- Most portable systems, such as cellular communication devices and laptop and palmtop computers, operate from a limited power supply, and the extension of battery-based operation time is a significant design goal. Therefore, it is very important to reduce the amount of power dissipated by the circuit in both DC and dynamic operation.
- The DC power dissipation of an inverter circuit can be calculated as the product of its power supply voltage and the amount of current drawn from the power supply during steady state.

$$P_{DC} = V_{DD} \cdot I_{DC} \quad \text{...(4.3)}$$

- The DC current drawn by the inverter circuit depends on the input and output voltage levels. If the input voltage level corresponds to logic "0" during 50% of the operation time and to logic "1" during the other 50%. Then the overall DC power consumption is given by

$$P_{DC} = \frac{V_{DD}}{2} \cdot [I_{DC}(V_{\text{in}} = \text{low}) + I_{DC}(V_{\text{in}} = \text{high})] \quad \text{...(4.4)}$$

- The DC power dissipation of different inverter designs varies significantly from each other, and that these differences may become a very important factor in the selection of a particular circuit type (e.g., CMOS vs. depletion-load NMOS) for a given design task.

4.3.8 Area Considerations

- Reduce the area of the MOS transistors used is necessary to reduce the chip area occupied by the inverter circuit.
- An MOS transistor has minimum area (WL) when both of the gate (channel) dimensions are made as small as possible within the constraints of the particular technology.
- For achieving the minimum transistor area, the ratio of the gate width to gate length (W/L) should also be as close to unity as possible.

4.3.9 Logic Gate Design Considerations

The following points have to be considered before designing the logic gates:

- Ideal logic gate is a highly non-linear device that attempts to quantize the input signal into two discrete output levels high or low. But in the actual gate possible ways to minimize the width of the undefined input voltage range and increasing the noise margin levels should be incorporated.
- Logic gates should the unidirectional. The voltage changes in the output of a gate should not affect the input given to a gate. The inputs should be able to control the output to produce a well defined logic function.
- The gate output should be capable of driving the inputs of more than one gate. This is discussed as the fan-in and fan-out capability of a gate.
- As the signal passes through the gate, the logic levels must be regenerated. That is, the voltage levels at the output of one gate must be compatible with the input voltage levels of the same or similar logic gates.
- The power and area consumed by a logic gate should be as small as possible to meet the basic design considerations of IC's.

4.4 THE BASIC MOS INVERTER

- The nMOS logic family is one of the major technologies used to realize digital circuits. The only component used in fabrication is the nMOS transistor, which can function as a controlled switch or as a resistor.
- Usually small chip area is required for nMOS transistor fabrication and this result in high packing density in VLSI system also this simple circuit configuration can be used to study a variety of functions of digital circuits.
- The simplest form of an nMOS logic circuit is an inverter as shown in Fig. 4.9. The basic inverter circuit requires a transistor with source connected to ground and a load resistor of some sort connected from the drain to the positive supply V_{DD}. The output is taken from the drain and input applied between gate and ground.
- There are various types of MOS inverters available. The classification of MOS inverter is the based on the type of pull up device used. In MOS inverters the pull down device is always NMOS enhancement FET.

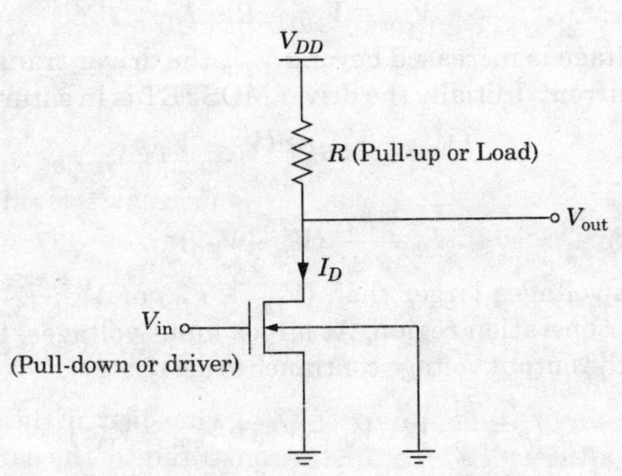

Fig. 4.9 Basic inverter circuit using MOSFET.

4.4.1 Resistive-Load Inverter

- The basic structure of the resistive-load inverter circuit is shown in Fig. 4.10. It is used an enhancement-type nMOS transistor and input is applied to the gate of the nMOS transistor which is also known as driver.
- The load consists of a simple linear resistor R_L which is connected between the drain and the power supply voltage V_{DD}.

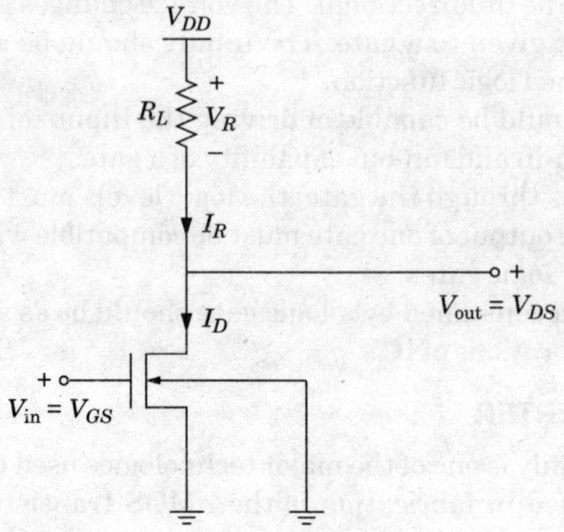

Fig. 4.10 Resistive-load inverter circuit.

- To simplify the calculations, the channel-length modulation effect will be neglected ($\lambda = 0$) in the following. The source and the substrate terminals of the driver transistor are both connected to the ground. Since $V_{SB} = 0$ the threshold voltage of the driver transistor is always equal to V_{T0} (V_T at $V_{SB} = 0$). We have already discussed that the drain current I_D of the driver MOSFET is equal to the load current (i.e, $I_D = I_R$) in DC steady-state operation.
- The transistor is in cut-off when the input voltages smaller than the threshold voltage V_{T0}. Since the voltage drop across the load resistor is equal to zero, the output voltage must be equal to the power supply voltage, V_{DD}.
- The output voltage V_{out} is given by

$$V_{out} = V_{DD} - R_L \cdot I_R \qquad \text{...(4.5)}$$

- As the input voltage is increased beyond V_{T0}, the driver transistor starts conducting a nonzero drain current. Initially the driver MOSFET is in saturation, since

$$(V_{out} = V_{DS}) > (V_{in} - V_{Tn})$$

Thus,

$$I_R = \frac{k_n}{2} \cdot (V_{in} - V_{T0})^2 \qquad \text{...(4.6)}$$

- When the input voltages larger than ($V_{DS} + T_{T0}$) or $V_{out} + V_{T0}$, the driver transistor enters the linear operation region. At larger input voltages, the transistor remains in linear mode, as the output voltage continues to decrease.

$$I_R = \frac{k_n}{2} \cdot [2 \cdot (V_{in} - V_{T0}) \cdot V_{out} - V_{out}^2] \qquad \text{...(4.7)}$$

- The voltage transfer characteristic of a typical resistive-load inverter circuit is shown in Fig. 4.11. The various operating regions of the driver transistor and the corresponding input-output conditions are given in the Table 4.1.

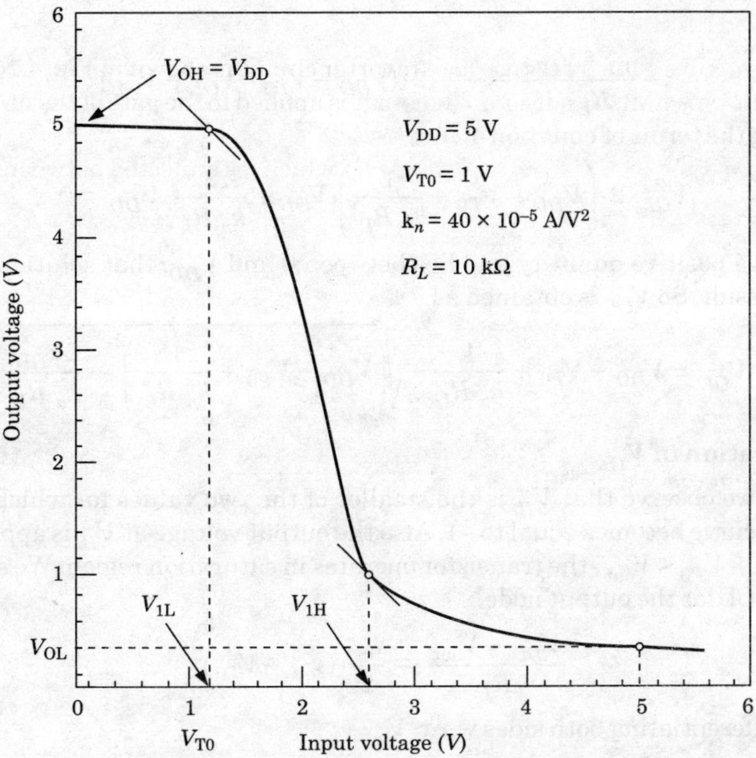

Fig. 4.11 Typical VTC of a resistive-load inverter circuit.

Table 4.1 Operating regions of the driver transistor in the resistive-load inverter.

Input Voltage Range	*Operating Mode*
$V_{in} < V_{T0}$	cut-off
$V_{T0} \leq V_{in} < V_{out} + V_{T0}$	saturation
$V_{in} \geq V_{out} + V_{T0}$	linear

Referring to the 4.1 and Fig. 4.11 the five critical voltage points are calculated as follows:

4.4.1.1 Calculation of V_{OH}

- When the input voltage V_{in} is low (smaller than the threshold voltage) then the driver transistor is cut-off. So the drain current of the driver transistor is equal to the load current ($I_R = I_D = 0$).
- It follows that the output voltage of the inverter under these conditions is:

$$V_{OH} = V_{DD} \quad \text{...(4.8)}$$

4.4.1.2 Calculation of V_{OL}

We assume that the input voltage is high ($V_{in} = V_{OH} = V_{DD}$). Since $V_{in} - V_{T0} > V_{out}$ and $V_{out} = V_{OL}$, the driver transistor operates in the linear region. Also note that the load current I_R is

$$I_R = \frac{V_{DD} - V_{out}}{R_L} \quad \text{...(4.9)}$$

and

$$\frac{V_{DD}-V_{OL}}{R_L} = \frac{k_n}{2}\cdot[2\cdot(V_{DD}-V_{T0})\cdot V_{OL}-V_{OL}^2] \qquad ...(4.10)$$

Rearranging the terms of equation 4.10.

$$V_{OL}^2\cdot 2\cdot\left(V_{DD}-V_{T0}+\frac{1}{k_n R_L}\right)\cdot V_{OL}+\frac{2}{k_n R_L}\cdot V_{DD}=0 \qquad ...(4.11)$$

Since V_{OL} is a positive quantity and lies between 0 and V_{DD}, that solution is chosen which gives a correct result. So V_{OL} is obtained as

$$V_{OL}=V_{DD}-V_{T0}+\frac{1}{k_n R_L}-\sqrt{\left(V_{DD}-V_{T0}+\frac{1}{k_n R_L}\right)^2-\frac{2V_{DD}}{k_n R_L}} \qquad ...(4.12)$$

4.4.1.3 Calculation of V_{IL}

From Fig. 4.11, we observe that V_{IL} is the smaller of the two values for which the slope of the voltage transfer curve becomes equal to –1. Also the output voltage at V_{IL} is approximately equal to V_{OH}. Since $V_{in}-V_{T0}<V_{out}$, the transistor operates in saturation region. We start our analysis by writing the KCL for the output node.

$$\frac{V_{DD}-V_{out}}{R_L}=\frac{k_n}{2}\cdot(V_{in}-V_{T0})^2 \qquad ...(4.13)$$

And after differentiating both sides w.r.t. V_{in}

$$-\frac{1}{R_L}\cdot\frac{dV_{out}}{dV_{in}}=k_n\cdot(V_{in}-V_{T0}) \qquad ...(4.14)$$

Since the derivative of the output voltage with respect to the input voltage is equal to (–1) at VIL, we can substitute $dV_{out}/dV_{in}=-1$ in (4.14)

$$-\frac{1}{R_L}\cdot(-1)=k_n\cdot(V_{IL}-V_{T0}) \qquad ...(4.15)$$

and

$$V_{IL}=V_{T0}+\frac{1}{k_n R_L} \qquad ...(4.16)$$

Hence, the value of output voltage is given by

$$V_{out}(V_{in}=V_{IL})=V_{DD}-\frac{k_n R_L}{2}\cdot\left(V_{T0}\frac{1}{k_n R_L}-V_{T0}\right)^2$$

$$=V_{DD}-\frac{1}{2k_n R_L} \qquad ...(4.17)$$

4.4.1.4 Calculation of V_{IH}

From Fig. 4.11, we can see that V_{IH} is the larger of the two voltage points on VTC at which the slope is equal to (–1). Also the output voltage (V_{OL}) at V_{IH} is slightly larger than the output low voltage V_{OL}. Hence, $V_{out}<V_{in}-V_{T0}$, and the driver transistor operates in the linear region. The KCL equation for the output node is given below.

$$\frac{V_{DD}-V_{out}}{R_L}=\frac{k_n}{2}\cdot[2\cdot(V_{in}-V_{T0})\cdot V_{out}-V_{out}^2] \qquad ...(4.18)$$

Differentiating both sides of (4.18) with respect to V_{in}, we obtain

$$-\frac{1}{R_L}\cdot\frac{dV_{out}}{dV_{in}} = \frac{k_n}{2}\cdot\left[2\cdot(V_{in}-V_{T0})\cdot\frac{dV_{out}}{dV_{in}} + 2V_{out} - 2V_{out}\cdot\frac{dV_{out}}{dV_{in}}\right] \qquad ...(4.19)$$

Substituting $dV_{out}/dV_{in} = -1$ and $V_{in} = V_{IH}$ in (4.19)

$$-\frac{1}{R_L}\cdot(-1) = k_n\cdot[(V_{IH}-V_{T0})\cdot(-1) + 2V_{out}] \qquad ...(4.20)$$

and

$$V_{IH} = V_{T0} + 2V_{out} - \frac{1}{k_n R_L} \qquad ...(4.21)$$

Thus, we obtain two algebraic equations, (4.18) and (4.21), for two unknowns, V_{IH} and V_{out}. Since from equations (4.21) and (4.18).

$$\frac{V_{DD}-V_{out}}{R_L} = \frac{k_n}{2}\cdot\left[2\cdot\left(V_{T0} + 2V_{out} - \frac{1}{k_n R_L} - V_{T0}\right)\cdot V_{out} - V_{out}^2\right] \qquad ...(4.22)$$

The positive solution of this second-order equation gives the output voltage V_{out} when the input is equal to V_{IH}.

$$V_{out}(V_{in} = V_{IH}) = \sqrt{\frac{2}{3}\cdot\frac{V_{DD}}{k_n R_L}} \qquad ...(4.23)$$

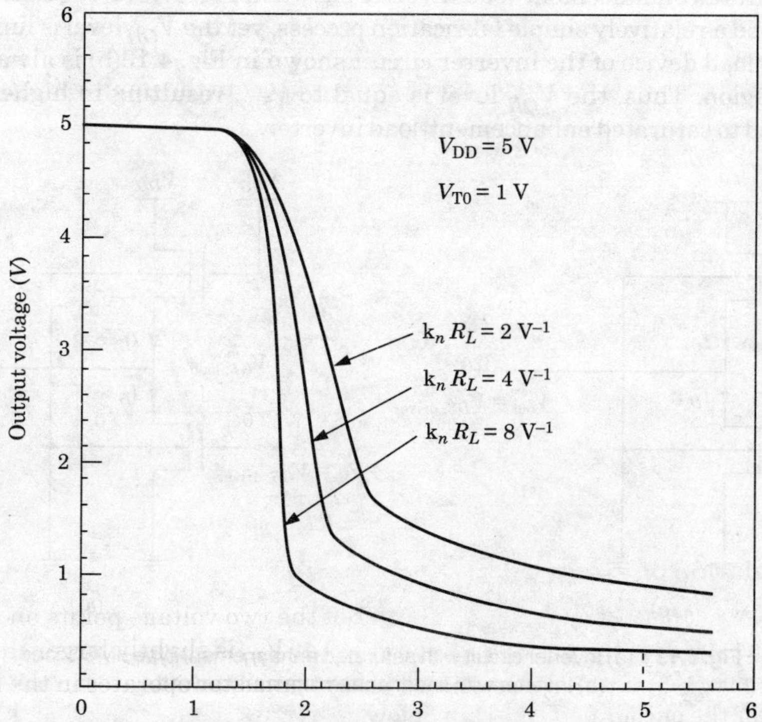

Fig. 4.12 Voltage transfer characteristics of the resistive-load inverter.

Finally, V_{IH} can be found by substituting (4.23) into (4.21), as follows.

$$V_{IH} = V_{T0} + \sqrt{\frac{8}{3} \cdot \frac{V_{DD}}{k_n R_L}} - \frac{1}{k_n R_L} \quad \text{...(4.24)}$$

- The four critical voltage points V_{OL}, V_{OH}, V_{IL}, and V_{IH} can now be used to determine the low and high noise margins of the resistive-load inverter circuit. Also the various parameters of the MOSFET and the load resistance R_L can be varied to see the effect on the VTC.
- Assuming that parameters such as the power supply voltage V_{DD} and the driver MOSFET threshold voltage V_{T0} are dictated by system- and processing-related constraints, the term $(K_n R_L)$ remains as the only design parameter which can be adjusted by the circuit designer to achieve certain design goals.
- The output high voltage V_{OH} is determined primarily by the power supply voltage. The other three critical voltage points, the adjustment of V_{OL} receives primary attention, while V_{IL} and V_{IH} are usually treated as secondary design variables. Figure 4.12 shows the VTC of a resistive-load inverter for different values of $(K_n R_L)$.
- The simple resistive-load inverter circuit is not a suitable candidate for most digital VLSI system applications, primarily because of the large area occupied by the load resistor.

4.4.2 Enhancement-Load nMOS Inverter

- The circuit configurations of two inverters with enhancement-type load devices are shown in Fig. 4.13. The enhancement-type load transistor can be operated either in the saturation region or in the linear region based on the bias voltage applied to its gate terminal.
- The saturated enhancement-load inverter shown in Fig. 4.13(a) requires a single voltage supply and a relatively simple fabrication process, yet the V_{OH} level is limited to $V_{DD} - V_{T,load}$. The load device of the inverter circuit shown in Fig. 4.13(b) is always biased in the linear region. Thus, the V_{OH} level is equal to V_{DD}, resulting in higher noise margins compared to saturated enhancement-load inverter.

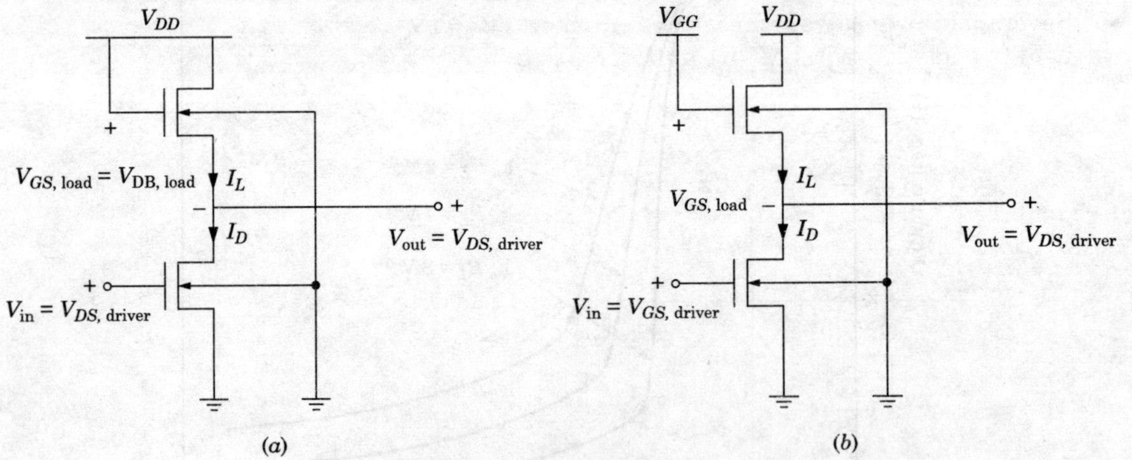

Fig. 4.13 (a) Inverter circuit with saturated enhancement-type nMOS load. (b) Inverter with linear enhancement-type load.

- MOSFETs with enhancement type driver and resistance as load are not well suited for VLSI circuit fabrication and in other application as resistance is a passive element and it is very difficult is fabricate a resistor.

- VTC also do not show sharper transition and better noise margins. So nMOS transistors as active loads poise a better candidature for inverter circuit. Also nMOS load occupies less chip area as compared to the linear resistor. Better performance was also achieved with nMOS loads.

4.4.2.1 Advantages of Enhancement-Load n-Type MOSFET as a Load

1. The main advantage of using a MOSFET as the load device is that the silicon area occupied by the transistor is usually smaller than that occupied by a comparable resistive load.
2. Inverter circuits with active loads can be designed to have better overall performance compared to that of passive-load inverters.
3. In a chronological view, the development of inverters with an enhancement-type MOSFET load precedes other active-load inverter types, since its fabrication process was perfected earlier.

4.4.2.2 Disadvantages of Enhancement-Load n-Type MOSFET as a Load

The most significant drawback of this configuration is the use of two separate power supply voltages. In addition, both types of inverter circuits shown in Fig. 4.13 suffer from relatively high stand-by (DC) power dissipation; hence, enhancement-load nMOS inverters are not used in any large-scale digital applications.

4.4.3 Depletion-Load nMOS Inverter

- The circuit diagram of the depletion-load inverter circuit is shown in Fig. 4.14(a), and a simplified view of the circuit consisting of a nonlinear load resistor and a non-ideal switch (driver) in shown in Fig. 4.14(b).
- The fabrication process for producing an inverter with an enhancement-type nMOS driver and a depletion-type nMOS load is slightly more complicated and requires additional processing steps, especially for the channel implant to adjust the threshold voltage of the load device.
- Several of the disadvantages of the enhancement-type load inverter can be avoided by using a depletion-type nMOS transistor as the load device.

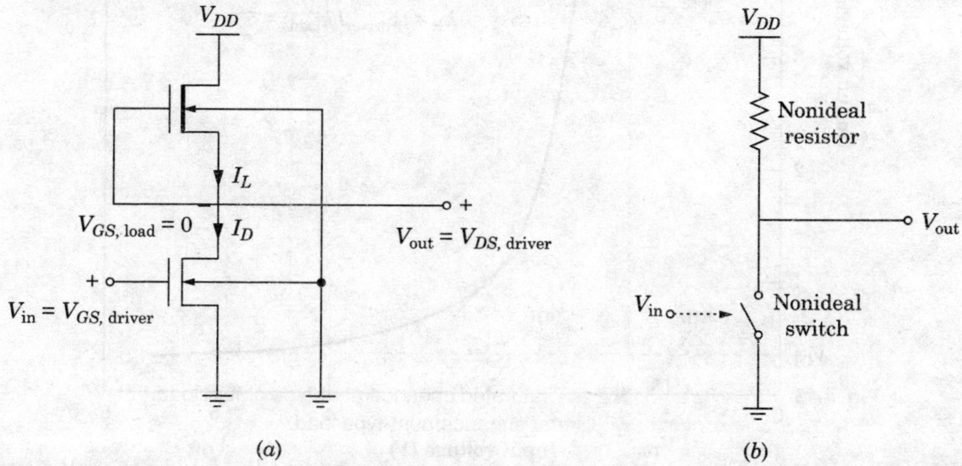

Fig. 4.14 (a) Inverter circuit with depletion-type nMOS load. (b) Simplified equivalent circuit consisting of a nonlinear load resistor and a non-ideal switch controlled by the input.

- The load is a depletion-type nMOS transistor so its threshold voltage is less than zero i.e., $V_{T0,\,load} < 0$. The driver device is an enhancement-type transistor so its threshold voltage is a positive quantity i.e., with $V_{T0,\,driver} > 0$.
- The gate and the source nodes of the load transistor are connected so that, $V_{GS,\,load} = 0$ and $V_{GS,\,load} > V_{T,\,load}$. So that the load device always has a conducting channel regardless of the input and output voltage levels.
- Both the driver transistor and the load transistor are built on the same *p*-type substrate, which is connected to the ground. So the source-to-substrate voltage of the load device is equal to output voltage ($V_{SB,\,load} = V_{out}$). The threshold voltage is also function of source-to-substrate voltage (V_{out}) and given by

$$V_{T,\,load} = V_{T0,\,load} + \gamma\left(\sqrt{|2\phi_F| + V_{out}} - \sqrt{|2\phi_F|}\right) \quad \text{...(4.25)}$$

The operating mode of the load transistor is determined by ($V_{GS} = 0$).

- In saturation region (i.e., when $V_{out} < V_{DD} + V_{T,\,load}$), the load current is given by

$$I_{D,\,load} = \frac{k_{n,\,load}}{2}[V_{GS,load} - V_{T,load}]^2$$

$$I_{D,\,load} = \frac{k_{n,\,load}}{2}\cdot\left[-V_{T,\,load}(V_{out})\right]^2 = \frac{k_{n,\,load}}{2}\cdot|V_{T,\,load}(V_{out})|^2 \quad \text{...(4.26)}$$

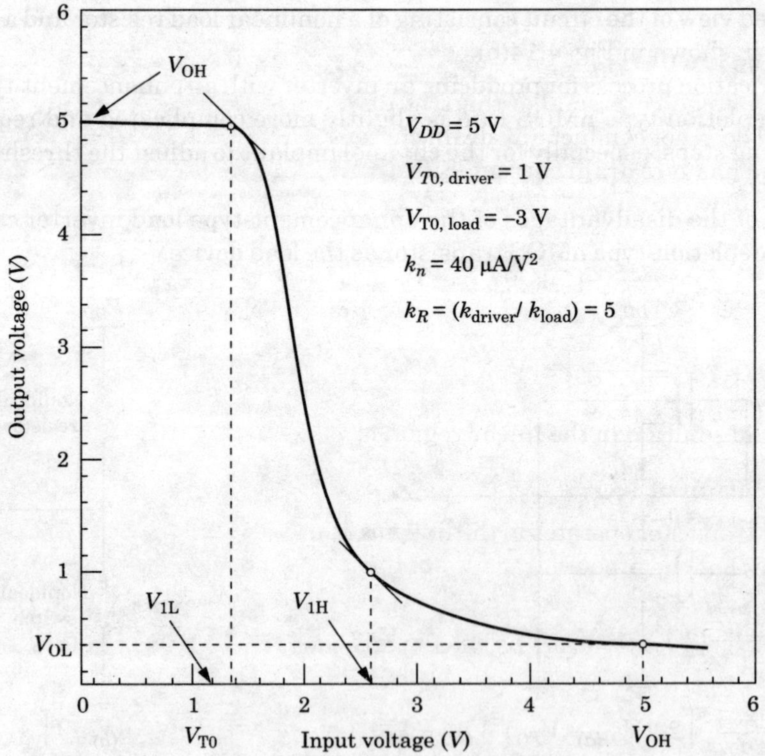

Fig. 4.15 Typical VTC of a depletion-load inverter circuit

- In linear region (i.e., when $V_{out} > V_{DD} + V_{T,\,load}$), the load current is given by

$$I_{D,\,load} = \frac{k_{n,\,load}}{2}[2(V_{GS,load} - V_{T,load})V_{DS,load} - V_{DS,load}^2]$$

$$I_{D,\,load} = \frac{k_{n,\,load}}{2}\cdot\left[2\,|V_{T,\,load}(V_{out})|\cdot(V_{DD} - V_{out}) - (V_{DD} - V_{out})^2\right] \quad ...(4.27)$$

- The voltage transfer characteristic (VTC) of this inverter can be constructed by setting $I_{D,driver} = I_{D,\,load}$, $V_{GS,driver} = V_{in}$ and $V_{DS,driver} = V_{out}$, and by solving the corresponding current equations for $V_{out} = f(V_{in})$. Figure 4.15 shows the VTC of a typical depletion-load inverter, with $k_{n,driver}' = k_{n,load}'$.
- We will consider the critical voltage points V_{OH}, V_{OL}, V_{IL} and V_{IH} for this inverter circuit. The operating regions and the voltage levels of the driver and the load transistors at these critical points are listed below.

Vin	*Vout*	*Drive operating region*	*Load operation region*
V_{OL}	V_{OH}	cut-off	linear
V_{IL}	$\approx V_{OH}$	saturation	linear
V_{IH}	samll	linear	saturation
V_{OH}	V_{OL}	linear	saturation

4.4.3.1 Calculation of V_{OH}

When the input voltage V_{in} is smaller than the driver threshold voltage V_{T0}, the driver transistor is turned off and does not conduct any drain current. Then the load device, which operates in the linear region and has zero drain current. Substituting V_{OH} for V_{out} in (4.27), and letting the load current $I_{D,\,load} = 0$, we obtain

$$I_{D,\,load} = \frac{k_{n,\,load}}{2}[2(V_{GS,load} - V_{T,load})V_{DS,load} - V_{DS,load}^2] = 0$$

$$I_{D,\,load} = \frac{k_{n,\,load}}{2}\cdot\left[2\,|\,V_{T,\,load}\,|\cdot(V_{DD} - V_{OH}) - (V_{DD} - V_{OH})^2\right] = 0 \quad ...(4.28)$$

The only valid solution in the linear region is $V_{OH} = V_{DD}$.

4.4.3.2 Calculation of V_{OL}

When the driver transistor operates in the linear region and the depletion-type load is in saturation than we assume that the input voltage V_{in} of the inverter is equal to $V_{OH} = V_{DD}$. Hence,

$$\frac{k_{driver}}{2}[2(V_{GS} - V_{T0})V_{DS,driver} - V_{DS,driver}^2] = \frac{k_{load}}{2}[V_{GS,load} - V_{T,load}(V_{OL})]^2$$

$$\frac{k_{driver}}{2}\cdot\left[2\cdot(V_{OH} - V_{T0})\cdot V_{OL} - V_{OL}^2\right] = \frac{k_{load}}{2}\cdot\left[-V_{T,\,load}(V_{OL})\right]^2 \quad ...(4.29)$$

After substituting $V_{in} = V_{OH}$ and $V_{out} = V_{OL}$ in above equation than its current solution is

$$V_{OL} = V_{OH} - V_{T0} - \sqrt{(V_{OH} - V_{T0})^2 - \left(\frac{k_{load}}{k_{driver}}\right) \cdot |V_{T,load}(V_{OL})|^2} \quad ...(4.30)$$

The actual value of V_{OL} is relatively small.

$$k_r = \frac{k_{load}}{k_{driver}} = \frac{[V_{T,load}(V_{OL})]^2}{[2(V_{OH} - V_T)V_{OL} - V_{OL}^2]}$$

4.4.3.3 Calculation of V_{IL}

V_{IL} is the smaller of the two values of the input voltage for which the slope of the VTC is equal to (–1). The driver transistor operates in saturation and the load transistor operates in the linear region when the input voltage is $V_{in} = V_{IL}$. Applying KCL for the output node, we obtain the following current equation:

$$\frac{k_{driver}}{2} \cdot (V_{in} - V_{T0})^2 = \frac{k_{load}}{2} \cdot \left[2\,|V_{T,load}(V_{out})| \cdot (V_{DD} - V_{out}) - (V_{DD} - V_{out})^2\right] \quad ...(4.31)$$

Differentiate both sides with respect to V_{in} we have

$$k_{driver} \cdot (V_{in} - V_{T0}) = \frac{k_{load}}{2} \cdot \left[2\left|V_{T,load}(V_{out})\right|\left(-\frac{dV_{out}}{dV_{in}}\right)\right.$$

$$\left. + 2(V_{DD} - V_{out})\left(-\frac{dV_{T,load}}{dV_{in}}\right) - 2(V_{DD} - V_{out})\left(-\frac{dV_{out}}{dV_{in}}\right)\right] \quad ...(4.32)$$

In general, we can assume that the term $(dV_{T,Load}/dV_{in})$ is negligible with respect to the others. Substituting V_{IL} for V_{in}, and letting $dV_{out}/V_{in} = 1$, we obtain V_{IL} as a function of the output voltage V_{out}.

$$V_{IL} = V_{T0} + \left(\frac{k_{load}}{k_{driver}}\right) \cdot \left[V_{out} - V_{DD} + \left|V_{T,load}(V_{out})\right|\right] \quad ...(4.33)$$

4.4.3.4 Calculation of V_{IH}

V_{IH} is the larger of the two voltage points on the VTC at which the slope is equal to (–1). Also when $V_{in} = V_{IH}$, output voltage is small. Since $V_{in} - V_{T0} > V_{out}$ the driver transistor is in the linear region and since $V_{out} - V_{T,load} < V_{DD}$ the load transistor is in saturation.

$$\frac{k_{driver}}{2} \cdot \left[2 \cdot (V_{in} - V_0) \cdot V_{out} - V_{out}^2\right] = \frac{k_{load}}{2} \cdot \left[-V_{T,load}(V_{out})\right]^2 \quad ...(4.34)$$

Differentiating both sides of (4.34) with respect to Vin, we obtain:

$$k_{driver} \cdot \left[V_{out} + (V_{in} - V_{T0})\left(\frac{dV_{out}}{dV_{in}}\right) - V_{out}\left(\frac{dV_{out}}{dV_{in}}\right)\right]$$

$$= k_{load}\left[-V_{T,load}(V_{out})\right] \cdot \left(\frac{dV_{T,load}}{dV_{out}}\right)\left(\frac{dV_{out}}{dV_{in}}\right) \quad ...(3.35)$$

Now, substitute $dV_{out}/dV_{in} = -1$ into (4.35), and solve for $V_{in} = V_{IH}$,

$$V_{IH} = V_{T0} + 2V_{out} + \left(\frac{k_{load}}{k_{driver}}\right) \cdot \left[-V_{T,\,load}(V_{out})\right] \cdot \left(\frac{dV_{T,\,load}}{dV_{out}}\right) \quad \text{...(3.36)}$$

The derivative of the load threshold voltage with respect to the output voltage cannot be neglected in this case.

$$\frac{dV_{T,\,load}}{dV_{out}} = \frac{\gamma}{2\sqrt{|+2\phi_F| + V_{out}}} \quad \text{...(4.37)}$$

The actual values of V_{IH} and the corresponding output voltage V_{out} are determined by solving (4.37) together with the current equation (4.34) and with the threshold voltage expression, using numerical iterations.

- It is seen from this discussion that the critical voltage points, the general shape of the inverter VTC, and ultimately, the noise margins, are determined essentially by the threshold voltages of the driver and the load devices, and by the driver-to-load ratio (K_{driver}/K_{load}).
- Notice that with $K'_{n,\,driver} = K'_{n,\,load}$, the driver-to-load ratio is solely determined by the (W/L) ratios of the driver and the load transistors. Figure 4.16 shows the VTCs of depletion-load inverter circuits with different driver-to-load ratios $K_R = (K_{driver}/K_{load})$.
- One important observation is that, unlike in the enhancement-load inverter case, a sharp VTC transition and larger noise margins can be obtained with relatively small driver-to-load ratios. Thus, the total area occupied by a depletion-load inverter circuit with an acceptable circuit performance is expected to be much smaller than the area occupied by a comparable resistive-load or enhancement-load inverter.

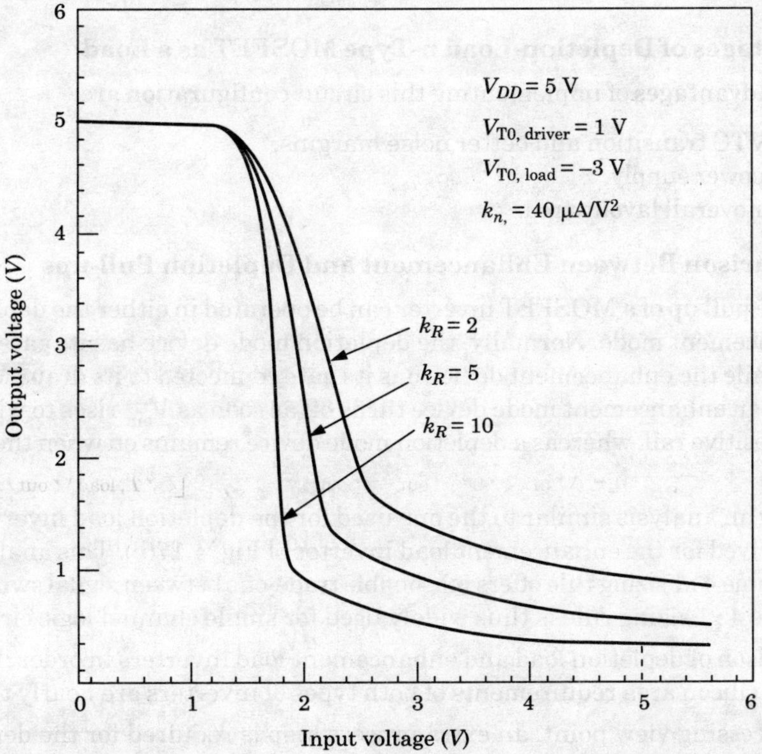

Fig. 4.16 Voltage transfer characteristics of depletion-load inverters, with different driver-to- load ratios.

4.4.3.5 Calculation of K_R

- To calculate the driver-to-load ratio, we can rearrange the equation (4.29) that achieves a target V_{OL} value. Here assume that the power supply voltage and the threshold voltage values are set previously by independent design and processing constraints.

$$k_R = \frac{k_{\text{driver}}}{k_{\text{load}}} = \frac{|V_{T,\text{load}}(V_{OL})|^2}{2(V_{OH} - V_{T0})V_{OL} - V_{OL}^2} \quad ...(4.38)$$

Here, the driver-to-load ratio is given by

$$k_R = \frac{k'_{n,\text{driver}} \cdot \left(\frac{W}{L}\right)_{\text{driver}}}{k'_{n,\text{load}} \cdot \left(\frac{W}{L}\right)_{\text{load}}} \quad ...(4.39)$$

Since the channel doping densities and, consequently, the channel electron mobilities of the enhancement-type driver transistor and the depletion-type load transistor are not equal, we shall expect that $K'_{n,\text{driver}} \neq K'_{n,\text{load}}$ in general. Only if $K'_{n,\text{driver}} = K'_{n,\text{load}}$, can be the driver-to-load ratio be reduced to

$$k_R = \frac{\left(\frac{W}{L}\right)_{\text{driver}}}{\left(\frac{W}{L}\right)_{\text{load}}} \quad ...(4.40)$$

4.4.3.6 Advantages of Depletion-Load n-Type MOSFET as a Load

The immediate advantages of implementing this circuit configuration are:

(i) Sharp VTC transition and better noise margins,
(ii) Single power supply,
(iii) Smaller overall layout area.

4.4.3.7 Comparison Between Enhancement and Depletion Pull-ups

1. The active pull up of a MOSFET inverter can be operated in either the depletion mode or in the enhancement mode. Normally, the depletion mode device has its gate connected to its source, while the enhancement device has its gate connected to its drain. When the output goes high an enhancement mode device turns off as soon as V_{out} rises to within a threshold value of positive rail, whereas a depletion mode device remains on when the output voltages goes high.
2. Following an analysis similar to the one used for the depletion load inverter, a sizing rule can be derived for the enhancement load inverter of Fig. 4.17(b). This analysis would show that the same 4 : 1 sizing rule offers reasonable trade-offs between signal swing and switching speed. The 4 : 1 sizing rule is thus widely used for single channel logic circuits.
3. A comparison of depletion load and enhancement load inverters in order. From a size view point, the silicon area requirements of both types of inverters are nearly the same.
4. From processing view point, an extra process step is required for the depletion implant, raising the cost of depletion mode logic relative to enhancement mode logic.

5. In terms of switching speed, the depletion mode inverter is considerably faster than an enhancement mode inverter in going from a low output voltage to a high output voltage. The reason for this variation is the large difference in equivalent resistance for the two transistors.

Depletion load inverters were found to be preferable to enhancement load inverters for most applications.

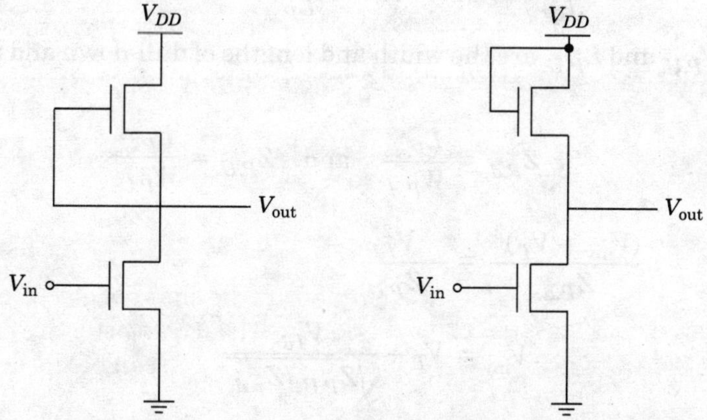

(*a*) Depletion load (*b*) Enhancement load

Fig. 4.17 Basic NMOS inverters.

4.5 DETERMINATION OF PULL-UP TO PULL-DOWN RATIO ($Z_{P.U.}/Z_{P.D.}$) FOR AN NMOS INVERTER DRIVEN BY ANOTHER NMOS INVERTER [UPTU EXAM: 2003-04, 2007-08, 2011-12]

Consider the arrangement shown in Fig. 4.18 in which an inverter is driven from the output of another similar inverter. Consider the depletion mode transistor for which $V_{GS} = 0$ under all conditions, and further assume that in order to cascade inverter without degradation of levels, we aim to meet the requirement:

$$V_{in} = V_{out} = V_{in}$$

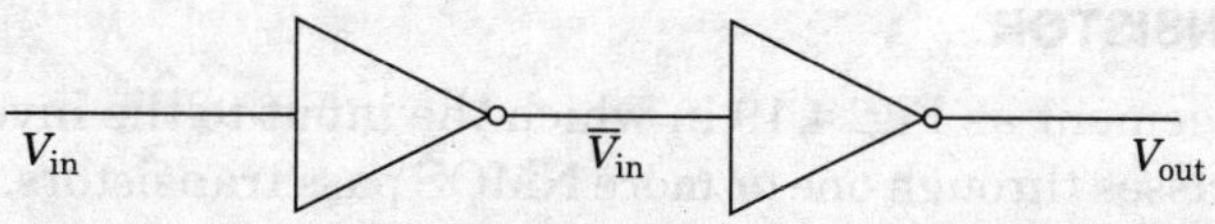

Fig. 4.18 NMOS inverter driven directly by another inverter.

For equal margins around the inverter threshold, we set $V_{GS} = V_{in} = 0.5\, V_{DD}$ for Enhancement Mode. At this point, both transistor are in saturation and

$$I_D = \frac{K_n'}{2} \cdot \frac{W}{L}(V_{GS} - V_T)^2$$

In depletion mode,

$$I_D = \frac{K_n'}{2} \cdot \frac{W_{P.U.}}{L_{P.U.}}\left(0 - V_{Td}^2\right)$$

where V_{Td} is the threshold voltage in depletion mode.

In Enhancement mode,

$$I_D = \frac{K'_n}{2} \cdot \frac{W_{p.d}}{L_{p.d}} (V_{\text{inv}} - V_T)^2$$

After equating the above equation

$$\frac{W_{P.d}}{L_{P.d}} (V_{\text{inv}} - V_T)^2 = \frac{W_{P.d}}{L_{P.d}} (0 - V_{Td}^2)$$

where, $W_{p.d}$, $L_{p.d.}$, $W_{P.U.}$ and $L_{P.U.}$ are the width and lengths of dull-down and pull-up transistors respectively.

Now, $$Z_{P.d.} = \frac{L_{P.d.}}{W_{P.d}} \quad \text{and} \quad Z_{P.U.} = \frac{L_{P.U.}}{W_{P.U.}}$$

so, $$\frac{(V_{\text{inv}} - V_T)^2}{Z_{P.d.}} = -\frac{V_{Td}^2}{Z_{P.U.}}$$

Simplifying $$V_{\text{inv}} = V_T - \frac{V_{Td}}{\sqrt{Z_{P.U.}/Z_{p.d.}}} \quad ...(4.41)$$

Here, we consider some typical values are follows:

$$V_T = 0.2\ V_{DD},\ V_{Td} = 0.6\ V_{DD} \quad \text{and} \quad V_{\text{inV}} = \frac{V_{DD}}{2}$$

so $$0.5 = 0.2 + \frac{0.6}{\sqrt{Z_{P.U.}/Z_{P.d.}}}$$

$$\Rightarrow \quad \sqrt{\frac{Z_{P.U.}}{Z_{P.d.}}} = 2 \quad \Rightarrow \quad Z_{P.U.} = 4\ Z_{p.d.}$$

For an inverter directly driven by another inverter.

4.6 PULL-UP TO PULL-DOWN RATIO FOR AN NMOS INVERTER DRIVEN THROUGH ONE OR MORE PASS TRANSISTOR

We consider the arrangement as Fig. 4.19 in which the input to the inverter 2 comes from the output inverter 1 but passes through one or more NMOS pass transistors. The connection of pass transistors in series will degrade the logic 1 level into inverter 2 so that the output will not be a proper logic 0 level. If the point A is at 0 V, and B point is thus at V_{DD}, but the voltage into inverter 2 at point C is now reduced from V_{DD} by the threshold voltage of the series pass transistor. However, there can be no voltage drop in the channels, since no static current flows through the pass transistors. With all pass transistors connected to V_{DD}, there is a loss of V_{Tn}.

Fig. 4.19 Pull-up to pull-down ratio for inverting logic coupled bypass transistors

Therefore, the input voltage to inverter 2 is

$$V_{in\,2} = V_{DD} - V_{Tn}$$

We must ensure that for this input voltage, we get the same voltage as would be the case for inverter 1 driven with input = V_{DD}.

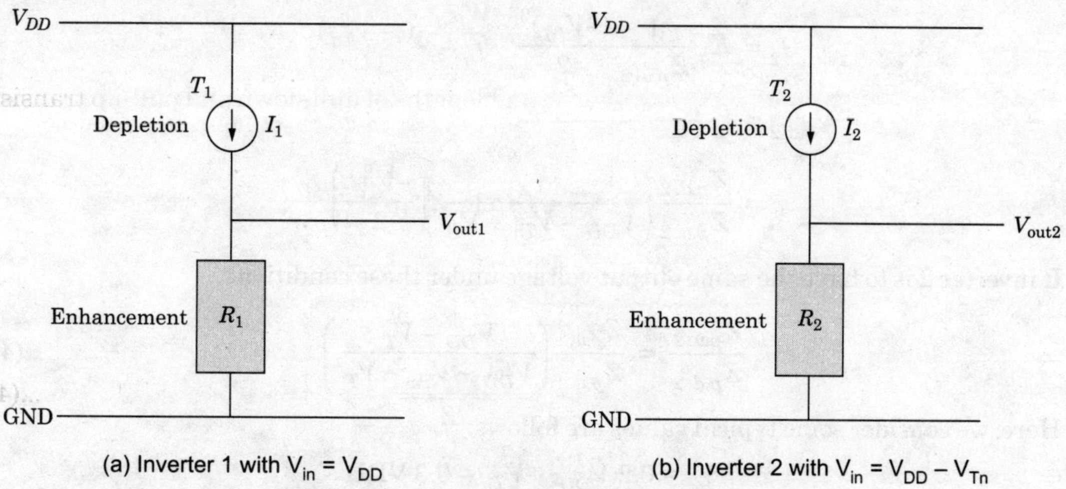

Fig. 4.20 Equivalent circuit of inverters 1 and 2.

Consider inverter 1 as shown in Fig. 4.20(a). If the input is at V_{DD}, then the pull-down transistor T_2 is conducting but with a low voltage across it, therefore, it is in its resistive region represented by R_1. Meanwhile, the pull-up transistor T_1 is in saturation and is represented as a current source.

For the pull-down transistor,

$$I_{DS} = K_n \frac{W_{P.d\,1}}{L_{P.d.1}}\left[(V_{DD} - V_T)V_{DSL} - \frac{V_{DS1}^2}{2}\right]$$

$$\therefore \quad R_1 = \frac{V_{DS,1}}{I_{DS,1}} = \frac{1}{K'_n}\cdot\frac{L_{P.d.1}}{W_{P.d.1}}\left[\frac{V_{DS,1}}{(V_{DD} - V_T)V_{DS\,1} - \frac{V_{DS,1}^2}{2}}\right]$$

if $V_{DS\,1}$ is very small, so $\frac{V_{DS}}{2} \approx 0$.

Thus $$R_1 = \frac{1}{K_n} Z_{p.d.1}\left(\frac{1}{V_{DD} - V_T}\right)$$

Now, for the depletion mode pull-up transistor in saturation, with $V_{GS} = 0$

$$I_1 = I_{DS} = K_n \frac{W_{P.U.1}}{L_{P.U.1}}\left(\frac{-V_{Td}^2}{2}\right)$$

$$V_{out\,1} = I_1 R_1 = \frac{Z_{p.d1}}{Z_{p.u.1}}\left(\frac{1}{V_{DD} - V_T}\right)\frac{V_{Td}^2}{2}$$

Consider inverter 2, when $V_{in} = V_{DD} - V_{Tn}$

then for inverter 2,

$$R_2 = \frac{1}{K_n} Z_{p.d.2} \frac{1}{[(V_{DD} - V_{Tn}) - V_T]}$$

$$I_2 = K_n \frac{1}{Z_{p.u.2}} \frac{(-V_{Td})^2}{2}$$

so $$V_{out\,2} = I_2 R_2$$

$$= \frac{Z_{p.d.2}}{Z_{p.u.2}} \left(\frac{1}{V_{DD} - V_{Tn} - V_T} \right) \left(\frac{-V_{Td}^2}{2} \right)$$

If inverter 2 is to have the same output voltage under these conditions.

$$\Rightarrow \quad \frac{Z_{p.u.2}}{Z_{p.d.2}} = \frac{Z_{p.u.1}}{Z_{p.d.1}} \left(\frac{V_{DD} - V_T}{V_{DD} - V_{Tn} - V_T} \right) \quad ...(4.42)$$

Here, we consider some typical values are follows:

$$V_T = 0.2\ V_{DD}, \quad V_{Tn} = 0.3\ \text{VDD}$$

$$\frac{Z_{p.u.2}}{Z_{p.d.2}} = \frac{Z_{p.u.1}}{Z_{p.d.1}} \left(\frac{0.8}{0.5} \right)$$

$$\therefore \quad \frac{Z_{p.u.2}}{Z_{p.d.2}} = 2 \left(\frac{Z_{p.u.1}}{Z_{p.d.1}} \right) = \frac{8}{1}$$

Summarizing for an NMOS transistor:

- An inverter driven directly from the output of another should have a

$$\frac{Z_{p.u.}}{Z_{p.d.}} \text{ ratio} \geq \frac{4}{1}$$

- An driven through one or more pass transistors should have a inverter

$$\frac{Z_{p.u.}}{Z_{p.d.}} \text{ ratio} \geq \frac{8}{1}$$

4.7 CMOS INVERTER **[M.Imp.]**

- CMOS inverter consists of an enhancement-type nMOS driver transistor and an enhancement-type pMOS transistor working in complementary mode as shown in Fig. 4.21.
- The circuit topology is complementary push-pull mode such that for high input nMOS transistor is ON and it drives (pulls down) the output node and the pMOS transistor acts as the load. For low input the pMOS transistor drives (pulls up) the output node and the nMOS transistor acts as the load.
- Hence, both devices (transistors) contribute equally to the circuit operation characteristics.

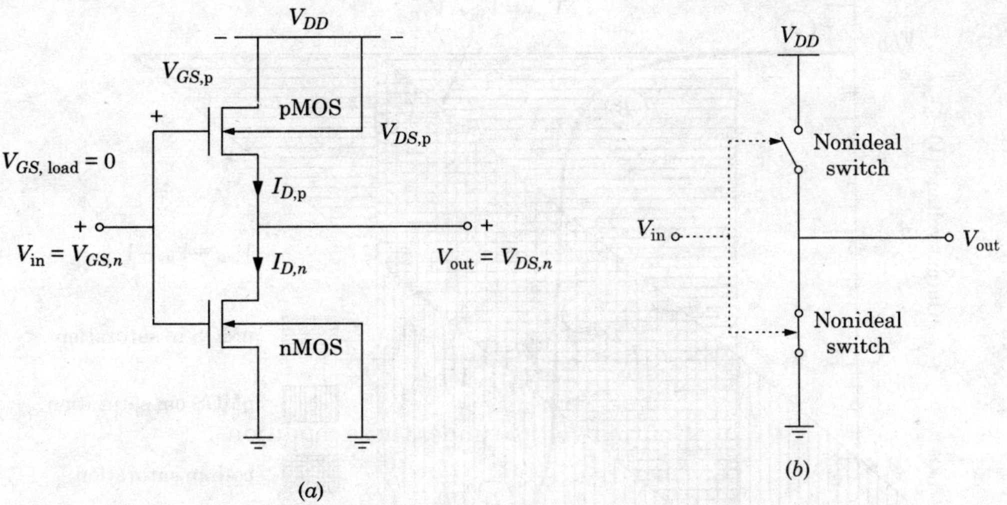

Fig. 4.21 (a) CMOS inverter circuit. (b) Simplified view of the CMOS inverter, consisting of two complementary non-ideal switches.

4.7.1 Circuit Operation

From Fig. 4.21 we can see that the input voltage is connected to the gate terminals of both the nMOS and the pMOS transistors. The substrate of the nMOS transistor is connected to the ground, while the substrate of the pMOS transistor is connected to the power supply voltage, V_{DD}. Since the effective source substrate-bias $VSB = 0$ for both devices. It can be seen from the circuit diagram in Fig. 4.21 that

$$V_{GS,\,n} = V_{in}$$

$$V_{DS,\,n} = V_{out} \qquad \text{...(4.43)}$$

and

$$V_{GS,\,p} = -(V_{DD} - V_{in})$$

$$V_{DS,\,p} = -(V_{DD} - V_{out}) \qquad \text{...(4.44)}$$

A typical CMOS inverter voltage transfer characteristic is shown in Fig. 4.22.

Here, we identify five distinct regions, labeled *A* through *E*, each corresponding to a different set of operating conditions. The table below lists these regions and the corresponding critical input and output voltage levels.

Region	*Vin*	*Vout*	*nMOS*	*pMOS*
A	$< V_{T0,\,n}$	V_{OH}	cut-off	linear
B	V_{IL}	high $\approx V_{OH}$	saturation	linear
C	V_{th}	V_{th}	saturation	saturation
D	V_{IH}	low $\approx V_{OL}$	linear	saturation
E	$> (V_{DD} + V_{T0,\,p})$	V_{OL}	Linear	cut-off

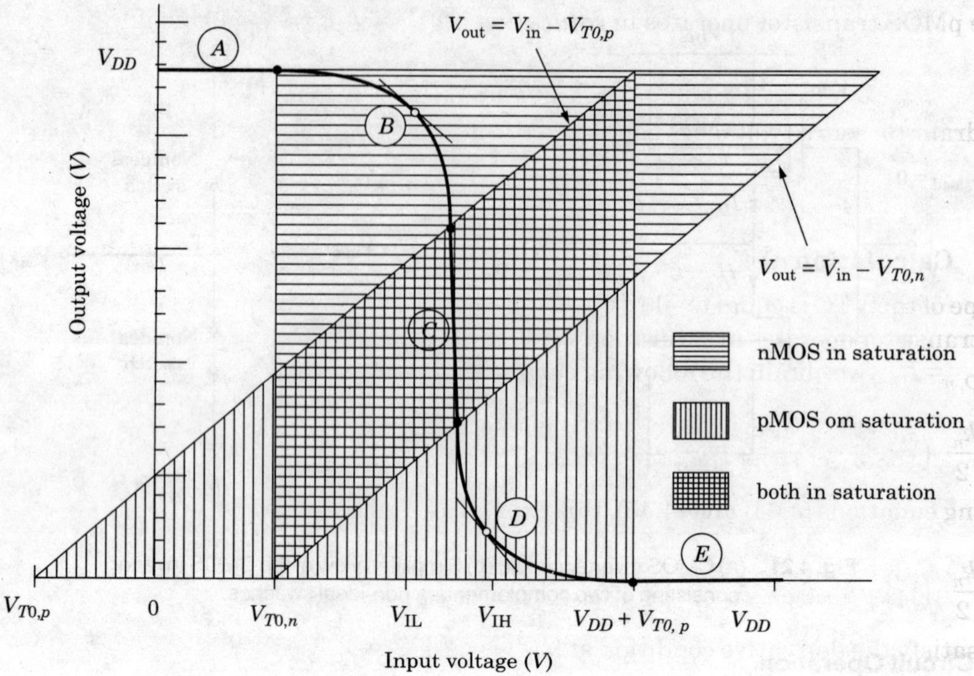

Fig. 4.22 VTC of CMOS inverter.

4.7.1.1 Calculation of V_{OH}

When the input voltage is smaller than the nMOS threshold voltage ($V_{in} < V_{T0,\,n}$), the nMOS transistor is cut-off and the pMOS transistor operates in the linear region. The drain currents of both transistors are approximately equal to zero (except for small leakage currents).

$$I_{D,\,n} = I_{D,\,p} = 0 \qquad \text{...(4.45)}$$

So the drain-to-source voltage of the pMOS transistor is also equal to zero, and the output voltage V_{OH} is equal to the power supply voltage.

$$V_{\text{out}} = V_{OH} = V_{DD} \qquad \text{...(4.46)}$$

Since pMOS is in linear region we have

$$I_{DP} = \frac{k_p}{2}\cdot\left[2\cdot(V_{GS,\,p} - V_{T0,\,p})\cdot V_{DS,\,p} - V_{DS,\,p}^2\right] = 0 \qquad \text{...(4.47)}$$

$$\frac{k_p}{2}\cdot\left[2\cdot(V_{\text{in}} - V_{DD} - V_{T0,\,p})\,(V_{out} - V_{DD}) - (V_{out} - V_{DD})^2\right] = 0 \qquad \text{...(4.48)}$$

The valid solution is $V_{OH} = V_{DD}$

4.7.1.2 Calculation of V_{OL}

When the input voltage exceeds $V_{DD} + V_{T0,\,p}$ the pMOS transistor is turned off and the nMOS transistor is operating in the linear region. The nMOS transistor operates in saturation if $V_{\text{in}} > V_{T0,n}$ and if the following condition is satisfied.

$$V_{DS,\,n} \geq V_{GS,\,n} - V_{T0,\,n} \;\Leftrightarrow\; V_{\text{out}} \geq V_{\text{in}} - V_{T0,\,n} \qquad \text{...(4.49)}$$

The pMOS transistor operates in *saturation* if $V_{in} < (V_{DD} + V_{T0,p})$, and if:

$$V_{DS,p} \le V_{GS,p} - V_{T0,p} \quad \Leftrightarrow \quad V_{out} \le V_{in} - V_{T0,p} \qquad \text{...(4.50)}$$

So drain-to- source voltage is equal to zero and $V_{out} = V_{on}$. The output voltage of the circuit is

$$V_{out} = V_{OL} = 0 \qquad \text{...(4.51)}$$

4.7.1.3 Calculation of V_{IL}

The slope of the VTC is equal to –1 ($dV_{out}/dV_{in} = -1$). When the input voltage is low ($V_{in} = V_{IL}$) the nMOS transistor operates in saturation while the pMOS transistor operates in the linear region. From $I_{D,n} = I_{D,p}$ we obtain the following current equation:

$$\frac{k_n}{2}\cdot\left(V_{GS,n} - V_{T0,n}\right) = \frac{k_p}{2}\cdot\left[2\cdot\left(V_{GS,p} - V_{T0,p}\right)\cdot V_{DS,p} - V_{DS,p}^2\right] \qquad \text{...(4.52)}$$

Using equations (4.44) and (4.45), this expression can be rewritten as

$$\frac{k_n}{2}\cdot\left(V_{in} - V_{T0,n}\right)^2 = \frac{k_p}{2}\cdot\left[2\cdot\left(V_{in} - V_{DD} - V_{T0,p}\right)\cdot\left(V_{out} - V_{DD}\right) - \left(V_{out} - V_{DD}\right)^2\right] \qquad \text{...(4.53)}$$

To satisfy the derivative condition at V_{IL} we differentiate both sides of (4.54) with respect to V_{in}.

$$K_n\cdot(V_{in} - V_{T0}, n) = k_p\cdot\left[\left(V_{in} - V_{DD} - V_{T0,p}\right)\cdot\left(\frac{dV_{out}}{dV_{in}}\right) + (V_{out} - V_{DD}) - (V_{out} - V_{DD})\cdot\left(\frac{dV_{out}}{dV_{in}}\right)\right] \qquad \text{...(4.54)}$$

Substituting $V_{in} = V_{IL}$ and $(dV_{out}/dV_{in}) = -1$ in (4.55), we obtain

$$k_n\cdot\left(V_{IL} - V_{T0,n}\right) = k_p\cdot\left(2V_{out} - V_{IL} + V_{T0,p} - V_{DD}\right) \qquad \text{...(4.55)}$$

The critical voltage V_{IL} can now be found as a function of the output voltage V_{out}, as follows:

$$V_{IL} = \frac{2V_{out} + V_{T0,p} - V_{DD} + k_R V_{T0,n}}{1 + k_R} \qquad \text{...(4.56)}$$

Where K_R is defined as

$$k_R = \frac{k_n}{k_p}$$

V_{IL} is obtained numerically by solving equation (4.56) and (4.51). Also the value of output voltage can be obtain the numerical solution.

4.7.1.4 Calculation of V_{IH}

When the input voltage is equal to V_{IH} the nMOS transistor operates in the linear region and the pMOS transistor operates in saturation. Applying KCL to the output node, we obtain

$$\frac{k_n}{2}\cdot\left[2\cdot\left(V_{GS,n} - V_{T0,N}\right)\cdot V_{DS,n} - V_{DS,n}^2\right] = \frac{k_p}{2}\cdot\left(V_{GS,p} - V_{T0,p}\right)^2 \qquad \text{...(4.57)}$$

Using equations (4.43) and (4.44), this expression can be rewritten as

$$\frac{k_n}{2}\cdot\left[2\cdot(V_{\text{in}}-V_{T0,n})\cdot V_{\text{out}}-V_{\text{out}}^2\right]=\frac{k_p}{2}\cdot(V_{\text{in}}-V_{DD}-V_{T0,p})^2 \quad ..(4.58)$$

Now, differentiate both sides of (4.58) with respect to V_{in}.

$$k_n\cdot\left[(V_{\text{in}}-V_{T0,n})\cdot\left(\frac{dV_{\text{out}}}{dV_{\text{in}}}\right)+V_{\text{out}}-V_{\text{out}}\cdot\left(\frac{dV_{\text{out}}}{dV_{\text{in}}}\right)\right]=k_p\cdot(V_{\text{in}}-V_{DD}-V_{T0,p}) \quad ...(4.59)$$

Substituting, $V_{\text{in}} = V_{IH}$ and $(dV_{\text{out}}/dV_{\text{in}}) = -1$ in (4.59), we obtain

$$k_n\cdot(-V_{IH}+V_{T0,n}+2V_{\text{out}}) = k_p\cdot(V_{IH}-V_{DD}-V_{T0,p}) \quad ...(4.60)$$

The critical voltage V_{IH} can now be found as a function of V_{out} as follow:

$$V_{IH}=\frac{V_{DD}+V_{T0,p}+k_R\cdot(2V_{\text{out}}+V_{T0,n})}{1+k_R} \quad ...(4.61)$$

Again, this equation must be solved simultaneously with the KCL equation (4.58) to obtain the numerical values of V_{IH} and V_{out}.

4.7.1.5 Calculation of V_{TH}

The inverter threshold voltage is defined as that voltage for which $V_{\text{th}} = V_{\text{in}} = V_{\text{out}}$. For $V_{\text{in}} = V_{\text{out}}$, both transistors are expected to be in saturation mode. So we can write the following KCL equation.

$$\frac{k_n}{2}\cdot(V_{GS,n}-V_{T0,n})^2=\frac{k_p}{2}\cdot(V_{GS,p}-V_{T0,p})^2 \quad ...(4.62)$$

Replacing $V_{GS,n}$ and $V_{GS,p}$ in (4.62) according to (4.43) and (4.44), we obtain

$$\frac{k_n}{2}\cdot(V_{\text{in}}-V_{T0,n})^2=\frac{k_p}{2}\cdot(V_{\text{in}}-V_{DD}-V_{T0,p})^2 \quad ...(4.63)$$

The correct solution for V_{in} for this equation is

$$V_{in}\cdot\left(1+\sqrt{\frac{k_p}{k_n}}\right)=V_{T0,n}+\sqrt{\frac{k_p}{k_n}}\cdot(V_{DD}+V_{T0,p}) \quad ...(4.64)$$

Finally, the inverter threshold (switching threshold) voltage V_{th} is found as

$$V_{\text{in}}=V_{\text{th}}=\frac{V_{T0,n}+\sqrt{\frac{1}{k_R}}\cdot(V_{DD}+V_{T0,p})}{\left(1+\sqrt{\frac{1}{k_R}}\right)} \quad ...(4.65)$$

- The variation of the inversion (switching) threshold voltage V_{th} as a function of the transconductance ratio K_R, and for fixed values of V_{DD}, $V_{T0,n}$ and $V_{T0,p}$ is shown in Fig. 4.23.

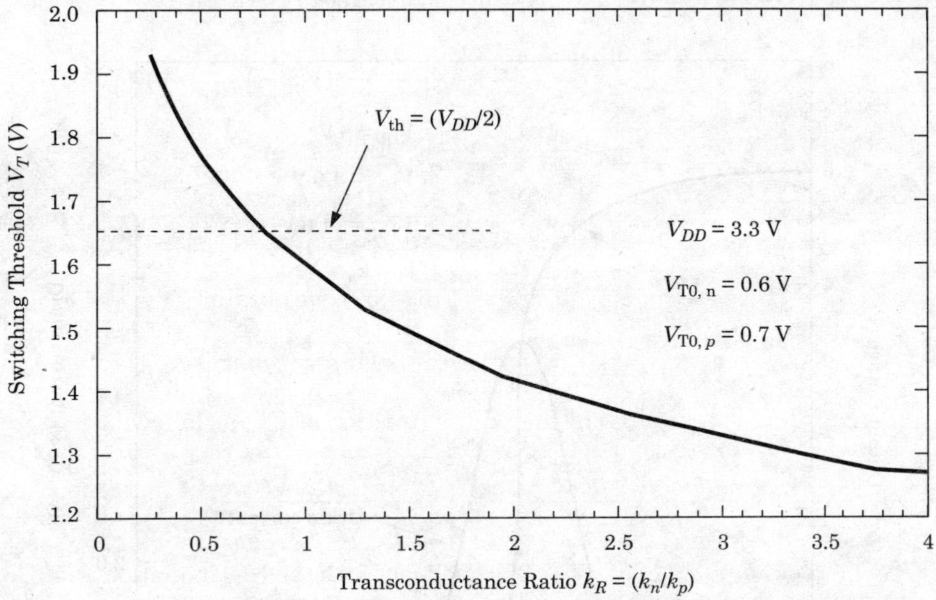

Fig. 4.23 Variation of the inversion threshold voltage as a function of K_R.

- The inverter threshold voltage is defined as $V_{th} = V_{in} = V_{out}$. However, when the input voltage is equal to V_{th}, we find that the output voltage can actually attain any value between $(V_{th} - V_{T0,n})$ and $(V_{th} - V_{T0,p})$, without violating the voltage conditions used in this analysis. This is due to the fact that the VTC segment corresponding to Region *C* in Fig. 4.22 becomes completely vertical if the channel-length modulation effect is neglected, i.e., if $\lambda = 0$. In more realistic cases with $\lambda > 0$, the VTC segment in Region *C* exhibits a finite, but very large, slope.
- It has already been established that the CMOS inverter does not draw any significant current from the power source, except for small leakage and sub-threshold currents, when the input voltage is either smaller than $V_{T0,n}$ or larger than $(V_{DD} + V_{T0,p})$. The nMOS and the pMOS transistors conduct a nonzero current, on the other hand, using low-to-high and high-to-low transitions, i.e., in Regions *B*, *C*, and *D*. It can be shown that the current being drawn from the power source during transition reaches its peak value when $V_{in} = V_{th}$. In other words, the maximum current is drawn when both transistors are operating in saturation mode. Figure 4.24 shows the voltage transfer characteristic of a typical CMOS inverter circuit and the power supply current, as a function of the input voltage.

4.7.1.6 Advantages of Depletion-Load *n*-Type MOSFET as a Load

1. The steady-state power dissipation of the CMOS inverter circuit is virtually negligible, except for small power dissipation due to leakage currents.
2. The voltage transfer characteristic (VTC) exhibits a full output voltage swing between 0 V and V_{DD}, and that the VTC transition is usually very sharp. So the VTC of the CMOS inverter resembles that of an ideal inverter.
3. Sharp transition between states, which increases the noise margins.

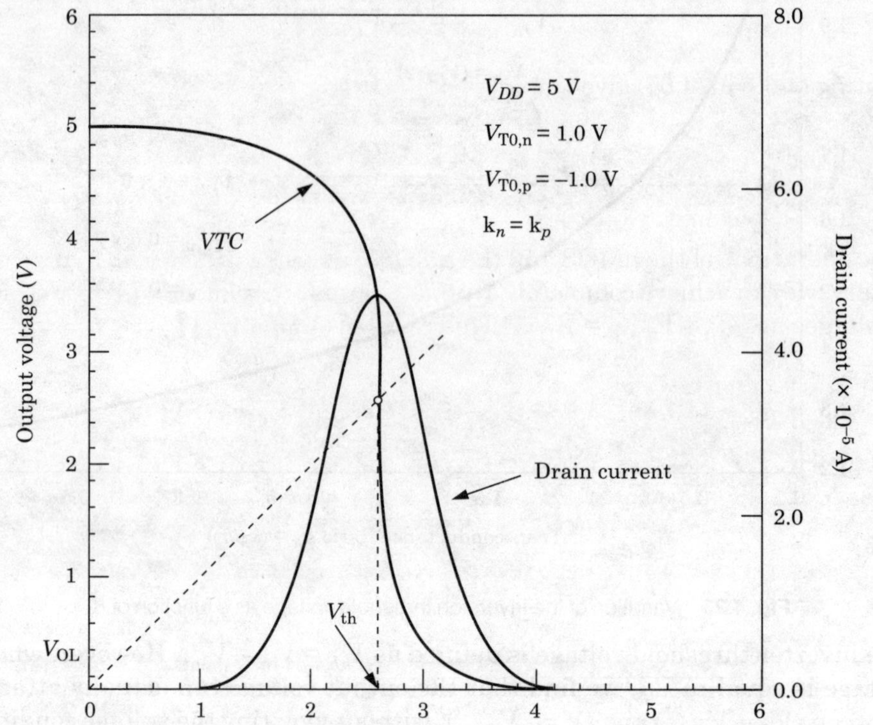

Fig. 4.24 Typical VTC and the power supply current of a CMOS inverter circuit.

4.7.2 Design Parameters of CMOS Inverters

The inverter threshold voltage V_{th} was identified as one of the most important parameters that characterize the steady-state input-output behavior of the CMOS inverter circuit. The CMOS inverter can, by virtue of its complementary push-pull operating mode, provide a full output voltage swing between 0 and V_{DD}, and therefore, the noise margins are relatively wide. Thus, the problem of designing a CMOS inverter can be reduced to setting the inverter threshold to a desired voltage value.

Given the power supply voltage V_{DD}, the nMOS and the pMOS transistor threshold voltages, and the desired inverter threshold voltage V_{th}, the corresponding ratio K_R can be found as follows. Reorganizing (4.65) yields

$$\sqrt{\frac{1}{k_R}} = \frac{V_{th} - V_{T0,n}}{V_{DD} + V_{T0,p} - V_{th}} \quad ...(4.66)$$

Now solve for K_R that is required to achieve the given V_{th}.

$$k_R = \frac{k_n}{k_p} = \left(\frac{V_{DD} + V_{T0,p} - V_{th}}{V_{th} - V_{T0,n}} \right)^2 \quad ...(4.67)$$

Recall that the switching threshold voltage of an *ideal* inverter is defined as

$$V_{th,\ \text{ideal}} = \frac{1}{2} \cdot V_{DD} \qquad \text{...(4.68)}$$

Substituting (4.68) in (4.67) gives

$$\left(\frac{k_n}{k_p}\right)_{\text{ideal}} = \left(\frac{0.5\ V_{DD} + V_{T0,\,p}}{0.5\ V_{DD} - V_{T0,\,n}}\right)^2 \qquad \text{...(4.69)}$$

Since the operations of the nMOS and the pMOS transistors of the CMOS inverter are fully complementary, we can achieve completely symmetric input-output characteristics by setting the threshold voltages as $V_{T0} = V_{T0,\,n} = V_{T0,\,p}$. This reduces (4.69) to:

$$\left(\frac{k_n}{k_p}\right)_{\substack{symmetric\\ thverter}} = 1 \qquad \text{...(4.70)}$$

$$\frac{k_n}{k_p} = \frac{u_n C_{ox} \cdot \left(\frac{W}{L}\right)_n}{\mu_p C_{ox} \cdot \left(\frac{W}{L}\right)_p} = \frac{\mu_n \cdot \left(\frac{W}{L}\right)_n}{\mu_p \left(\frac{W}{L}\right)_p} \qquad \text{...(4.71)}$$

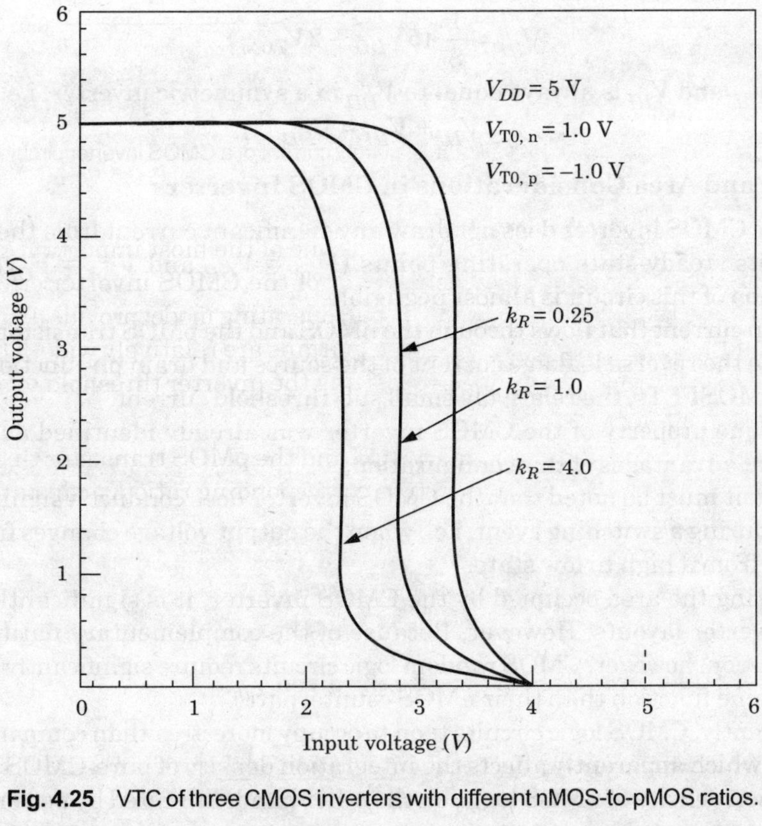

Fig. 4.25 VTC of three CMOS inverters with different nMOS-to-pMOS ratios.

Assume that the gate oxide thickness t_{ox}, and hence, the gate oxide capacitance C_{ox} have the same value for both nMOS and pMOS transistors. The unity-ratio condition (4.70) for the ideal symmetric inverter requires that

$$\frac{\left(\frac{W}{L}\right)_n}{\left(\frac{W}{L}\right)_p} = \frac{\mu_p}{\mu_n} \approx \frac{230\ \text{cm}^2/\text{V.s}}{580\ \text{cm}^2/\text{V.s}} \quad ...(4.72)$$

Hence,

$$\left(\frac{W}{L}\right)_p \approx 2.5\left(\frac{W}{L}\right)_n \quad ...(4.73)$$

The VTCs of three CMOS inverter circuits with different K_R ratios are shown in Fig. 4.24. It can be seen clearly that the inverter threshold voltage V_{th} shifts to lower values with increasing K_R ratio.

4.7.2.1 Symmetric CMOS Inverters

For a Symmetric CMOS inverter with $V_{T0,n} = V_{T0,p}$ and $K_R = 1$, therefore, the input-low level voltage is found from equation (4.56) as below:

$$V_{IL} = \frac{1}{8}\cdot(3V_{DD} + 2V_{T0,n}) \quad ...(4.74)$$

Also, the critical voltage V_{IH} is found as

$$V_{IH} = \frac{1}{8}\cdot(5V_{DD} - 2V_{T0,n}) \quad ...(4.75)$$

The sum of V_{IL} and V_{IH} is always equal to V_{DD} in a symmetric inverter. i.e,

$$V_{IL} + V_{IH} = V_{DD} \quad ...(4.76)$$

4.7.2.2 Power and Area Considerations in CMOS Inverters

1. Since the CMOS inverter does not draw any significant current from the power source in both of its steady-state operating points ($V_{out} = V_{OH}$ and $V_{out} = V_{OL}$), the *DC* power dissipation of this circuit is almost negligible.
2. The drain current that flows through the nMOS and the pMOS transistors in both cases is limited to the reverse leakage current of the source and drain pn-junctions, and in short-channel MOSFETs, the relatively small sub-threshold current.
3. This unique property of the CMOS inverter was already identified as one of the most important advantages of this configuration.
4. However, it must be noted that the CMOS inverter does conduct a significant amount of current during a switching event, i.e., when the output voltage changes from a low to high state, or from a high to low state.
5. Considering the area occupied by the CMOS inverter, it is significantly less compared other inverter layouts. However, Because of the complementary nature of this circuit configuration, however, CMOS random logic circuits require significantly more transistors for the same function than their nMOS counterparts.
6. Consequently, CMOS logic circuits tend to occupy more area than comparable nMOS logic circuits, which apparently affects the integration density of pure-CMOS logic. The actual integration density of nMOS logic, on the other hand, is limited by power dissipation and heat generation problems.

4.8 SWITCHING CHARACTERISTICS OF IN CMOS INVERTERS (UTU-2012)

4.8.1 Delay-Time Definitions

- In this topic we discussed about some commonly used delay time definitions. The input and output voltage waveforms of a typical inverter circuit are shown in Fig. 4.26(a). The propagation delay times τ_{PHL} and τ_{PLH} determine the input-to-output signal delay during the high-to-low and low-to-high transitions of the output, respectively.
- By definition, τ_{PHL} is the time delay between the $V_{50\%}$-transition of the *rising* input voltage and the $V_{50\%}$ -transition of *the falling* output voltage. Similarly, τ_{PLH} is defined as the time delay between the $V_{50\%}$ -transition of *the falling* input voltage and the $V_{50\%}$-transition of the *rising* output voltage.
- For the analysis of delay, we assumed that the input voltage waveform is an ideal step pulse with zero rise and fall times. Under this assumption, τ_{PHL} becomes the time required for the output voltage to fall from V_{OH} to the $V_{50\%}$ level, and τ_{PLH} becomes the time required for the output voltage to rise from V_{OL} to the $V_{50\%}$ level. The voltage point $V_{50\%}$ is defined as follows.

$$V_{50\%} = V_{OL} + \frac{1}{2}(V_{OH} - V_{OL}) = \frac{1}{2}(V_{OL} + V_{OH}) \qquad ...(4.77)$$

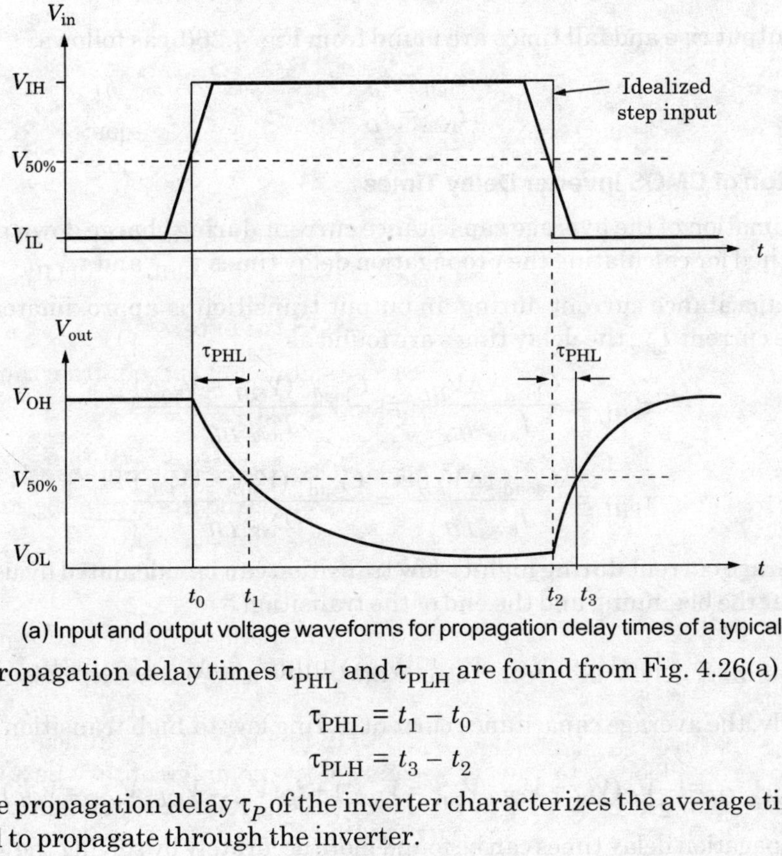

Fig. 4.26 (a) Input and output voltage waveforms for propagation delay times of a typical inverter.

Thus, the propagation delay times τ_{PHL} and τ_{PLH} are found from Fig. 4.26(a) as

$$\tau_{PHL} = t_1 - t_0$$

$$\tau_{PLH} = t_3 - t_2 \qquad ...(4.78)$$

The average propagation delay τ_P of the inverter characterizes the average time required for the input signal to propagate through the inverter.

$$\tau_P = \frac{\tau_{PHL} + \tau_{PLH}}{2} \qquad ...(4.79)$$

For the definition of output voltage rise and fall times, we refer to Fig. 4.26(b). The rise time τ_{rise} is defined here as the time required for the output voltage to rise from the $V_{10\%}$ level to $V_{90\%}$ level. Similarly, the fall time τ_{fall} is defined here as the time required for the output voltage to drop from the $V_{90\%}$ level to $V_{10\%}$ level. The voltage levels $V_{10\%}$ and $V_{90\%}$ are defined as

$$V_{10\%} = V_{OL} + 0.1 \cdot (V_{OH} - V_{OL}) \quad \text{...(4.80)}$$

$$V_{90\%} = V_{OL} + 0.9 \cdot (V_{OH} - V_{OL}) \quad \text{...(4.81)}$$

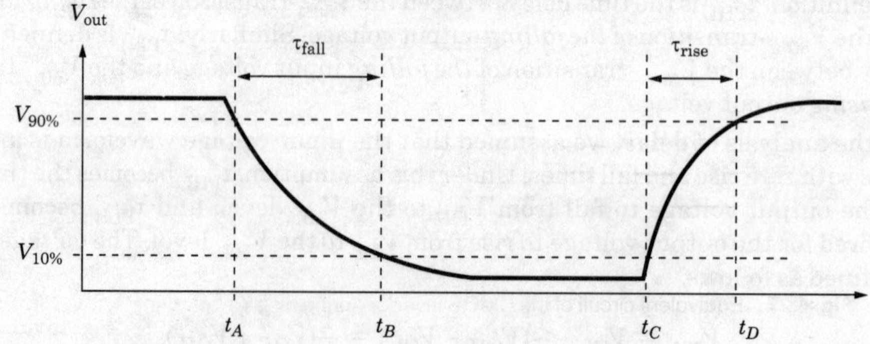

Fig. 4.26 (b) Output voltage rise and fall times.

Thus, the output rise and fall times are found from Fig. 4.26(b) as follows.

$$\tau_{fall} = t_B - t_A$$

$$\tau_{rise} = t_D - t_C \quad \text{...(4.82)}$$

4.8.2 Estimation of CMOS Inverter Delay Times

- The estimation of the average capacitance current during charge down and charge up is the method for calculating the propagation delay times τ_{PHL} and τ_{PLH}.
- If the capacitance current during an output transition is approximated by a constant average current I_{avg} the delay times are found as

$$\tau_{PHL} = \frac{C_{load} \cdot \Delta V_{HL}}{I_{avg,\,HL}} = \frac{C_{load} \cdot (V_{OH} - V_{50\%})}{I_{avg.\,HL}} \quad \text{...(4.83)}$$

$$\tau_{PHL} = \frac{C_{load} \cdot \Delta V_{LH}}{I_{avg,\,LH}} = \frac{C_{load} \cdot (V_{50\%} - V_{OL})}{I_{avg,\,LH}} \quad \text{...(4.84)}$$

- The average current during high-to-low transition can be calculated by using the current values at the beginning and the end of the transition.

$$I_{avg,\,HL} = \frac{1}{2}[i_C(V_{in} = V_{OH}, V_{out} - V_{OH}) + i_C(V_{in} = V_{OH}, V_{out} - V_{50\%})] \quad \text{...(4.85)}$$

- Similarly, the average capacitance current during low-to-high transition is

$$I_{avg,\,LH} = \frac{1}{2}[i_C(V_{in} = V_{OL}, V_{out} = V_{50\%}) + i_C(V_{in} = V_{OL}, V_{out} = V_{OL})] \quad \text{...(4.86)}$$

- The propagation delay times can be found more accurately by solving the state equation of the output node in the time domain. The differential equation associated with the output node is given below. Note that the capacitance current is also a function of the output voltage.

$$C_{\text{load}} \frac{dV_{\text{out}}}{dt} = i_C = i_{D,p} - i_{D,n} \quad \text{...(4.87)}$$

- First, we consider the rising-input case for a CMOS inverter. Initially, the output Voltage is assumed to be equal to V_{OH}. When the input voltage switches from low (V_{OL}) to high (V_{OH}), the nMOS transistor is turned on and it starts to discharge the load capacitance. At the same time, the pMOS transistor is switched off; thus,

$$i_{D,p} \approx 0 \quad \text{..(4.88)}$$

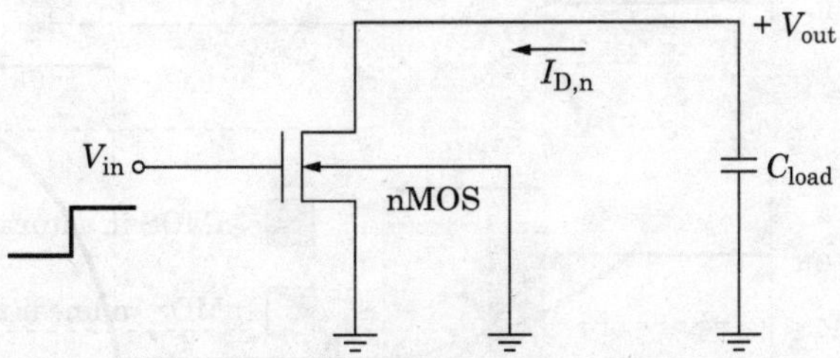

Fig. 4.27 Equivalent circuit of the CMOS inverter during high-to-low output transition.

$$C_{\text{load}} \frac{dV_{\text{out}}}{dt} = -i_{D,n} \quad \text{...(4.89)}$$

- In other types of inverter circuits, such as the resistive-load inverter or the depletion-load inverter, the load device continues to conduct a nonzero current when the input is switched from low to high. However, the load current is usually negligible in comparison to the driver current.

4.9 CMOS DYNAMIC CHARACTERISTICS (DYNAMIC RESPONSE OF CMOS INVERTER)

4.9.1 Calculation for τ_{PHL}

The, input and output voltage waveforms during this high-to-low transition are illustrated in Fig. 4.28. When the nMOS transistor starts conducting, it initially operates in the saturation region. When the output voltage falls below ($V_{DD} - V_{T,n}$), the nMOS transistor starts to conduct in the linear region. These two operating regions are also shown in Fig. 4.28(a).

First, consider the nMOS transistor operating in saturation.

$$i_{D,n} = \frac{k_n}{2}(V_{\text{in}} - V_{T,n})^2$$

$$= \frac{k_n}{2}(V_{OH} - V_{T,n})^2, \quad \text{for} \quad V_{OH} - V_T, n < V_{\text{out}} \le V_{OH} \quad \text{...(4.90)}$$

Since the saturation current is practically independent of the output voltage (neglecting channel-length modulation), the solution (4.90) in the time interval between t_0 and t_1, can be found as

$$\int_{t=t_0}^{t=t_1'} dt = -C_{\text{load}} \int_{V_{\text{out}}=V_{OH}}^{V_{\text{out}}=V_{OH}-V_{T,n}} \left(\frac{1}{i_{D,n}}\right) dV_{\text{out}} = \frac{2\,C_{load}}{k_n(V_{OH} - V_{T,n})^2} \int_{V_{\text{out}}=V_{OH}}^{V_{out}=V_{OH}-V_{T,n}} dV_{\text{out}} \quad \text{...(4.91)}$$

Evaluating this simple integral yields

$$t'_1 - t_0 = \frac{2C_{\text{load}}\,V_{T,n}}{k_n(V_{OH} - V_{T,n})^2} \quad \text{...(4.92)}$$

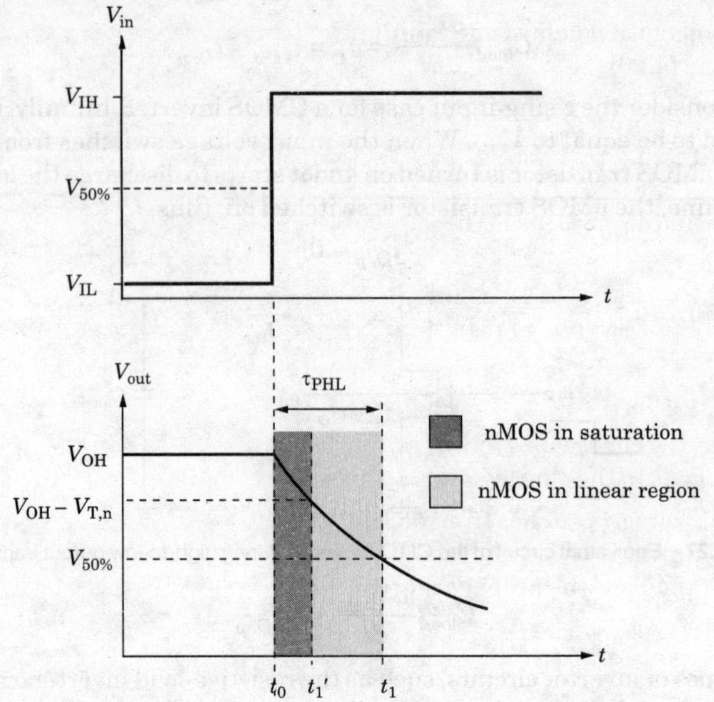

Fig. 4.28 (a) Input and output voltage waveforms during high-to-low transition.

At $t = t'_1$, the output voltage will be equal to $(V_{DD} - V_{Tn})$ and the transistor will be at the saturation-linear region boundary. Next, consider the nMOS transistor operating in the linear region.

$$i_{D,n} = \frac{k_n}{2}\left[2(V_{in} - V_{T,n})V_{out} - V_{out}^2\right]$$

$$= \frac{k_n}{2}\left[2(V_{OH} - V_{T,n})V_{out} - V_{out}^2\right], \quad \text{for} \quad V_{out} \le V_{OH} - V_{T,n} \quad \text{...(4.93)}$$

The solution of (4.89) in the time interval between t', and t can be found as

$$\int_{t=t_1'}^{t=t_1} dt = -C_{load} \int_{V_{out} - V_{OH} - V_{T,n}}^{V_{out} = V_{50\%}} \left(\frac{1}{i_{D,n}}\right) dV_{out}$$

$$= -2C_{load} \int_{V_{out} - V_{OH} - V_{T,n}}^{V_{out} = V_{50\%}} \left(\frac{1}{k_n\left[2(V_{OH} - V_{T,n})V_{out} - V_{out}^2\right]}\right) dV_{out} \quad \text{...(4.94)}$$

Evaluating this integral yields

$$t_1 - t'_1 = \frac{-2\,C_{load}}{k_n} \cdot \frac{1}{2(V_{OH} - V_{T,n})} \ln\left(\frac{V_{out}}{2(V_{OH} - V_{T,n}) - V_{out}}\right)\Bigg|_{V_{out} = V_{OH} - V_{T,n}}^{V_{out} = V_{50\%}} \quad \text{...(4.95)}$$

$$t_1 - t'_1 = \frac{C_{load}}{k_n(V_{OH} - V_{T,n})} \ln\left(\frac{2(V_{OH} - V_{T,n}) - V_{50\%}}{V_{50\%}}\right) \quad \text{...(4.96)}$$

Finally, the propagation delay time for high-to-low output transition (τ_{PHL}) can be found by combining (4.92) and (4.96):

$$\tau_{PHL} = \frac{C_{load}}{k_n(V_{OH} - V_{T,n})}\left[\frac{2V_{T,n}}{V_{OH} - V_{T,n}} + \ln\left(\frac{4(V_{OH} - V_{T,N})}{V_{OH} + V_{OL}} - 1\right)\right] \quad ...(4.97)$$

For $V_{OH} = V_{DD}$ and $V_{OL} = 0$, as is the case for the CMOS inverter, (4.97) becomes:

$$\tau_{PHL} = \frac{C_{load}}{k_n(V_{DD} - V_{T,n})}\left[\frac{2V_{T,n}}{V_{DD} - V_{T,n}} + \ln\left(\frac{4(V_{DD} - V_{T,n})}{V_{DD}} - 1\right)\right] \quad ...(4.98)$$

4.9.2 Calculation for τ_{PLH}

In a CMOS inverter, the charge up event of the output load capacitance for falling input transition is completely analogous to the charge-down event for rising input.

When the input voltage switches from high (V_{OH}) to Low (V_{OL}), the nMOS transistor is cutoff, and the load capacitance is being charged up through the pMOS transistor. Following a very similar derivation produces, the propagation delay time τ_{PLH} can be found with the Fig. 4.28(b).

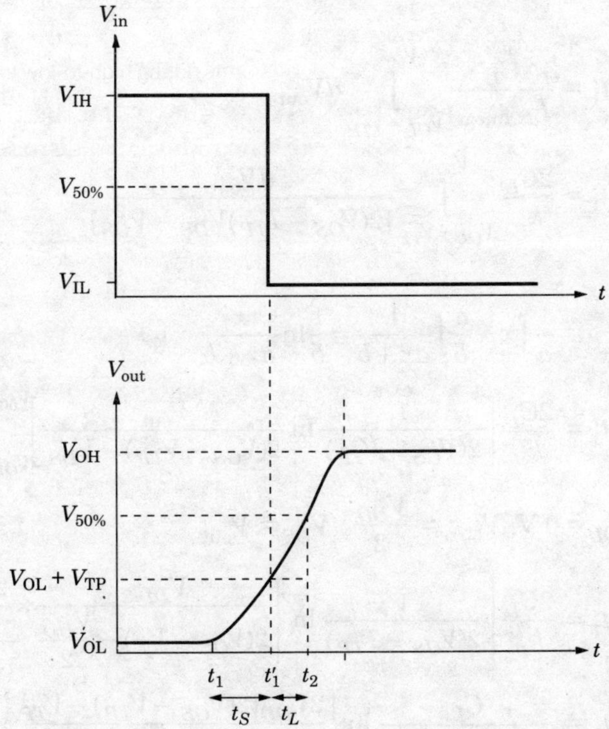

Fig. 4.28 (b) Input and output voltage waveforms during Low to High transition

When the output voltage grow over than ($V_{OL} + V_T$), the pMOS transistor starts to conduct in the linear region and when the output voltage stands between ($V_{OL} + V_T$) to V_{DL}, the pMOS transistor works in saturation region.

$$\tau_{PLH} = (t_2 - t_1') + (t_1' - t_1)$$

$\Downarrow$ Due to Linear $\quad \Downarrow$ Due to Saturation

We know that, the current across the load capacitor is given by

$$I = C_L \frac{dV}{dt}$$

$$I\,dt = C_L\,dV$$

Case I: When saturation Region.

$$\int_{t_1}^{t_1'} dt = \frac{C_L}{I_{ds}} \int_{V_{OL}}^{V_{OL}+V_{TP}} dV_{out}$$

$$t_1' - t_1 = \frac{2C_L}{k_p(V_{GS}-V_T)} \int_{V_{DL}}^{V_{OL}+V_{TP}} dV_{out}$$

$$t_1' - t_1 = \frac{2C_L\,V_T}{k_p(V_{GS}-V_{TP})^2} \quad \text{...(4.99)}$$

Similarly,

$$t_2 - t_1' = \frac{C_L}{I_{ds,\text{linear}}} \int_{V_{OL}+V_{TP}}^{V_{50\%}} dV_{out}$$

$$t_2 - t_1' = \frac{2C_L}{k_p} \int_{V_{OL}+V_{TP}}^{V_{50\%}} \frac{dV_{out}}{[2(V_{GS}-T_{TP})V_{DS} - V_{DS}^2]}$$

We know that

$$\int \frac{dx}{ax^2+bx} = \frac{1}{a}\int x - \frac{a}{b}\int \frac{1}{ax+b} = \frac{1}{b}\ln\frac{x}{ax+b}$$

so, $$t_2 - t_1' = \frac{2C_L}{k_p}\left[\frac{1}{2(V_{GS}-V_{TP})}\ln\frac{V_{out}}{2(V_{GS}-V_{TP})-V_{DS}}\right]_{V_{OL}+V_T}^{V_{50\%}}$$

Here, $$V_{OL} = 0\text{ V},\ V_{50\%} = \frac{V_{DD}}{2},\ V_{DS} = V_{out}$$

$$t_2 - t_1' = \frac{2C_L}{k_p}\left[\frac{1}{2(V_{GS}-T_{TP})}\ln\left\{\frac{V_{DD/2}}{2(V_{GS}-V_T)-\dfrac{V_{DD}}{2}} \times \frac{2(V_{GS}-V_{TP})-V_{TP}}{V_{TP}}\right\}\right]$$

$$t_2 - t_1' = \frac{C_L}{k_p(V_{GS}-V_{TP})}\ln\left\{\frac{2V_{DD}[2(V_{GS}-V_{TP})-V_{TP}}{2V_T[4(V_{GS}-V_{TP})-V_{DD}}\right\}$$

$$t_2 - t_1' = \frac{C_L}{k_p(V_{GS}-V_{TP})}\ln\left[\frac{V_{DD}\cdot V_T\left[2\left(\dfrac{V_{GS}-V_{TP}}{V_{TP}}\right)\right]-1}{V_T\cdot V_{DD}\left[4\left(\dfrac{V_{GS}-V_{TP}}{V_{DD}}\right)-1\right]}\right]$$

$$t_2 - t_1' = \frac{C_L}{k_p(V_{GS} - V_{TP})} \ln \left[\frac{2\left(\frac{V_{GS} - V_{TP}}{V_{TP}}\right) - 1}{4\left(\frac{V_{GS} - V_{TP}}{V_{DD}}\right) - 1} \right] \quad ...(4.100)$$

From Equation (4.99) and (4.100)

$$\tau_{PLH} = t_2 - t_1' + t_1' + t_1$$

$$\tau_{PLH} = \frac{C_L}{k_p(V_{GS} - V_{TP})} \left[\ln \left\{ \frac{2\left(\frac{V_{GS} - V_{TP}}{V_{TP}}\right) - 1}{4\left(\frac{V_{GS} - V_{TP}}{V_{DD}}\right) - 1} \right\} + \frac{2V_T}{(V_{GS} - V_{TP})} \right]$$

Put $V_{GS} = V_{DD}$

$$\tau_{PLH} = \frac{C_L}{k_p(V_{DD} - V_{TP})} \left[\frac{2V_{TP}}{(V_{DD} - V_{TP})} + \ln \left\{ 4\left(\frac{V_{DD} - V_{TP}}{V_{DD}}\right) - 1 \right\} \right] \quad ...(4.101)$$

$$\left(\text{if } \frac{V_{GS} - V_{TP}}{V_P} \approx 0.5 \right)$$

Comparing the delay expression (4.101) with (4.99), we can see that the sufficient conditions for *balanced* propagation delays, i.e., for $\tau_{PHL} = \tau_{PLH}$ in a CMOS inverter are:

$$V_{T,n} = |V_{T,p}| \quad \text{and} \quad k_n = k_p \quad (\text{or } W_p/W_n = m_p/u_n)$$

- **The calculation of τ_{PLH} in different types of inverters depends on the load device and its operation and the analysis procedure is very similar to the CMOS case.**
- **We will consider the nMOS depletion-load inverter case as an example in the following. When the input voltage switches from high to low, the enhancement-type nMOS driver transistor is turned off. The output load capacitance is then being charged up through the depletion-type load transistor. The differential equation describing this event is**

$$C_{\text{load}} \frac{dV_{\text{out}}}{dt} = i_{D,\text{load}}(V_{\text{out}}) \quad ...(4.102)$$

- **Note that the load device is initially in saturation, and enters the linear region when the output voltage rises above ($V_{DD} + V_{T,\text{load}}$), where $V_{T,\text{load}} < 0$.**

$$i_{Dl,\text{load}} = \frac{k_{n,\text{load}}}{2}\left(|V_{T,\text{load}}|\right)^2, \quad \text{for} \quad V_{\text{out}} \le V_{DD} - |V_{T,\text{load}}| \quad ...(4.103)$$

$$i_{D,\text{load}} = \frac{k_{n,\text{load}}}{2}\left[2|V_{T,\text{load}}|(V_{DD} - V_{\text{out}}) - (V_{DD} - V_{\text{out}})^2\right],$$

$$\text{for} \quad V_{\text{out}} > V_{DD} - |V_{T,\text{load}}| \quad ...(4.104)$$

The delay time τ_{PLH} can be found as follows:

$$\tau_{PLH} = C_{\text{load}} \left[\int_{V_{\text{out}} = V_{OL}}^{V_{\text{out}} - V_{DD} - |V_{T,load}|} \left(\frac{dV_{\text{out}}}{i_{D,\text{load}}(\text{sat})} \right) + \int_{V_{\text{out}} = V_{DD} - |V_{T,\text{load}}|}^{V_{\text{out}} = V_{50\%}} \left(\frac{dV_{\text{out}}}{i_{D,\text{load}}(\text{linear})} \right) \right] \quad ...(4.105)$$

$$\tau_{PLH} = \frac{C_{load}}{k_{n,load}|V_{T,load}|}\left[\frac{2(V_{DD}-|V_{T,load}|-V_{OL})}{|V_{T,load}|} + \ln\left(\frac{2|V_{T,load}|-(V_{DD}-V_{50\%})}{V_{DD}-V_{50\%}}\right)\right] \quad ...(4.106)$$

- In this section, we assume that the input signal waveform is a step pulse with zero rise and fall times for the delay time derivations.
- Now, we consider the case where the input voltage waveform is not an ideal (step) pulse waveform, but has finite rise and fall times, τ_f and τ_r.
- The exact calculation of the output voltage delay times is more complicated under this more realistic assumption, since both the nMOS transistor and the pMOS transistor conduct current during the charge-up and charge-down events. To simplify the estimation of the actual propagation delays, we can utilize the propagation delay times calculated under the step-input assumption, using the following empirical expressions:

$$\tau_{PHL}(\text{actual}) = \sqrt{\tau_{PHL}^2(\text{step input}) + \left(\frac{t_r}{2}\right)^2} \quad ...(4.107)$$

$$\tau_{PLH}(\text{actual}) = \sqrt{t_{PLH}^2(\text{step input}) + \left(\frac{t_f}{2}\right)^2} \quad ...(4.108)$$

- Here, τ_{PHL}*(step input)* and τ_{PLH}*(step input)* denote the propagation delay time values calculated assuming a step pulse input waveform at the input, i.e., using (4.100) and (4.101). While the expressions given above are purely empirical, they provide a simple estimation of how much the propagation delays are *increased as* a result of non-zero input rise and fall times.

4.10 INVERTER DESIGN WITH DELAY CONSTRAINTS

- The delay (timing) specifications is one of the most fundamental issues in digital circuit design which ultimately determine the overall performance of complex systems. In the following, we will consider some of the fundamental aspects of CMOS inverter design based on timing constraints.
- In the most general case, we must take into account that the combined load capacitance of the inverter is also a function of transistor sizes as shown in Fig. 4.29. The load capacitance C_{load} consists of intrinsic components (inter connect, wiring capacitance and fan-out capacitance which are usually independent of transistor dimensions of the inverter under consideration).
- If C_{load} mainly consist of extrinsic components and this load-capacitance can be estimated accurately and independently of transistor dimensions, then the problem of inverter design can be reduced with a required delay value of τ^*_{PHL}, then (*W*/*L*) ratio of the nMOS transistor can be found as:

$$\left(\frac{W}{L}\right)_n = \frac{C_{load}}{\tau^*_{PHL}\mu_n C_{ox}(V_{DD}-V_{T,n})}\left[\frac{2V_{T,n}}{V_{DD}-V_{T,n}} + \ln\left(\frac{4(V_{DD}-V_{T,n})}{V_{DD}}-1\right)\right] \quad ...(4.109)$$

Similarly,

The (*W*/*L*) ratio of the pMOS transistor to satisfy a given target value of τ^*_{PLH} can be calculated as:

$$\left(\frac{W}{L}\right)_p = \frac{C_{load}}{\tau^*_{PLH}\mu_p C_{ox}(V_{DD}-|V_{T,p}|)}\left[\frac{2|V_{T,p}|}{V_{DD}-|V_{T,p}|} + \ln\left(\frac{4(V_{DD}-|V_{T,P}|}{V_{DD}}-1\right)\right] \quad ...(4.110)$$

- In most cases, the transistor sizes found from delay requirements must also meet other design criteria such a noise margin and logic inversion threshold. The extrinsic components in not very sensitive to device dimensions.
- Yet in most cases, we have to take into account that intrinsic components of C_{load} are increasing functions of the device dimensions, W_n and W_p. We know that the components of the output load capacitance becomes.

$$C_{load} = C_{gd,n}(W_n) + C_{gd,p}(W_p) + C_{db,n}(W_n) + C_{db,P}(W_P) + C_{int} + C_g$$

$$C_{load} = f(W_n, W_p) \quad \text{...(4.111)}$$

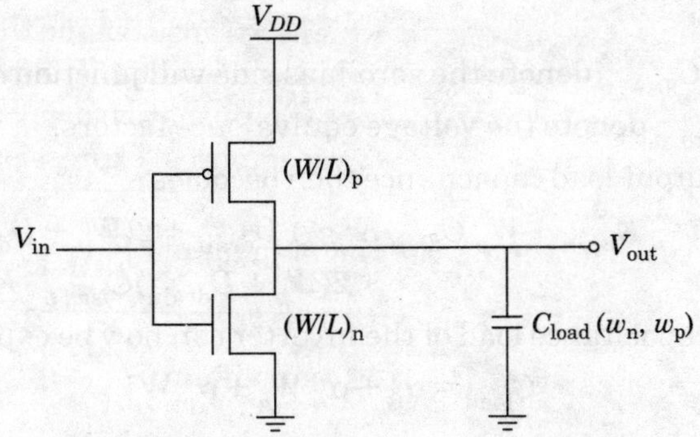

Fig. 4.29 General circuit structure of inverter.

- The fan-out capacitance C_g is also a function of the device dimensions in the next-stage gate(s). The fan-out capacitance C_g will be treated as a constant parameter which is independent of the device dimensions to driving gate. Any effort to increase the channel width of nMOS and pMOS transistors in order to reduce delay will increase the intrinsic components of the load capacitance.
- To gain some insight into the transistor sizing problem under delay constraint and to analyze the interaction of various design parameters, we will consider the simplified mask layout of the CMOS inverter is shown in Fig. 4.30.

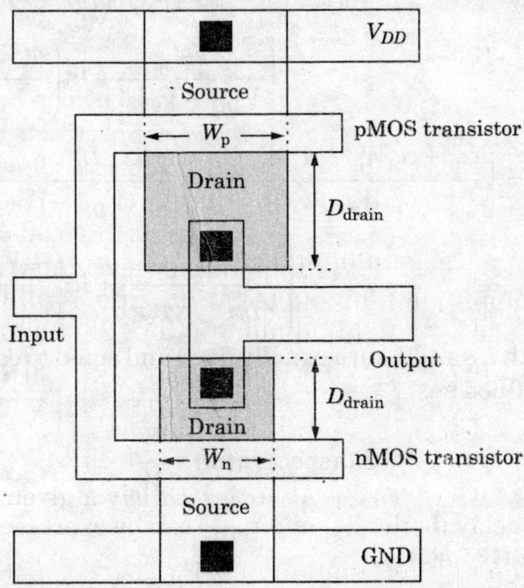

Fig. 4.30 Simplified CMOS inverter mask layout used for delay analysis.

- The diffusion areas of both nMOS and pMOS transistors have a simple rectangular geometry and the drain region length D_{drain} is assumed to be same for both devices. The relatively small gate-to-drain capacitance $C_{gd,\,n}$ and $C_{gd,\,p}$ will be neglected in this analysis. The drain parasitic capacitances can be found as:

$$C_{db,\,n} = W_n\, D_{drain}\, C_{j0,\,n}\, K_{eq,\,n} + 2(W_n + D_{drain})\, C_{jsw,\,n}\, K_{eq,\,n} \quad ...(4.112)$$

$$C_{db,\,p} = W_p\, D_{drain}\, C_{j0,\,p}\, K_{eq,\,p} + 2(W_p + D_{drain})\, C_{jsw,\,p}\, K_{eq,\,p} \quad ...(4.113)$$

where

⇒ $C_{j0,\,n}$ and $C_{j0,\,p}$ denote the zero-bias junction capacitance for n-type and p-type diffusion regions.

⇒ $C_{j0sw,\,n}$ and $C_{j0sw,\,p}$ denote the zero-bias side will junction capacitances.

⇒ $K_{eq,\,n}$ and $K_{eq,\,p}$ denote the voltage equivalence factors.

- The combined output load capacitance then becomes

$$C_{load} = (Wn\, C_{j0,\,n}\, K_{eq,\,n} + W_p\, C_{j0,\,p}\, K_{eq,\,p})\, D_{drain} + 2(W_n + D_{drain})C_{jsw,\,n}\, K_{eq,\,n} + 2(W_p + D_{drain})C_{jsw,\,p} \cdot K_{eq,\,p} + C_{int} + C_g \quad ...(4.114)$$

Thus, the total capacitance load of the inverter can now be expressed as

$$C_{load} = \alpha_0 + \alpha_n\, W_n + \alpha_p\, W_p$$

where,

$$\alpha_0 = 2\,\text{Drain}(C_{jsw,\,n}\, K_{eq,\,n} + C_{jsw,\,p}\, K_{eq,\,p}) + C_{int} + C_g$$

$$\alpha_n = K_{eq,\,n}(C_{j0,\,n}\, D_{drain} + 2\, C_{jsw,\,n})$$

$$\alpha_p = K_{eq,\,p}(C_{j0,\,p}\, D_{drain} + 2\, C_{jsw,\,p})$$

Using (4.114), the propagation delay expression (4.98) and (4.101) for falling and using output transitions are rewritten as,

$$\tau_{PHL} = \left(\frac{\alpha_0 + \alpha_n W_n + \alpha_p W_p}{W_n}\right)\left(\frac{L_n}{\mu_n\, C_{ox}(V_{DD} - V_{T,\,n})}\right) \times \left[\frac{2V_{T,\,n}}{V_{DD} - V_{T,\,n}} + \ln\left(\frac{4(V_{DD} - V_{T,\,n})}{V_{DD}} - 1\right)\right] \quad ...(4.115)$$

$$\tau_{PLH} = \left(\frac{\alpha_0 + \alpha_n W_n + \alpha_p W_p}{W_n}\right)\left(\frac{L_p}{\mu_p\, C_{ox}(V_{DD} - |V_{T,\,p}|)}\right) \times \left[\frac{2\,|V_{T,\,P}|}{V_{DD} - |V_{T,\,P}|} + \ln\left(\frac{4(V_{DD} - V_{T,\,P})}{V_{DD}} - 1\right)\right] \quad ...(4.116)$$

- The channel length L_N and L_P are usually fixed and equal to each other. Let that transistor aspect ratio be defined as;

$$R(\text{aspect ratio}) \equiv \frac{W_p}{W_n} \quad ...(4.117)$$

- The propagation delay (both τ_{PLH} and τ_{PHL}) can be expressed as a function of the single design parameter (W_n or W_p).

$$\tau_{PHL} = \Gamma_n\left(\frac{\alpha_n + (\alpha_n + R\alpha_p)W_n}{W_n}\right) \quad ...(4.118)$$

$$\tau_{PLH} = \Gamma_n\left(\frac{\alpha_n + \left(\frac{\alpha_n}{R} + \alpha_p\right)W_p}{W_p}\right) \quad ...(4.119)$$

where,

$$\Gamma_n = \left[\frac{L_n}{\mu_n C_{ox}(V_{DD} - V_{T,n})}\right]\left[\frac{2\,V_{T,n}}{V_{DD} - V_{T,n}} + \ln\left(\frac{4(V_{DD} - V_{T,n})}{V_{DD}} - 1\right)\right] \quad ...(4.120)$$

$$\Gamma_p = \left[\frac{L_p}{\mu_n C_{ox}(V_{DD} - |V_{T,p}|)}\right]\left[\frac{2\,V_{T,p}}{V_{DD} - (V_{T,p})} + \ln\left(\frac{4(V_{DD} - |V_{T,p}|)}{V_{DD}} - 1\right)\right] \quad ...(4.121)$$

- Given target delay value τ^*_{PHL} and τ^*_{PLH}. It can be seen that increasing W_n and W_p to reduce the propagation delay times will have a diminishing influence upon delay beyond certain values and the delay values will approach a limit value for value W_n and W_p. The limit delay vales can be found as;

$$\tau_{PHL}^{limit} = \Gamma_n(\alpha_n + R\alpha_p) \quad ...(4.122)$$

$$\tau_{PLH}^{limit} = \Gamma_p\left(\alpha_p + \frac{\alpha_n}{R}\right) \quad ...(4.123)$$

- The propagation delay times of a CMOS inverter cannot be reduced beyond these limit values. The propagation delay limit is independent of the extrinsic — capacitance components, C_{int} and C_g.
- If the extrinsic components dominate that total load capacitance, then delay reduction can be achieved for wider range of W_n and W_p.
- If the intrinsic component is dominant, then the speed limit is reached for smaller values of W_n and W_p.

4.11 CMOS INVERTER LAYOUT DESIGN

- In this topic, we will discuss about the mask layout design of a CMOS inverter. Here, we consider the design of inverter with minimum-size transistors.
- The width of the active area is determined by the minimum diffusion contact size and the minimum separation from diffusion contact to both active area edges. The width of the polysilicon line over the active area (which is the gate of the transistor) is typically taken as the minimum poly width (Fig. 4.31). Then, the minimum overall length of the active area is determined by the following sum: (minimum polysilicon width) + 2 × (minimum poly-to-contact spacing) + 2 × (minimum contact size) + 2 × (minimum spacing from contact to active area edge).
- The pMOS transistor must be placed in an n-well region, and the minimum size of the n-well is dictated by the pMOS active area and the minimum n-well overlap over n^+.
- The distance between the nMOS and the pMOS transistor is determined by the minimum separation between the n^+ active area and the n-well (as Fig. 4.32). The poly-silicon gates of the nMOS and the pMOS transistors are usually aligned.
- All the signals are preferably connected by the metal lines and metal-poly-silicon contacts are used to provide the electrical connection between the two layers.

- The last step for the mask layout is the local interconnections in metal for the output node, V_{DD} and G_{ND} contacts (as Fig. 4.33).
- The dimensions of metal lines in a mask layout are usually dictated by the minimum metal width and the minimum metal separation. The n-well region has a V_{DD} contact for proper biasing.

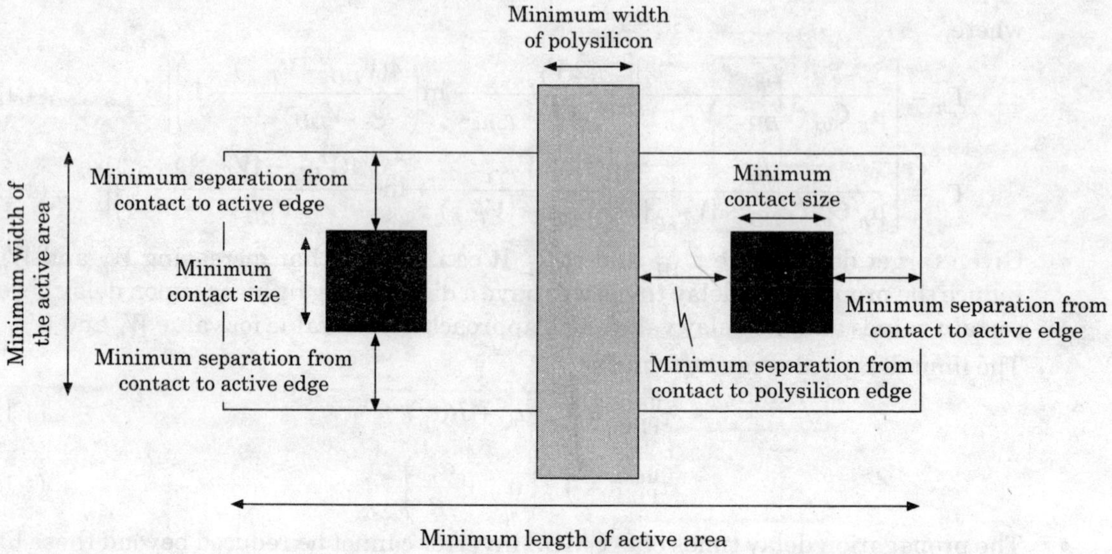

Fig. 4.31 Design rules which determine the dimensions of a minimum-size transistor.

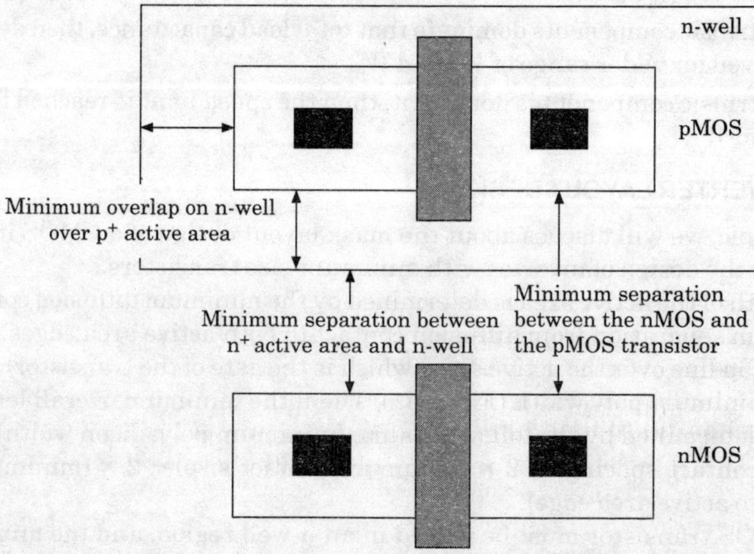

Fig. 4.32 Design rules which determine the separation between the nMOS and the pMOS transistor of the CMOS inverter.

- The layout design rules having a set of limitations for the mask geometry. The full-custom layout design process still allows a large number of variations in terms of device sizing, the placement of individual devices, and the routing of interconnections between the devices.

- Depending on the dominant design criteria and design constraints (minimization of overall silicon area, minimization of delay times, placement of input/output pins, etc.), one can choose a certain mask layout design over other alternatives. The number of layout possibilities also tends to increase with circuit complexity, i.e., with the number of transistors involved in the design.

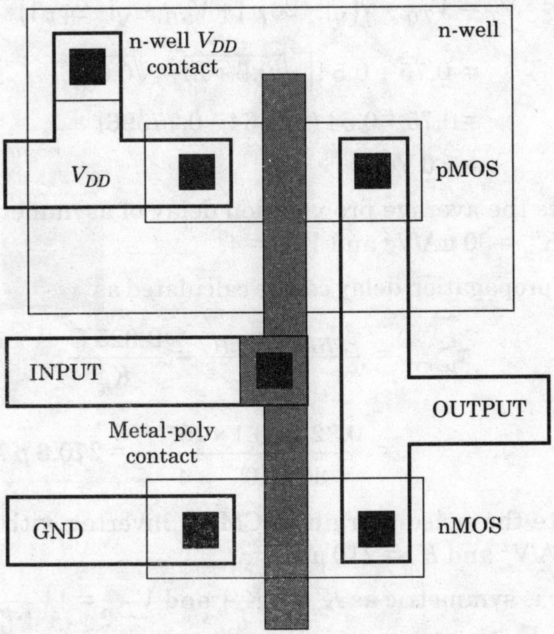

Fig. 4.33 Complete mask layout of the CMOS inverter.

POINTS TO BE REMEMBER

1. The inverter threshold voltage is $V_{Th} = V_{in} = V_{out}$.
2. Noise immensity of the digital circuit increases with noise margin.
3. The slope of the voltage transfer characteristics is equal to –1.
4. Guard rings prevent the formation of parasitic transistors and contact cut.
5. In nMOS inverter with resistors pull-up or resistive load inverter, high current required for fast switching, which can be achieved by smaller value of R_L. But for large noise margin, larger value of resistance is required. To achieve, this it is suggested to use current source as pull-up in place of resistance. This is trade-off between delay and power, for fast operation one needs small resistance whereas for low power, larger resistance in required.
6. In all calculations of V_{IH}, V_{IL}, V_{OL} and V_{Th}, we are not considering the body effect in any transistor. If we are considering the short channel advices then current will be saturated due to velocity saturation effect and we have to consider the velocity saturated current in both transistors.
7. The mobility of the respective transistors depends on the surface doping concentration temperature, electric field and also substrate concentrate the capacitor.
8. The gate response is determined by the lime it takes to charge the capacitor.

SOLVED PROBLEMS

Example 4.1. A MOS transistor has a threshold voltage of 0.75 V while the body effect coefficient equal to 0.54. Compute the threshold voltage for $V_{SB} = 5V$ and $2\phi_F = -0.6$ V.

Solution: We know that

$$V_T = V_{T0} + \gamma\left(\sqrt{|-2\phi_F| + V_{SB}} - \sqrt{|-2\varphi_F|}\right)$$

$$= 0.75 + 0.54\left[\sqrt{(0.6+5)} - \sqrt{0.6}\right]$$

$$= 0.75 + 0.54\,(2.3564 - 0.77596)$$

$$= 1.6 \text{ Volts}$$

Example 4.2. Calculate the average propagation delay of asymmetrical CMOS inverter with $C = 0.1$ pF. Given that $K'_n = 30\ \mu A/V^2$ and $W/L = 4$. **(UTU-2012)**

Solution: The average propagation delay can be calculated as

$$\tau_{p,\,avg} = \frac{\tau_{PHL} + \tau_{PLH}}{2} = \frac{0.325\,C}{K_n}$$

$$= \frac{0.325 \times 0.1 \times 10^{-12}}{30 \times 10^{-6} \times 4} = 270.8\ p \text{ sec.}$$

Example 4.3. Calculate the noise margin of CMOS inverter with $V_{DD} = 5$ V, $V_{T0,\,n} = 1$ V, $V_{T,\,p} = -1$ V, $K_n = 210\ \mu A/V^2$ and $K_p = 210\ \mu A/V^2$. **(UTU-2012)**

Solution: The inverter is symmetric as $K_n = |K_p|$ and $V_{T,\,n} = |V_{T,\,p}|$

So, the noise margin is given by

$$NM_L = V_{IL} - V_{OL} = V_{IL}$$

$$NM_H = V_{OH} - V_{IH} = V_{DD} - V_{IH}$$

and
$$V_{IL} = \frac{1}{8}(3V_{DD} + 2V_{T,n}) = \frac{1}{8}(3 \times 5 + 2 \times 1) = \frac{17}{8} \text{ Volt}$$

$$V_{IH} = \frac{1}{8}(5V_{DD} - 2V_{T,n}) = \frac{1}{8}(5 \times 5 - 2 \times 1) = \frac{23}{8} \text{ Volt}$$

so the noise margin,

$$NM_L = V_{IL} = \frac{17}{8} = 2.125 \text{ Volts}$$

$$NM_H = V_{DD} - V_{IH} = 5 - 2.875 \text{ Volts} = 2.125 \text{ Volt}$$

Example 4.4. A logic gate is defined by the following voltage levels of $V_{OH} = 6$ V, $V_{OL} = 0.3$ V, $V_{IH} = 2.7$ V and $V_{IL} = 0.6$ V. Find the noise margin of this gate.

Solution: The low noise margin is given by

$$NM_L = V_{IL} - V_{OL} = 0.6 - 0.3 = 0.3 \text{ Volt}$$

and the high noise margin is given by

$$NM_H = V_{OH} - V_{IH} = 6 - 2.7 = 3.3 \text{ Volt}$$

Example 4.5. Consider a CMOS inverter circuit with the following parameter: $V_{DD} = 3.3$ V, $V_{Tn} = 0.6$ V, $V_{Tp} = -0.7$ V, $K_n = 200\ \mu A/V^2$ and $K_p = 80$ mA/V^2.

Calculate the noise margins of the circuit.

Solution: The CMOS inverter being considered her has $K_R = 2.5$ and $V_{Tn} \neq |V_{Tp}|$. Here it is not a symmetric inverter.

Here, $\quad V_{OL} = 0$ V, $\quad V_{DH} = V_{DD} = 3.3$ Volt

We have
$$V_{IL} = \frac{2V_{out} + V_{TP} - V_{DD} + K_R \cdot V_{Tn}}{1 + K_R}$$

$$V_{IL} = \frac{2V_{out} + 0.7 - 3.3 + 2.5 \times 0.6}{1 + 2.5} = 0.57\ V_{out} - 0.71$$

We known that,

$$\frac{K_n}{2}(V_{GSn} - V_{Tn})^2 = K_p\left[(V_{GSp} - V_{Tp}\left(V_{DS,p} - \frac{V_{DS,p}^2}{2}\right)\right]$$

$$\Rightarrow \quad \frac{K_n}{2}(V_{IL} - V_{Tn})^2 = K_p\left[(V_{IL} - V_{DD} - V_{Tp})(V_{out} - V_{DD}) - \frac{(V_{out} - V_{DD})^2}{2}\right]$$

Put the value of V_{IL}.

$$\Rightarrow \quad \frac{K_n}{2}(0.57V_{out} - 0.71 - 0.6)$$

$$= K_p\left[(0.57V_{out} - 0.71 - 3.3 + 0.7)(V_{out} - 3.3) - \frac{(V_{out} - 3.3)^2}{2}\right]$$

$$\Rightarrow \quad \frac{200}{2}(0.57V_{out} - 1.31) = 80\left[(0.57V_{out} - 3.4)(V_{out} - 3.3) - \frac{(V_{out} - 3.3)^2}{2}\right]$$

$$\Rightarrow \quad 0.66V_{out}^2 + 0.05V_{out} - 6.65 = 0$$

$$V_{out} = 3.14 \text{ Volt}$$

so
$$V_{IL} = 0.57 \times 3.14 - 0.71 = 1.08 \text{ Volt}$$

To Find V_{IL}:

$$V_{IH} = \frac{V_{DD} - V_{TP} + K_R(2V_{out} + V_{Tn})}{1 + K_R}$$

$$V_{IH} = \frac{3.3 - 0.7 + 2.5(2V_{out} + 0.6)}{1 + 2.5}$$

$$V_{IH} = 1.43\ V_{out} + 1.17$$

We know the relation,

$$\frac{K_n}{2}\left[(V_{GSn} - V_{Tn})V_{DSn} - \frac{V_{DSn}^2}{2}\right] = \frac{K_p}{2}(V_{GSp} - V_{T,p})^2$$

$$\Rightarrow \quad \frac{K_n}{K_p}\left[(V_{th} - V_{Tn})V_{out} - \frac{V_{out}^2}{2}\right] = \frac{1}{2}(V_{in} - V_{DD} - V_{T,p})^2$$

$$\Rightarrow \quad \frac{2K_n}{K_p}\left[(V_{IH} - V_{Tn})V_{out} - \frac{V_{out}^2}{2}\right] = (V_{IH} - V_{DD} - V_{T,p})^2$$

$$\Rightarrow \quad \frac{2\times 200}{80}\left[(V_{IH} - 0.6)V_{out} - \frac{V_{out}^2}{2}\right] = (\text{VIH} - 3.3 + 0.7)^2$$

$$\Rightarrow \quad 2.61\, V_{out}^2 + 6.94\, V_{out} - 204 = 0$$

$$\Rightarrow \quad V_{out} = \frac{-6.94 \pm \sqrt{(6.94)^2 - 4\times 2.61\times 204}}{2\times 2.61}$$

$$\Rightarrow \quad V_{out} = 0.27 \text{ Volt}$$

so $$V_{IH} = 1.43 \times 0.27 + 1.17 = 1.55 \text{ Volt}$$

The low level and high level noise margins can be calculated using the formula,

$$NM_L = V_{IL} - V_{OL} = 1.08 - 0$$

$$= 1.08 \text{ V}$$

$$NM_H = V_{OH} - V_{IH} = 3.3 - 1.55$$

$$= 1.75 \text{ V}$$

Example 4.6. Design the *(W/L)* ratio of the nMOS transistor used in resistive load-inverter with $V_{T0} = 1$ V, $V_{DD} = 5$ V and the permissible power dissipation is 0.5 mW, $K_n = 25\ \mu\text{A/V}^2$.

Solution: If the transistor is cutoff then input is $V_{OL} < V_{T0}$. If $V_{T0} = 1$ V for normal design value of V_{OL} is in the range of 25 to 50% of the V_{T0}. So assume that $V_{OL} = 0.25$ V.

Hence power dissipation,

$$P = V_{DD} \times I_{DD}$$

$$\Rightarrow \quad 0.5 \times 10^{-3} = 5 \times I_{DD}$$

$$\Rightarrow \quad I_{DD} = 10^{-4} \text{ Amp}$$

For output to V_{OL} input must be $V_{OH} = 5\text{ V} = V_{GS}$. If $V_{GS} - V_{T0} > V_{DS}$ then transistor will operate in linear region so,

$$I_{DD} = K_n\left(\frac{W}{L}\right)\left[V_{GS} - V_{T0} - \frac{V_{DS}}{2}\right]V_{DS}$$

$$10^{-4} = 25\times 10^{-6}\left(\frac{W}{L}\right)\left[5 - 1 - \frac{0.25}{2}\right](0.25)$$

$$\Rightarrow \quad \frac{W}{L} = 4.13$$

Example 4.7. Let $V_{DD} = 3$ V and the transistor parameter be $K_p = 30\ \mu A/V^2$ and $V_{T0} = 0.7$ V, Design the inverter circuit [(*W*/*L*)n and R_L] in Fig. 4.34. To achieve a noise margin $N_{MH} = 1$ V and a static power dissipation of 100 μW (assume the out put voltage $V_{out} = 3$ V during 50% of the operation time and $V_{out} = 0.2$ during the remaining 50%). You may use

$$V_{IH} = V_{Ton} + \sqrt{\frac{2V_{DD}}{3K_nR_L}} - \frac{1}{K_nR_L}$$

Solution:

$$N_{MH} = V_{OH} - V_{IH} = 1 \text{ V}$$

$$V_{IH} = 3 - 1 = 2 \text{ V}$$

$$\Rightarrow \quad V_{IH} = 3 - 1 = 2\text{V}$$

using

$$V_{IH} = V_{Ton}t_2\sqrt{\frac{2V_{DD}}{3K_nP_L}} - \frac{1}{K_nR_L}$$

$$2 - 0.5 = 2\sqrt{\frac{2 \times \not{3}}{3\,K_nR_L}} - \frac{1}{K_nR_L}$$

$$\Rightarrow \quad 1.5 = \sqrt{\frac{8}{K_nR_L}} - \frac{1}{K_nR_L}$$

Let us $K_n R_L = 1/x$

$$\Rightarrow \quad 1.5 = \sqrt{8x} - x$$

$$\Rightarrow \quad 8x = (1.5 + x)^2$$

$$\Rightarrow \quad 8x = x^2 + 2.25 + 3x$$

$$\Rightarrow \quad x^2 + 2.25 - 5x = 0$$

$$\Rightarrow \quad x = \frac{+5 \pm \sqrt{25 - 9}}{2} = 0.5,\ 4.5 \text{ V}$$

To determine the correct value of $x = \dfrac{1}{K_nR_L}$, evaluate V_{out}

$$V_{IH} = V_{T0n} + 2\,V_{out} - \frac{1}{K_nR_L}$$

$$\Rightarrow \quad V_{out} = \frac{V_{IH}}{2} + \frac{1}{2K_nR_L} - \frac{V_{T0n}}{2}$$

$$\Rightarrow \quad V_{out} = \frac{1}{2}[2 - V_{T0n} + x]$$

At $\quad x = 0.5 \Rightarrow V_{out} = 1\text{ V} = \frac{1}{2}(2 - V_{T0n} + x)$

$\quad x = 4.5 \Rightarrow V_{out} = 3\text{ V} = \frac{1}{2}(2 - V_{T0n} + x)$

Hence $\quad x = 0.5$ V

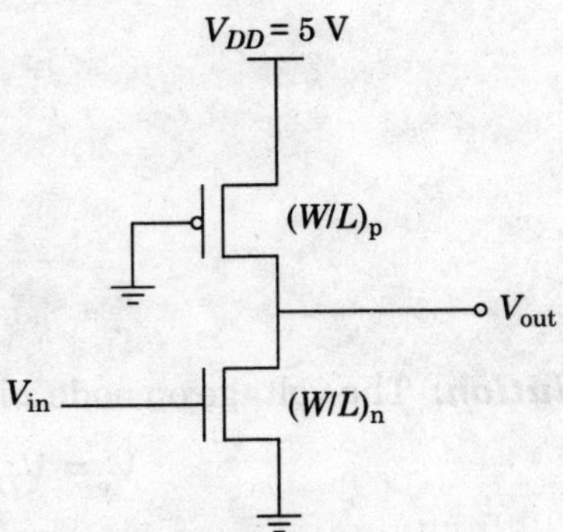

Fig. 4.34

For the power dissipation,

$$100\,\mu\text{W} = \frac{V_{DD}}{2}\left(\frac{V_{DD} - V_{OL}}{R_L}\right)$$

$$\Rightarrow \quad 10^{-4} = \frac{3}{2}\left(\frac{3 - 0.2}{R_L}\right)$$

$$\Rightarrow \quad \frac{2}{3} \times 10^{-4} \times \frac{1}{2.8} = \frac{1}{R_L}$$

$$\Rightarrow \quad R_L = 4.2 \times 10^4\ \Omega.$$

$$\frac{1}{K_n R_L} = 0.5\text{ V}$$

$$\Rightarrow \quad K_n = \frac{1}{0.5 R_L} = \frac{1}{0.5 \times 4.2 \times 10^4} = 47.62\ \mu\text{A/V}^2$$

$$K_n = K_p\left(\frac{W}{L}\right)_n$$

$$\Rightarrow \quad \left(\frac{W}{L}\right)_n = \frac{K_n}{K_p} = \frac{47.62}{30} = 1.59$$

Example 4.8. Consider the following Fig. 4.35 NMOs inverter. Assume that the bulk terminal of NMOS inverter. Assume that the bulk terminals of all NMOS device are connected to GND.

Assume that the point I_N has a 0 V to 2.5 swing. Set up the equation to compute the voltage on node X. Assume $\gamma = 0.5$.

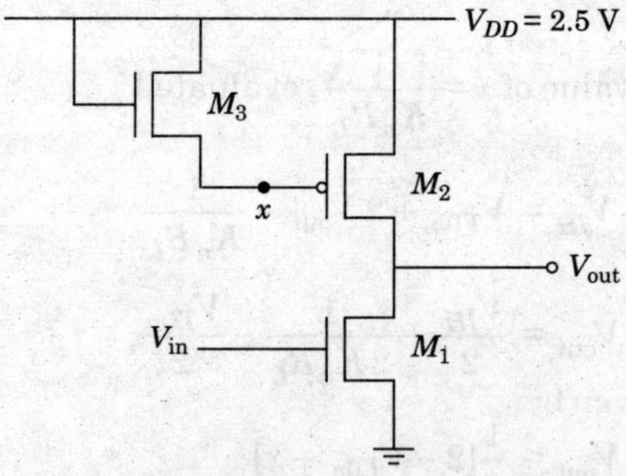

Fig. 4.35

Solution: The voltage on node x is the set to one threshold value V_T below V_{DD}. So

$$V_x = V_{DD} - V_T$$

$$V_x = V_{DD} - \left[V_{T0} + \gamma\left(\sqrt{V_{SB} + 1 - 2\phi_F}\right) - \sqrt{(-2\phi_F)}\right]$$

$$V_x = 2.5 - \left[0.43 + 0.5\left(\sqrt{V_x + 0.6} - \sqrt{0.6}\right)\right]$$

$$V_x = 2.07 + 0.39 - \sqrt{V_x + 0.6}(0.5)$$

$$V_x = 2.46 - 0.5\sqrt{V_x + 0.6}$$

$$(V_x - 2.46)^2 = 0.25(V_x + 0.6)$$

$\Rightarrow$ After solving the above equation,

$$V_x = 1.7014 \quad \text{Volt}$$

Example 4.9. Show that the aspect ratio of PMOS load device must be 1/3 if the aspect ratio (*L*/*W*) of nMOS driver is 1/1 in CMOS inverter to have symmetrical propagation delay and noise margins.

Solution: (a) The transconductance ratio K_R can be gives by

$$K_R = \frac{\mu_n (W/L)_n}{\mu_p (W/L)_p}$$

For symmetric inverter,

$$K_R = 1$$

so
$$\frac{(W/L)_n}{(W/L)_p} = \frac{\mu_p}{\mu_n}$$

We know that the typical values of μ_p and μ_n are 460 cm^2/V sec and 1350 cm^2/VS respectively.

$\therefore$
$$\frac{(W/L)_n}{(W/L)_p} = \frac{\mu_p}{\mu_n} = \frac{460}{1350}$$

$\Rightarrow$
$$\frac{1}{(W/L)_p} = \frac{1}{3}$$

$\Rightarrow$
$$(W/L)_p = 3$$

Hence, the aspect ratio for pMOS $\dfrac{L}{W} = \dfrac{1}{3}$

(*b*) The input low voltage V_{IL} for the the symmetrical inverter can be calculated

as
$$V_{IL} = \frac{1}{8}(3V_{DD} + 2V_{T0})$$

and input high voltage V_{IH} is given by

$$V_{IH} = \frac{1}{8}(5V_{DD} - 2\,V_{T0})$$

$\therefore$
$$V_{IH} + V_{IL} = V_{DD}$$

Noise margin can be given as,

$$N_{MH} = V_{OH} - V_{IL} = V_{DD} - V_{IH}$$

$$N_{ML} = V_{IL} - V_{OL} = V_{IL}$$

Example 4.10. Consider an inverter circuit that has FET aspect ratios of $(W/L)_n = 6$ and $(W/L)_p = 8$ in a process where $K'_n = 150\ \mu A/V^2$, $V_{Tn} = +0.70$ V, $K'_p = 62\ \mu A/V^2$, $V_{Tp} = -0.85$, $V_{DD} = 3.3$ V and $C_{out} = 150$ pF. Compute rise and fall times.

Solution: The value of transconductance factor K_n and K_p for NFET and PFET respectively.

$$\therefore \quad K_n = \left(\frac{W}{L}\right)_n \cdot K'_n$$

$$K_n = 6 \times 150\ \mu A/V^2 = 900\ \mu A/V^2$$

$$K_p = \left(\frac{W}{L}\right)_p \cdot K'_p$$

$$\therefore \quad K_p = 8 \times 62\ \mu A/V^2 = 496\ \mu A/V^2$$

the value resistance offered by NFET and PFET.

$$\therefore \quad R_{eqp} = \frac{1}{K_p(V_{DD} - |V_{TP}|)}$$

$$= \frac{1}{496\ \mu A/V^2\ (3.3\ V - 0.85\ V)}$$

$$\therefore \quad R_{eqp} = 822.9\ \Omega$$

For resistance of NFET,

$$R_{eqn} = \frac{1}{K_n(V_{DD} - |V_{Tn}|)}$$

$$= \frac{1}{900\ \mu A/V^2\ (3.3\ V - 0.70)}$$

$$R_{eqn} = 235.6\ \Omega$$

Propagation delay due to PFET can be calculated as,

$$\tau_p = R_{eqp} \cdot C_{out}$$

$$= 822.9\ \Omega \times 150\ pF$$

$$\therefore \quad \tau_p = 123.43\ ns$$

$\therefore$ Rise time t_r is given by,

$$t_r = 2.2\ \tau_p$$

$$\therefore \quad \tau_r = 2.2 \times 123.43\ ns$$

$$= 271.55\ ns$$

Now the propagation delay due to NFET can be calculated as,

$$\tau_n = R_{eqn} \cdot C_{out}$$

$$= 235.6\ \Omega \times 150\ ns$$

$$\therefore \quad \tau_n = 35.34\ ns$$

$\therefore$ fall time t_f is given by,

$$t_f = 2.2\ t_n$$

$$\therefore \quad t_f = 2.2 \times 35.34\ ns$$

$$= 77.78\ ns$$

Rise time $t_r = 271.55$ ns

Fall time $t_f = 77.748$ ns

Example 4.11. Design a CMOS inverter by determining the W_n and W_p of the NMOS and PMOS transistors to meet the following specifications.

1. $V_{TH} = 2$ V for $V_{DD} = 5$ V
2. Delay time of 2 ns for V_{out} transistion from 4 V to 1 V with $C_L = 1.0$ pF
3. $K'_n = 30$ µA/V2
4. $K'_p = 10$ mA/V^2
5. $L_n = L_p = 1.0$ µm
6. $V_{Ton} = 1$ V
7. $V_{Top} = -1.5$ V
8. $W_{min} = 2$ µm

Solution: It satisfy the delay constraint τ_{PHL} from 4 V to 1 V. For high-to-low transistion, the NMOS transistor must be in linear.

i.e. $$V_{in} = V_{OH} \text{ and } 1 \leq V_{out} \leq 4\text{ V} \Rightarrow \text{NMOS } L_{in}$$

$$C_L \frac{dV_{out}}{dt} = -\frac{dV_{out}}{dt} = -\frac{m_n C_{ox}}{2}\frac{W_n}{L_n}[2(V_{OH} - V_{Ton})V_{out} - V_{out}^2]$$

$$\text{tdelay} = 2\times10^{-9}\text{s} = -2C_L \frac{1}{\mu_n C_{ox}\frac{W_n}{L_n}} \int_{V_{out}=4}^{V_{out}=1} \frac{dV_{out}}{[2(V_{OH} - V_{Ton})V_{out} - V_{out}^2]}$$

$$= -C_L \frac{1}{\mu_n C_{ox}\frac{W_n}{L_n}} \frac{1}{2(V_{OH} - V_{Ton})} \ln\left[\frac{V_{out}}{2(V_{OH} - V_{Ton}) - V_{out}}\right]\Bigg|_{V_{out}=4}^{V_{out}=1}$$

$$= \frac{-C_L}{\mu_n C_{ox}\frac{W_n}{L_n}(V_{DD} - V_{Ton})}\left\{\ln\left[\frac{1}{2(5-1)-1}\right] - \ln\left[\frac{4}{2(5-1)-4}\right]\right\}$$

$$2.0\times10^{-9} = \frac{-1\times10^{-12}}{(30\times10^{-6})\frac{W_n}{L_n}(5-1)}\left\{\ln\left[\frac{1}{7}\right] - \ln\left[\frac{4}{4}\right]\right\}$$

$$\frac{W_n}{L_n} = \frac{1\times10^{-12}}{(2\times10^{-9})(30\times10^{-6}\times4)}(\ln(7))$$

$$\frac{W_n}{L_n} = 8.108$$

$$W_n = 8.108\, L_n$$

$$\therefore \quad W_n = 8.108 \text{ µm} \qquad \because L_n = 1 \text{ µm}$$

Now to satisfy V_{TH} constraint

$$V_{TH} = \frac{V_{\text{Ton}} + \sqrt{\frac{1}{K_R}}(V_{DD} + V_{\text{Top}})}{\left(1 + \sqrt{\frac{1}{K_R}}\right)}$$

$$2 = \frac{1.0 + \sqrt{\frac{1}{K_R}}(5 + (-1.5))}{\left(1 + \sqrt{\frac{1}{K_R}}\right)}$$

$$\therefore \quad K_R = \frac{9}{4}$$

$$K_R = \frac{\mu_n\, C_{ox}\,(W/L)_n}{\mu_p\, C_{ox}\,(W/L)_p} = \frac{30\, W_p}{W_p}$$

$$\frac{9}{4} = 3\frac{W_n}{W_p}$$

$$\therefore \quad W_p = \frac{4}{9}(3)\, W_n \qquad \because \quad L = 1\ \mu\text{m}$$

$$W_p = 10.8\ \mu\text{m}$$

Example 4.12. Consider a CMOS inverter with $C_{\text{load}} = 1.2$ pF, where the IV characteristics of the nMOS transistor driver are specified as follows:

$$V_{GS_n} = 5\text{ V and } V_{DS_n} \geq 4\text{ V} \Rightarrow T_{Dn} = T_{\text{Dnsat}} = 5\text{ mA}$$

Assume V_{in} is step pulse that switches instantaneosuly from 0 to to 5 V. Calculate the delay time necessary for the inverter output to fall from its initial value of 5 V to 2.5 V.

Solution: $\quad V_{50\%} = 0.5\,[V_{OL} + V_{OH}] = 0.5\,[0 + 5] = 2.5\text{ V}$

From I-V Data

at $\quad$ SAT $V_{DS_n} = 5\text{ V} - V_{\text{Ton}} = 4\text{ V} \Rightarrow V_{\text{Ton}} = 1\text{ V}$

Calculate transconductance $K_n \rightarrow$

$$K_n = \frac{2I_{\text{Dnsat}}}{(V_{GS} - V_{\text{Ton}})} = \frac{10\text{ mA}}{(4\text{ V})^2}$$

$$= 0.625 \times 10^3\ \text{A/V}^2$$

For $t_0 < t < t_1'$.

Since $\quad I_{Dn} = I_{\text{Dnsat}} = 5\text{ mA}$

$$V_{OH} - V_{\text{Ton}} = 4\text{ V}$$

$$V_{OH} = 5\text{ V}$$

$$\int_{t=t_0}^{t=t_1'} dt = -C_{\text{load}} \int_{V_{\text{out}}=V_{OH}}^{V_{\text{out}}=V_{OH}-V_{\text{Ton}}} \left(\frac{1}{I_{Dn}}\right) dV_{\text{out}}$$

$$t_1' - t_0 = -\frac{C_{\text{load}}}{I_{\text{Dnsat}}} \int_{V_{\text{out}}=5\text{ V}}^{V_{\text{out}}=4\text{ V}} dV_{\text{out}} = -\frac{1.2\text{ pF}}{5\text{ mA}}(-1\text{ V}) = 0.24\text{ ns}$$

For $t_1' < t < t_1$:

$$t_1 - t_1' = \frac{C_{\text{load}}}{K_n(V_{OH} - V_{\text{Ton}})} \ln\left[\frac{2(V_{\text{OH}} - V_{\text{Ton}}) - V_{50\%}}{V_{50\%}}\right]$$

$$= \frac{1.2\text{ pF}}{(0.625 \times 10^{-3}\text{ A/V}^2)(5-1)\text{ V}} \ln\left[\frac{2(5-1)\text{ V} - 2.5\text{ V}}{2.5\text{ V}}\right]$$

$$= 0.3784\text{ ns}$$

Hence,

$$\tau_{\text{PHL}} = 0.24 + 0.3784$$

$$= 0.6184\text{ ns}$$

Example 4.13. Consider a CMOS inverter with $C_{\text{load}} = 1.2$ pF and $V_{DD} = 5$ V, where the I-V characteristics of the nMOS transistor driver are specified as follows:

$$K_n' = \mu_n C_{ox} = 20\ \mu\text{A/V}^2,\ (W/L)_n = 10 \text{ and } V_{\text{Ton}} = 1\text{ V}$$

Use both the average – current method and the differential equation method to calcualte τ_{fall} (time elapsed between the time $V_{\text{out}} = V_{90\%} = 4.5$ V to the time at which $V_{\text{out}} = V_{10\%} = 0.5$ V)

Solution: For the calculation of $I_{\text{avg, fall}}$

$$I_{\text{avg, fall}} = \frac{1}{2}[i_c(V_{\text{in}} = V_{OH}, V_{\text{out}} = V_{90\%}) + i_c\,(V_{\text{in}} = V_{OH}, V_{\text{out}} = V_{10\%})]$$

$$= \frac{1}{2}[i_c(V_{\text{in}} = 5\text{ V}; V_{\text{out}} = 4.5\text{ V}) + i_c\,(V_{\text{in}} = 5\text{ V}, V_{\text{out}} = 0.5)]$$

$$= \frac{1}{2}\left[\frac{1}{2}K_n(V_{\text{in}} - V_{\text{Ton}})^2 + \frac{1}{2}K_n(2(V_{\text{in}} - V_{\text{Ton}})V_{\text{out}} - V_{\text{out}}^2)\right]$$

$$= \frac{1}{4}K_n[(V_{OH} - V_{\text{Ton}})^2 + (2(V_{OH} - V_{\text{Ton}})\,V_{10\%} - V_{10\%}^2)]$$

$$= \frac{1}{4}20 \times 10^{-6} 10[(5-1)^2 + (2(5-1)\,0.5 - (0.5)^2)]$$

$$= 0.9875\text{ mA}$$

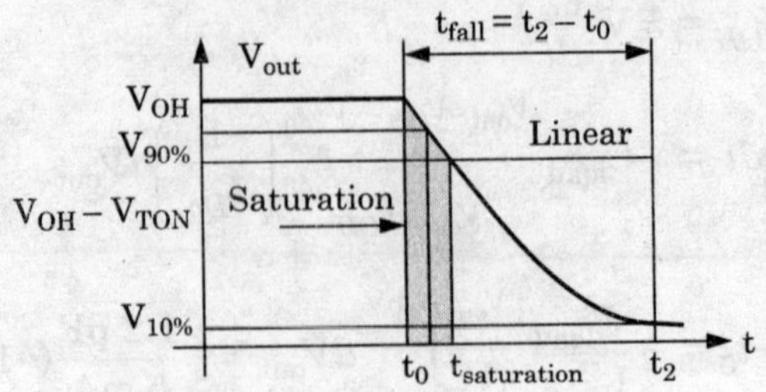

Fig. 4.36

Now calculate t_{fall}

$$\tau_{fall} = \frac{C_{load}\ \Delta V}{I_{avg,}\ \text{fall}} = \frac{1.2 \times 10^{-12}\ \text{F}\ (4.5 - 0.5)\ \text{V}}{0.9875 \times 10^{-3}\text{A}}$$

$$= 4.86\ \text{ns}$$

SAT for $4.0\ \text{V} \le V_{out}\ 4.5\ \text{V}$

$$C_{load}\frac{dV_{out}}{dt} = -\frac{1}{2}K_n(V_{in} - V_{Ton})^2$$

$$\frac{dV_{out}}{dt} = \frac{K_n}{2C_{load}}(V_{in} - V_{Ton})^2$$

$$= -\frac{K_n}{2C_{load}}(V_{OH} - V_{Ton})^2$$

$$= -\frac{K_n}{2C_{load}}(5 - V_{Ton})^2$$

$$= -\frac{20 \times 10^{-6}}{2 \times (1.2 \times 10^{-10})}(5-1)^2 = -0.33 \times 10^9\ \text{V/S}$$

$$\int_{t=t_0}^{t=t_{sat}} dt = \frac{-1}{0.133 \times 10^9}\int_{V_{out}=4.5\ \text{V}}^{V_{out}=4\ \text{V}} dV_{out}$$

$$t_{sat} - t_0 = -\frac{1}{0.133 \times 10^9} V_{out}\Big|_{V_{out}=4.5}^{V_{out}=4\ \text{V}}$$

$$= -\frac{(4-4.5)\ \text{V}}{0.133 \times 10^9\ \text{V/s}}$$

$$t_{sat} - t_0 = +3.759\ \text{ns}$$

$$t_{sat} = 3.759\ \text{ns}$$

LIN for $0.5\ \text{V} \le V_{out} < 4.0\ \text{V}$

$$C_{load} = \frac{dV_{out}}{dt} = -\frac{1}{2}K_n(2(V_{in} - V_{Ton})V_{out} - V_{out}^2)$$

$$\int_{t=t_{sat}}^{t+t_2} dt = -C_{log}\int_{V_{out}=4\text{ V}}^{V_{out}=0.5\text{ V}} \frac{dV_{out}}{K_n(2V_{in} - V_{Ton})V_{out} - V_{out}^2)}$$

$$t_2 - t_{sat} = \frac{C_{load}}{K_n}\frac{1}{(V_{in} - V_{Ton})}\ln\left[\frac{2(V_{in} - V_{Ton}) - V_{10\%}}{V_{10\%}}\right]$$

$$= \frac{C_{load}}{K_n}\frac{1}{(V_{OH} - V_{Ton})}\ln\left[\frac{2(V_{OH} - V_{Ton}) - V_{10\%}}{V_{10\%}}\right]$$

$$= \frac{1.2\times10^{-12}\text{F}}{20\times10^{-6}\text{A/V}^2(5-1)\text{ V}}\ln\left[\frac{2(5-1) - 0.5}{0.5}\right]$$

$$= 40.62 \text{ ns}$$

$$\tau_{fall} = (t_2 - t_{sat}) + (t_{sat} - t_0)$$

$$= 3.759 + 40.62 = 44.379 \text{ ns}$$

SHORT ANSWERS TYPE QUESTIONS

Q. 1. Discuss the drawbacks of the resistive load inverter.

Ans: Even through the resistive load inverter yield decent transfer characteristics and an optimal V_{OH} of V_{DD}. Still there were big issues like larger static power consumption, rise/fall times, and fabrication in efficiency. It is observed that the resistor pull-up dissipated more power and also, a resistor that is too large can increase the rise time, while a resistor that is too small can increase the fall time. Another drawback was that V_M was not close to $V_{DD/2}$, so there would be a greater chance for errors and V_{OL} is larger than 0 means reduced noise margin. Lastly, during the fabrication of an I_C, the resistor would add unnecessary are to the chip.

Q. 2. Derive the expression for overall dc power consumption of depletion load inverter.

Ans: For depletion load inverter, steady state dc power consumption is determined by calculating the current drawn from the power supply when input is low and when input is high. When input is low, the driver transistor is in the cut-off region and the current flowing through the driver transistor is zero. Since load current and driver currents must be equal to each other, for low input, no significant current is drawn from power supply either by load or driver transistor. As a result, the inverter does not dissipate dc power for low input voltage if subthreshold condition is assumed to be zero.

When input in high, the driver transistor turns ON. Then, both transistors conduct a significant current and are given by

$$k_{L,n}(-V_{THL})^2 = k_{d,n}[2(V_{in} - V_{Tho})V_{out} - V^2_{out}]$$

where $V_{out} = V_{OL}$

Let us assume that the input voltage is low during 50% of the operation time and high for other 50%. Under this situation, the overall dc power consumption is given as

$$P_{dc} = \frac{V_{DD}}{2}\cdot\frac{k_{L,n}}{2}(-V_{Th,L})^2$$

where $V_{out} = V_{OL}$

Q. 3. Why stitching threshold voltage is undesirable?

Ans: When $V_{in} = V_M$, both transistors operate in the saturation region. If means both are ON at the same time and hence a direct conducting path will be established between VDD and ground. This causes a large current to flow in the CMOS inverter and hence considerable power consumption. Therefore, such situation should be avoided, where both transistors are ON to save power.

Q. 4. Why a circular chain consisting of an even number of inverters cannot be used as a ring oscillator?

Ans: Because the last output in this case is same as the input. However, this configuration of inverter feedback can be used as a storage element, which is a building block of SRAM.

Q. 5. Why it is required to keep equal rise/fall time of output signal and input signal in CMOS inverter?

Ans: It was found that for a loaded gate, if output signals and input signals have equal rise/fall times, the short circuit power can be reduced by 20% of total power.

Q. 6. What are the applications of ring oscillator?

Ans: The ring oscillator frequency measurements are routinely utilized to characterize a particular design and/or a new fabrication process. Ring oscillator can be used as a simple pulse generator where the output waveform is utilized as a simple master clock signal generated on the chip.

Q. 7. What is CMOS inverter?

Ans: It is a series connection of P and *n*-devices.

Q. 8. Define stage ratio.

Ans: The ratio by which each stage is increased in size is called stage ratio.

Q. 9. Give one important advantage of CMOS inverter over others inverter configurations.

Ans: The steady state power dissipation of the CMOS inverter circuit is negligible.

REVIEW QUESTIONS

1. What is the (W/L) ratio of the drive switching transistor of a saturated load inverter to obtain $V_{OL} = 0.20$ V? Assume that $I_{DS} = 60$ µA. Also calculated the (W/L) ratio of the saturated load.
2. Compare the nMOS load device characteristic (load current vs output voltage) for:
   (a) saturated load
   (b) linear load
   (c) depletion load with or without body effect.
3. Determine an inverter with a resistive load for $V_{DD} = 3.3$ V, $V_{OL} = 0.15$ V. Assume $I_D = 30$ µA, $K'_n = 60$ µA/V^2 and $V_{Th} = 1.25$ Volt.
4. What are the noise margins for CMOS inverter having $\left(\frac{W}{L}\right)_n = \left(\frac{W}{L}\right)_p$ operating with $V_{DD} = 5$V and $V_{th} = -V_{T,p} = 1$ V? Find the same for $V_{DD} = 3.3$ Volt.
5. Explain CMOS inverter *VTC* with a neat diagram. What is the criteria for voltage threshold for high and low level in inverter characteristics? **[UPTU 2005-06]**
6. Explain the characteristics of CMOS Inverter.
7. Describe the nMOS NAND gate ratio determination.
8. Write the advantages and disadvantages of nMOS depletion load inverter.
9. How you can minimize the parasitic effects occurs in CMOS inverter?
10. Describe the operation of basic MOS Inverter. Derive the expression for pull-up to pull-down ratio for an NMOS inverter driven by another NMOS inverter. **[UPTU 2003-04, 2007-08, 2011-12]**

11. Draw the different type of MOS inverter circuits and their transfer-characteristics compare their relative advantages and disadvantages. **[UPTU 2005-06]**

12. Does the inverter with lower V_{OL} always have the shorter high to low switching-time? Justify your answer.

13. Derive the expression for V_{IH}, V_{IL}, NM_L and NM_H for CMOS inverter. **[UPTU 2011-12, 2007-08]**

14. What do you understand by the symmetric CMOS inverter? Given the condition for it.

15. Calculate the critical voltages (V_{OL}, V_{OH}, V_{IL} and V_{IH}) and noise margin for depletion load inverter with following data $V_{DD} = 5$ V, $K_n = K_p = 25$ μA/V^2, $V_{T0,\,n} = 1$ V, $V_{T0,\,p} = -3.0$ V, $R_L = 200$ KΩ, $(W/L)_n = 2$, $(W/L)_p = 1/3$, l = 0.4 V$^{1/2}$ and $\phi_F = -0.3$ V.

16. Compare voltage and noise margin for bipolar and CMOS logic family.

17. Explain MOS technology families. How is CMOS different from NMOS?

18. Explain briefly the design considerations of CMOS circuits.

19. Draw and explain nMOS inverter with enhancement mode pull up and its transfer charactgeristics. Why dedepletion load is preferred compared to enhancement load. **[UPTU 2010-11]**

20. Explain the CMOS static transfer characteristics. **(UTU-2012)**

❑❑❑

5 Chapter
Power Consumption for CMOS Circuits

Chapter Overview

- Introduction
- Overview Of Power Consumption
- Needs For Low Power VLSI Chips
- Low-power Design Through Voltage Scaling

5.1 INTRODUCTION

The art of power analysis and optimization of integrated circuits used to be a narrow speciality in analog circuit design. It is now appearing in the mainstream digital design community affecting all aspects of the design process. The increase in low power chips and systems are driven by both business and technical needs. New generations of semiconductor processing technologies present more requirements. In the power dissipation to digital chips due to increased device density, speed and complexity.

"The rate at which energy is delivered or exchanged is known as Power." An analogy of energy and power is water and its flow rate. In a VLSI chip, electric energy is converted to heat energy during operations. The rate at which energy is takes from the source and converted into heat is the power dissipation.

The VLSI low power design problems can be broadly classified into two analysis and optimization.

Analysis problems are concerned about the accurate estimation of the power circuitry dissipation at different phases of the design process. Analysis techniques differ in their accuracy and efficiency. The accuracy of analysis depends on the availability of design information. Analysis techniques also serve as the foundation for design optimization.

Optimization is the process of generating the best design given an optimization goal, without violating design specifications. An automatic design optimization algorithm requires a fast analysis engine to evaluate the merits of the design choices. Manual optimization also depends a reliable analysis tool to provide accurate estimation of dower dissipation.

- A decision to apply a particular low-power technique often involves trade of from different sources putting in various directions.
- Major criteria to be considered are the impact to the circuit delay, which affects the performance and throughput of the chip, and the chip area, which directly translates to manufacturing costs.

5.2 NEEDS FOR LOW POWER VLSI CHIPS

- As the scale of integration improves, more transistors, faster and smaller than their predecessors, are being packed into a chip. This leads to the steady growth of the operating frequency and processing capacity per chip, resulting in increased power dissipation.
- A need for low power VLSI chips arises from various evolution forces of integration circuits.
- The requirements for low power consumption must be met along with equally demanding goals of high chip density and high throughput.
- The need for low-power design is also becoming a major issue in high performance digital systems, such as DSP's, microprocessors and other applications.
- High performance computing system characterized by large power dissipation also drives the low power needs.
- Power dissipation has a direct impact on the packaging cost of the chip and the cooling cost of the system.
- Another major demand for low power chips and systems comes from environmental concerns. Computers are the fastest-growing electricity loads in the commercial sector. Since electricity generation is a major source of air pollution, inefficiency energy usage in computing equipment indirectly contributes to environmental pollution.
- Device characteristics (e.g. threshold voltage), device geometries and interconnect proper are significant factors in lowering the power consumption. Circuit level measures such as the proper choice of circuit design styles, reduction of the voltage swing and clock strategies can be used to reduce power dissipation at the transistor level.
- The power consumed by the system can be reduced by a proper selection of the data processing algorithms, specifically to minimize the number of switching events for a given task.

5.3 OVERVIEW OF POWER CONSUMPTION

In this section, we will study the various sources (components) of time-averaged power consumption on in CMOS circuits. The average power consumption in conventional CMOS digital circuits can be expressed as the sum of three main components, namely

(*i*) the dynamic (switching) power consumption.

(*ii*) the short-circuit power consumption.

(*iii*) the leakages power consumption.

If the system or chip includes circuits other than conventional CMOS gates that have continuous current paths between the power supply and the ground, a fourth (static) power component should also be considered.

5.3.1 Dynamic (Switching) Power Dissipation (UTU-2012)

Dynamic power dissipation is due to the switching activities of the CMOS circuits. A higher operating frequency leads to more frequent switching activities in the circuits and results is increased power dissipation.

In digital CMOS circuits, switching power is dissipated when energy is drawn from the power supply to charge up the output node capacitance (OR). The most significant source of dynamic power dissipation in CMOS circuits is the charging and discharging of capacitance.

Most digital CMOS circuits do not require capacitors for their intended operations. The capacitance forms due to parasitic effects of interconnection wires and transistors. Such parasitic capacitance cannot be avoided and it has a significant impact on the power dissipation of the circuits. The estimation and analysis of parasitic capacitance is a crucial subject matter not only for signal delay, but also for power.

The equivalent circuit of charging and discharging output capacitance of a CMOS logic gate is shown in Fig. 5.1. Here, we use a switch that flips up and down to model the charging and discharging cycles.

From the Fig. 5.1, V_{DD} is an ideal constant voltage source and R_c, R_d is the resistance of charging, discharging circuitry, either conventional or parasitic.

According to the law of physics, the voltage $V_o(t)$ and the current $I_c(t)$ of a capacitance C_L at time t are given by

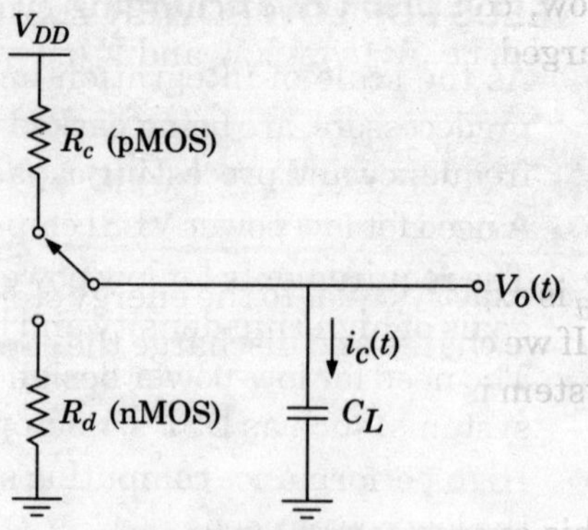

Fig. 3.1 Equivalent circuit for charging and discharging a capacitor.

$$i_c(t) = C_L \frac{dV_o(t)}{dt} \quad ...(5.1)$$

During the charging cycle from time t_0 to t_1, the energy E_g drawn from the voltage source is

$$E_g = \int_{t_0}^{t_1} V_{DD} \cdot i_c(t)dt \quad ...(5.2)$$

Initially the capacitor contains no charge and the voltage across its terminals is zero. i.e., $V_c(t_0) = 0$.

Assume that the capacitor is fully charged at the end of the charging cycle, i.e., $V_c(t_1) = V_{DD}$.

So from equations (5.1) and (5.2), we have

$$E_g = C_L V_{DD} \int_{t_2}^{t_1} \frac{dV_0(t)}{dt} \cdot dt = C_L V_{DD} \int_{0}^{V_{DD}} dV_0$$

$$E_g = C_L\, V_{DD}^2 \quad ...(5.3)$$

Part of the electrical energy E_g drawn from the voltage source is stored in the capacitor and the rest is dissipated as heat energy in the resistor R_C.

The energy E_{cap} stored in the capacitor at the end of charging cycle is

$$E_{cap} = \int_{t_0}^{t_1} V_0(t)\, i_c(t)\, dt = C_L \int_{t_0}^{t_1} V_0(t) \frac{dV_0(t)}{dt} dt$$

$$= C_L \int_{0}^{V_{DD}} V_0\, dV_0 = \frac{1}{2} C_L V_{DD}^2 \quad ...(5.4)$$

From equation (5.3) and (5.4), the energy E_C dissipated at R_C during charging is given by

$$E_C = E_g - E_{cap} = \frac{1}{2}\, C_L V_{DD}^2 \quad ...(5.5)$$

Now, consider the discharging cycle from t_1 to t_2, we assume that the capacitor is fully discharged, i.e., $V_o(t_1) = V_{DD}$ and $V_o(t_2) = 0$. The energy E_d dissipated in the discharge resistor R_d is

$$E_d = \int_{t_1}^{t_2} V_o(t)\, i_c(t)\, dt = -C_L \int_{V_{DD}}^{0} V_o(t)\, dV = \frac{1}{2} C_L V_{DD}^2 \quad ...(5.6)$$

E_d is exactly equal to the energy stored in the capacitance at the beginning at the discharging cycle. If we charge and discharge the capacitance at the frequency of f_{CLK}, the power dissipation of the system is

$$P_{avg} = E_d \cdot f = C_L V^2 f_{CLK} \quad ...(5.7)$$

This average switching power dissipation of a CMOS gate is essentially independent of all transistor characteristics and transistor sizes as long as a full voltage swing in achieved.

We now discuss the generality and applicability of the $P_{avg} = C_L V^2 f_{CLK}$ equation by reviewing the derivation procedure. We observe that during charging, $C_L V_{DD}^2$ energy is drawn from the energy source, half of which is dissipated in the charging resistance R_c and the other half is stored in the capacitor. During discharging the energy stored in the capacitor is dissipated as heat is the discharging resistor R_d.

The result is independent of the charging and discharging circuitry R_c, R_d, the length of charging and discharging cycle, the voltage or current wave from $V_o(t)$, $i_c(t)$ etc.

Equation (5.7) is only the power dissipation caused by a single capacitor C_L. In general, the total power should be summed over each capacitance C_i in a circuit yielding

$$P_{avg} = \sum_{i=0}^{n} C_i V_i^2 f_i \quad ...(5.8)$$

For CMOS circuits, V_{DD} is typically the same for all capacitance C_i and assume that f_{CLK} is constant. Then, the power dissipation

$$P_{avg} = V_{DD}^2 \cdot f \sum_i C_i = C_{Total} V_{DD}^2 f_{CLK} \quad ...(5.9)$$

where, C_{Total} is the sum of all capacitance, f_{CLK} is the average frequency and V_{DD} is the voltage swing.

The analysis of switching power dissipation presented above is based on the assumption that the output node of a CMOS gate undergoes one power-consuming transition (0 to V_{DD}) in each clock cycle. This assumption, however, is not always correct; the node transition rate can be slower than the clock rate depending on the circuit topology, logic style and the input signal statistics. To better represent this behaviour, we will introduce α_T (node transition factor), which is effective number of power-consuming voltage transitions experienced per clock cycle. Then, the average switching power consumption becomes

$$P_{avg} = \alpha_T C_{total} \cdot V_{DD}^2 \cdot f_{CLK} \quad ...(5.10)$$

5.3.1.1 Observations on Switching Power Reduction

The average switching power dissipation of CMOS logic gates suggest that we have several different means for reducing the power consumption. The switching power dissipation is also a linear function of the clock frequency, yet simply reducing the frequency would significantly diminish the overall system performance.

These measures includes

(i) **Reduction of power supply voltage:** The reduction of power supply voltage is one of the most widely practical measures for low-power design.

- The $P = C_L V_{DD}^2 f_{CLK}$ equation consists of three terms voltage, capacitance and frequency. Due to the quadratic effect of the voltage term reducing the switching voltage can achieve dramatic savings.
- The easiest method to achieve this is to reduce the operation voltage of the CMOS circuits.
- Other methods seek to reduce voltage swing by using well-known circuit techniques such as charge sharing, transistor threshold voltage etc.
- There are many trade-offs to be considered in voltage reduction.
- Performance is lost because MOS transistor become slower at lower operating voltages.
- In particular, we need to consider that reducing the power supply voltage leads to an increase of delay.
- The input and output signal level of a low-voltage circuit or module should be made compatible with the peripheral circuitry, in order to maintain correct signal transmission.

(ii) **Reduction of switching (clock) frequency:**

- The reduction of clock frequency would be a viable option only in cases where the overall throughput of the system can be maintained by other means.
- The reduction of switching activity requires a detailed analysis of signal transition probabilities and implementation of various circuit-level and system level measures such as logic optimization, use of gated clock signals and prevention of glitches.
- Reduction of switching frequency also has the side effect of improving the reliability of a chip as some failure mechanism is related to the switching frequency.
- One effective method of reducing switching frequency is to eliminate logic switching that is not necessary for computation.

(iii) **Reduction of load capacitance:**

- Reducing parasitic capacitance in digital design has always been a good way to improve performance as well as power.
- The real goal is to reduce the product of capacitance and its switching frequency.
- Signals with high switching frequency should be routed with minimum parasitic capacitance to conserve power.
- The load capacitance can be reduced by using certain circuit design styles and by proper transistor sizing.
- Capacitance reduction can be achieved at most design abstraction levels; material process technology. Physical design (Floor-planning, placement and routing), circuit techniques, transistor sizing, logic restructuring, architecture transformation and alternative computation algorithms.

(iv) **Reduction of leakage and static current:**

- Leakage current, whether reverse biased junction or subthreshold current, is generally not very useful in digital design. So, designers often have very little control over the leakage current of the digital circuit.
- The leakage power dissipation of a CMOS digital circuit is several orders of magnitude smaller than the dynamic power.

- The leakage power problem mainly appears in very low frequency circuits or ones with "sleep modes" where dynamic activities are suppressed.
- Most leakage reduction techniques are applied at low-level design abstraction such as process in very low frequency device and circuit design.
- Static current can be reduced by transistor sizing, layout techniques and careful circuit design.
- Static current depends on the logic state of its output and we can consider eversion the signal polarity to minimize the probability of static current flow.

5.3.2 Short-Circuit Power Consumption

The switching power dissipation is purely due to the energy required to charge up the parasitic load capacitances in the circuit and the switching power is independent of the rise and fall times of the input signals.

If a CMOS inverter (or a logic gate) is driven with input voltage waveforms with finite rise and fall times, both the nMOS and pMOS transistors in the circuit may conduct simultaneously for a short amount of time during switching, forming a direct current path between the power supply and ground as shown in Fig. 5.2.

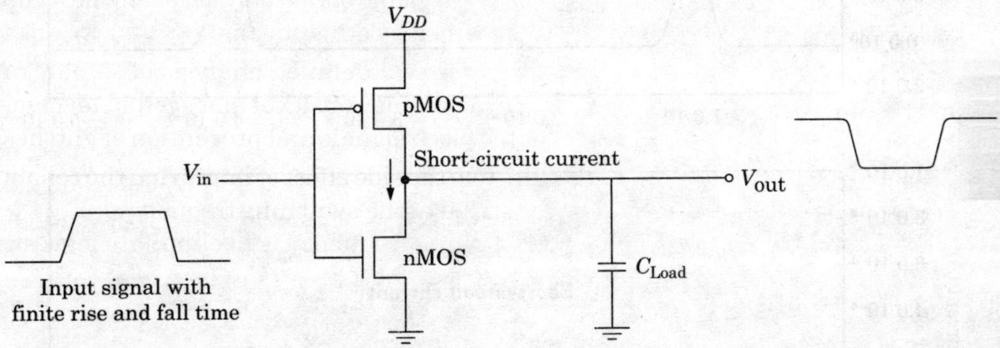

Fig. 5.2 Both nMOS and pMOS transistor may conduct a short-circuit current during switching.

The current component which passes through both the nMOS and pMOS devices during switching does not contribute to the charging of the capacitances in the circuit, hence it is known as the **short-circuit current** component. This component is prevalent if the output load capacitance is small, and/or if the input signal rise and fall time are large, as shown in Fig. 5.3. Here, the input/output voltage waveforms and the components of the current drawn from the power supply are illustrated for a symmetrical CMOS inverter with small capacitive load. The nMOS transistor in the circuit starts conducting when the rising input voltage exceeds the threshold voltage $V_{T,n}$. The pMOS transistor remains an until the input reaches the voltage level $(V_{DD} - |V_{T,p}|)$.

So there is a time window during which both transistor are turned on. As the output capacitance is discharged through the nMOS transistor, the output voltage start to fall. The drain to source voltage drop of the pMOS transistor becomes non-zero; which allows the pMOS transistor to conduct as well. The short circuit current is terminated when the input voltage transition is completed and the pMOS transistor is turned off. A similar event is responsible for the short-circuit current component during the falling input transition, when the output voltage starts rising while both transistors are on.

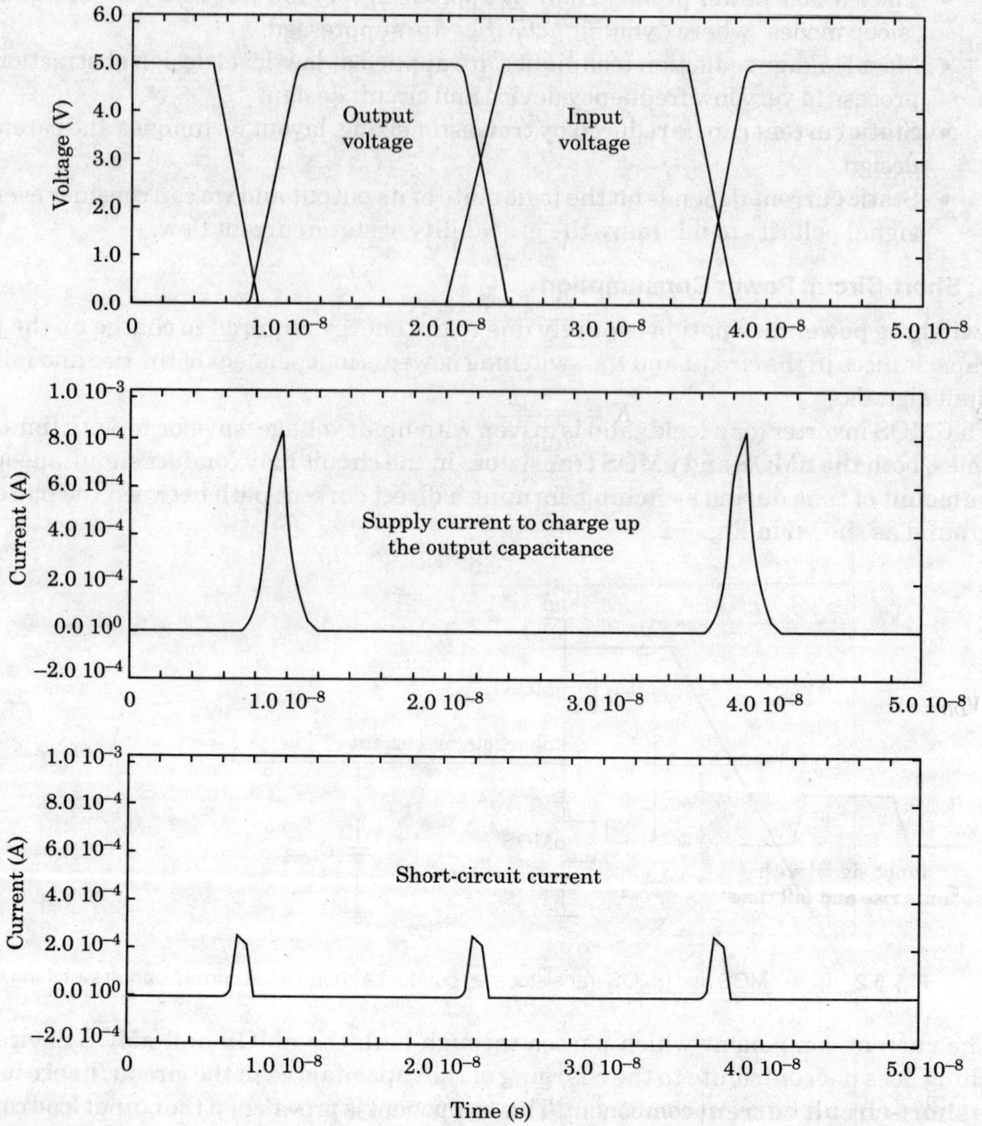

Fig. 5.3 Different waveforms- input-output voltage waveforms, the supply current, the short-circuit current.

For an unloaded inverter, the short-circuit power dissipation is given by

$$P_{sc} = I_{mean} \cdot V_{DD} \quad \text{...(5.11)}$$

For the input waveform shown in Fig. 5.3(a), which depicts the short circuit Fig. 5.3(b) in an unloaded inverter,

$$I_{mean} = 2\left[\frac{1}{T}\int_{t_1}^{t_2} I(t)\,dt + \frac{1}{T}\int_{t_2}^{t_3} I(t)dt\right] \quad \text{...(5.12)}$$

assuming that $V_{T,n} = -V_{Tp}$ and $\beta_n = \beta_p = \beta$ and that the behaviour is symmetrical around t_2.

$$I_{\text{mean}} = 2\times\frac{2}{T}\int_{t_1}^{t_2}\frac{\beta}{2}(V_{\text{in}}(t)-V_T)^2\,dt$$

with $$V_{\text{in}}(t) = \frac{V_{DD}}{t_r}\cdot t$$

$$t_1 = \frac{V_T}{V_{DD}}\cdot t_r \quad \text{and} \quad t_2 = \frac{t_r}{2}.$$

$$I_{\text{mean}} = \frac{1}{12}\frac{K\tau f_{\text{CLK}}}{V_{DD}}(V_{DD}-2V_T)^3 \qquad \text{...(5.13)}$$

where $$K = \frac{\beta}{f_{\text{CLK}}\cdot T_\tau}$$

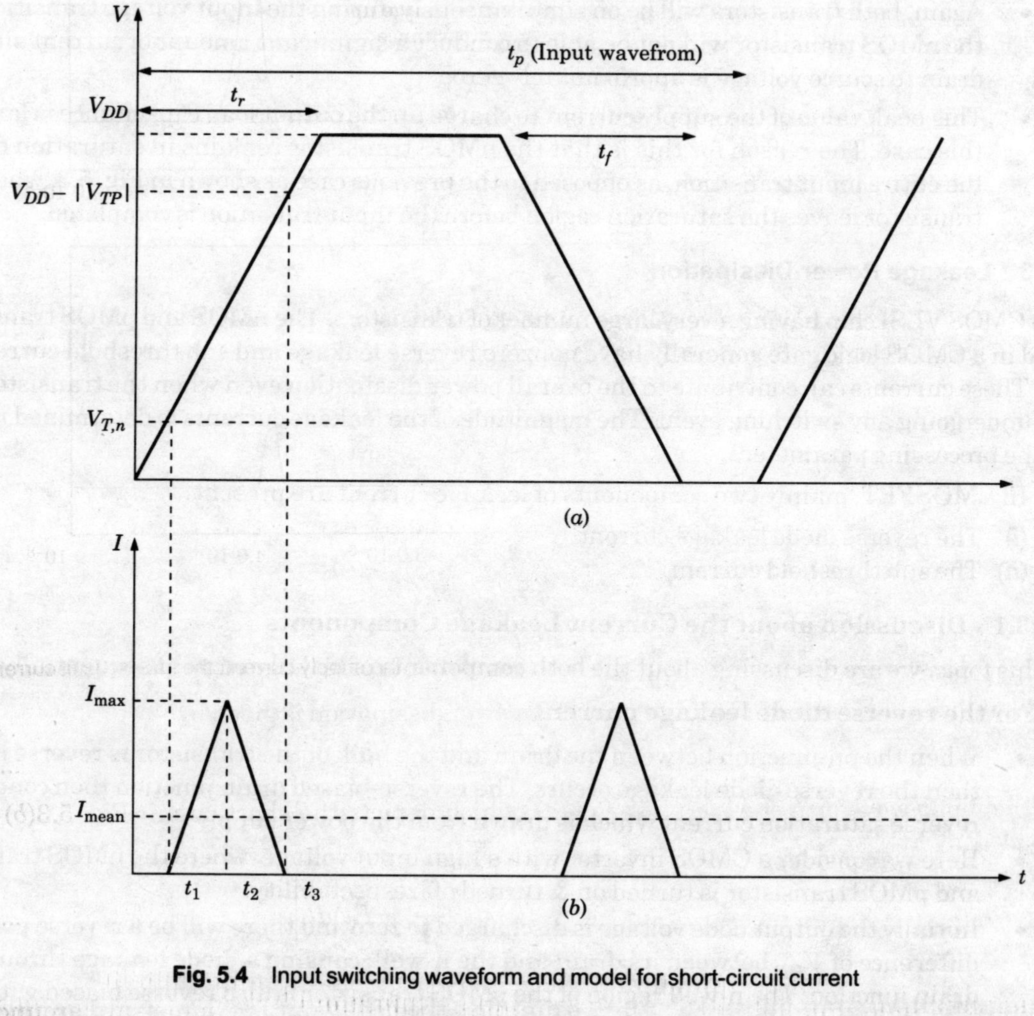

Fig. 5.4 Input switching waveform and model for short-circuit current

Hence, the short-circuit power dissipation becomes

$$\left.\begin{aligned} P_{SC} &= \frac{1}{12} K \cdot \tau \cdot f_{\text{CLK}}\,(V_{DD} - 2V_T)^3 \\ P_{SC} &= \frac{\beta}{12} (V_{DD} - 2V_T)^3 \,\frac{t_{\text{rf}}}{T} \end{aligned}\right\} \quad \text{...(5.14)}$$

The short-circuit power dissipation is linearly proportional to the input signal rise and fall times, and also to the trans-conductance of the transistor.

Reducing the input transition times will decrease the short circuit current component.

- Now we consider the same CMOS inverter with a larger output load capacitance and smaller input transition times. During the rising input transition the output voltage will effectively remain at V_{DD} until the input voltage completes its swing and the output will start to drop only after the input has reached its final value.
- Here, both nMOS and pMOS transistors are on simultaneously during the transition, the pMOS transistor cannot conduct a significant amount of current since the voltage drop between its source and drain terminals is almost zero.
- Again, both transistors will be on simultaneously during the input voltage transition, yet the nMOS transistor will not be able to conduct a significant amount of current since its drain to source voltage is approximately zero.
- This peak value of the supply current to charge up the output load capacitance is larger in this case. The reason for this is that the pMOS transistor remains in saturation during the entire input transition, as opposed to the previous case as shown in Fig. 5.3, where the transistor leaves the saturation region before the input transition is completed.

5.3.3 Leakage Power Dissipation

The CMOS VLSI chip having a very large number of transistors. The nMOS and pMOS transistors used in a CMOS logic gate generally have nonzero reverse leakage and sub threshold currents.

These currents can contribute to the overall power dissipation even when the transistors are not undergoing any switching event. The magnitude of the leakage currents is determined mainly by the processing parameters.

In a MOSFET, mainly two components of leakage current are present.

(i) The reverse diode leakage current

(ii) The sub threshold current

5.3.3.1 Discussion about the Current Leakage Components

In this topic, we are discussing about the both components of leakage current sequentially.

1. For the reverse diode leakage current

- When the pn-junction between the drain and the bulk of the transistor is reverse biased, then the reverse diode leakage occurs. The reverse-biased drain junction then conducts a reverse saturation current which is drawn from the power supply.
- Here we consider a CMOS inverter with a high input voltage, where the nMOS transistor and pMOS transistor is turned on & turned off respectability.
- Initially the output node voltage is discharged to zero and there will be a reverse potential difference of V_{DD} between its drain and the n-well, causing a diode leakage through the drain junction. The n-well region of the pMOS transistor is also reverse biased with V_{DD},

with respect to the p-type substrate. Therefore, another significant leakage current component exists because of the n-well junction (Fig. 5.5).

- A similar situation can be observed when the input voltage is zero, and the output voltage is charged up to V_{DD} through the pMOS transistor. Then, the reverse potential difference between the nMOS drain region and the p-type substrate causes a reverse leakage current which is also drawn from the power supply (through the pMOS transistor).

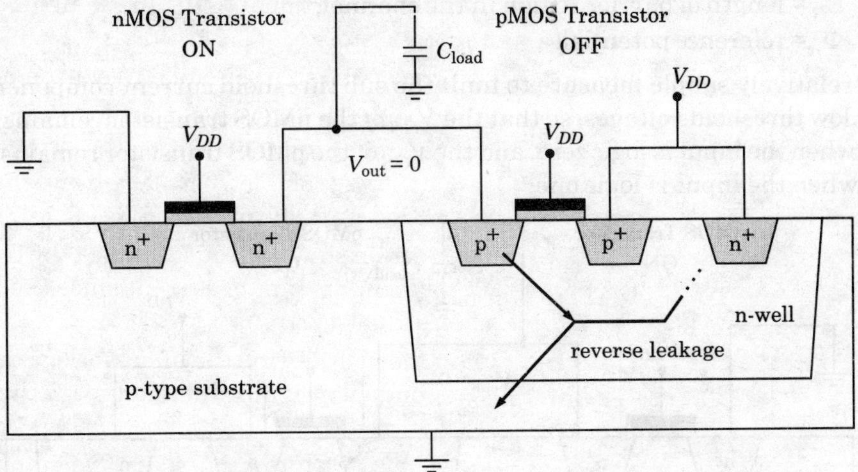

Fig. 5.5 Reverse leakage current paths in a CMOS inverter with high input voltage.

- The reverse leakage current of a pn-junction is expressed by

$$I_{\text{reverse}} = A \cdot J_S \left(e^{\frac{qV_{\text{bias}}}{kT}} - 1 \right) \quad \text{...(5.15)}$$

where, V_{bias} = reverse bias voltage across the junction,
J_S = reverse saturation current density,
A = junction area.

- The typical reverse saturation current density is 1–5 pA/μm^2, and it increases quite significantly with temperature. The reverse leakage occurs even during the stand-by operation when no switching takes place.

2. For the Sub threshold current

- The sub threshold current is another component of leakage currents in CMOS circuits. This component is due to carrier diffusion between the source and the drain regions of the transistor in weak inversion.
- The behaviour of an MOS transistor in the sub threshold operating region is similar to a bipolar device.
- The sub threshold current is exponentially dependent on the gate voltage. The amount of the sub threshold current may become significant when the $V_{GS} \leq V_{TH}$.
- In this case, the power dissipation due to sub threshold leakage can become comparable in magnitude to the switching power dissipation of the circuit. The device structure for sub threshold leakage current is shown in Fig. 5.6.
- The sub threshold leakage current can occur even when there is no switching activity in the circuit and that this component must be carefully considered for estimating the total power dissipation in the stand-by operation mode.

- The sub threshold current expression is given below:

$$I_D\,(\text{subthreshold}) = \frac{qD_n W x_c n_0}{L_B} \cdot e^{\frac{q\phi_r}{kT}} \cdot e^{\frac{q}{kT}(A \cdot V_{GS} + B \cdot V_{DS})} \quad ...(5.16)$$

Where, X_c = sub threshold channel length,
D_n = electron diffusion coefficient,
L_B = length of barrier region in the channel,
Φ_r = reference potential.

- One relatively simple measure to limit the sub threshold current component is to avoid very low threshold voltages, so that the V_{GS} *of* the nMOS transistor remains safely below V_{Tn} when the input is logic zero, and the V_{GS} of the pMOS transistor remains safely below V_{TP} when the input is logic one.

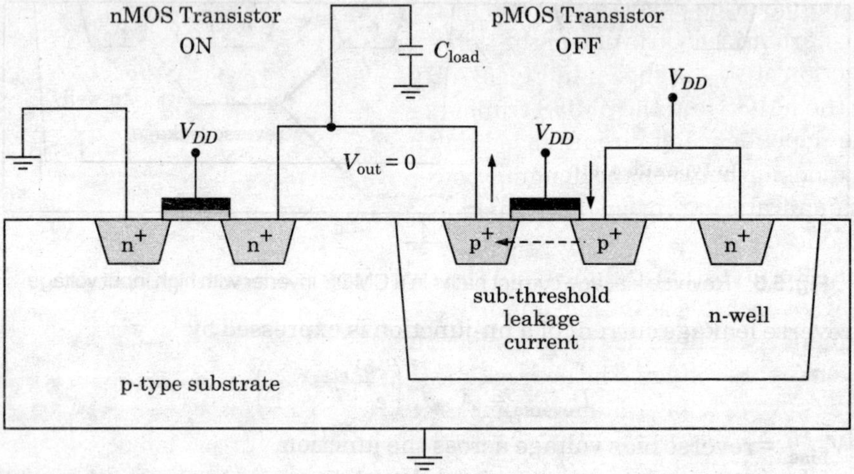

Fig. 5.6 Subthreshold leakage current path in a CMOS inverter with high input voltage.

Here, we discussed about three major sources of power consumption in CMOS digital integrated circuits. Some chips may also contain circuits which consume static power. One example is the pseudo-nMOS logic circuits which utilize a pMOS transistor as the pull-up device, which was shown that the circuit exhibits a non zero static power dissipation due to the load current. The presence of such circuits should be taken into account when estimating the overall power dissipation.

The total power dissipation in CMOS digital circuits can be expressed as the sum of four components,

$$P_{total} = \alpha_T \cdot C_{load} \cdot V_{DD}^2 \cdot f_{CLK} + V_{DD}(I_{short\text{-}circuit} + I_{leakage} + I_{static}) \quad ...(5.17)$$

where, $I_{short\text{-}circuit}$ = the average short-circuit current,
$I_{leakage}$ = the reverse leakage and sub threshold leakage currents,
I_{static} = the current component drawn from the power supply.

5.4 LOW-POWER DESIGN THROUGH VOLTAGE SCALING

The average switching power dissipation is proportional to the square of the power supply voltage, Here we discuss about the effects of reducing the power supply voltage V_{DD} upon switching power consumption and dynamic performance of the gate.

5.4.1 Influence of Voltage Scaling on Power and Delay

Due to the reduction of power supply voltage, reduces the dynamic power dissipation and the inevitable design trade-off is the increase of delay. We know that the propagation delay expressions for the CMOS inverter circuit is given by

$$\tau_{PHL} = \frac{C_{load}}{k_n(V_{DD} - V_{T,n})}\left[\frac{2V_{T,n}}{V_{DD} - V_{T,n}} + ln\left(\frac{4(V_{DD} - V_{T,n})}{V_{DD}} - 1\right)\right] \quad ...(5.18a)$$

$$\tau_{PLH} = \frac{C_{load}}{k_p(V_{DD} - V_{T,p})}\left[\frac{2\,|V_{T,p}|}{V_{DD} - |V_{T,p}|} + ln\left(\frac{4(V_{DD} - |V_{T,p}|)}{V_{DD}} - 1\right)\right] \quad ...(5.18b)$$

- If the power supply voltage is scaled down while all other parameters are kept constant, the propagation delay time would increase.
- The normalized variation of the delay *and* the average switching power dissipation as a function of V_{DD} *is* shown in Fig. 5.7. Here our assumption is that the threshold voltages for the nMOS and the pMOS transistor are $V_{Tn} = 0.8$ V and $V_{TP} = -0.8$ V, respectively.
- The dependence of circuit speed on the power supply voltage may also influence the relationship between the dynamic power dissipation and the supply voltage. We knew that the quadratic improvement (reduction) of power consumption as the power supply voltage is reduced.
- This interpretation assumes that the switching frequency remains constant. If the circuit is always operated at the maximum frequency allowed by its propagation delay, the operating frequency will drop as the propagation delay becomes larger with the reduction of the power supply voltage. The net result is that the dependence of switching power dissipation on the power supply voltage becomes stronger than a simple quadratic relationship, shown in Fig. 5.7.

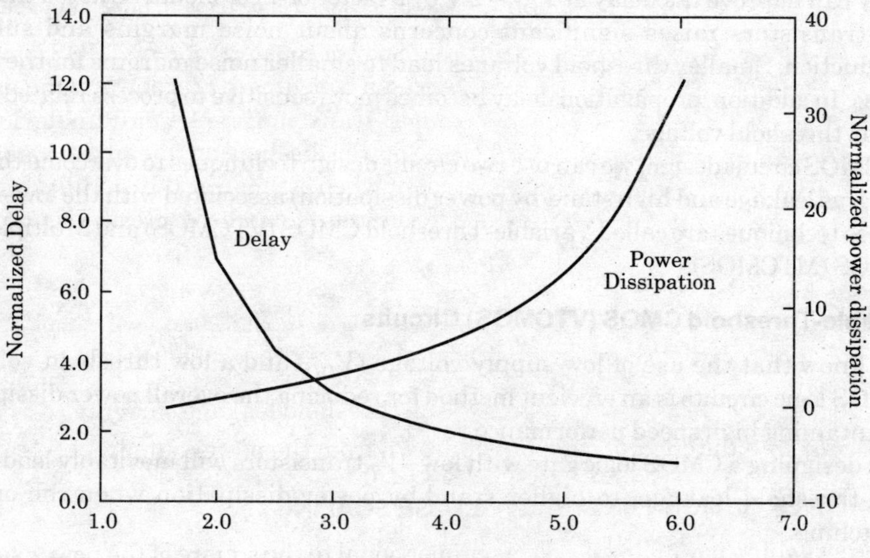

Fig. 5.7 Normalized propagation delay and average switching power dissipation of a CMOS inverter

- In this voltage scaling, the power supply voltage as well as the critical device dimensions (channel length, gate oxide thickness) and doping densities are scaled by the same factor. The variation of the propagation delay of a CMOS inverter as a function of the power supply voltage for different threshold voltage values is shown in Fig. 5.8.

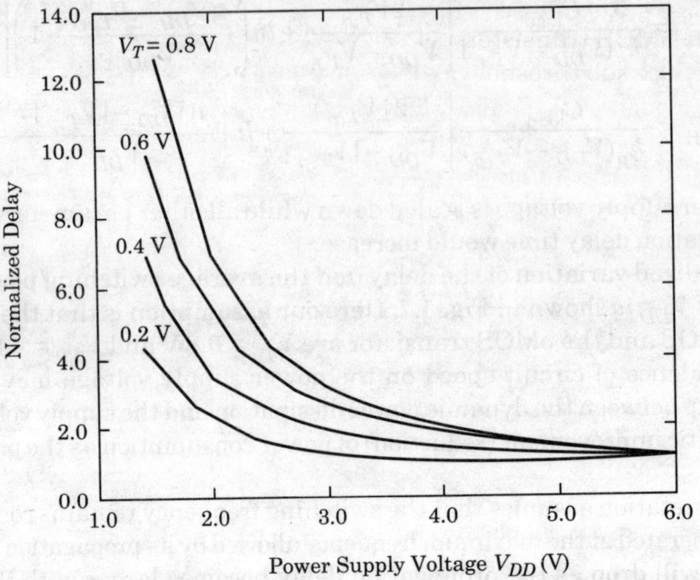

Fig. 5.8 Variation of the normalized propagation delay of a CMOS inverter w.r.t. Power supply voltage and threshold voltage

- According to the Fig. 5.8, we can see that the reducing of threshold voltage from 0.8 V to 0.2 V can improve the delay at V_{DD} = 2 V by a factor of 2. It should be noted that using low-VT transistors raises significant concerns about noise margins and sub threshold conduction. Smaller threshold voltages lead to smaller noise margins for the CMOS logic gates. In addition, propagation delay becomes more sensitive to process related fluctuations of the threshold voltage.
- In CMOS circuit design, we can use two circuit design techniques to overcome the difficulties (such as leakage and high stand-by power dissipation) associated with the low – V_T circuits. These techniques are called Variable-Threshold CMOS (VTCMOS) and Multiple–Threshold CMOS (MTCMOS).

5.4.2 Variable-Threshold CMOS (VTCMOS) Circuits

- We know that the use of low supply voltage (V_{DD}) and a low threshold voltage (V_T) in CMOS logic circuits is an efficient method for reducing the overall power dissipation, while maintaining high speed performance.
- But designing a CMOS logic gate with low – V_T transistors will inevitably lead to increased sub-threshold leakage, to higher stand-by power dissipation when the output is not switching.
- This problem is reducing by the adjustment of threshold voltages of the transistors in order to avoid leakage in the stand-by mode, by changing the substrate bias.
- In conventional CMOS logic circuits, the substrate terminals of all nMOS transistors are connected to ground potential, while the substrate terminals of all pMOS transistors are

connected to V_{DD}. This ensures that the source and drain diffusion regions always remain reverse-biased with respect to the substrate, and that the threshold voltages of the transistors are not significantly influenced by the body (back gate-bias) effect.

- In VTCMOS circuit technique, the transistors are designed inherently with a low threshold voltage, and the substrate bias voltages of nMOS and pMOS transistors are generated by a variable substrate bias control circuit, as shown in Fig. 5.9. The threshold voltages of nMOS and pMOS transistors are increased by adjusting the substrate bias voltage, in order to reduce sub threshold leakage currents in the stand-by mode.

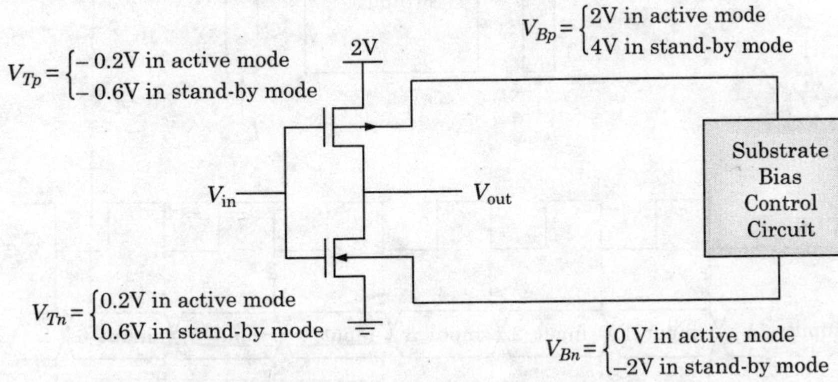

Fig. 5.9 A variable-threshold CMOS (VTCMOS) inverter circuit.

- When the inverter circuit in Fig. 5.9 is operating in its *active* mode, the substrate bias voltage of the nMOS transistor is Von = 0 and the substrate bias voltage of the pMOS transistor is $V_{BP} = V_{DD}$. Thus, the inverter transistors do not experience any back gate-bias effect.
- When the inverter circuit is in the stand-by mode, the substrate bias control circuit generates a lower substrate bias voltage for the nMOS transistor and a higher substrate bias voltage for the pMOS transistor. As a result, the magnitudes of the threshold voltages V_{Tn} and V_T both increase in the stand-by mode, due to the back gate bias effect. Since the sub threshold leakage current drops exponentially with increasing threshold voltage, the leakage power dissipation in the stand-by mode can be significantly reduced with this technique.
- The VTCMOS technique can also be used to automatically control the threshold voltages of the transistors in order to reduce leakage currents, and to compensate for process-related fluctuations of the threshold voltages. This approach is also called the Self-Adjusting Threshold-Voltage Scheme (SATS).
- The variable-threshold CMOS circuit design techniques are very effective for reducing the sub-threshold leakage currents and for controlling threshold voltage values in low V_{DD} – low V_T applications.
- This technique usually requires twin-well or triple-well CMOS technology in order to apply different substrate bias voltages to different parts of the chip.

5.4.3 Pipelining Approach

- Firstly, we consider a single functional block which implements a logic function F(INPUT) of the input vector, INPUT. This architectural design is used for reducing power dissipation. Here, both the input and the output vectors are sampled through register arrays and driven by a clock signal CLK.

- The critical path in this logic block (at a power supply voltage of V_{DD}) allows a maximum sampling frequency of f_{CLK} is equal to or less than $T_{CLK} = l/f_{CLK}$. *The* functional block and simplified timing diagram of the circuit is shown in Fig. 5.10.

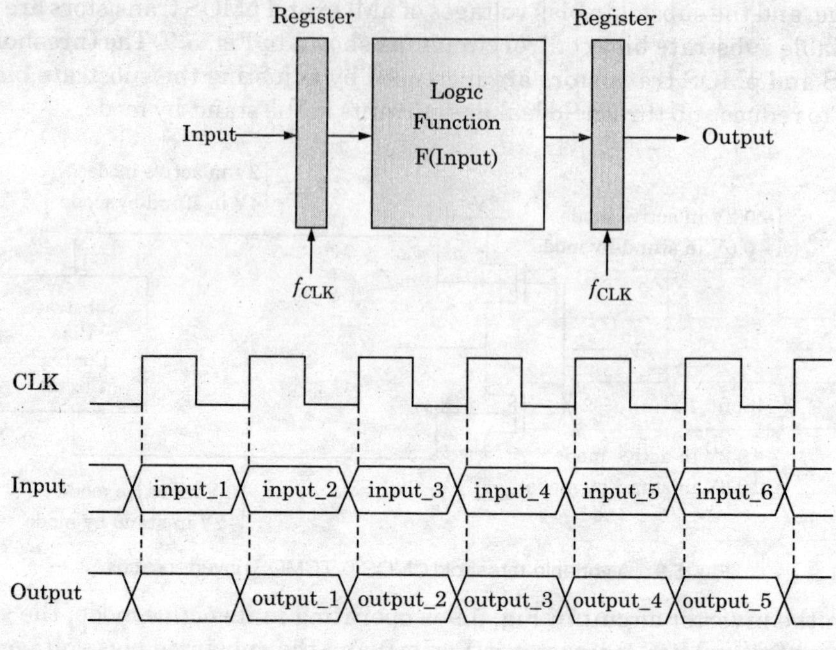

Fig. 5.10 Single-stage implementation of a logic function and its simplified timing diagram.

We assume that the total capacitance switched every clock cycle (C_{total}) *consists* of:

(i) the capacitance switched in the input register array,

(ii) the capacitance switched to implement the logic function,

(iii) the capacitance switched in the output register array.

So, the dynamic power consumption of this structure can be found as;

$$P_{\text{reference}} = C_{\text{total}} \cdot V_{DD}^2 \cdot f_{\text{CLK}} \quad \text{...(5.19)}$$

- Now we consider an N-stage pipelined structure for implementing the same logic function, as shown in Fig. 5.11. The logic function F (INPUT) has been partitioned into successive stages, and a total of $(N-1)$ register arrays have been introduced to create the pipeline.
- All registers are clocked at the original sample rate, f_{CLK}. If all stages of the partitioned function have approximately equal delays of

$$\tau_P(\text{pipeline} - \text{stage}) = \frac{\tau_{P,\max}(\text{input - to - output})}{N} = T_{\text{CLK}} \quad \text{...(5.20)}$$

- Then the logic blocks between two successive registers can operate N-times slower while *maintaining the same functional throughput* as before. This implies that the power supply voltage can be reduced to a value of V_{DD}, *new, to* effectively slow down the circuit by a factor of N.
- The maximum pipeline stage delay is equal to the clock period, and the latency is N clock cycles.

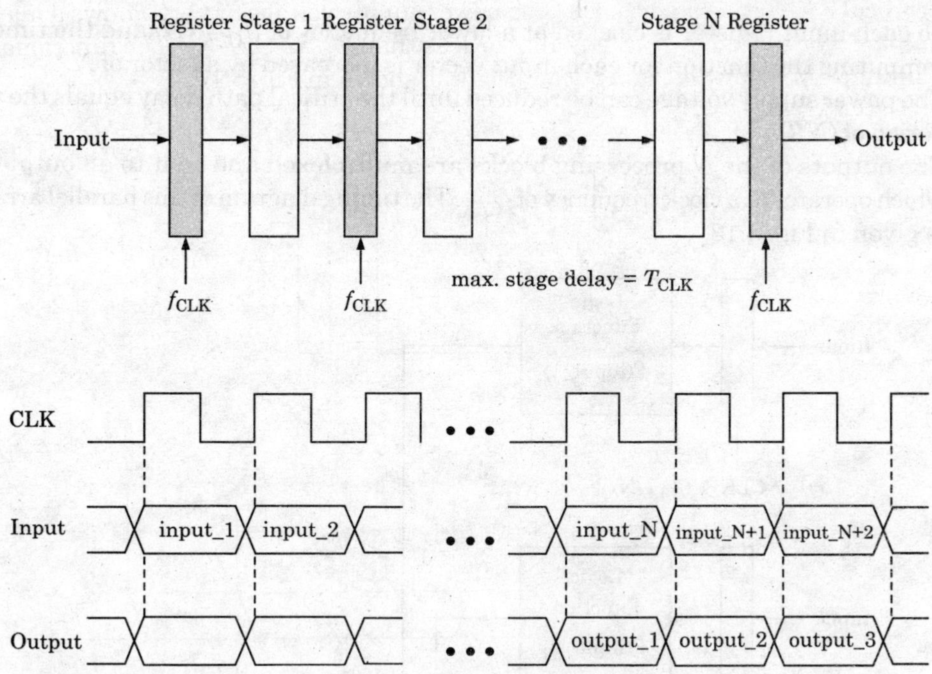

Fig. 5.11 *N*-stage pipeline structure realizing the same logic functions as in Fig. 5.10.

The dynamic power consumption of the N-stage pipelined structure with a lower supply voltage and with the same functional throughput as the single-stage structure can be approximated by

$$P_{\text{pipeline}} = [C_{\text{total}} + (N-1)\, C_{\text{reg}}] \cdot V_{DD,\text{new}}^2 \cdot f_{\text{CLK}} \qquad \text{...(5.21)}$$

Where Creg represents the capacitance switched by each pipeline register.

Then, the power reduction factor achieved in a N-stage pipeline structure is

$$\frac{P_{\text{pipeline}}}{P_{\text{reference}}} = \frac{[C_{\text{total}} + (N-1)\, C_{\text{reg}}] \cdot V_{DD,\text{new}}^2 \cdot f_{\text{CLK}}}{C_{\text{total}} \cdot V_{DD}^2 \cdot f_{\text{CLK}}} = \left[1 + \frac{C_{\text{reg}}}{C_{\text{total}}}(N-1)\right] \frac{V_{DD,\text{new}}^2}{V_{DD}^2} \qquad \text{...(5.22)}$$

- The architectural modification described here has a relatively small area overhead. A total of $(N-1)$ register arrays have to be added to convert the original single-stage structure into a pipeline. While trading off area for lower power, this approach also increases the latency from one to N clock cycles. Yet in many applications such as signal processing and data encoding, latency is not a very significant concern.

5.4.4 Parallel Processing Approach (Hardware Replication)

- This approach could be useful especially when the logic function to be implemented is not suitable for pipelining. The N-block parallel structure realizing the same logic function is shown in Fig. 5.12.
- Assume that the consecutive input vectors arrive at the same rate as in the single stage case. The input vectors are routed to all the registers of the N processing blocks.
- Gated clock signals, each with a clock period of (NT_{CLK}), are used to load each register every N clock cycles. This means that the clock signals to each input register are skewed by T_{CLK}, so each one of the N consecutive input vectors is loaded into a different input register.

- So each input register is clocked at a lower frequency of (f_{CLK}/N) and the time allowed computing the function for each input vector is increased by a factor of N.
- The power supply voltage can be reduced until the critical path delay equals the new clock period of $(N\,T_{CLK})$.
- The outputs of the N processing blocks are multiplexed and sent to an output register which operates at a clock frequency of f_{CLK}. The timing diagram of this parallel arrangement is given in Fig. 5.13.

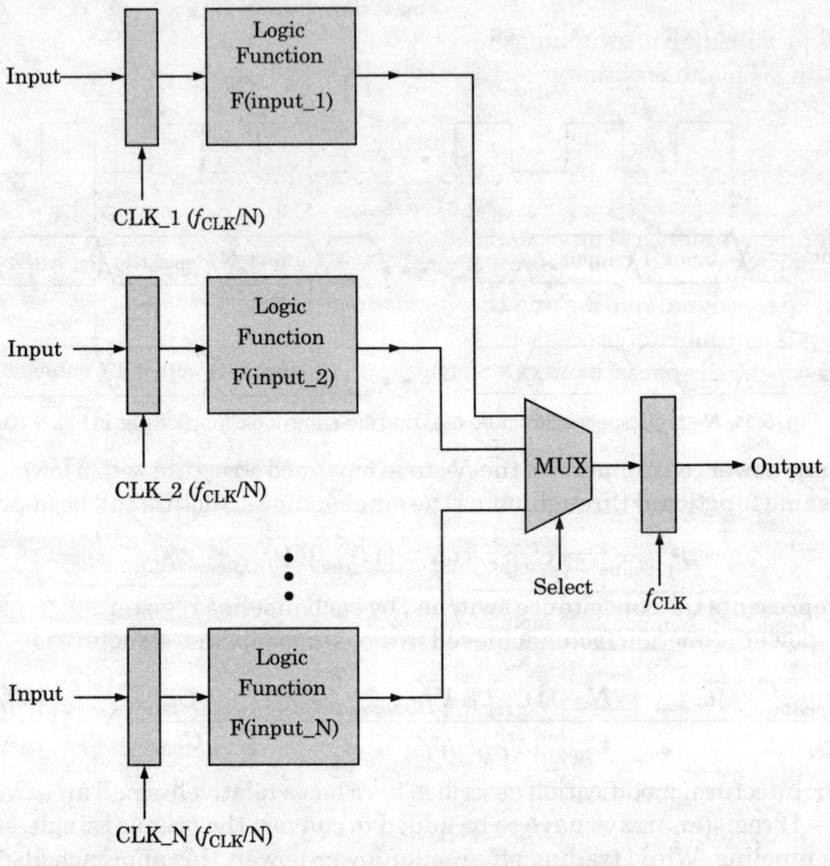

Fig. 5.12 N-block parallel structures realizing the same logic function as in Fig. 5.10.

- The time allowed computing the function for each input vector is increased by a factor of N; the power supply voltage can be reduced to a value of V_{DD}, new, to effectively slow down the circuit.
- The total dynamic power dissipation of the parallel structure (neglecting the dissipation of the multiplexor) is found as the sum of the power dissipated by the input registers and the logic blocks operating at a clock frequency of (f_{CLK}/N), and the output register operating at a clock frequency of f_{CLK}.

$$P_{\text{parallel}} = N \cdot C_{\text{total}} \cdot V_{DD,\text{new}}^2 \cdot \frac{f_{\text{CLK}}}{N} + C_{\text{reg}} \cdot V_{DD,\text{new}}^2 \cdot f_{\text{CLK}}$$

$$= \left(1 + \frac{C_{\text{reg}}}{C_{\text{total}}}\right) \cdot C_{total} \cdot V_{DD,\text{new}}^2 \cdot f_{\text{CLK}} \quad \text{...(5.23)}$$

- Note that there is also an additional overhead which consists of the input routing capacitance, the output routing capacitance and the capacitance of the output multiplexor structure, all of which are increasing functions of N. If this overhead is neglected, the amount of power reduction achievable in an N-block parallel implementation is

$$\frac{P_{\text{parallel}}}{P_{\text{reference}}} = \frac{V_{DD,\text{new}}^2}{V_{DD}^2} \cdot \left(1 + \frac{C_{\text{reg}}}{C_{\text{total}}}\right) \quad \text{...(5.24)}$$

- The lower bound of switching power reduction realizable with architecture-driven voltage scaling is found, assuming zero threshold voltage, as

$$\frac{P_{\text{parallel}}}{P_{\text{reference}}} \geq \frac{1}{N^2} \quad \text{...(5.25)}$$

- A total of N identical processing blocks must be used to slow down the operation speed by a factor of N.
- Due to the signal routing and the overhead circuitry, the silicon area will grow even faster than the number of processors.
- The pipelined approach offers a more efficient alternative for reducing the power dissipation.

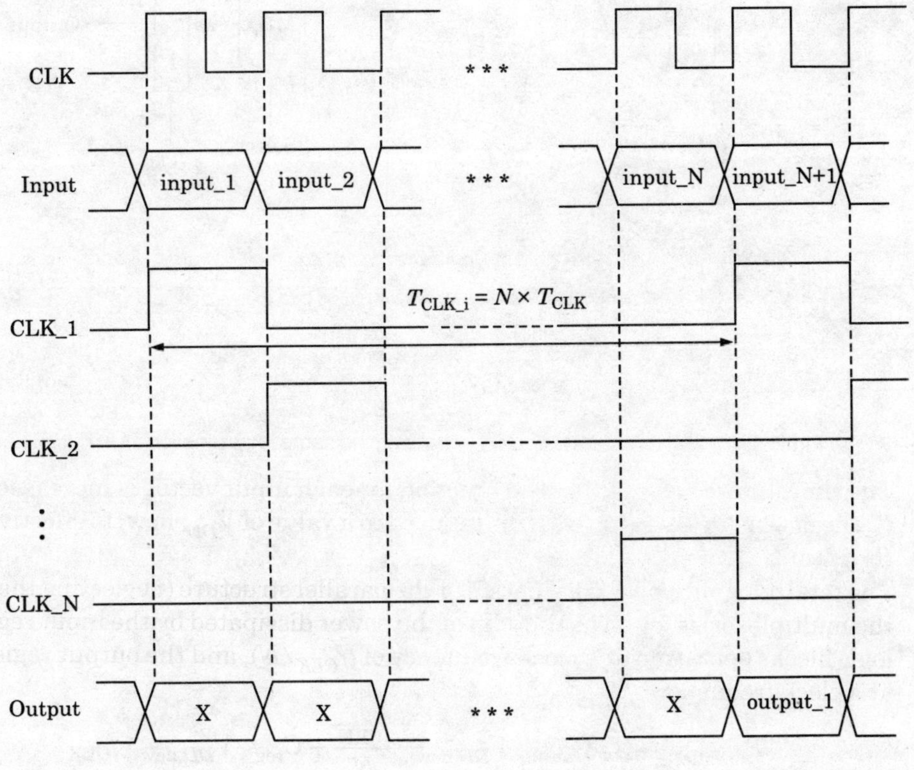

Fig. 5.13 Simplified timing diagram of the N-block parallel structure shown in Fig. 5.12.

REVIEW QUESTIONS

1. Design and adiabatic full adder circuit. Compare its transistor count with
   (*a*) the conventional CMOS full adder circuit and
   (*b*) the pass-transistor full adder circuit.
2. Discuss about the low-power design through voltage scaling.
3. Explain the following circuits in detail
   (*a*) Variable Threshold CMOS circuits
   (*b*) Multiple threshold CMOS circuits
4. What do you understand by the leakage power dissipation
5. Explain the different types of power dissipation.
6. Describe: (*a*) Pipelining Approach.
   (*b*) Step-wise charging circuits.
7. Explain adiabatic logic circuits in detail.
8. Describe: (*a*) Dynamic Power Dissipation.
   (*b*) Short-circuit Power Dissipation.
   (*c*) Leakage Power Dissipation.
9. What are the various sources of power dissipation in CMOS circuits? **(UPTU–2011–12)**
10. Derive the total power consumption in CMOS inverter. **(UTU–2012)**

❑❑❑

6 Combinational MOS Logic Circuits

Chapter

Chapter Overview

- Introduction
- MOS Logic Circuits with Depletion nMOS Loads
- Ratioed Logic
- Complementary Pass-Transistor Logic (CPL)
- Static CMOS Design
- CMOS Logic Circuits
- CMOS Transmission Gate (Pass-Transistor)

6.1 INTRODUCTION

A *combinational circuit* is one where the output at any time depends only on the present combination of inputs at that point of time with total disregard to the past state of the inputs. The logic gate is the most basic building block of combinational logic. The logical function performed by a combinational circuit is fully defined by a set of Boolean expressions. There is no internal connection from outputs back to inputs.

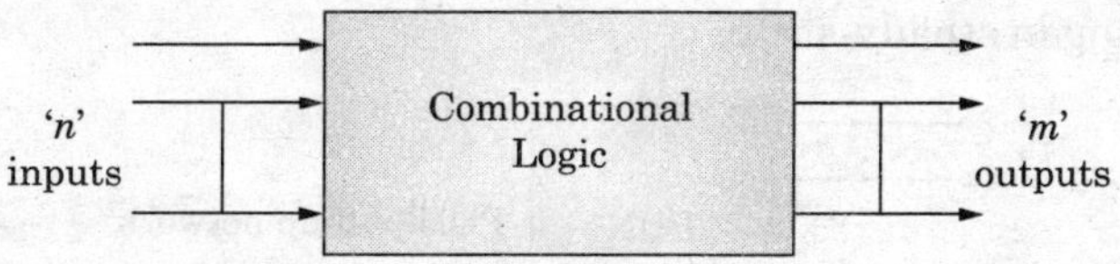

Fig. 6.1 Block schematic representation of a generalized combinational circuit

Figure 6.1 shows the block schematic representation of a generalized combinational circuit having *n* input variables and *m* output variables or simply outputs. Since the number of input variables is *n*, there are 2^n possible combinations of bits at the input. Each output can be expressed in terms of input variables by a Boolean expression, with the result that the generalized system of Fig. 6.1 can be expressed by *m* Boolean expressions. The common design metrics by which a gate is evaluated are area, speed, energy and power. In addition to these metrics, robustness to noise and reliability are also very important considerations. We will see that certain logic styles can significantly improve performance, but they usually are more sensitive to noise. In this chapter, we will examine the static and dynamic characteristics of various combinational MOS logic circuits.

6.2 STATIC CMOS DESIGN

The most widely used logic is static complementary CMOS. The primary advantage of the CMOS structure is robustness, good performance and low power consumption with no static power dissipation.

- The static behavior of a circuit is discussed in term of its functionality and robustness. The static CMOS inverter has excellent properties in many areas like low sensitivity to noise and process variation, excellent speed and low power consumption. So these properties are carried over to more complex logic gates implemented using the same topology.
- Simple gates using CMOS were discussed in previous chapter but increasing the number of gates inputs resulted in increased chip area and larger propagation delays.
- Design styles other than complementary design are revised like radioed and pass transistor logic styles that can address the issues like chip area and delay. These two design style are also static in nature and at every point in time each gate output is connected either to V_{DD} or GND via a low resistance path. The outputs of the gates also assume that at a all times the value of the Boolean function implement by the circuit.
- The rationed logic style consists of an active pull down (or up) network that is connected to a load. So the design complexity is much reduced but static power consumption increases substantially. Also transistor sizings (W/L) have to be done carefully to maintain large noise margins.
- Pass transistor logic uses a logic gate as simple switch network. So some logic functions can be very easily implemented. But it long chains of switches are used then propagation delay also increases. Also static power consumption may increase and noise margins may be reduced.

6.2.1 Complementary Logic

A static CMOS gate is a combination of two networks-the pull-up network (PUN) and the pull-down (PDN) as shown in Fig. 6.2. The function of the PUN is to provide a connection between the output and V_{DD} anytime the output of the logic gate is meant to be 1. Similarly, the function of the PDN is to connect the output to V_{SS} when the output of the logic gate is meant to be 0. The PUN and PDN networks are constructed in a mutually exclusive such that one and only one of the networks are conducting in steady-state.

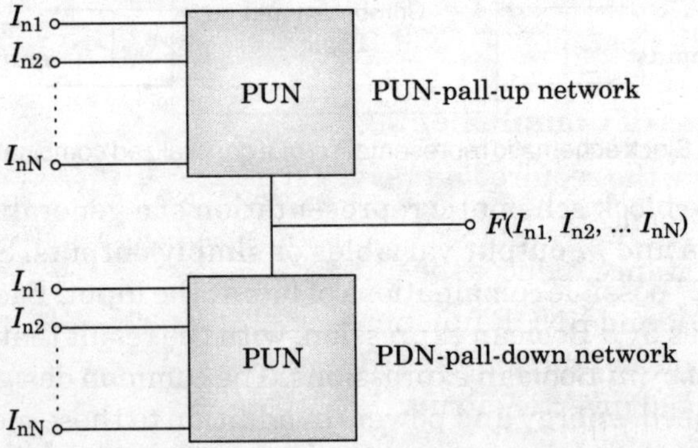

Fig. 6.2 Complementary logic gates as combination of pull-down and pull-up networks.

In constructing the PDN and PUN networks, the following points should be noted:

- A transistor can be through of as a switch controlled by its gate signal. An NMOS switch assumes positive logic and PMOS switch assumes negative logic for their operation.
- The PDN is constructed using NMOS devices while PMOS transistors are used in the PUN. This is due to the fact that NMOS transistors produce 'strong 0s' and PMOS device generate 'strong 1s'.

6.2.1.1 Static Properties of Complementary Logic Gates

- They exhibit rail-to-rail swing with $V_{OH} = V_{DD}$ and V_{OL} = GND.
- The circuits have no static power dissipation, since the circuits are designed such that the pull-down and pull-up networks are mutually exclusive.
- The analysis of dc voltage transfer characteristics and the noise margins is more complicated than for the inverter, as these parameters depend upon the data input pattern applied to the gate.
- The main advantage is that it is a very robust and simple approach for implementing logic gates.

6.2.1.2 Problems with Complementary Logic Gates

Two major problems associated with using this style as the complexity of the gate (fan-in) increases are:

1. The number of transistors required to implement an N fan-in gate is $2N$. This can result in a significantly large implementation area.
2. The propagation delay of a complementary CMOS gate deteriorates rapidly as a function of the fan-in.

6.2.1.3 Advantages of Complementary Logic

1. Easy to design
2. Simple to analyze
3. Robust operation
4. Static circuit
5. Ratio less logic
6. Maximum logic swing from V_{DD} to ground
7. High noise margin
8. No static power dissipation

6.2.1.4 Disadvantages of Complementary Logic

1. For given n inputs, the required number of transistor is $2n$. It means occupying larger area on chip the MOS logic.
2. High input capacitance reduces the ultimate performance of the circuit.
3. Only NAND, NOR and XNOR functions can be implemented whereas the realization of a non-inverting Boolean function like OR, AND and XOR in a single stage is not possible and needs the additional inverter circuit.

6.3 MOS LOGIC CIRCUITS WITH DEPLETION NMOS LOADS

6.3.1 Two-Input NOR Gate

- The circuit diagram, the logic symbol, and the corresponding truth table of the two-input NOR gate is shown in Fig. 6.3.

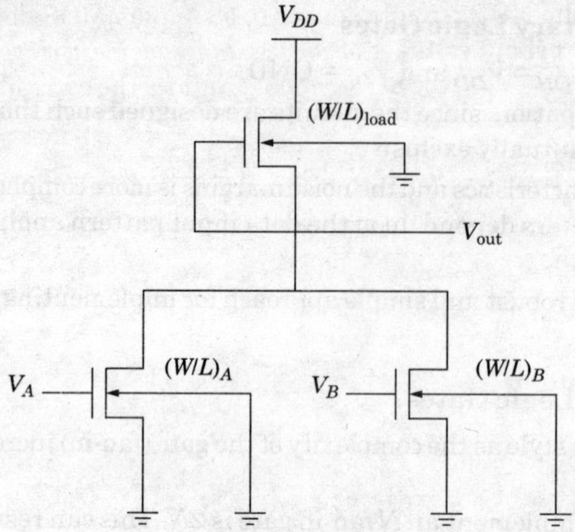

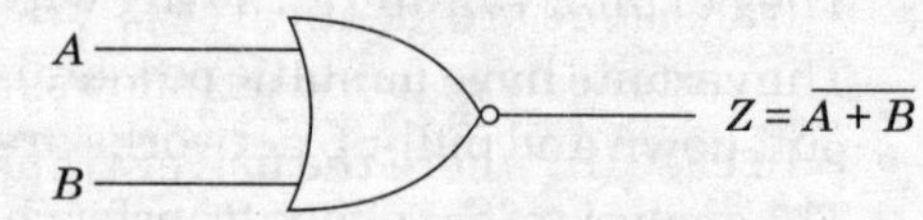

V_A	V_B	V_{out}
low	low	high
low	high	low
high	low	low
high	high	low

Fig. 6.3 A two-input depletion-load NOR gate, its logic symbol.

- The Boolean OR operation is performed by the parallel connection of the two enhancement-type nMOS driver transistors.
- When both V_A and V_B are high, then the corresponding driver transistor turns on and provides a conducting path between the output node and the ground. Hence, the output voltage becomes low. In this case, the circuit operates like a depletion-load inverter with respect to its static behavior.
- If both V_A and V_B are low, both driver transistors remain cutoff. The output node voltage is pulled to a logic-high level by the depletion-type nMOS load transistor.
- The DC analysis of the circuit can be simplified significantly by considering the structural similarities between this circuit and the simple nMOS depletion-load inverter. In the following, the calculation of output low and output high voltages will be examined.

6.3.1.1 Calculation of V_{OH}

- When both input voltages V_A and V_B are lower than the corresponding driver threshold voltage, the driver transistors are turned off and conduct no drain current. So, the load device operates in the linear region, also has zero drain current.
- Its linear region current equation becomes

$$I_{D,\,\text{load}} = \frac{k_{n,\text{load}}}{2} \cdot [2\,|V_{T,\text{load}}(V_{OH})|\cdot(V_{DD} - V_{OH}) - (V_{DD} - V_{OH})^2] = 0 \quad \text{...(6.1)}$$

- The solution of this equation gives $V_{OH} = V_{DD}$.

6.3.1.2 Calculation of V_{OL}

For finding the output low voltage V_{OL}, we must consider three different cases (three different input voltage combinations), which produce a conducting path from the output node to the ground. These cases are:

(i) $V_A = V_{OH}$ $\quad V_B = V_{OL}$

(ii) $V_A = V_{OL}$ $\quad V_B = V_{OH}$

(iii) $V_A = V_{OH}$ $\quad V_B = V_{OH}$

- For the first two cases, (i) and (ii), the NOR circuit reduces to a simple nMOS depletion load inverter. Assuming that the threshold voltages of the two enhancement-type driver transistors are identical ($V_{T0,A} = V_{T0,B} = V_{T0}$), the driver-to-load ratio of the corresponding inverter can be found as follows.
- In case (i), where the driver transistor A is on, the ratio is

$$k_R = \frac{k_{driver,A}}{k_{load}} = \frac{k'_{n,driver}\left(\frac{W}{L}\right)_A}{k'_{n,load}\left(\frac{W}{L}\right)_{load}} \quad \text{...(6.2)}$$

- In case (ii), where the driver transistor B is on, the ratio is

$$k_R = \frac{k_{driver,B}}{k_{load}} = \frac{k'_{n,driver}\left(\frac{W}{L}\right)_B}{k'_{n,load}\left(\frac{W}{L}\right)_{load}} \quad \text{...(6.3)}$$

The output low voltage level V_{OL} in both cases is given by:

$$V_{OL} = V_{OH} - V_{T0} - \sqrt{(V_{OH} - V_{T0})^2 - \left(\frac{k_{load}}{k_{driver}}\right) \cdot |V_{T,load}(V_{OL})|^2} \quad \text{...(6.4)}$$

Note that if the (W/L) ratios of both drivers are identical, i.e., $(W/L)_A = (W/L)_B$ the output low voltage (V_{OL}) values calculated for case (i) and case (ii) will be identical.

- In case (iii), where both driver transistors are turned on, the saturated load current is the sum of the two linear-mode driver currents.

$$I_{D,\,load} = I_{D,\,driver\,A} + I_{D,\,driver\,B} \quad \text{...(6.5)}$$

$$\frac{k_{load}}{2}|V_{T,load}(V_{OL})|^2 = \frac{k_{driver,A}}{2}[2(V_A - V_{T0})V_{OL} - V_{OL}^2] + \frac{k_{driver,B}}{2}[2(V_B - V_{T0})V_{OL} - V_{OL}^2] \quad \text{...(6.6)}$$

Since the gate voltages of both driver transistors are equal ($V_A = V_B = V_{OH}$), we can devise an equivalent driver-to-load ratio for the NOR structure:

$$k_R = \frac{k_{driver,A} + k_{driver,B}}{k_{load}} = \frac{k'_{n,driver}\left[\left(\frac{W}{L}\right)_A + \left(\frac{W}{L}\right)_B\right]}{k'_{n,load}\left(\frac{W}{L}\right)_{load}} \quad \text{...(6.7)}$$

- Thus, the NOR gate with both of its inputs tied to a logic-high voltage is replaced with an nMOS depletion-load inverter circuit with the driver-to-load ratio given by (6.7).
- The output voltage level in this case is

$$V_{OL} = V_{OH} - V_{T0} - \sqrt{(V_{OH} - V_{T0})^2 - \left(\frac{k_{load}}{k_{driver,A} + k_{driver,B}}\right) |V_{T,load}(V_{OL})|^2} \quad \text{...(6.8)}$$

- We have to achieve a certain maximum V_{OL} for the worst case, i.e., when only one input is high. Thus, we assume that one input (either V_A or V_B) is logic-high and determine the driver to- load ratio of the resulting inverter using (6.4). Then set

$$k_{\text{driver}, A} = k_{\text{driver}, B} = k_R \, k_{\text{load}} \quad \text{...(6.9)}$$

- When both inputs are logic-high, the output voltage is even lower than the required maximum V_{OL} thus the design constraint is satisfied.

6.3.2 Generalized NOR Structure with Multiple Inputs

- In this topic, we discussed about the analysis to generalize n-input NOR gates, which consist of *n* parallel driver transistors. The circuit arrangement for the generalize *n*-input NOR gates is shown in Fig. 6.4.
- The combined current I_D in this circuit is supplied by the driver transistors which are turned on, i.e., transistors which have gate voltages higher than the threshold voltage V_{T0}.
- The combined pull-down current can then be expressed as follows:

$$I_D = \sum_{k(on)} I_{D,k} = \begin{cases} \displaystyle\sum_{k(on)} \frac{\mu_n C_{ox}}{2}\left(\frac{W}{L}\right)_k \left[2(V_{GS,k} - V_{T0})\, V_{\text{out}} - V_{\text{out}}^2\right] & \text{linear} \\ \displaystyle\sum_{k(on)} \frac{\mu_n C_{ox}}{2}\left(\frac{W}{L}\right)_k (V_{GS,k} - V_{T0})^2 & \text{saturation} \end{cases} \quad \text{...(6.10)}$$

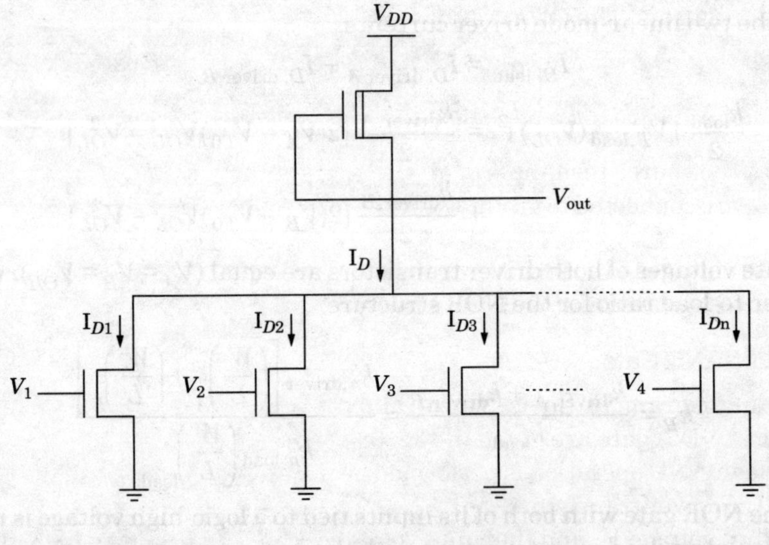

Fig. 6.4 Generalized *n*-input NOR gate.

Assuming that the input voltages of all driver transistors are identical,

$$V_{GS,\,k} = V_{GS} \quad \text{for} \quad k = 1, 2, \ldots, n \quad \text{...(6.11)}$$

The pull-down current expression can be rewritten as;

$$I_D = \begin{cases} \dfrac{\mu_n C_{ox}}{2}\left(\sum_{k(on)}\left(\dfrac{W}{L}\right)_k\right)[2(V_{GS} - V_{T0})\,V_{out} - V_{out}^2] & \text{linear} \\ \dfrac{\mu_n C_{ox}}{2}\left(\sum_{k(on)}\left(\dfrac{W}{L}\right)_k\right)(V_{GS} - V_{T0})^2 & \text{saturation} \end{cases} \quad ...(6.12)$$

Thus, the multiple-input NOR gate can also be reduced to an equivalent inverter, shown in Fig. 6.5.

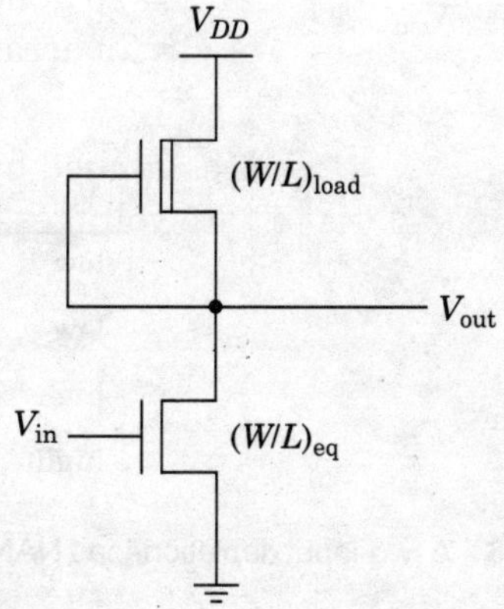

Fig. 6.5 Equivalent inverter circuit corresponding to the n-input NOR gate.

- For static analysis. The (W/L) ratio of the driver transistor here is

$$\left(\frac{W}{L}\right)_{equivalent} = \sum_{k(on)}\left(\frac{W}{L}\right)_k \quad ...(6.13)$$

- Note that the source terminals of all enhancement-type nMOS driver transistors in the NOR gate are connected to ground. Thus, the drivers do not experience any substrate-bias effect.
- The depletion-type nMOS load transistor is subject to substrate-bias effect, since its source is connected to the output node, and its source-to-substrate voltage is $V_{SB} = V_{out}$.

6.3.3 Two-Input NAND Gate

- The circuit diagram, the logic symbol, and the corresponding truth table of the two-input NAND (NAND2) gate are given in Fig. 6.6.
- The Boolean AND operation is performed by the series connection of the two enhancement type nMOS driver transistors.
- When input voltage V_A and the input voltage V_B are equal to logic-high, then a conducting path is introduced between the output node and the ground.
- When both of the series-connected drivers are turned on, the output voltage will be low, which is the complemented result of the AND operation. Otherwise, either one or both of the driver transistors will be off, and the output voltage will be pulled to a logic-high level by the depletion-type nMOS load transistor.
- Figure 6.6 shows that all transistors except the one closest to the ground are subject to substrate-bias effect, since their source voltages are larger than zero.

- We have to consider this fact in detailed calculations. For all of the three input combinations which produce a logic-high output voltage, the corresponding V_{OH} value can easily be found as $V_{OH} = V_{DD}$.

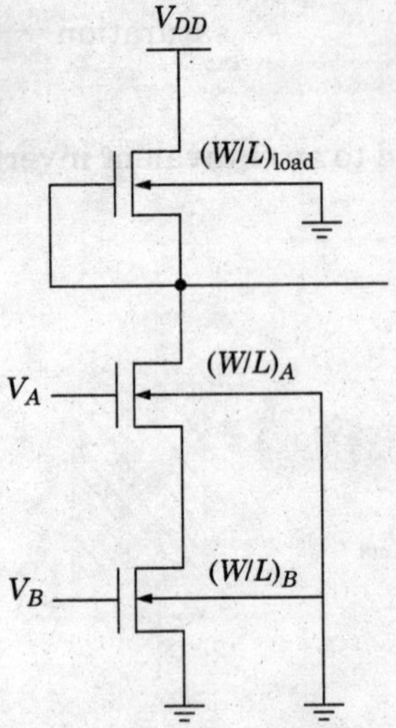

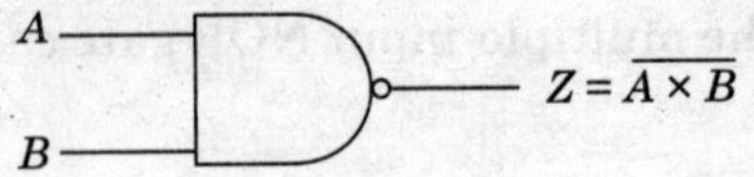

V_A	V_B	V_{out}
low	low	high
low	high	high
high	low	high
high	high	low

Fig. 6.6 A two-input depletion-load NAND gate.

- Consider the NAND2 gate with both of its inputs equal to V_{OH} as shown in Fig. 6.7. It can easily be seen that the drain currents of all transistors in the circuit are equal to each other.

$$I_{D,\text{ load}} = I_{D,\text{ driver }A} = I_{D,\text{ driver }B} \quad \text{...(6.14)}$$

$$\frac{k_{\text{load}}}{2} \,|\, V_{T,\text{load}}(V_{OL}) \,|^2 = \frac{k_{\text{driver},A}}{2}\left[2(V_{GS,A} - V_{T,A})V_{DS,A} - V_{DS,A}^2\right]$$

$$= \frac{k_{\text{driver},B}}{2}\left[2(V_{GS,B} - V_{T,B})V_{DS,B} - V_{DS,B}^2\right] \quad \text{...(6.15)}$$

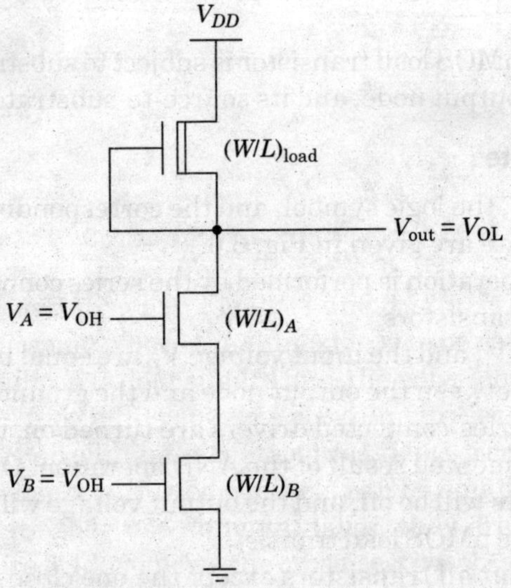

Fig. 6.7 The NAND2 gate with both of its inputs at logic-high level.

- The gate-to-source voltages of both driver transistors can be assumed to be approximately equal to V_{OH}.
- For simplicity, the substrate-bias effect for driver transistor A, and assume $V_{TA} = V_{TB} = V_{T0}$, since the source-to-substrate voltage of driver A is relatively low.
- The drain-to-source voltages of both driver transistors can then be solved from (6.15) as

$$V_{OL} \text{ as } V_{DS,A} = V_{OH} - V_{T0} - \sqrt{(V_{OH} - V_{T0})^2 - \left(\frac{k_{total}}{k_{driver,A}}\right) \cdot |V_{T,load}(V_{OL})|^2} \quad ...(6.16)$$

$$V_{DS,B} = V_{OH} - V_{T0} - \sqrt{(V_{OH} - V_{T0})^2 - \left(\frac{k_{load}}{k_{driver,B}}\right) \cdot |V_{T,load}(V_{OL})|^2} \quad ...(6.17)$$

- Let the two driver transistors be identical, i.e., $k,_{driver\,A} = k,_{driver\,B} = k,_{driver}$. Noting that the output voltage V_{OL} is equal to the sum of the drain-to-source voltages of both drivers, we obtain

$$V_{OL} \approx 2\left(V_{OH} - V_{T0} - \sqrt{(V_{OH} - V_{T0})^2 - \left(\frac{k_{load}}{k_{driver}}\right) \cdot |V_{T,load}\,(V_{OL})|^2}\right) \quad ...(6.18)$$

- The following analysis gives a better and more accurate view of the operation of two series-connected driver transistors. Consider the two identical enhancement-type nMOS transistors with their- gate terminals connected.
- At this point, the only simplifying assumption will be $V_{TA} = V_{TB} = V_{T0}$. When both driver transistors are in the linear region the drain currents can be written as;

$$I_{D,A} = \frac{k_{driver}}{2}\,[2(V_{GS,A} - V_{T0})V_{DS,A} - V_{DS,A}^2] \quad ...(6.19)$$

Since, $I_{D,A} = I_{D,B}$, this current can also be expressed as;

$$I_{D,B} = \frac{k_{driver}}{2}\,[2(V_{GS,B} - V_{T0})V_{DS,B} - V_{DS,B}^2] \quad ...(6.20)$$

Using $V_{GS,A} = V_{GS,B} - V_{DS,B}$,

$$I_D = I_{D,A} = I_{D,B} = \frac{I_{D,A} + I_{D,B}}{2} \quad ...(6.21)$$

Now let $V_{GS} = V_{GS,B}$ and $V_{DS} = V_{DS,B} + V_{DS,A}$. The drain-current expression can then be written as follows.

$$I_D = \frac{k_{driver}}{4}\,[2(V_{GS,B} - V_{T0})\,(V_{DS,A} + V_{DS,B}) - (V_{DS,A} + V_{DS,B})^2] \quad ...(6.22)$$

$$I_D = \frac{k_{driver}}{4}\,[2(V_{GS} - V_{T0})\,V_{DS} - V_{DS}^2] \quad ...(6.23)$$

Thus, two nMOS transistors connected in series and with the same gate voltage behave like one nMOS transistor with $K_{eq} = 0.5\,K_{driver}$.

6.3.2.1 Generalized NAND Structure with Multiple Inputs

- At this point, we can discuss about generalized n-input NAND gates, which consist of n series-connected driver transistors, as shown in Fig. 6.8.
- Neglecting the substrate bias effect, and assuming that the threshold voltages of all transistors are equal to V_{T0}., the driver current I_D in the linear region can be derived as in Eq. (6.24) whereas I_D in saturation is taken as its extension.

$$I_D = \frac{\mu_n C_{ox}}{2}\left(\frac{1}{\sum\limits_{k(on)} \frac{1}{\left(\frac{W}{L}\right)_k}}\right) \cdot \begin{cases} [2(V_{in} - V_{T0})\,V_{out} - V_{out}^2] & \text{linear} \\ (V_{in} - V_{T0})^2 & \text{saturation} \end{cases} \quad ...(6.24)$$

Hence, the (W/L) ratio of the equivalent driver transistor is

$$\left(\frac{W}{L}\right)_{equivalent} = \frac{1}{\sum\limits_{k(on)} \frac{1}{\left(\frac{W}{L}\right)_k}} \quad ...(6.25)$$

If the series-connected transistors are identical, i.e., $(W/L)_1 = (W/L)_2 = ... = (W/L)$, the width-to-length ratio of the equivalent transistor becomes

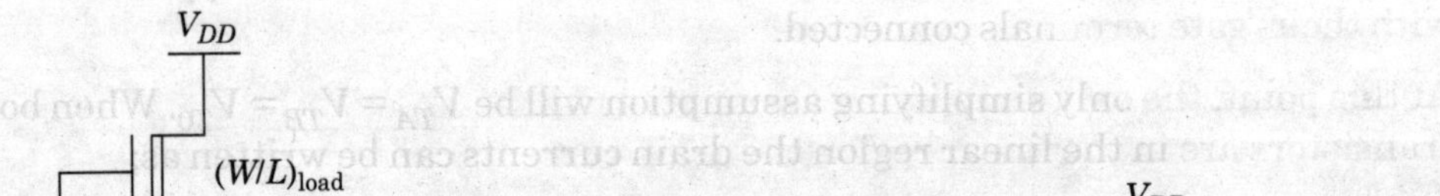

Fig. 6.8 The generalized NAND structure and its inverter equivalent.

$$\left(\frac{W}{L}\right)_{equivalent} = \frac{1}{n}\left(\frac{W}{L}\right) \quad ...(6.26)$$

For a two-input NAND gate, this means that each driver transistor must have a (W/L) ratio twice that of the equivalent inverter driver. If the area occupied by the depletion-type load transistor is negligible, the resulting NAND2 structure will occupy approximately four times the area occupied by the equivalent inverter which has the same static characteristics.

6.4 CMOS LOGIC CIRCUITS

6.4.1 CMOS NOR2 (Two-Input NOR) Gate

- The design and analysis of CMOS combinational logic circuits can be based on the basic principles developed for the nMOS depletion-load logic circuits.
- The circuit diagram of a two-input CMOS NOR gate is shown in Fig. 6.9. The circuit consists of a parallel-connected n-net and a series-connected complementary p-net. The input voltages V_A and V_B are applied to the gates of one nMOS and one pMOS transistor.
- When either one or both inputs are high (when the n-net creates a conducting path between the output node and the ground) then the p-net is cut-off.
- If both input voltages are low (the n-net is cut-off), then the p-net creates a conducting path between the output node and the supply voltage V_{DD}. Thus, the dual or complementary circuit structure allows that, for any given input combination, the output is connected either to V_{DD} or to ground via a low-resistance path.
- A DC current path between the V_{DD} and ground is not established for any of the input combinations.

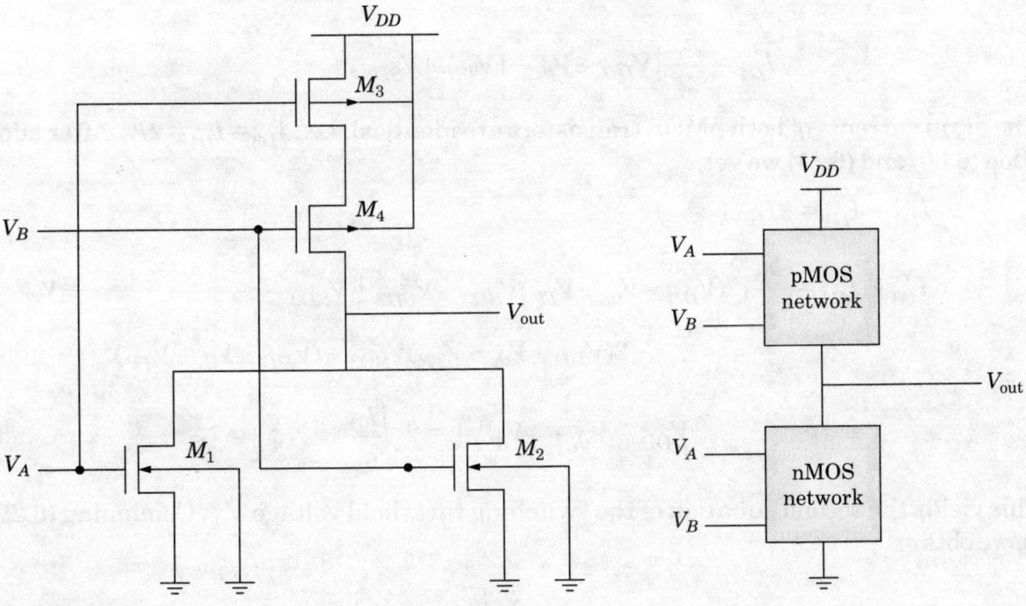

Fig. 6.9 A CMOS NOR2 gate and its complementary operation.

- The output voltage of the CMOS NOR2 gate will attain a logic-low voltage of $V_{OL} = 0$ and a logic-high voltage of $V_{OH} = V_{DD}$.

For circuit design purposes, we assume that

- The switching threshold voltage V_{th} for the both input voltages switch simultaneously $(V_A = V_B)$.
- The device sizes in each block are identical $((W/L)_{n,A} = (W/L)_{n,B}$ and $(W/L)_{p,A} = (W/L)_{p,B})$
- The substrate-bias effect for the pMOS transistors is neglected for simplicity.

By definition, the output voltage is equal to the input voltage at the switching threshold.

$$V_A = V_B = V_{out} = V_{th} \qquad ...(6.27)$$

It is obvious that the two parallel nMOS transistors are saturated at this point, because $V_{GS} = V_{DS}$. The combined drain current of the two nMOS transistors is

$$I_D = k_n(V_{th} - V_{T,n})^2 \qquad ...(6.28)$$

Thus, we obtain the first equation for the switching threshold V_{th}.

$$V_{th} = V_{T,n} + \sqrt{\frac{I_D}{k_n}} \qquad ...(6.29)$$

Examination of the p-net in Fig. 6.9 shows that the pMOS transistor M_3 operates in the linear region, while the other pMOS transistor, M_4, is in saturation for $V_{in} = V_{out}$. Thus

$$I_{D3} = \frac{k_p}{2}[2(V_{DD} - V_{th} - |V_{T,P}|)V_{SD3} - V_{SD3}^2] \qquad ...(6.30)$$

$$I_{D4} = \frac{k_p}{2}(V_{DD} - V_{th} - |V_{T,P}|V_{SD3})^2 \qquad ...(6.31)$$

The drain currents of both pMOS transistors are identical, i.e., $I_{D3} = I_{D4} = 2I_D$. After addition, equation (6.30) and (6.31) we get,

$$I_{D3} + I_{D4} = 2I_D$$

$$I_{D3} + I_{D4} = \frac{k_p}{2}[2(V_{DD} - V_{th} - V_{TP})V_{SD3} - V_{SD3}^2 + V_{SD3}^2$$
$$- 2(V_{DD} - V_{th} - V_{TP})V_{SD3} + (V_{DD} - V_{th} - V_{TP})^2$$

$$V_{DD} - V_{th} - |V_{T,P}| = 2\sqrt{\frac{I_D}{k_p}} \qquad ...(6.32)$$

This yields the second equation of the switching threshold voltage V_{th}. Combining (6.29) and (6.32), we obtain

$$V_{th}(\text{NOR2}) = \frac{V_{T,n} + \frac{1}{2}\sqrt{\frac{k_p}{k_n}}(V_{DD} - |V_{T,P}|)}{1 + \frac{1}{2}\sqrt{\frac{k_p}{k_n}}} \qquad ...(6.33)$$

Now compare this expression with the switching threshold voltage of the CMOS inverter, which was derived earlier.

$$V_{th}(\text{INR}) = \frac{V_{T,n} + \sqrt{\frac{k_p}{k_n}}(V_{DD} - |V_{T,P}|)}{1 + \sqrt{\frac{k_p}{k_n}}} \quad ...(6.34)$$

If $K_n = K_p$ and $V_{T,n} = V_{T,p}$, the switching threshold of the CMOS inverter is equal to $V_{DD}/2$. Using the same parameters, the switching threshold of the NOR2 gate is

$$V_{th}(\text{NOR2}) = \frac{V_{DD} + V_{T,n}}{3} \quad ...(6.35)$$

Which is not equal to $V_{DD}/2$.

For example, when $V_{DD} = 5\text{V}$ and $V_{T,n} = V_{T,p} = 1$ V, the switching threshold voltages of the NOR2 gate and the inverter are:

$$V_{th}(\text{NOR2}) = 2 \text{ V}$$

$$V_{th}(\text{INR}) = 2.5 \text{ V}$$

$$V_{th}(\text{NOR2}) = \frac{V_{T,n} + \sqrt{\frac{k_p}{4k_n}}(V_{DD} - |V_{T,P}|)}{1 + \sqrt{\frac{k_p}{4k_n}}} \quad ...(6.36)$$

Which is identical to (6.33), From (6.33), we can easily derive simple design guidelines for the NOR2 gate. For example, in order to achieve a switching threshold voltage of $V_{DD}/2$ for simultaneous switching, we have to set $V_{T,n} = V_{T,p}$ and $K_p = 4K_n$.

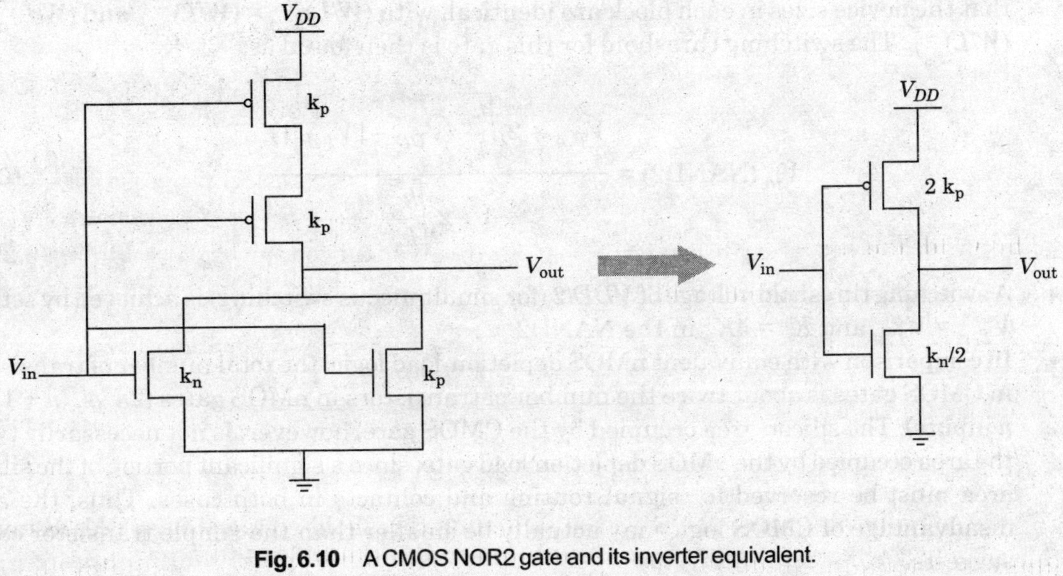

Fig. 6.10 A CMOS NOR2 gate and its inverter equivalent.

6.4.2 CMOS NAND2 (Two-Input NAND) Gate

- The circuit arrangement of two-input CMOS NAND (NAND2) gate is shown in Fig. 6.11.
- The n-net consisting of two series-connected nMOS transistors creates a conducting path between the output node and the ground only if both input voltages are logic-high.
- In this case, both of the parallel-connected pMOS transistors in the p-net will be off. For all other input combinations, either one or both of the pMOS transistors will be turned on, while the n-net is cut-off, thus creating a current path between the output node and the power supply voltage.

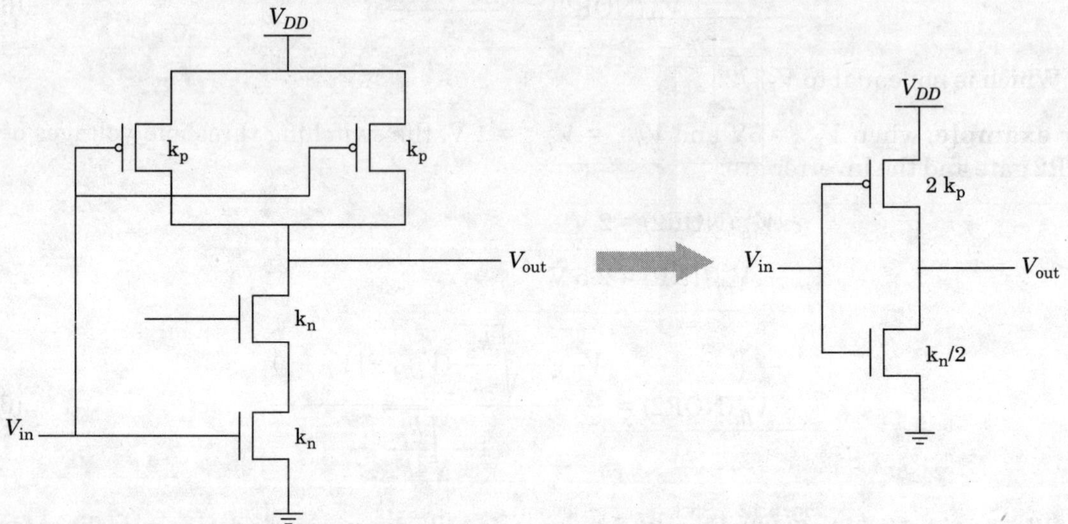

Fig. 6.11 A CMOS NAND2 gate and its inverter equivalent.

- For the calculation the switching threshold for the CMOS NAND2 gate. We will assume that the device sizes in each block are identical, with $(W/L)_{n,A} = (W/L)_{n,B}$ and $(W/L)_{p,A} = (W/L)_{p,B}$. The switching threshold for this gate is then found as;

$$V_{th}\,(\text{NAND2}) = \frac{V_{T,n} + 2\sqrt{\dfrac{k_p}{k_n}}(V_{DD} - |V_{T,P}|)}{1 + 2\sqrt{\dfrac{k_p}{k_n}}} \qquad \text{...(6.37)}$$

- A switching threshold voltage of $VDD/2$ (for simultaneous switching) is achieved by setting $V_{T,n} = V_{T,p}$ and $K_n = 4K_p$ in the NAND2.
- In comparison with equivalent nMOS depletion-load logic, the total number of transistors in CMOS gates is about twice the number of transistors in nMOS gates ($2n$ vs. $(n+1)$ for n inputs). The silicon area occupied by the CMOS gate, however, is not necessarily twice the area occupied by the nMOS depletion load gate, since a significant portion of the silicon area must be reserved for signal routing and contacts in both cases. Thus, the area disadvantage of CMOS logic may actually be smaller than the simple transistor count suggests.

6.4.2.1 Layout of Simple CMOS Logic Gates

- In this topic, we will examine simplified layout examples for CMOS NOR2 and NAND2 gates.
- The simplified (stick diagram) view of the CMOS NOR2 gate layout given in Fig. 6.12. Here, the diffusion areas are depicted by rectangles, the metal connections and contacts are represented by solid lines and circles, respectively, and the poly silicon columns are represented by crosshatched strips. The stick-diagram layout does not carry any information on the actual geometry relations of the individual features, but it conveys valuable information on the relative placement of the transistors and their interconnections.

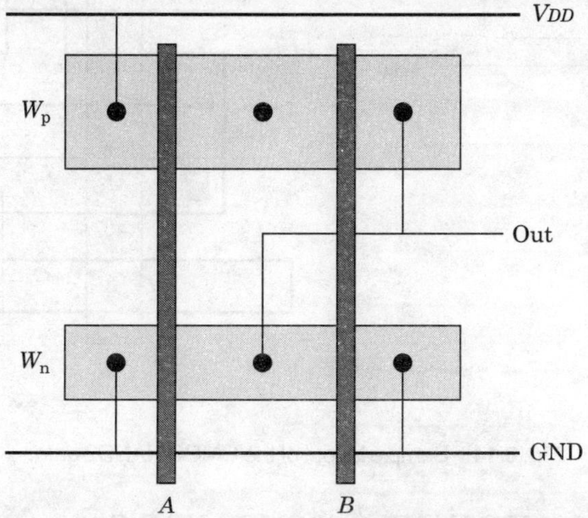

Fig. 6.12 Stick-diagram layout of the CMOS NOR2 gate.

- A sample layout of a CMOS NOR2 gate, using single layer metal and single-layer poly silicon is shown in Fig. 6.13. Here, the *p*-type diffusion area for pMOS transistors and the *n*-type diffusion area for nMOS transistors are aligned in parallel to allow simple routing of the gate signals via two parallel polysilicon lines running vertically.

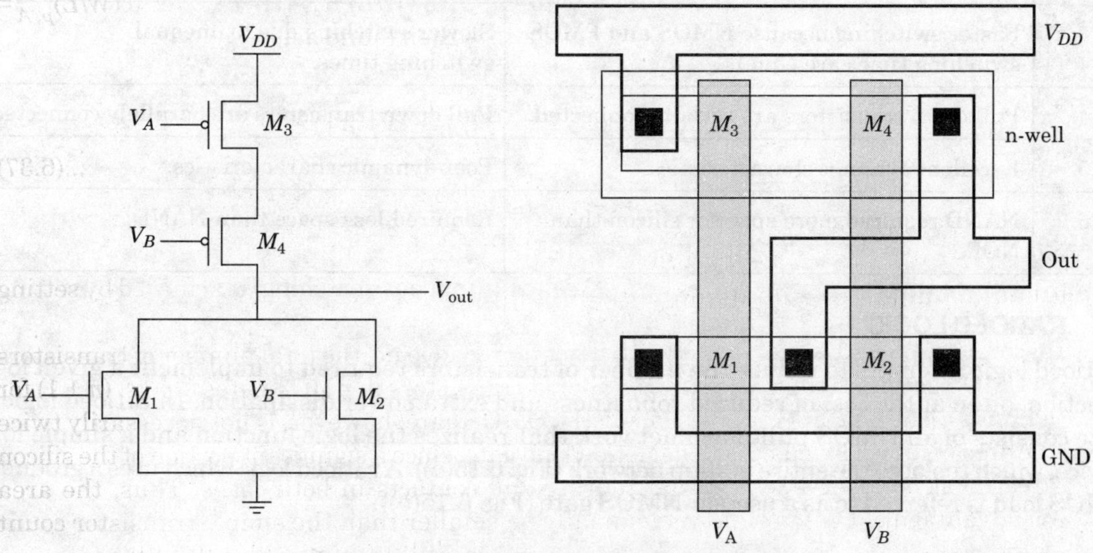

Fig. 6.13 Sample layout of the CMOS NOR2 gate.

- The layout of a CMOS NAND2 gate, using the same basic layout principles is shown in Fig. 6.14.

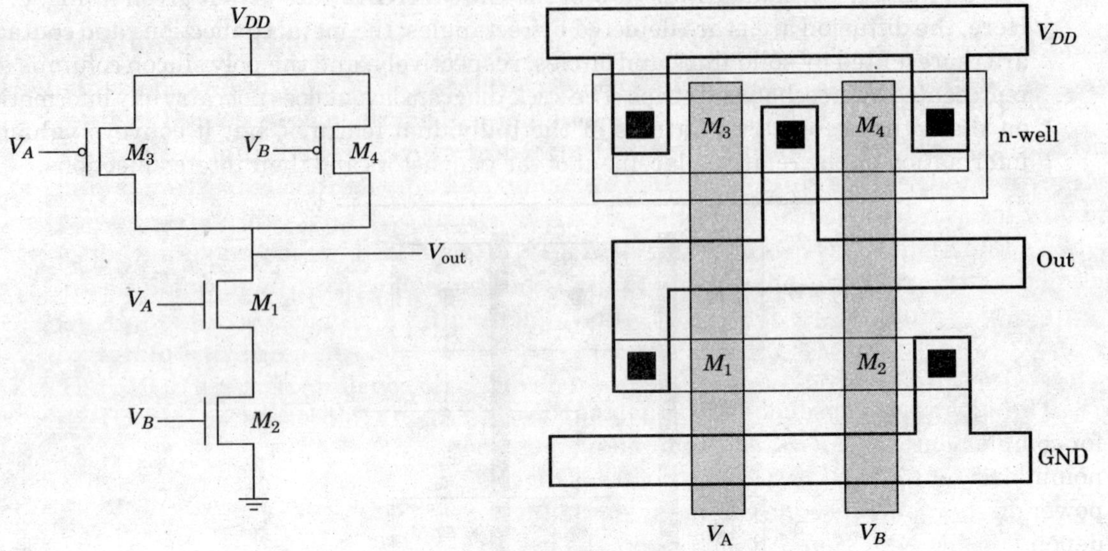

Fig. 6.14 Sample layout of the CMOS NAND2 gate.

6.4.2.2 Comparison of CMOS NAND and CMOS NOR Gates

S.No.	CMOS NAND	CMOS NOR
1.	Pull up transistors are in parallely connected.	Pull up transistors are in serially connected.
2.	Better noise margins than CMOS NOR gate.	Noise margine are poor.
3.	Faster switching because NMOS and PMOS switching times are equal.	Slower switching due to unequal switching times.
4.	Pull down transistors are serially connected.	Pull down transistors are parallely connected.
5.	Excellent dynamic characteristics.	Poor dynamic characteristics.
6.	NAND required more space or silicon than NOR.	Required less space than NAND.

6.5 RATIOED LOGIC

Ratioed logic attempts to reduce the number of transistors required to implement a given logic function, often at the cost of reduced robustness and extra power dissipation. In ratioed logic, a gate consists of an NMOS pull-down network that realizes the logic lunction and a simple load device, which replaces the entire pull-up network (Fig. 6.15(*a*)). A ratioed logic which uses a grounded PMOS load is referred to as a pseudo-NMOS gate (Fig. 6.15(*b*)).

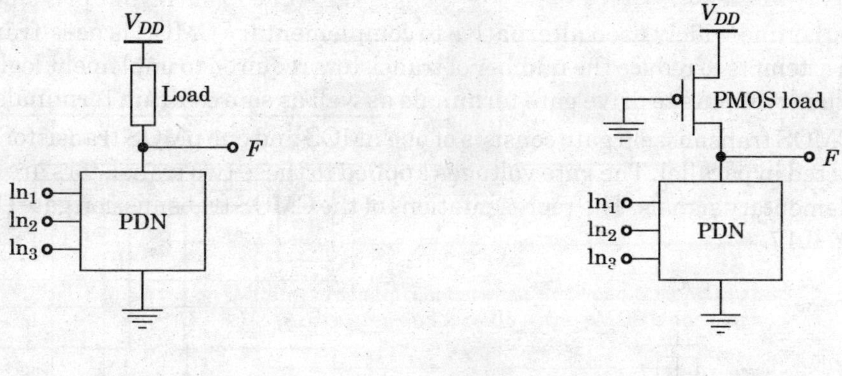

(a) Generic (b) Pseudo-NMOS

Fig. 6.15 Ratioed logic gate

The advantage of pseudo-NMOS gate is the reduced number of transistors $(N + 1)$ versus $2N$ for complementary CMOS. The nominal high output voltage (V_{OH}) for this gate is V_{DD}, but the nominal low output voltage (V_{OL}) is not 0 V. This results in reduced noise margins and more static power dissipation. Since the voltage swing on the output and the overall functionality of the gate depend on the NMOS and PMOS sizes, the circuit is called *ratioed*. This is in contrast to the ratioless logic such as complementary CMOS, where the low and high levels do not depend on transistor sizes.

The static power dissipation of pseudo-NMOS limits its use. When area is most important, however, its reduced transistor count compared with complementary CMOS is quite attractive. Pseudo-NMOS thus still finds occasional use in large fan-in circuits.

Differential Cascade Voltage Switch Logic (DCVSL). This logic family combines two concepts differential logic and positive feedback. A differential gate requires that each input is provided in complementary format, and it produces complementary outputs in turn. The feedback mechanism ensures that the logic device is turned off when not needed. This logic completely eliminates static currents and provides rail-to-rail swing (Fig. 6.16).

The pull-down networks PDN1 and PDN2 use NMOS devices and are mutually exclusive, *i.e.*, when PDN1 conducts, PDN2 is off, and when PDN1 is off, PDN2 conducts—such that the required logic function and its inverse are simultaneously implemented. The resulting circuit exhibits a rail-to-rail swing, and the static power dissipation is eliminated. The circuit is still ratioed since the sizing of the PMOS devices relative to the pull-down devices is critical to functionality, not just performance. In addition to the problem of increased design complexity, this circuit style has a power-dissipation problem due to crossover currents. During the transition, there is a period of time when PMOS and PDN are turned ON simultaneously, producing a short-circuit path.

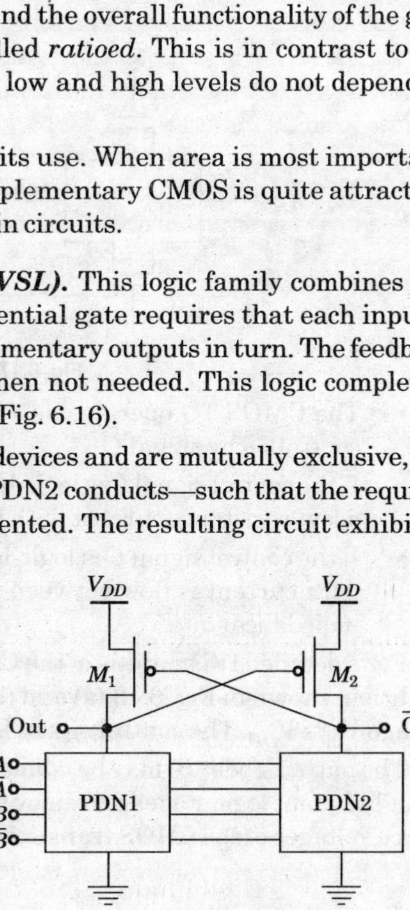

Fig. 6.16 DCVSL logic gate.

6.6 CMOS TRANSMISSION GATE (PASS-TRANSISTOR) (UTU-2012)

- A popular and widely used alternative to complementary CMOS is pass-transistor logic, which attempts to reduce the number of transistors required to implement logic by allowing the primary inputs to drive gate terminals as well as source-drain terminals.
- The CMOS transmission gate consists of one nMOS and one pMOS transistor and they are connected in parallel. The gate voltages applied to these two transistors are also set to be complementary signals. The representations of the CMOS transmission gate (TG) as shown in Fig. 6.17.

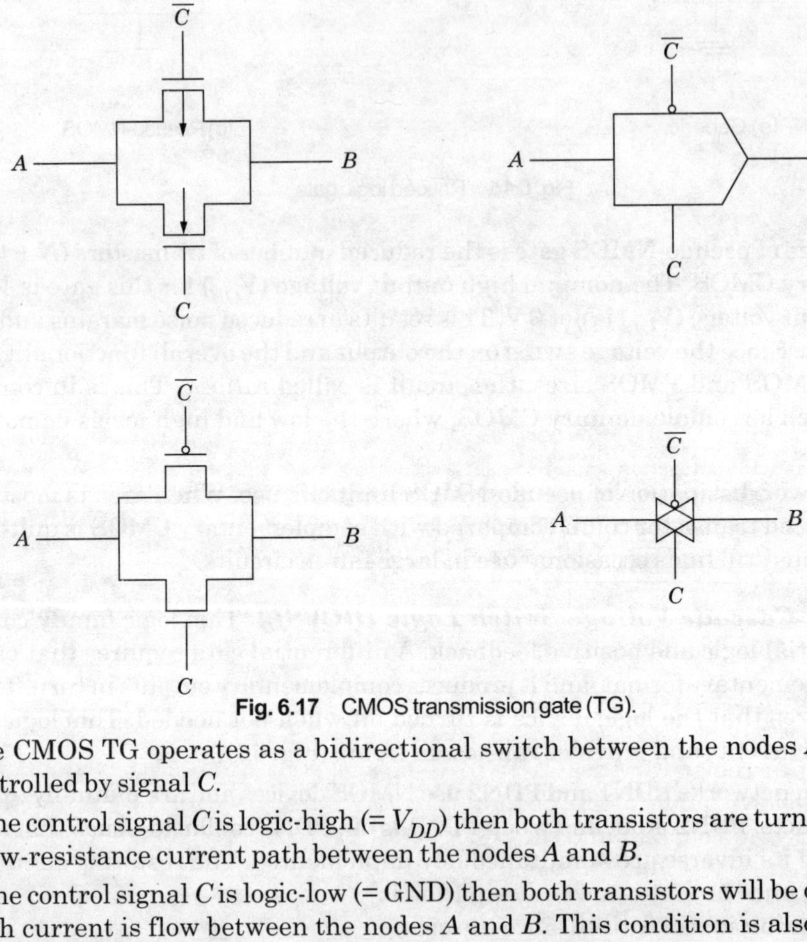

Fig. 6.17 CMOS transmission gate (TG).

- The CMOS TG operates as a bidirectional switch between the nodes *A* and *B* which is controlled by signal *C*.
- If the control signal *C* is logic-high (= V_{DD}) then both transistors are turned on and provide a low-resistance current path between the nodes *A* and *B*.
- If the control signal *C* is logic-low (= GND) then both transistors will be off and there is no path current is flow between the nodes *A* and *B*. This condition is also called the high-impedance state.

For a detailed DC analysis of the CMOS transmission gate, we will consider the following bias condition, shown in Fig. 6.18 (a) and (b). The input node (*A*) is connected to a constant logic-high voltage $V_{in} = V_{DD}$. The control signal is also logic-high. So that both transistors are turned on.

The output node (*B*) may be connected to a capacitor, which represents capacitive loading of the subsequent logic stages driven by the transmission gate. The drain-to-source and the gate-to-source voltages of the nMOS transistor are

$$V_{DS,\,n} = V_{DD} - V_{out}$$

$$V_{GS,\,n} = V_{DD} - V_{out} \qquad \text{...(6.38)}$$

Thus, the nMOS transistor will be turned off for $V_{out} > V_{DD} - V_{T,n}$ and will operate in the saturation mode for $V_{out} < V_{DD} - V_{T,n}$. The V_{DS} and V_{GS} voltages of the pMOS transistor are

$$V_{DS,p} = V_{out} - V_{DD}$$

$$V_{GS,p} = -V_{DD} \qquad ...(6.39)$$

The pMOS transistor is in saturation for $V_{out} < V_{T,p}$ and it is in linear region for $V_{out} > V_{T,p}$. The pMOS transistor is always turned on irrespective of the output voltage V_{out} level.

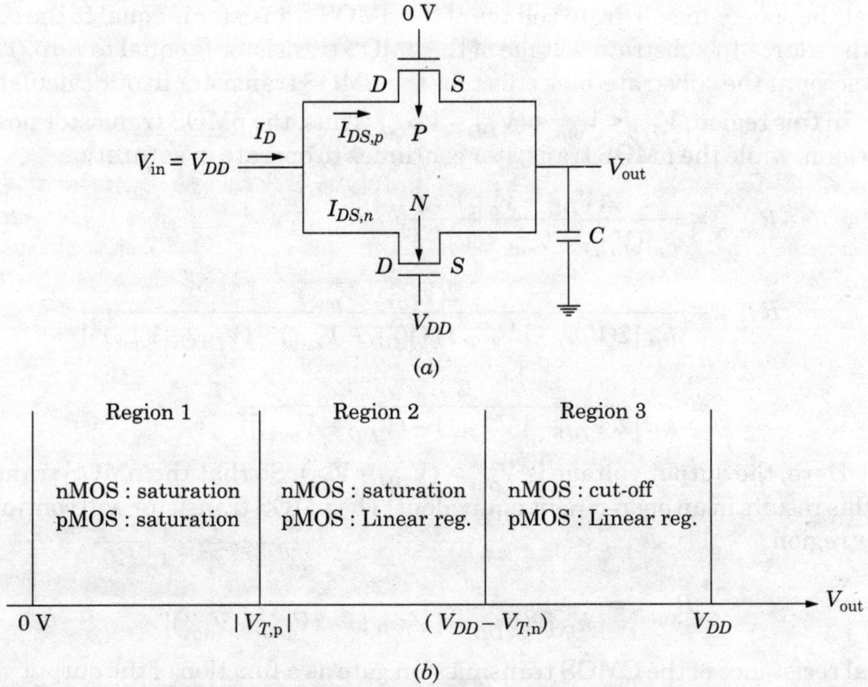

Fig. 6.18 (a) Bias of the CMOS transmission gate (b) Regions of the CMOS transmission gate.

Three operating regions for the CMOS transmission gate, which are depending on the output voltage level are shown in Fig. 6.18. The total current flowing through the transmission gate is the sum of the nMOS drain current and the pMOS drain current.

$$I_D = I_{DS,n} + I_{SD,P} \qquad ...(6.40)$$

At this point, the *equivalent resistance* for each transistor is given by

$$R_{eq,n} = \frac{V_{DD} - V_{out}}{I_{DS,n}}$$

$$R_{eq,P} = \frac{V_{DD} - V_{out}}{I_{DS,P}} \qquad ...(6.41)$$

The total equivalent resistance of the CMOS TG will then be the parallel equivalent of these two resistances, $R_{eq,n}$ and $R_{eq,P}$. Since, the equivalent resistance values of the transmission gate for the three operating regions can be calculated below.

Region 1: In this region, the output voltage is smaller than the absolute value of the pMOS transistor threshold voltage, i.e., $V_{out} < V_{TP}$. According to Fig. 6.18, both transistors are in saturation. We obtain the equivalent resistance of both devices as

$$R_{eq,\,n} = \frac{2(V_{DD} - V_{out})}{k_n(V_{DD} - V_{out} - V_{T,n})^2} \quad ...(6.42)$$

$$R_{eq,\,P} = \frac{2(V_{DD} - V_{out})}{k_P(V_{DD} - |V_{T,n}|)^2} \quad ...(6.43)$$

Note that the source-to-substrate voltage of the nMOS transistor is equal to the output voltage V_{out}, while the source-to-substrate voltage of the pMOS transistor is equal to zero. Thus, we have to take into account the substrate-bias effect for the nMOS transistor in our calculations.

Region 2: In this region, $V_{T,p} < V_{out} < (V_{DD} - V_{T,n})$. Thus, the pMOS transistor now operates in the linear region, while the nMOS transistor continues to operate in saturation.

$$R_{eq,\,n} = \frac{2(V_{DD} - V_{out})}{k_n\,(V_{DD} - V_{out} - V_{T,n})^2} \quad ...(6.44)$$

$$R_{eq,\,P} = \frac{2(V_{DD} - V_{out})}{k_P\,[2(V_{DD} - |V_{T,P}|)(V_{DD} - V_{out}) - (V_{DD} - V_{out})^2]} \quad ...(6.45)$$

$$= \frac{2}{k_P\,[2(V_{DD} - |V_{T,P}|) - (V_{DD} - V_{out})]}$$

Region 3: Here, the output voltage is $V_{out} > (V_{DD} - V_{Tn})$. So that the nMOS transistor will be turned off, this results in an open-circuit equivalent. The pMOS transistor will continue to operate in the linear region.

$$R_{eq,\,P} = \frac{2}{k_P\,[2(V_{DD} - |V_{T,P}|) - (V_{DD} - V_{out})]} \quad ...(6.46)$$

The total resistance of the CMOS transmission gate as a function of the output voltage V_{out} is shown in Fig. 6.19.

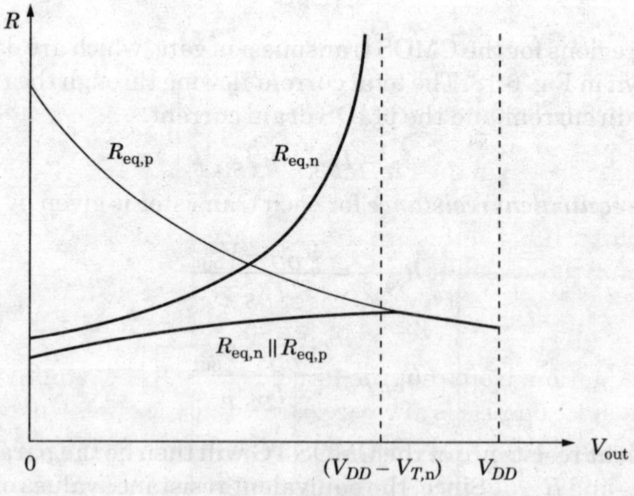

Fig. 6.19 Equivalent resistance of the CMOS transmission gate, plotted as a function of the output voltage.

6.6.1 Property of CMOS Transmission Gate

The total equivalent resistance of the TG remains relatively constant, i.e., its value is almost independent of the output voltage, whereas the individual equivalent resistances of both the nMOS and the pMOS transistors are strongly dependent on V_{out}.

The implementation of CMOS transmission gates in logic circuit design usually results in compact circuit structures which may even require a smaller number of transistors than their standard CMOS counterparts.

6.6.2 TG-Based Multiplexer

Transmission gate logic design has been used extensively in CMOS design for many years. The simplicity of the switching and the ability to transmit the entire range of voltages make it attractive for many applications.

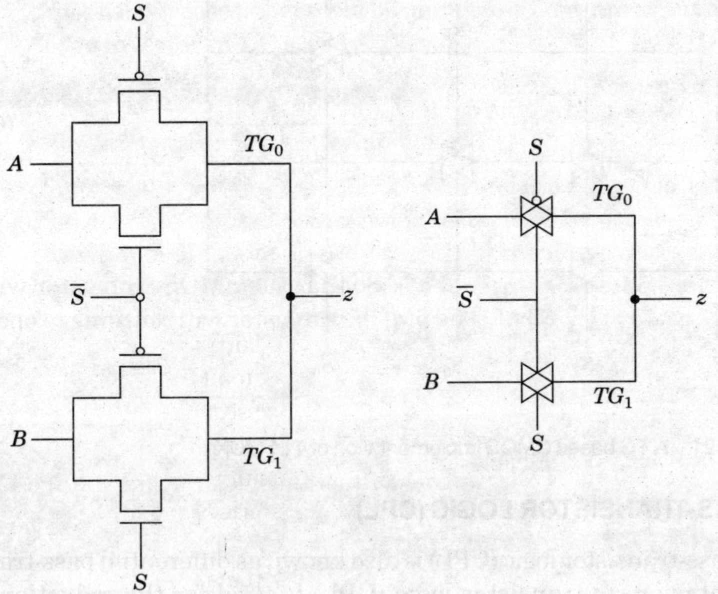

S	TG_0 TG_1	Z
0	Closed open	A
1	Open closed	B

Fig. 6.20 A TG based 2 : 1 multiplexer circuit and symbol

The 2 : 1 multiplexer is shown in the Fig. 6.20. The operation of the circuit is specified in the truth table. When the selector signal has a value $S = 0$, TG_0 is closed and TG_1 is open, so that A is transmitted to the output Z. If $S = 1$, the situation is reversed with TG_0 open and TG_1 closed, in this case $B = Z$. Combining these results gives

$$Z = A \cdot \overline{S} + B \cdot S$$

Which is required equation. Note that the use of a pair of TGs eliminates the possibility of having a floating output since one TG is always closed while the other will be open.

The 2 : 1 multiplexer can be modified to produce other useful functions such as XOR, XNOR, OR logic gates.)

The 2 : 1 mux architecture can be extended to 4 : 1 mux by using 2 bit selector (S_1, S_0). Each input line (A, B, C, D) will have two TGs in its path such that the output is

$$Z = A \cdot \overline{S}_1 \cdot \overline{S}_0 + B \cdot \overline{S}_1 \cdot S_0 + C \cdot S_1 \cdot \overline{S}_0 + D \cdot S_1 \cdot S_0$$

A 4 : 1 TG based CMOS multiplexer is shown in Fig. 6.21. The structure is same as for the NMOS NOR multiplexer except that the NMOS pass transistors have been replaced by CMOS transmission gates. The NMOS 4 : 1 multiplexer has a faster fall time, and the CMOS MUX has the faster rise time.

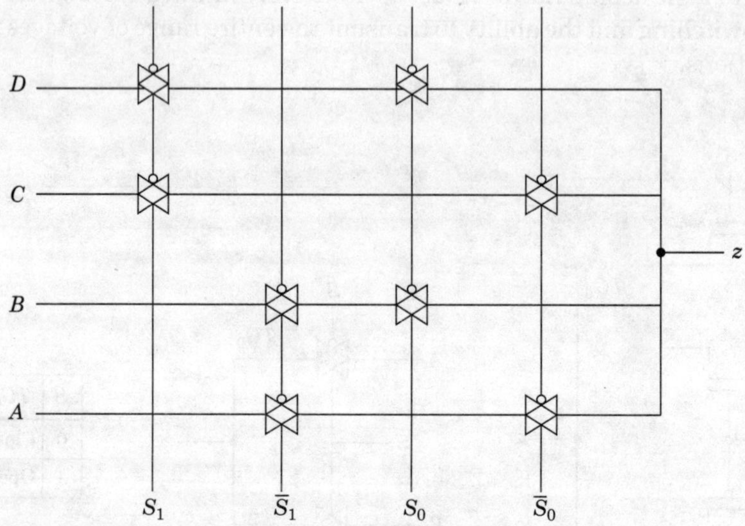

Fig. 6.21 A TG based CMOS implementation of 4 : 1 MUX

6.7 COMPLEMENTARY PASS-TRANSISTOR LOGIC (CPL)

- The complementary pass-transistor logic (CPL) is also known as differential pass-transistor logic. The complementary pass-transistor logic (CPL) is used for the reduction of the complexity of full-CMOS pass-gate logic circuits. It is also used for high performance design.
- In complementary pass-transistor logic (CPL), a purely nMOS pass-transistor network is used for the logic operations, instead of a CMOS TG network.
- All inputs are applied in complementary form, *i.e.*, every input signal and its inverse must be provided; the circuit also produces complementary outputs, to be used by subsequent CPL stages.
- The elimination of pMOS transistors from the pass-gate network reduces the parasitic capacitances. The operation speed is typically higher compared to a full-CMOS counterpart.
- The CPL design style is highly modular and a wide range of functions can be realized by using the same basic pass-transistor structures.
- The circuit diagrams of a CPL NOR2, a CPL NAND2 and a XOR2 gate are shown in Fig. 6.22.

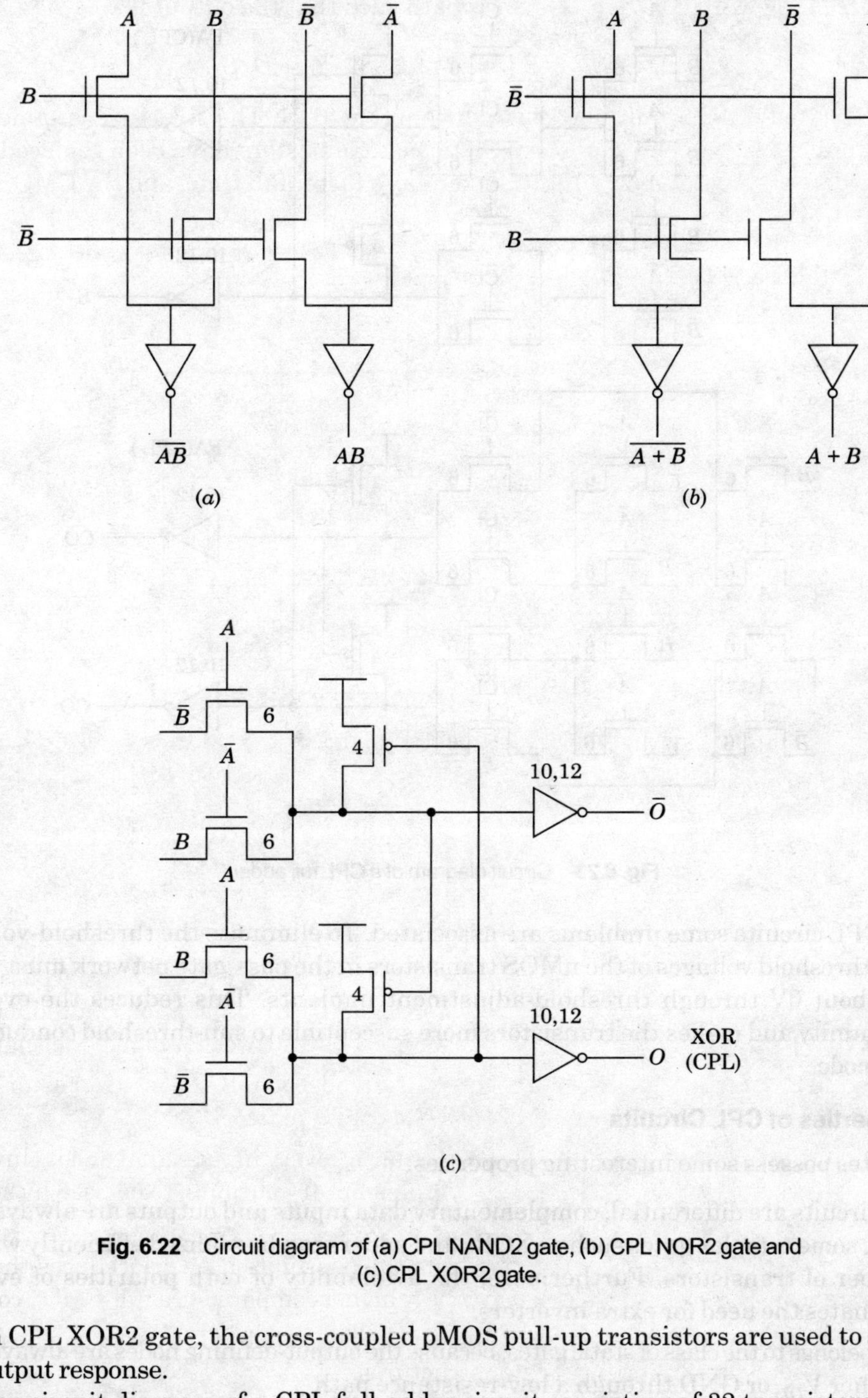

Fig. 6.22 Circuit diagram of (a) CPL NAND2 gate, (b) CPL NOR2 gate and (c) CPL XOR2 gate.

- In CPL XOR2 gate, the cross-coupled pMOS pull-up transistors are used to speed up the output response.
- The circuit diagram of a CPL full-adder circuit consisting of 32 transistors is shown in Fig. 6.23.

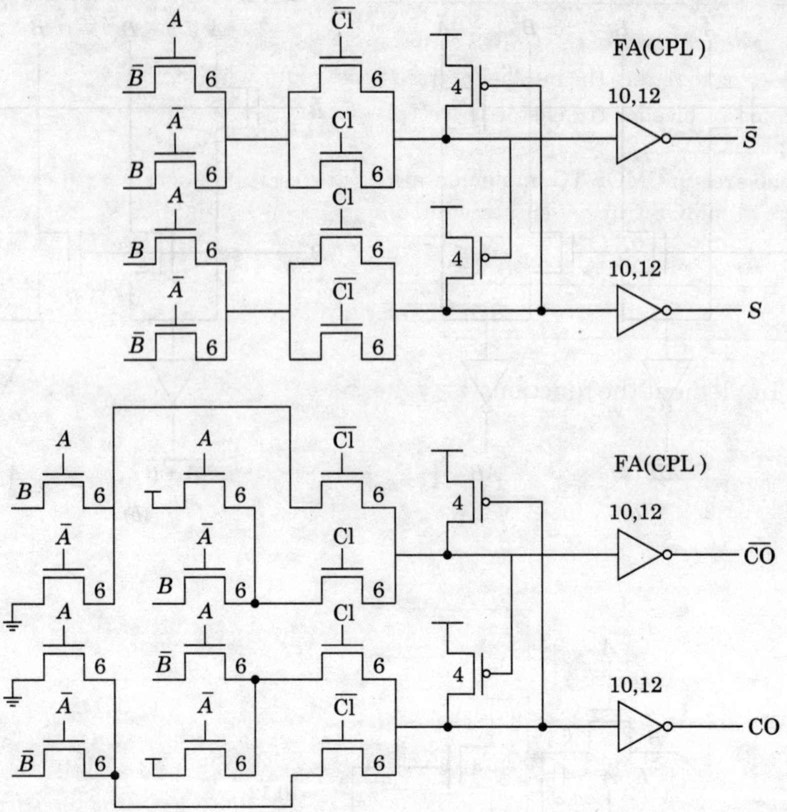

Fig. 6.23 Circuit diagram of a CPL full adder.

- In CPL circuits some problems are associated. To eliminate the threshold-voltage drop, the threshold voltages of the nMOS transistors in the pass-gate network must be reduced to about 0V through threshold-adjustment implants. This reduces the overall noise immunity and makes the transistors more susceptible to sub-threshold conduction in the off-mode.

6.7.1 Properties of CPL Circuits

The CPL gates possess some interesting properties:

1. The circuits are differential, complementary data inputs and outputs are always available. Thus, some complex gates such as XORs and adders can be realized efficiently with smaller number of transistors. Furthermore, the availability of both polarities of every signal eliminates the need for extra inverters.
2. CPL belongs to the class of static gates, because the output-defining nodes are always connected to either V_{DD} or GND through a low-resistance path.
3. The design is very simple. More complex gates can be built by cascading the standard pass-transistor modules.

POINTS TO BE REMEMBER

1. The primary advantage of the CMOS structure is robustness.
2. Ratioed logic is to reduce the number of transistors required to implement a given logic function.
3. The function of bleeder transistor is to compensate for the charge lost due to the pull-down leakage paths.
4. To save the area in CMOS TG implementation of logic, TG's can be laid out as separated nMOS pairs with all pMOS transistors placed in one single-well.

SOLVED EXAMPLES

Example 6.1: Implement the functions $Y = \overline{(A + B + C)\,(D + E)\,F}$ using CMOS logic based.

Solution:

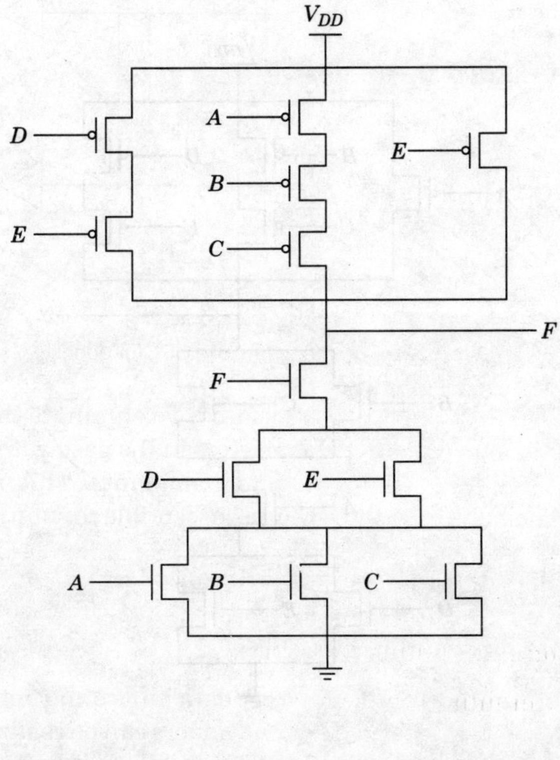

Example 6.2: Design the circuit described the function $Y = \overline{A\,(B + C)\,(D + E)}$ using CMOS logic. Also find the equivalent CMOS inverter circuit for simulator. Switching of all inputs assuming that $(W/L)_p = 5$ and $(W/L)_n = 2$.

Solution: The equivalent W/L ratios of the nMOS and PMOS network are determined by using the series parallel equivalence rules.

So,

$$\left(\frac{W}{L}\right)_{n,eq} = \frac{1}{\dfrac{1}{\left(\frac{W}{L}\right)_D + \left(\frac{W}{L}\right)_E} + \dfrac{1}{\left(\frac{W}{L}\right)_B + \left(\frac{W}{L}\right)_C} + \dfrac{1}{\left(\frac{W}{L}\right)_A}} = \frac{1}{\dfrac{1}{2+2} + \dfrac{1}{2+2} + \dfrac{1}{2}} = 1$$

$$\left(\frac{W}{L}\right)_{P,eq} = \frac{1}{\dfrac{1}{\left(\frac{W}{L}\right)_B} + \dfrac{1}{\left(\frac{W}{L}\right)_E}} + \frac{1}{\dfrac{1}{\left(\frac{W}{L}\right)_C} + \dfrac{1}{\left(\frac{W}{L}\right)_B}} + \frac{1}{\dfrac{1}{\left(\frac{W}{L}\right)_A}}$$

$$= \frac{1}{\frac{1}{5} + \frac{1}{5}} + \frac{1}{\frac{1}{5} + \frac{1}{5}} + \frac{1}{\frac{1}{5}} = 10$$

So we have the (*W*/*L*) ratio of the equivalent inverter circuit. The CMOS circuit for expression *Y* is given by.

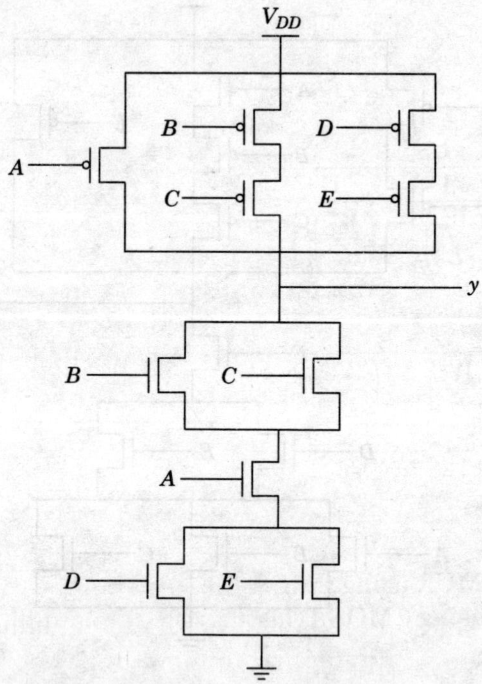

Example 6.3: Design the circuit described by the function $Y = \overline{(A + D + E)(B + C)}$ using CMOS logic. Also find the equivalent CMOS inverter circuit if $(W/L)_P = 15$ and $(W/L)_n = 10$.

Solution: The CMOS logic

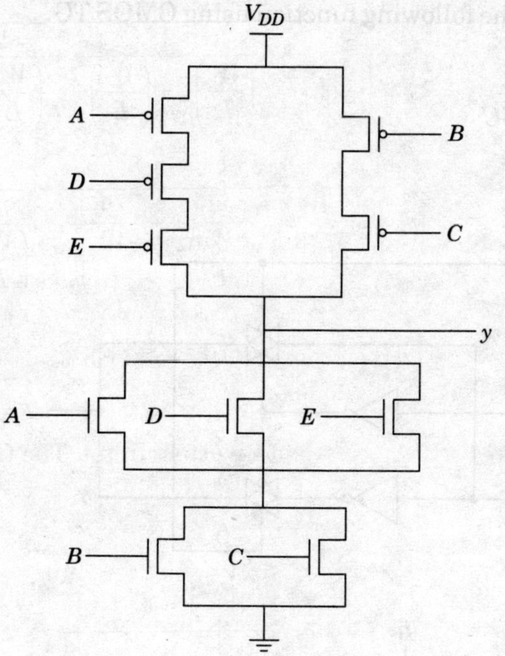

and

$$\left(\frac{W}{L}\right)_{P,eq} = \frac{1}{\dfrac{1}{\left(\frac{W}{L}\right)_D}+\dfrac{1}{\left(\frac{W}{L}\right)_A}+\dfrac{1}{\left(\frac{W}{L}\right)_E}} + \frac{1}{\dfrac{1}{\left(\frac{W}{L}\right)_C}+\dfrac{1}{\left(\frac{W}{L}\right)_B}} = \frac{1}{\frac{1}{15}+\frac{1}{15}+\frac{1}{15}} + \frac{1}{\frac{1}{15}+\frac{1}{15}} = 12.5$$

$$\left(\frac{W}{L}\right)_{eq,n} = \frac{1}{\dfrac{1}{\left(\frac{W}{L}\right)_B+\left(\frac{W}{L}\right)_C}+\dfrac{1}{\left(\frac{W}{L}\right)_A+\left(\frac{W}{L}\right)_D+\left(\frac{W}{L}\right)_E}} = \frac{1}{\dfrac{1}{\frac{1}{10}+\frac{1}{10}}+\dfrac{1}{\frac{1}{10}+\frac{1}{10}+\frac{1}{10}}}$$

$$= \frac{1}{\frac{1}{30}+\frac{1}{20}} = 12$$

Example 6.4: Design a 2 × 1 MUX using CMOS TG.
Solution: A two input MUX using CMOS TG is show below.

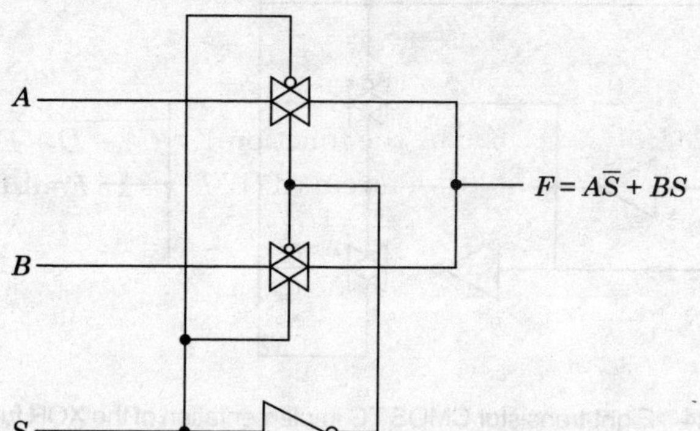

Example 6.5: Implement the following function using CMOS TG.

(a) $F = A\bar{B} + \bar{A}B$

(b) $F = AB + \bar{A}\bar{C} + A\bar{B}C$

Solution:

(a)

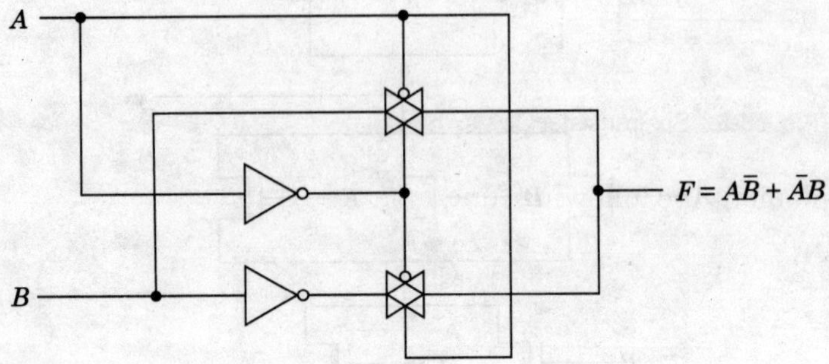

(b)

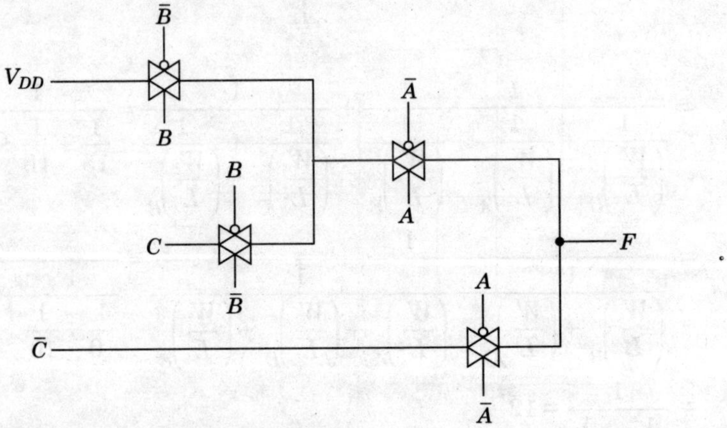

Examples 6.6: Design a XOR gate using CMOS TG. **(UTU-2012)**

Solution: An eight-transistor and only six transistors implementation of the logic XOR function, using CMOS TGs and CMOS inverters is shown in Fig. 6.24 and Fig. 6.25 respectively.

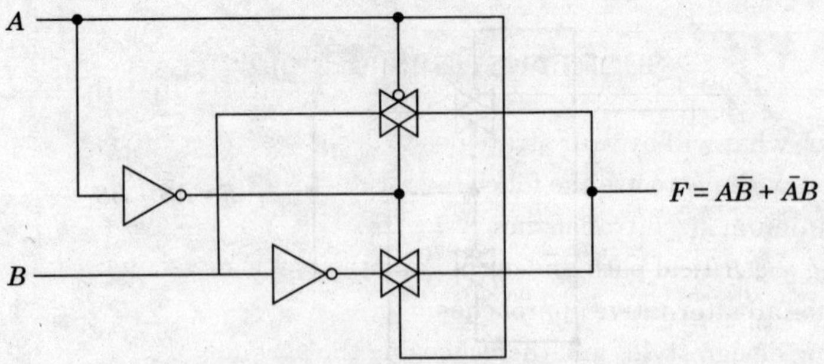

Fig. 6.24 Eight-transistor CMOS TG implementation of the XOR function.

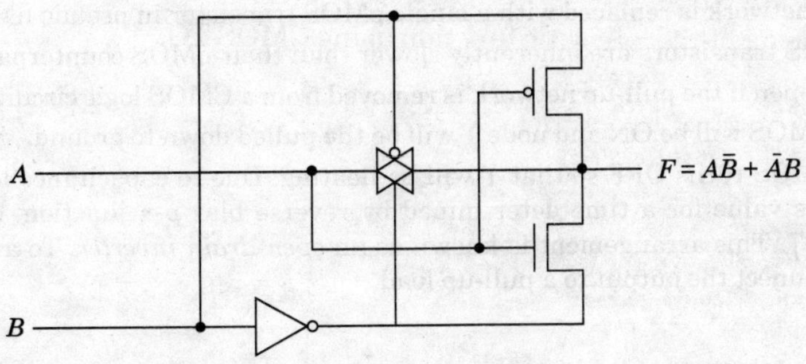

Fig. 6.25 Six-transistor CMOS TG implementation of the XOR function.

Examples 6.7: Implement the following functions using 8 transistors CMOS TG.
Solution:

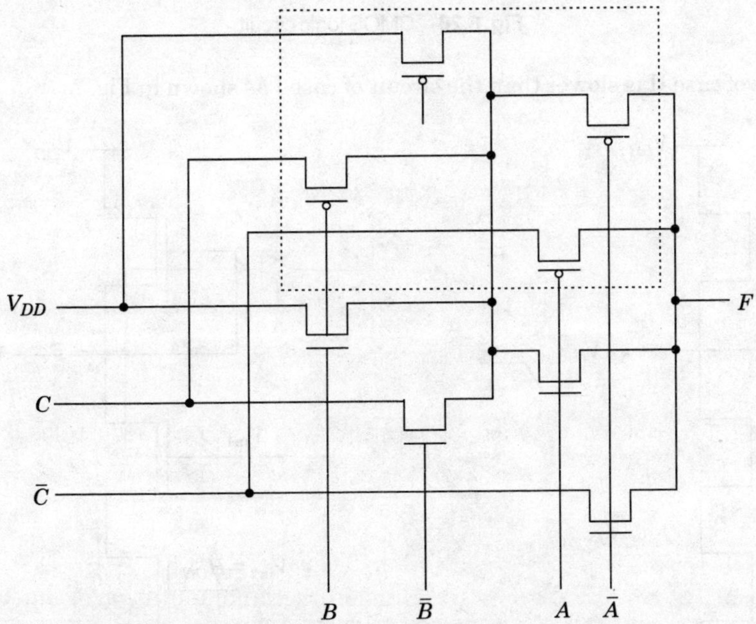

SHORT ANSWERS TYPE QUESTIONS

Q.1. As a designer, what will by your strategies to design complex circuits?

Ans. As a designer, we have to use the following strategy for design.

- Using minimum sized transistors.
- Analyzing the critical path and optimizing the transistor sizing.
- Looking for all alternative approaches.
- Comparing design styles and then choosing the best one.

Q.2. Why pull-up network is replaced with a single pMOS transistor in pseudo nMOS logic?

Ans. Because pMOS transistors are inherently slower than their nMOS counterpart.

Q.3. What will happen if the pull-up network is removed from a CMOS logic circuit as shown in Figure.

Ans. If A = high, nMOS will be ON and node Y will be the pulled down to ground.

If A = low, nMOS turns OFF so that Y will be floating. Due to capacitance load at Y, the output remains at its value for a time deter mined by reverse bias *p-n* junction, current and lumped capacitance C_L. This arrangement is known as an *open-drain inverter*. To avoid this circuit it is required to connect the output to a pull-up load.

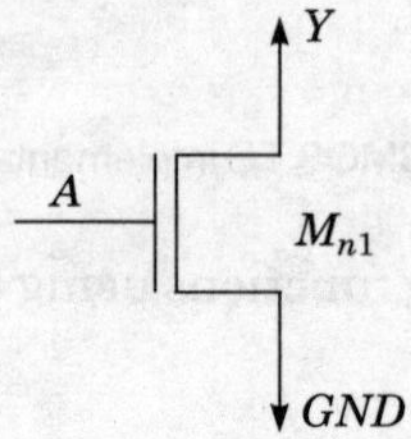

Fig. 6.26 CMOS logic circuit

Q.4. Why circuit of case II is slower than the circuit of case I as shown in Fig. ?

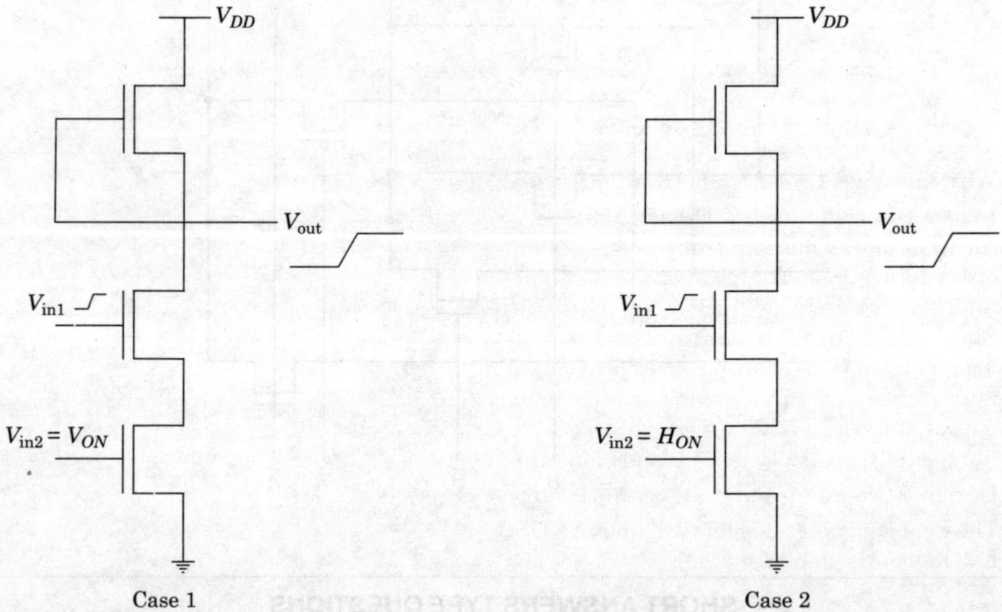

Ans. Case II has additional capacitance on the internal node which charges to $V_{DD} - V_{th}$ which takes some of the charging current normally going to G.

Q.5. What are the major drawbacks of PIL?

Ans. We can summarize the main disadvantages of pass transistor logic as follows:

- The threshold voltage drop through the nMOS transistors ($V_{out} = V_{DD} - V_{Tn}$) makes level restoration at the gate output necessary (more transistors for invertase at the output).
- To decuple gate inputs and outputs and for acceptable output driving capabilities, inverters are usually attached to the output.

- Because the MOS networks are connected to gate inputs rather than constant power lines, only one signal path through each network must be active at a time to avoid short circuits between inputs. Therefore, each pass transistors network must realize a multiplexer structure which limits the number of logic functions that can be implemented efficiently.
- Because pass-transistor multiplexers require complementary control signals, dual rail logic is usually used to provide all signals in complementary form (more transistors, more wires).
- Pass transistors logic with swing restoration circuitry is sensitive to voltage scaling and transistors sizing with respect to circuit robustness reduced noise margins.

Q.6. Design a 2-to-1 MUX using pass transistor logic and compare this design with TG implementation in terms of number of transistors used in each design.

Ans. The 2 : 1 MUX implementation using pass transistor logic is shown in Figure. The pass transistor logic implementation of 2 to 1 MUX cases only two transistors and 1 level of propagation delay, whereas in TG implementation we need 4 transistors. For the same MUX design using CMOS logic needs 8 numbers of transistors.

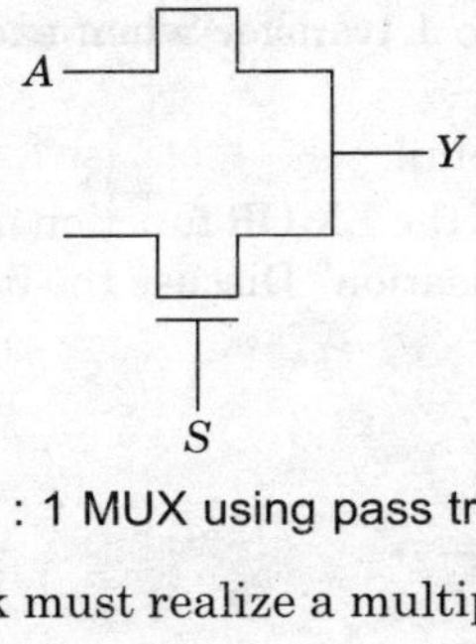

Fig. 6.27 2 : 1 MUX using pass transistor logic

Q.7. Why each pass transistor network must realize a multiplexer structure?

Ans. In pass transistor logic (PTL), the MOS networks are connected to variable gate inputs rather than constant power lines and only one signal path through each network must be active at a time in order to avoid shorts between inputs. Therefore, each pass transistor network must realize a multiplexer structure which limits the number of logic function that can be implemented efficiently. Since pass transistors multiplexer structure requires complementary control signals, dual rail logic is usually used in order to provide all signals in their complementary form. Due to these multiplexer structures, again two MOS networks are required in addition to swing restoration and output buffering circuits. These all extra requirements will annitilate the advantage of using less number of transistors in PTL and small input loads.

Q.8. Design a two input AND gate with pass transistors logic.

Ans. The Boolean expression for two input AND gate is : $F = A.B$. The circuit implemented using pseudo-nMOS logic is shown in Fig.

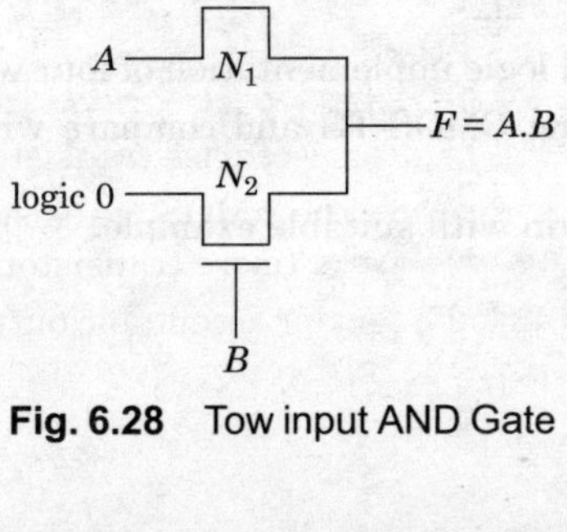

Fig. 6.28 Tow input AND Gate

The two input AND gate is implemented by using nMOS pass transistor only. When B = high, the transistors N_1 turns ON and transistors N_2 turns OFF. Therefore, N_1 will give low impedance path from input to output which result in $A.B$. When B = low, N_2 turns ON and N_1 turns OFF. Therefore, input 0 will transfer at the output and hence, the final result will be $A.B$. The realization of AND gate with pseudo-nMOS logic requires only 4 transistors whereas in complementary CMOS implementation the number of used transistors is six. The reduced number of transistors has the additional advantages of lower capacitance.

Q.9. What is the arrangement in CPL circuit to reduce the static current due to incomplete turn off the pMOSFET in the output inverters?

Ans. To minimize this circuit, a weak pMOS feedback device can be used in CPL circuit in order to pull the pass transistors output to full supply voltage at the cost of increased output node capacitance, which results in higher switching power dissipation and higher delay.

REVIEW QUESTIONS

1. Design a pass transistor network that implements the function

$$F = ABC + A\bar{B}\bar{C} + \bar{A}\bar{B}C + \bar{A}B\bar{C}$$

2. Find the node voltages during logic 1 transfer when each pass transistor is driving the gate of another pass transistor.
3. Explain the transmission gate in detail.
4. Draw the CMOS VLSI realisation of the EX-OR function using pass transistor logic for two signals. What is the advantages of this realisation? Discuss the design considerations of digital circuits. **(UPTU-2007-08)**
5. Write short notes on:
   (a) AOI Gate
   (b) Propagation delay
   (c) Noise Margin
   (d) Fanout and Fanin
6. Why we prefer CMOS transmission gates over other gates?
7. Draw the circuit of 8 × 1 MUX using T-G's
8. Consider the following circuit. What is the output voltage if only one input is high fall input are high?

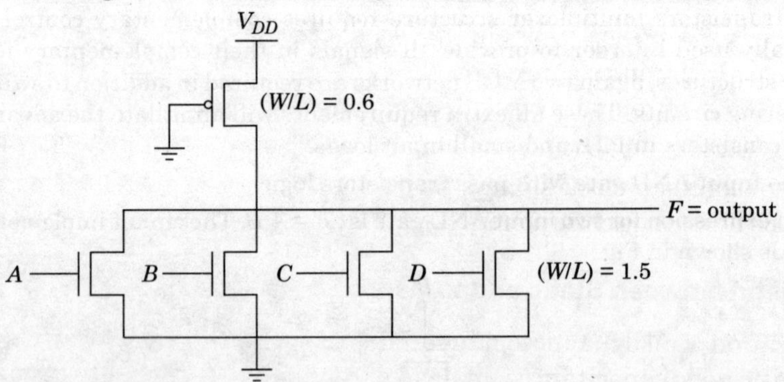

9. Design an NMOS switch based logic implementation of four way MUX layout. **(UPTU-2005-06)**
10. Design 2 input XOR gate using CMOS-TG and compare with pass transistor and CMOS logic circuit. **(UPTU-2011-12)**
11. What is pass transistor? Explain with suitable example. **(UTU-2012)**

□□□

7

Dynamic Logic Circuits

Chapter

Chapter Overview

- Introduction
- Voltage Bootstrapping
- Dynamic CMOS Circuit Design Techniques
- Basic Principles of Pass Transistor Circuits
- Synchronous Dynamic Circuit Techniques
- High-Performance Dynamic CMOS Circuits

7.1 INTRODUCTION

- Static logic circuits allow versatile implementation of logic functions based on static, or steady-state, behavior of simple nMOS or CMOS structures.
- A typical static logic gate generates its output corresponding to the applied input voltages after a certain time delay, and it can preserve its output level (or state) as long as the power supply is provided.
- Static logic circuits may require a large number of transistors to implement a function, and may cause a considerable time delay.
- The major objective of dynamic logic circuits is the high-performance digital implementations where reduction of circuit delay and silicon area is required.
- Dynamic logic circuits require periodic clock signals in order to control charge refreshing. The capability of temporarily storing a state, *i.e.*, a voltage level, at a capacitive node allows us to implement very simple sequential circuits with memory functions.
- The use of common clock signals throughout the system enables us to *synchronize* the operations of various circuit blocks.

7.1.1 Comparison between Static and Dynamic Logic Circuits

- The operation of all dynamic logic gates depends on temporary (transient) storage of charge in parasitic node capacitances, instead of relying on steady-state circuit behavior.
- The dynamic logic implementation of complex functions generally requires a smaller silicon area than the static logic implementation.
- The power consumption which increases with the parasitic capacitances, the dynamic circuit implementation in a smaller area will consume less power than the static counterpart, despite its use of clock signals.

7.2 BASIC PRINCIPLES OF PASS TRANSISTOR CIRCUITS

The fundamental building block of nMOS dynamic logic circuits, consisting of an nMOS pass transistor driving the gate of another nMOS transistor is shown in Fig. 7.1.

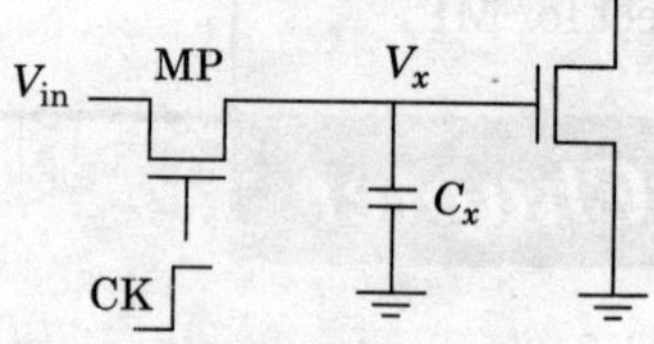

Fig. 7.1 The basic building block for nMOS dynamic logic.

Notice that the pass transistor MP provides the only current path to the intermediate capacitive node (soft node) X. When the clock signal becomes inactive ($CK = 0$), the pass transistor ceases to conduct and the charge stored in the parasitic capacitor C_x continues to determine the output level of the inverter. In the following, we will first examine the charge-up event.

(a) Logic "1" Transfer

- Assume that the soft node voltage is equal to 0 initially, *i.e.*, $V_x(t = 0) = 0$ V.
- A logic "1" level is applied to the input terminal, which corresponds to $V_{IN} = V_{OH} = V_{DD}$.
- The clock signal at the gate of the pass transistor goes from 0 to V_{DD} at $t = 0$. It can be seen that the pass transistor MP starts to conduct as soon as the clock signal becomes active and that MP will operate in saturation throughout this cycle since $V_{DS} = V_{GS}$. Consequently, $V_{DS} > V_{GS} - V_{Tn}$. The circuit to be analyzed for the logic "1" transfer event can be simplified into an equivalent circuit as shown in Fig. 7.2.

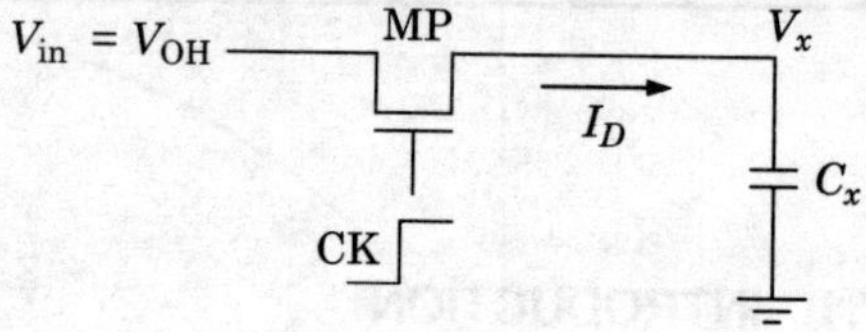

Fig. 7.2 Equivalent circuit for the logic "1" transfer event.

The pass transistor MP operating in the saturation region starts to charge up the capacitor C_x, thus,

$$C_x \frac{dV_x}{dt} = \frac{k_n}{2}(V_{DD} - V_x - V_{T,n})^2 \quad \text{...(7.1)}$$

Note that the threshold voltage of the pass transistor is actually subject to substrate bias effect and therefore, depends on the voltage level V_x. To simplify our analysis, we will neglect the substrate bias effect at this point. Integrating (7.1), we obtain

$$\int_0^t dt = \frac{2C_x}{k_n}\int_0^{V_x} \frac{dV_x}{(V_{DD} - V_x - V_{T,n})} = \frac{2C_x}{k_n}\left(\frac{1}{(V_{DD} - V_x - V_{T,n}}\right)\Bigg|_0^{V_x} \quad \text{...(7.2)}$$

$$t = \frac{2C_x}{k_n}\left[\left(\frac{1}{V_{DD} - V_x - V_{T,n}}\right) - \left(\frac{1}{V_{DD} - V_{T,n}}\right)\right] \quad \text{...(7.3)}$$

This equation can be solved for $V_x(t)$, as follows.

$$V_x(t) = (V_{DD} - V_{T,n}) - \frac{\left(\dfrac{k_n(V_{DD} - V_{T,n})}{2C_x}\right)}{1 + \left(\dfrac{k_n(V_{DD} - V_{T,n})}{2C_x}\right)} \quad \text{...(7.4)}$$

The graph between the variations of the node voltage V_x (according to (7.4)) and time is shown in Fig. 7.3. The pass transistor will turn off when $V_x = V_{max}$, so at this point, its gate-to-source voltage will be equal to its threshold voltage. Therefore, the voltage at node X can never attain the full power supply V_{max} voltage level of V_{DD} during the logic "1" transfer. The actual value of the maximum 1 possible voltage V_{max} at node X can be found by taking into account the substrate bias effect for MP.

$$V_{max} = V_x\Big|_{t \to \infty} = V_{DD} - V_{T,\,n}$$

$$= V_{DD} - V_{T0,\,n} - \gamma\left(\sqrt{|2\phi_F| + V_{max}} - \sqrt{|2\phi_F|}\right) \qquad \text{...(7.5)}$$

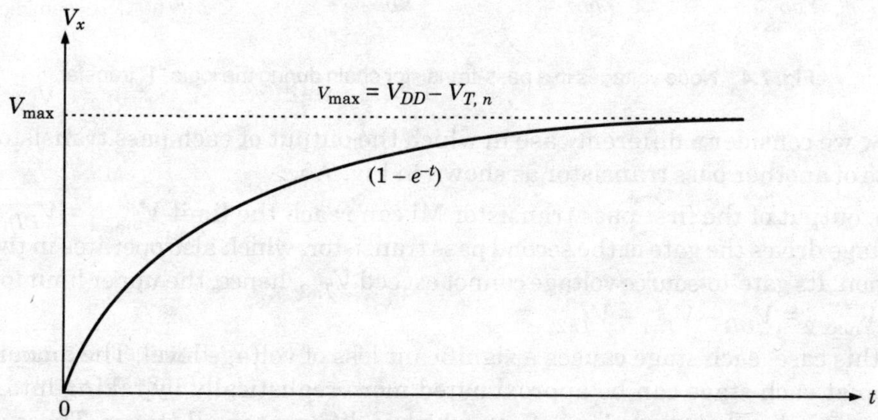

Fig. 7.3 Variation of V_x as a function of time during logic "1" transfer.

The rise time of the voltage V_x will be underestimated if the zero-bias threshold voltage V_{T0} is used in (eq. 7.3). In that case, the actual charge-up time will be longer than predicted by (eq. 7.3), because the drain current of the nMOS transistor is decreased due to the substrate bias effect. The node voltage V_x has an upper limit of $V_{max} = (V_{DD} - V_T)$ has a significant implication for circuit design.

- For example, we consider a circuit of pass-transistor chain during the logic "1" transfer. In this case a logic "1" at the input node ($V_{IN} = V_{DD}$) is being transferred through a chain of cascaded pass transistors as shown in Fig. 7.4.
- For analysis, we assume that initially all internal node voltages V_1 through V_4 are zero. The first pass transistor M_1 operates in saturation with $V_{DS1} > V_{GS1} - V_{Tn1}$. So, the voltage at node 1 cannot limit value $V_{max\,1} = V_{DD} - V_{Tn1}$.

$$V_{T,\,n1} = V_{T0,\,n} - \gamma\left(\sqrt{|2\phi_F| + V_{max\,1}} - \sqrt{|2\phi_F|}\right)$$

$$V_{T,\,n2} = V_{T0,\,n} - \gamma\left(\sqrt{|2\phi_F| + V_{max\,2}} - \sqrt{|2\phi_F|}\right) \qquad \text{...(7.6)}$$

- If all the pass transistors in this circuit are identical, than the second pass transistor $M2$ operates at the saturation boundary. So, the voltage at node 2 will be equal to $V_{max\ 2} = V_{DD} - V_{Tn2}$.
- Due to this we can easily be seen that with $V_{Tn1} = V_{Tn2} = V_{Tn3}, \ldots$, the node voltage at the end of the pass transistor will become one threshold voltage lower than V_{DD}, without effect of the number of pass transistors present in the chain.

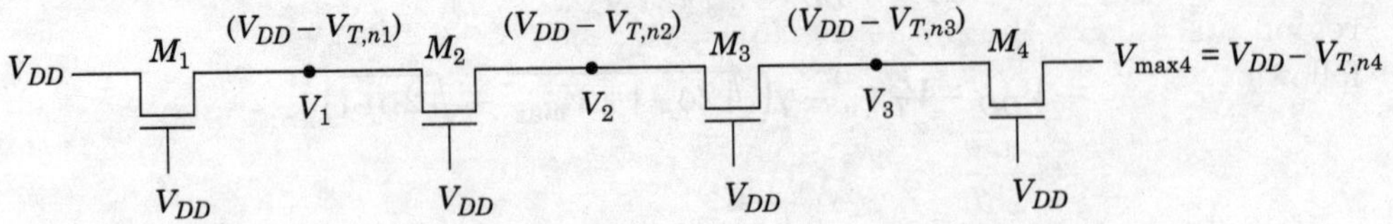

Fig. 7.4 Node voltages in a pass-transistor chain during the logic "1" transfer.

- Now we consider a different case in which the output of each pass transistor drives the gate of another pass transistor as shown in Fig. 7.5.
- The output of the first pass transistor Ml can reach the limit $V_{max\ 1} = V_{DD} - V_{Tn1}$. This voltage drives the gate of the second pass transistor, which also operates in the saturation region. Its gate-to-source voltage cannot exceed V_{Tn2}, hence, the upper limit for V_2 is found as $V_{max\ 2} = V_{DD} - V_{Tn1} - V_{Tn2}$.
- In this case, each stage causes a significant loss of voltage level. The amount of voltage drop at each stage can be approximated more realistically by taking into account the corresponding substrate bias effect, which is different in all stages. These analyses are helped to examine important characteristics of the logic "1" transfer event.

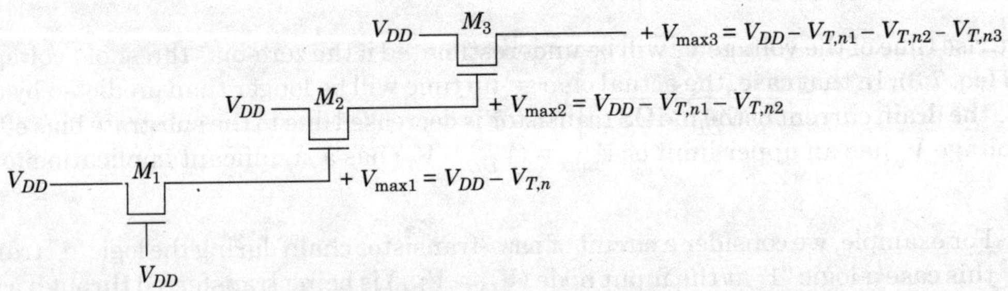

Fig. 7.5 Node voltages during the logic "1" transfer, when each pass transistor is driving another pass transistor.

(b) Logic "0" Transfer

- The circuit to be analyzed for the logic "0" transfer event can be simplified into an equivalent circuit as shown in Fig. 7.6. When the soft-node voltage V_x is initially equal to a logic "1" level, i.e., $V_x(t = 0) = V_{max} = (V_{DD} - V_{Tn})$. A logic "0" level is applied to the input terminal. Now, the clock signal at the gate of the pass transistor goes from 0 to V_{DD} at $t = 0$.
- The pass transistor MP starts to conduct as soon as the clock signal becomes active, and the direction of drain current flow through MP will be opposite to that during the charge-up (logic "1" transfer) event.

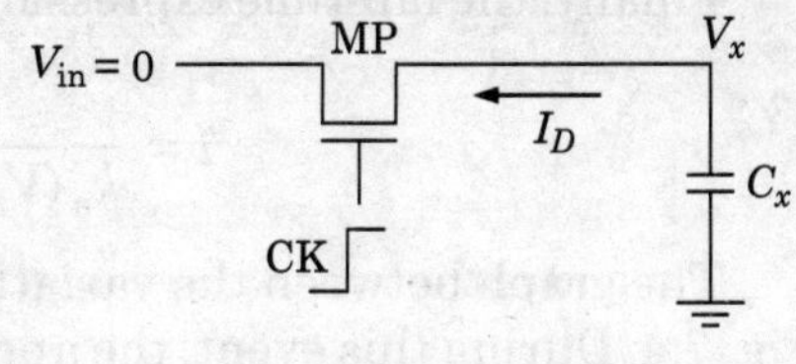

Fig. 7.6 Equivalent circuit for the logic "0" transfer event.

- The intermediate node X will now corresponds to the drain terminal of MP and that the input node will correspond to its source terminal.
- The pass transistor operates in the linear region, when $V_{GS} = V_{DD}$ and $V_{DS} = V_{max}$. So that $V_{DS} < V_{GS} - V_{Tn}$.
- The pass transistor MP operating in the linear region discharges the parasitic capacitor C_x, as follows:

$$-C_x \frac{dV_x}{dt} = \frac{k_n}{2}(2(V_{DD} - V_{T,n})V_x - V_x^2) \quad ...(7.7)$$

$$dt = -\frac{2C_x}{k_n} \cdot \frac{dV_x}{2(V_{DD} - V_{T,n})V_x - V_x^2} \quad ...(7.8)$$

- Here, the source voltage of the nMOS pass transistor is equal to 0 V; hence, there is no substrate bias effect for MP($V_{Tn} = V_{T0,n}$). But the initial condition: $V_x(t=0) = (V_{DD} - V_{Tn})$ contains the threshold voltage *with* substrate bias effect, because the voltage V_x is set during the preceding logic "1" transfer event.
- To simplify the expressions, we will use V_{Tn} in the following. From the above equation (7.8), we can find the expression for the fall time t.

$$\int_0^1 dt = -\frac{2C_x}{k_n} \int_{V_{DD}-V_{T,n}}^{V_x} \left(\frac{\frac{1}{2(V_{DD}-V_{T,n})} + \frac{1}{2(V_{DD}-V_{T,n})}}{2(V_{DD}-V_{T,n}) - V_x} \right) dV_x \quad ...(7.9)$$

$$t = \frac{C_x}{k_n(V_{DD} - V_{T,n})} \left[\ln\left(\frac{2(V_{DD} - V_{T,n}) - V_x}{V_x} \right) \right]_{V_{DD}-V_{T,n}}^{V_x} \quad ...(7.10)$$

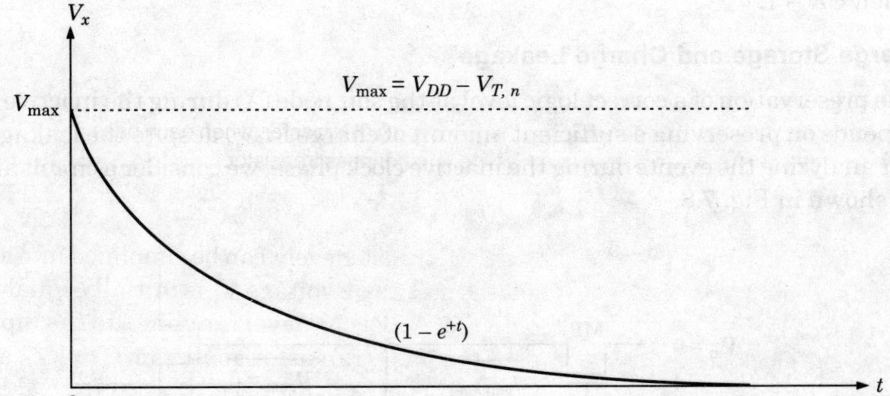

Fig. 7.7 Variation of V_x as a function of time during logic "0" transfer.

Finally, the fall-time expression for the node voltage V_x can be obtained as

$$t = \frac{C_x}{k_n(V_{DD} - V_{T,n})} \ln\left(\frac{2(V_{DD} - V_{T,n}) - V_x}{V_x}\right) \quad \text{...(7.11)}$$

The graph between the variations of the node voltage V_x and a function of time is shown in Fig. 7.4. During this event, the applied input voltage level (logic 0) can be transferred to the soft node without any modification.

Calculation of fall time

First we define the two time point's $t_{90\%}$ and $t_{10\%}$ as the times at which the node voltage is equal to 0.9 V_{max} and 0.1 V_{max}, respectively. These two time points can easily calculated by equation (7.11).

$$t_{90\%} = \frac{C_x}{k_n(V_{DD} - V_{T,n})} \ln\left(\frac{(2 - 0.9)(V_{DD} - V_{T,n})}{0.9(V_{DD} - V_{T,n})}\right)$$

$$= \frac{C_x}{k_n(V_{DD} - V_{T,n})} \ln\left(\frac{1.1}{0.9}\right) \quad \text{...(7.12)}$$

$$t_{10\%} = \frac{C_x}{k_n(V_{DD} - V_{T,n})} \ln\left(\frac{1.9}{0.1}\right) \quad \text{...(7.13)}$$

The fall time of the soft-node voltage V_x is by definition the difference between t $t_{10\%}$ and $t_{90\%}$, which is found as

$$\tau_{fall} = t_{10\%} - t_{90\%}$$

$$= \frac{C_x}{k_n(V_{DD} - V_{T,n})}[\ln(1.9) - \ln(1.22)]$$

$$= 2.74 \frac{C_x}{k_n(V_{DD} - V_{T,n})} \quad \text{...(7.14)}$$

- Hence, we have examined the transient charge-up and charge-down events which are responsible for logic "1" transfer and logic "0" transfer during the active clock phase, i.e., when $CK = 1$.

7.2.1 Charge Storage and Charge Leakage

- The preservation of a correct logic level at the soft node (X) during the inactive clock phase depends on preserving a sufficient amount of charge in C_x, despite the leakage currents.
- For analyzing the events during the inactive clock phase, we consider a circuit arrangement as shown in Fig. 7.8.

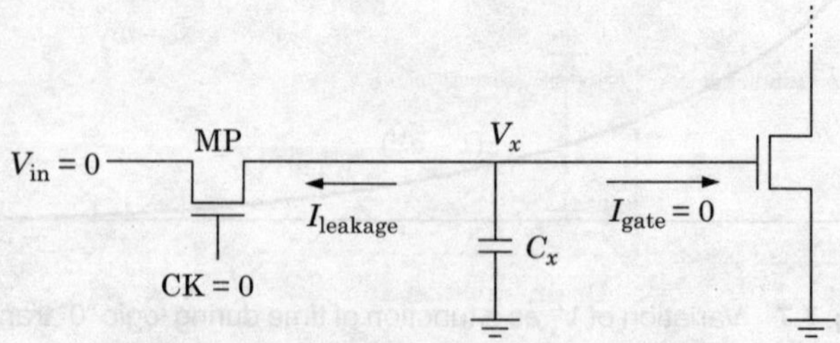

Fig. 7.8 Charge leakages from the soft node X.

- We will assume that a logic-high voltage level has been transferred to the soft node during the active clock phase and that now both the input voltage V_{IN} and the clock CK are equal to 0 V.
- The charge stored in C_x will gradually leak away due to the leakage currents associated with the pass transistor. The gate current of the inverter driver transistor is negligible for all practical purposes.
- A simplified cross-section of the nMOS pass transistor, together with the lumped node capacitance C_x is shown in Fig. 7.9.
- This leakage current has two main components, namely, the sub threshold channel current and the reverse conduction current of the drain-substrate junction.

$$I_{leakage} = I_{subthreshold(MP)} + I_{reverse(MP)} \quad ...(7.15)$$

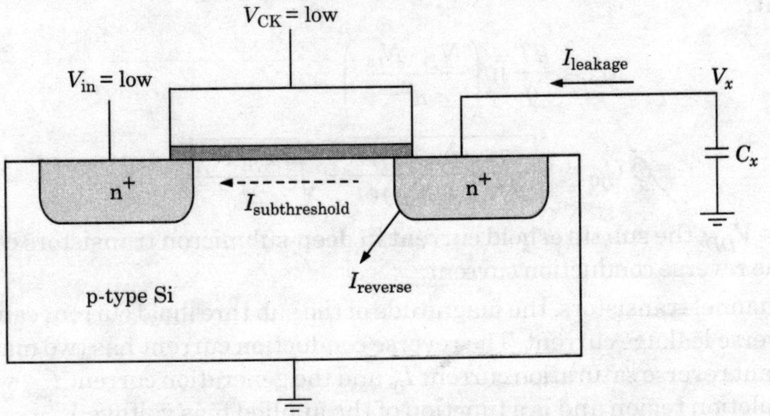

Fig. 7.9 Simplified cross-section of the nMOS pass transistor.

- In this analysis, mainly two portions of soft-node capacitance C_x are present. In which one component of C_x can be considered constants, which are primarily due to oxide-related parasitic and other portion of the total soft-node capacitance C_x is due to the reverse biased drain-substrate junction, which is also a function of the soft-node voltage V_x.
- These constant capacitance components will be represented by C_{in} as shown in Fig. 7.10. So the total charge stored in the soft node as the sum of two main components, as follows.

$$Q = Q_j(V_x) + Q_{in} \quad \text{where} \quad Q_{in} = C_{in} \cdot V_x$$

$$C_{in} = C_{gb} + C_{poly} + C_{metal} \quad ...(7.16)$$

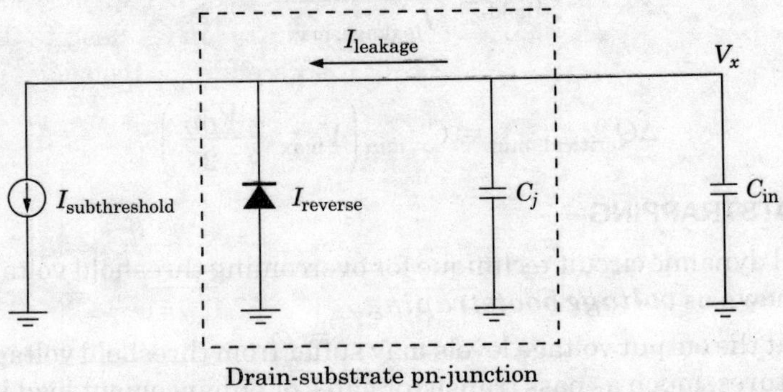

Fig. 7.10 Equivalent circuit used for analyzing the charge leakage process.

The total leakage current can be expressed as the time derivative of the total soft-node charge Q.

$$I_{\text{leakage}} = \frac{dQ}{dt} = \frac{dQ_j(V_x)}{dt} + \frac{dQ_{in}}{dt}$$

$$= \frac{dQ_j(V_x)}{dV_x}\frac{dV_x}{dt} + C_{\text{in}} \cdot \frac{dV_x}{dt} \quad \text{...(7.17)}$$

Where,

$$\frac{dQ_j(V_x)}{dV_x} = C_j(V_x) = \frac{A \cdot C_{j0}}{\sqrt{1 + \frac{V_x}{\phi_0}}} = A \cdot \sqrt{\frac{q\varepsilon_{Si} N_A}{2(\phi_0 + V_x)}} \quad \text{...(7.18)}$$

We know that,

$$\phi_0 = \frac{kT}{q} \ln\left(\frac{N_D \cdot N_A}{n_i^2}\right) \quad \text{...(7.19)}$$

$$C_{j0} = \sqrt{\frac{q\,\varepsilon_{Si} N_A\, N_D}{2(N_A + N_D)\phi_0}} - \sqrt{\frac{q\,\varepsilon_{Si}\, N_A}{2\phi_0}} \quad \text{...(7.20)}$$

- For $V_{DS} = V_{DD}$, the sub threshold current in deep-submicron transistors can significantly exceed the reverse conduction current.
- In long-channel transistors, the magnitude of the sub threshold current can be comparable to the reverse leakage current. The reverse conduction current has two main components, the constant reverse saturation current I_0, and the generation current I_{gen} which originates in the depletion region and is a function of the applied bias voltage V_x.
- The minimum combined soft-node capacitance is given as

$$C_{x,\,\min} = C_{\text{gb}} + C_{\text{poly}} + C_{\text{metal}} + C_{db,\,\min} \quad \text{...(7.21)}$$

Where $C_{db,\,\min}$ represents the minimum junction capacitance (At $V_x = V_{\max}$).

7.2.2 Holding Time

"The holding time (t_{hold}) is the shortest time required for the soft-node voltage to drop from its initial logic-high value to the logic threshold voltage due to leakage". The holding time (t_{hold}) is given by

$$t_{\text{hold}} = \frac{\Delta Q_{\text{critical, min}}}{I_{\text{leakage, max}}} \quad \text{...(7.22)}$$

Where,

$$\Delta Q_{\text{critical, min}} = C_{x,\,\min}\left(V_{\max} - \frac{V_{DD}}{2}\right) \quad \text{...(7.23)}$$

7.3 VOLTAGE BOOTSTRAPPING

- A very useful dynamic circuit technique for overcoming threshold voltage drops in digital circuits is known as ***voltage bootstraping.***
- We know that the output voltage levels may suffer from threshold voltage drops in several circuit structures, much as pass transistor gates or enhancement load inverters and logic gates.

- Dynamic voltage bootstrapping techniques offer a simple yet effective way to overcome threshold voltage drops which occur in most situations. Consider the circuit shown in Fig. 7.11, where the voltage V_x is equal to or smaller than the power supply voltage, $V_x < V_{DD}$. Consequently, the enhancement-type nMOS transistor M_2 will operate in saturation.
- When the input voltage V is low, the maximum value that the output voltage can attain is limited by

$$V_{out}(\max) = V_x - V_{T_2}(V_{out}) \quad ...(7.24)$$

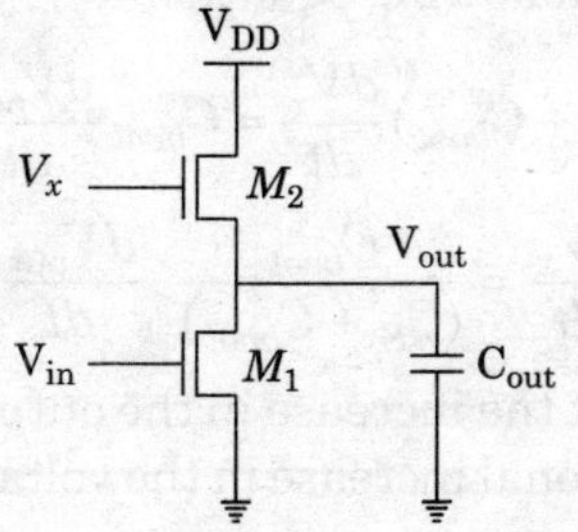

Fig. 7.11 Enhancement-type circuit in which the output node is weakly driven

- To overcome the threshold voltage drop and to obtain a full logic-high level (V_{DD}) at the output node, the voltage V_x must be increased. Now consider the circuit shown in Fig. 7.12, where a third transistor M_3 has been added to the circuit. The two capacitors C_s and C_{boot}, seen in the circuit diagram represent the capacitances which dynamically couple the voltage V_x, to the ground and to the output, respectively. We will see that this circuit can produce a high V_x during switching, so that the threshold voltage drop can be overcome at the output node.

$$V_x \leq V_{DD} + V_{T2}(V_{out}) \quad ...(7.25)$$

- Initially assume that the input voltage V_{in} is logic-high, so that M_1 and M_2 have a nonzero drain current and that the output voltage is low. At this point, M_1 is in the linear region and M_2 is in saturation. Since $I_{D3} = 0$, the initial condition for the voltage V_d can be found as

$$V_x = V_{DD} - V_{T_3}(V_x) \quad ...(7.26)$$

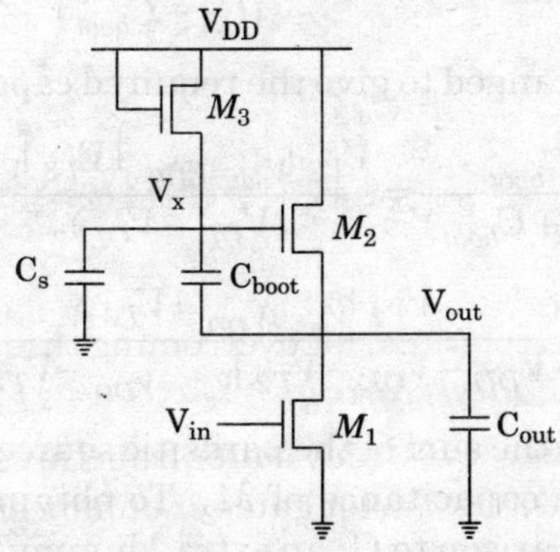

Fig. 7.12 Dynamic bootstrapping arrangement to boost VK during switching

- Now, assume that the input switches from its logic-high level to 0 V at $t = 0$. As a result, the driver transistor M_1 will turn off and the output voltage $\mathbf{V_{OUT}}$, will start to rise. This change in the output voltage level will now be coupled to V_x through the bootstrap capacitor, C_{boot} Let i_{Cboot} represent the transient current flowing through the capacitor C_{boot} during this charge-up event, and let is be the current through C_s. Assuming that the two current components are approximately equal, we obtain

$$i_{Cs} = i_{Cboot} \Leftrightarrow C_S \frac{dV_x}{dt} \approx C_{boot} \frac{d(V_{out} - V_x)}{dt} \qquad \text{...(7.27)}$$

Reorganizing (7.27) yield the following equation

$$(C_S + C_{boot}) \frac{dV_x}{dt} = C_{boot} \frac{dV_{out}}{dt} \qquad \text{...(7.28)}$$

$$\frac{dV_x}{dt} = \frac{C_{boot}}{(C_S + C_{boot})} \cdot \frac{dV_{out}}{dt} \qquad \text{...(7.29)}$$

- It can be seen from (7.29) that the increase in the output voltage V_x, during this switching event will generate a proportional increase in the voltage level V_x. Integrating both sides of (7.29), we obtained

$$\int_{V_{DD}-V_{T3}}^{V_x} dV_x = \frac{C_{boot}}{(C_S + C_{boot})} \cdot \int_{V_{OL}}^{V_{DD}} dV_{out} \qquad \text{...(7.30)}$$

$$V_x = (V_{DD} + V_{T_3}) + \frac{C_{boot}}{(C_S + C_{boot})} (V_{DD} - V_{OL}) \qquad \text{...(7.31)}$$

If the capacitor C_{boot} is must larger than C_s ($C_{boot} >> C_s$) the maximum value of V_x, can be approximated as

$$V_x(\max) = 2V_{DD} - V_{T_3} - V_{OL} \qquad \text{...(7.32)}$$

which proves that voltage bootstrapping can significantly boost the voltage leave V_x. Now remember that in order to overcome the threshold voltage drop at the output, the minimum required voltage level V_x is

$$V_x(\min) = V_{DD} + V_{T2}\,|_{V_{out}V_{DD}}$$
$$= (V_{DD} - V_{T3}(V_x)) + \frac{C_{boot}}{(C_S + C_{boot})}(V_{DD} - V_{OL}) \qquad \text{...(7.33)}$$

This equation can be rearranged to give the required capacitance ratio, as follows,

$$\frac{C_{boot}}{(C_S + C_{boot})} = \frac{V_{T2}\,|_{V_{out}=V_{DD}} + V_{T3}\,|_{V_x}}{(V_{DD} - V_{OL})} \qquad \text{...(7.34)}$$

$$\frac{C_{boot}}{C_S} = \frac{V_{T2}\,|_{V_{out}=V_{DD}} + V_{T3}\,|_{V_x}}{V_{DD} - V_{OL} - V_{T2}\,|_{V_{out}=V_{DD}} - V_{T3}\,|_{V_x}} \qquad \text{...(7.35)}$$

- Note that Cs is essentially the sum of the parasitic source-to-substrate capacitance of M_3 and the gate-to-substrate capacitance of M_2. To obtain a sufficiently large bootstrap capacitance Cboot in comparison to C_S an extra "dummy" transistor is typically added to the circuit, as shown in Fig. 7.13.

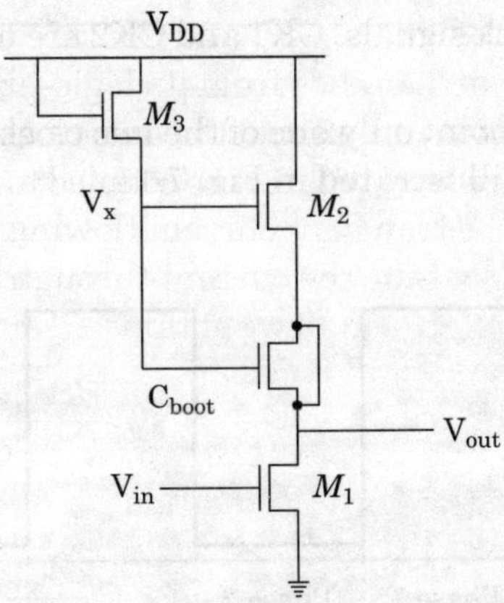

Fig. 7.13 Realization of the bootstrapping capacitor with a dummy MOS device.

Since its drain and source terminals are connected together, the dummy transistor simply acts as an MOS capacitor between V_{out} and V_x. Although this circuit arrangement contains two additional transistors to achieve voltage bootstraping, the resulting circuit performance improvement is usually well worth the extra silicon area used for the bootstrapping devices.

7.4 SYNCHRONOUS DYNAMIC CIRCUIT TECHNIQUES

- The basic concepts associated with temporary storage of logic levels in capacitive circuit nodes, but here, our attention to digital circuit design techniques, which take advantage of this simple yet effective principle.
- In this topic, we will discuss different examples of synchronous dynamic circuits implemented using depletion-load nMOS, enhancement-load nMOS, and CMOS building blocks.

7.4.1 Dynamic Pass Transistor Circuits

- The generalized view of a multi-stage synchronous circuit is shown in Fig. 7.14. The circuit consists of cascaded combinational logic stages, which are interconnected through nMOS pass transistors.
- All inputs of each combinational logic block are driven by a single clock signal. Individual input capacitances are not shown in this figure for simplicity, but the operation of the circuit depends on temporary charge storage in the parasitic input capacitances.

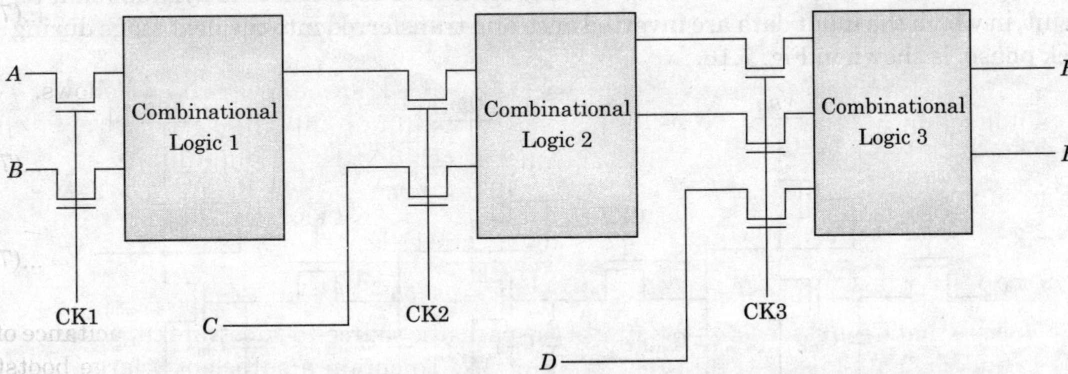

Fig. 7.14 Multi-stage pass transistor logic driven by two non over lapping clocks.

- Two non overlapping clock signals, CK1 and CK2 are used in this system for drive the pass transistors.
- Here, at any given time point only one of the two clock signals can be active due to its non overlapping property as illustrated in Fig. 7.15.

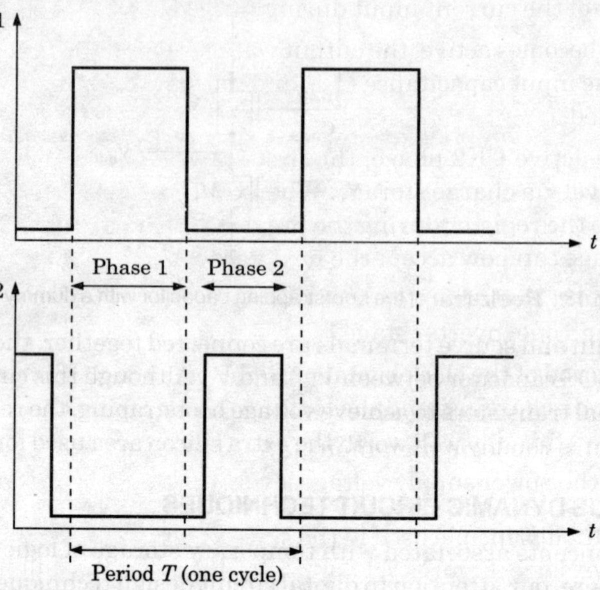

Fig. 7.15 *Non* overlapping clock signals used for two-phase synchronous operation.

- When clock CK1 is active, the input levels of Stage 1 (and Stage 3) are applied through the pass transistors, but the input capacitances of Stage 2 retains their previously set logic levels.
- When clock CK2 is active the input levels of Stage 2 will be applied through the pass transistors, but the input capacitances of Stage 1 and Stage 3 retain their logic levels.
- This allows us to incorporate the simple dynamic memory function at each stage input, and at the same time, to facilitate synchronous operation by controlling the signal flow in the circuit using the two periodic clock signals. This signal timing scheme is also called *two-phase clocking* and is one of the most widely used timing strategies.

We have not made any specific assumptions about the internal structure of the combinational logic stages, when we used the two-phase clocking scheme. A depletion-load dynamic shift register circuit, in which the input data are inverted once and transferred into the next stage during each clock phase, is shown in Fig. 7.16.

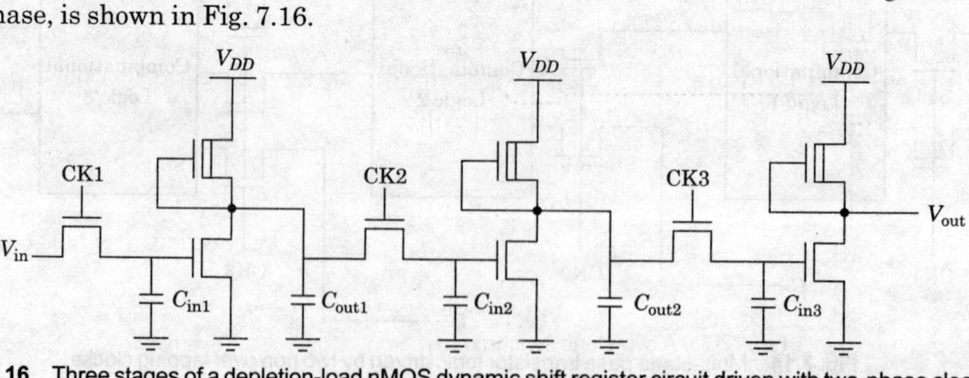

Fig. 7.16 Three stages of a depletion-load nMOS dynamic shift register circuit driven with two-phase clocking.

Operations of shift register circuit

- During the active phase of CK1 the input voltage level V_{IN} is transferred into the input capacitance $C_{in\,1}$. Thus, the valid output voltage level of the first stage is determined as the inverse of the current input during this cycle.
- When CK2 becomes active, the output voltage level of the first stage is transferred into the second stage input capacitance C_{in2}, and the valid output voltage level of the second stage is determined.
- During the active CK2 phase, the first-stage input capacitance continues to retain its previous level via charge storage. When CK1 becomes active again, the original data bit *written* into the register during the previous cycle is transferred into the third stage, and the first stage can now accept the next data bit.
- The maximum clock frequency of this circuit is determined by the signal propagation delay through one inverter stage.
- One half-period of the clock signal must be enough to allow the input capacitance C_{in}, for charging or discharging, and the logic level to propagate to the output by charging C_{out}.
- In this circuit, the logic-high input level of each inverter stage is one threshold voltage lower than the power supply voltage level.

The same operation principle used in the simple shift register circuit can easily be extended to synchronous complex logic. Figures 7.17 and 7.18 show a two-stage circuit example implemented using depletion-load nMOS complex logic gates.

In a complex logic circuit such as the one shown in Fig. 7.18, we see that the signal propagation delay of each stage may be different. Thus, in order to guarantee that correct logic levels are propagated during each active clock cycle, the half-period length of the clock signal must be longer than the largest single-stage signal propagation delay found in the circuit.

Next, we consider a dynamic enhancement-load shift register implementation. In which, at each stage, the input pass transistor and the load transistor are driven by the same clock phase (as Fig. 7.19).

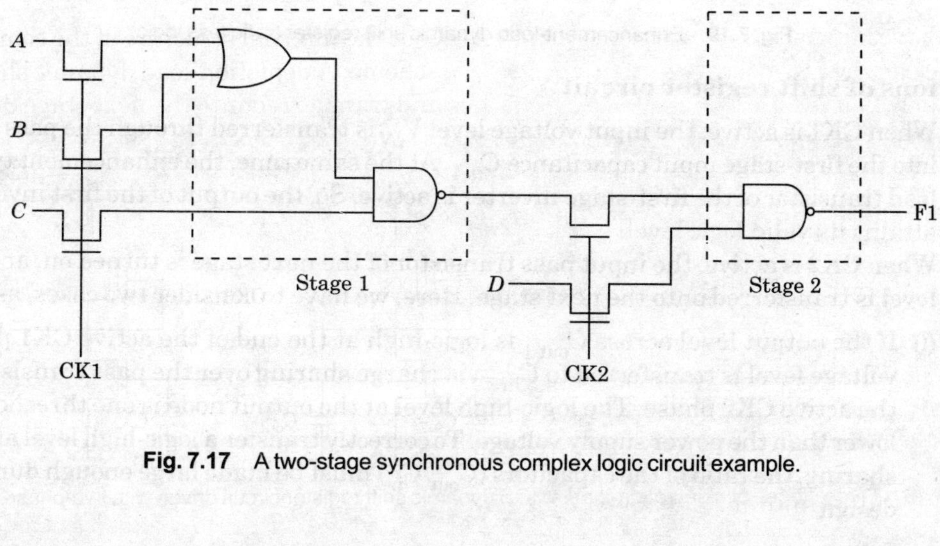

Fig. 7.17 A two-stage synchronous complex logic circuit example.

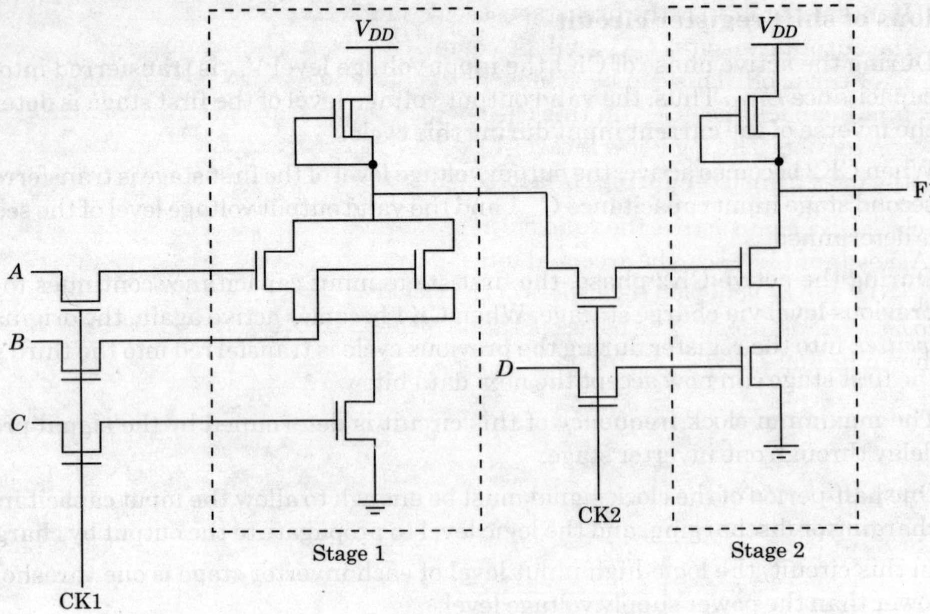

Fig. 7.18 Depletion-load nMOS implementation of synchronous complex logic.

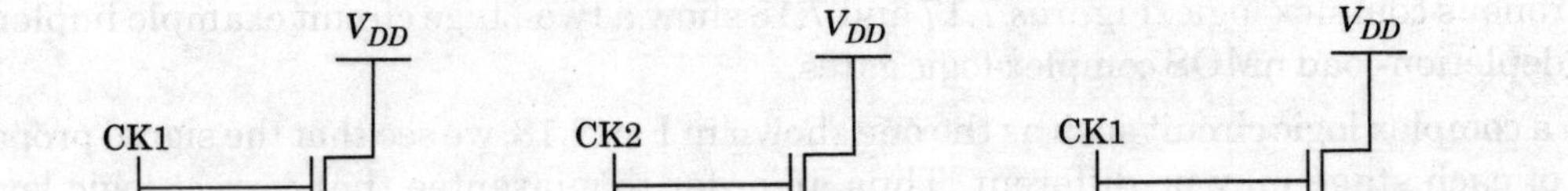

Fig. 7.19 Enhancement-load dynamic shift register (ratio less logic).

Operations of shift register circuit

- When CK1 is active, the input voltage level V_{IN} is transferred through the pass transistor into the first-stage input capacitance $C_{in\,1}$. At the same time, the enhancement type nMOS load transistor of the first-stage inverter is active. So, the output of the first inverter stage attains its valid logic level.
- When CK2 is active, the input pass transistor of the next stage is turned on, and the logic level is transferred onto the next stage. Here, we have to consider two cases, as follows.
  (i) If the output level across $C_{out\,1}$ is logic-high at the end of the active CK1 phase, this voltage level is transferred to $C_{in\,2}$ via charge sharing over the pass transistor during the active CK2 phase. The logic-high level at the output node is one threshold voltage lower than the power supply voltage. To correctly transfer a logic-high level after charge sharing, the ratio of the capacitors (C_{out}/C_{in}) must be made large enough during circuit design.

(ii) If the output level of the first stage is logic-low at the end of the active CK1 phase, then the output capacitor $C_{out\ 1}$ will be completely drained to a voltage of $V_{OL} = 0$ V when CK1 turns off. This can be achieved because a logic-high level is being stored in the input capacitance $C_{in\ 1}$ in this case, which forces the driver transistor to remain in conduction. The logic-low level of $V_{OL} = 0$ V is also transferred into the next stage via the pass transistor during the active CK2 phase.

When clock CK1 again active, the valid output level across $C_{out\ 2}$ is determined and transferred into $C_{in\ 3}$. A new input level can be accepted into $C_{in\ 1}$, during this phase. Since the valid logic-low level of $V_{OL} = 0$ V can be achieved regardless of the driver-to-load ratio, this circuit arrangement is called *ratio less dynamic logic.* The basic operation principle can be extended to arbitrary complex logic, as shown in Fig. 7.20.

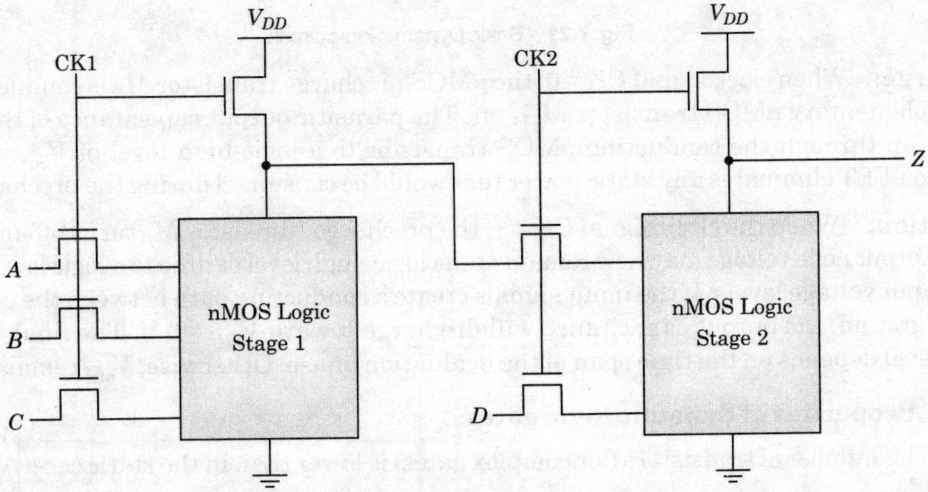

Fig. 7.20 General circuit structure of ratio less synchronous dynamic logic.

7.5 DYNAMIC CMOS CIRCUIT DESIGN TECHNIQUES

A variety of approaches including pseudo-NMOS, pass-transistor, etc., were presented to reduce the number of transistors required to implement a given logic function. The pseudo-NMOS logic style requires only $N + 1$ transistor to implement an N input logic gate, but unfortunately it has static power dissipation. In this section, an alternative logic style called dynamic logic is presented that obtains a similar result, while avoiding static power consumption. With the addition of a clock input, it uses a sequence of precharge and conditional evaluation phases.

7.5.1 Dynamic Logic: Basic Principles

A dynamic CMOS circuit technique which allows us to significantly reduce the number of transistors used to implement any logic function. The basic construction of an (n-type) dynamic logic gate is shown in Fig. 7.21. The PDN is constructed exactly as in complementary CMOS. The operation of this circuit is divided into two phases-precharge and evaluation-with the mode of operation determined by clock signal, CK.

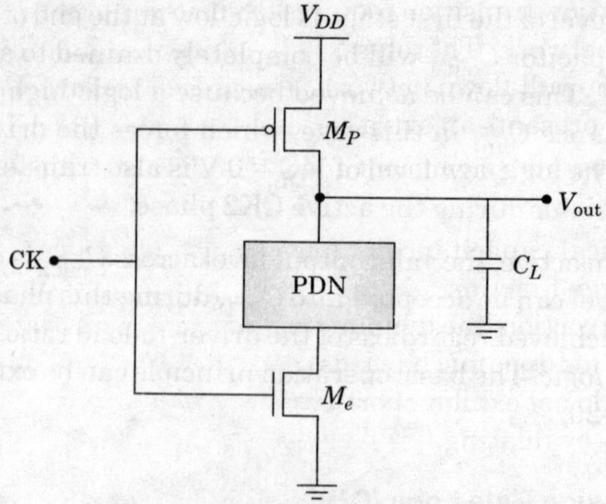

Fig. 7.21 Basic Dynamic logic circuit.

Precharge: When clock signal CK = 0, the pMOS precharge transistor M_P is conducting, while the complementary nMOS transistor M_e is off. The parasitic output capacitance of the circuit is charged up through the conducting pMOS transistor to a logic-high level of $V_{out} = V_{DD}$. The evaluation FET eliminates any static power that would be consumed during the precharge period.

Evaluation: When the clock signal CK = 1, the precharge transistor M_P turns off and M_e turns on. The output node voltage may now remain at the logic- high level or drop to a logic low, depending on the input voltage levels. If the input signals create a conducting path between the output node and the ground, the output capacitance will discharge toward $V_{OL} = 0$ V. The final discharged output level depends on the time span of the evaluation phase. Otherwise, V_{out} remains at V_{DD}.

7.5.1.1 Properties of dynamic logic gates:

1. The number of transistors (for complex gates) is lower than in the static case: $N + 2$ versus 2N.
2. The logic function is implemented by nMOS pull- network.
3. It only consumes dynamic power. Ideally, no static current path ever exists between V_{DD} and GND. The total power dissipation, however, can be significantly higher copared with a static logic gate.
4. The logic gates have faster switching speeds.

7.5.1.2 Limitations of domino logic circuits:

1. Number of inverting static logic stages in cascade must be even, so the inputs of the next domino CMOS stage experience only 0 to ⊥ transition curing evaluation.
2. Only non-inverting structures can be implemented using domino CMOS. If necessary inversion must be carried out using conventional CMOS logic.
3. Charge sharing between the dynamic stage output node and the inter mediate nodes of the nMOS logic block during the evaluation phase may cause erroneous output.

7.5.2 Speed and Power Dissipation of Dynamic Logic

- The main advantages of dynamic logic are increased speed and reduced area. Fewer devices to implement a given logic function implies that the overall load capacitance is much smaller. After precharge phase, the output is high. For a low input-signal, no additional switching occurs. So $t_{PLH} = 0$.

- When the high to low transition requires the discharging of the output capacitance through the pull-down network. Therefore, t_{PHL} is proportional to C_x and the current sinking capabilities of the pull-down network. The presence of evolution transistor slows the gate somewhat, as it presents an extra series resistance.

The dynamic logic presents a significant advantage from a power perspective, due to the following reasons:

- First, the physical capacitance is lower since dynamic logic uses fewer transistors to implement a given function.
- Second, by construction, the dynamic logic gates can have at most one transistor per clock cycle. Glitching (or dynamic hazards) does not occur in dynamic logic.
- Dynamic gates do not exhibit short-circuit power since the pull-up path is not turned ON when the gate is evaluating.

7.5.3 CMOS Transmission Gate Logic/CMOS Shift Register

- CMOS transmission gate logic is based on the principle of the two-phase synchronous logic circuit.
- In this, individual logic blocks are cascaded via clock-controlled switches. Here, static CMOS gates are used for implementing the logic blocks, and CMOS transmission gates are used for transferring the output levels of one stage to the inputs of the next stage.
- Each transmission gate is controlled by the clock signal and its complement.

7.5.3.1 Operation of CMOS Transmission Logic

1. The fundamental building block of a dynamic CMOS transmission gate shift register is shown in Fig. 7.22. It consists of a CMOS inverter, which is driven by a CMOS transmission gate.

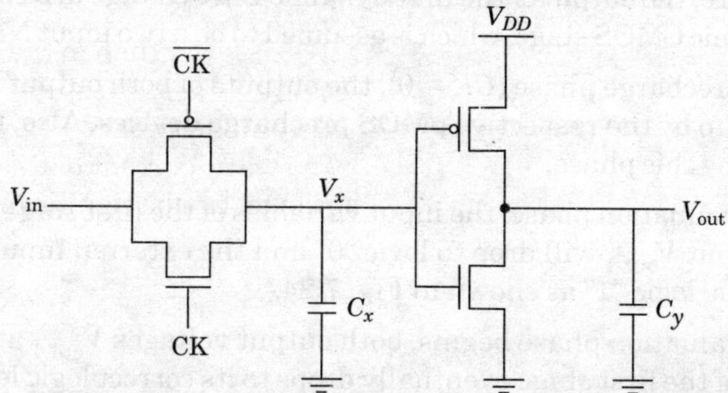

Fig. 7.22 Basic building block of a CMOS transmission gate dynamic shift register.

2. When the clock phase $CK = 1$, the input voltage V_{in} is transferred onto the parasitic input capacitance C_X, via the transmission gate. The CMOS transmission gate gives smaller transfer time compared to only nMOS switches due to its low on-resistance. There is no threshold voltage drop across the CMOS transmission gate.
3. When the clock phase $CK = 0$, the CMOS transmission gate turns off and the voltage level across C_x, can be preserved until the next cycle.
4. Figure 7.23 shows a single-phase CMOS shift register, which is built by cascading identical units as in Fig. 7.22 and by driving each stage alternately with the clock signal and its complement.

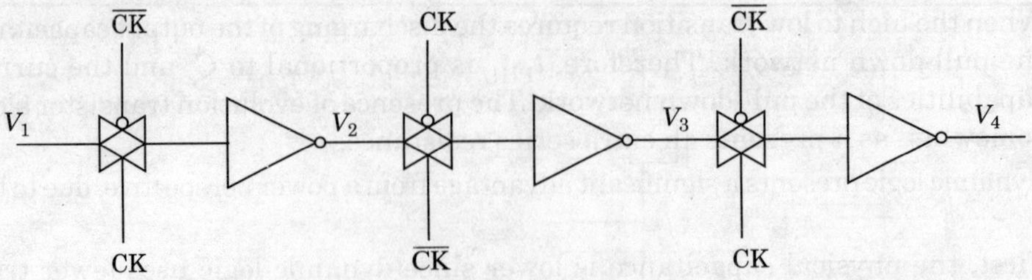

Fig. 7.23 Single-phase CMOS transmission gate dynamic shift register.

- When $CK = 1$, the transmission gates of the odd-numbered stages would conduct and the transmission gates of the even-numbered stages are off, so that the cascaded inverter stages in the chain are alternately isolated. So that the inputs are permitted in alternating half cycles.
- In practice, however, the clock signal and its complement do not constitute a truly non overlapping signal pair, since the clock voltage waveform has finite rise and fall times. Therefore, true two-phase clocking with two non overlapping clock signals (ϕ_1 and ϕ_2) *and* their complements is usually preferred over single-phase clocking in dynamic CMOS transmission gate logic.

7.5.4 Cascading Dynamic CMOS Logic

- The operation of the single-stage dynamic CMOS logic gate is quite straightforward. There is one major problem that complicates the design of dynamic circuits during straight forward cascading of dynamic gates to create multilevel logic structures.
- The problem is best illustrated with two-stage cascaded nMOS structure as shown in Fig. 7.24. Here, the output of the first dynamic CMOS stage drives one of the inputs of the second dynamic CMOS stage, which is assumed to be a two input NAND gate for simplicity.
- During the precharge phase ($CK = 0$), the outputs of both output voltages $V_{out\,1}$ and $V_{out\,2}$ are pulled up by the respective pMOS precharge devices. Also, the external inputs are applied during this phase.
- During the evaluation phase, the input variables of the first stage are assumed to be such that the output $V_{out\,1}$ will drop to logic "0" and the external input of the second-stage is assumed to be logic "1" as shown in Fig. 7.24.
- When the evaluation phase begins, both output voltages $V_{out\,1}$ and $V_{out\,2}$ are logic-high. The output of the first stage eventually drops to its correct logic level after a certain time delay.
- When the evaluation in the second stage is start concurrently with the high value of $V_{out\,1}$ at the beginning of the evaluation phase, the output voltage $V_{out\,2}$ at the end of the evaluation phase will be *erroneously* low. Although the first stage output subsequently assumes its correct output value once the stored charge is drained, the correction of the second-stage output is not possible.
- With the help of this example, we can see that the dynamic CMOS logic stages cannot be cascaded directly when driven by the same clock signal. Alternative clocking schemes and circuit structures must be developed to overcome this problem.

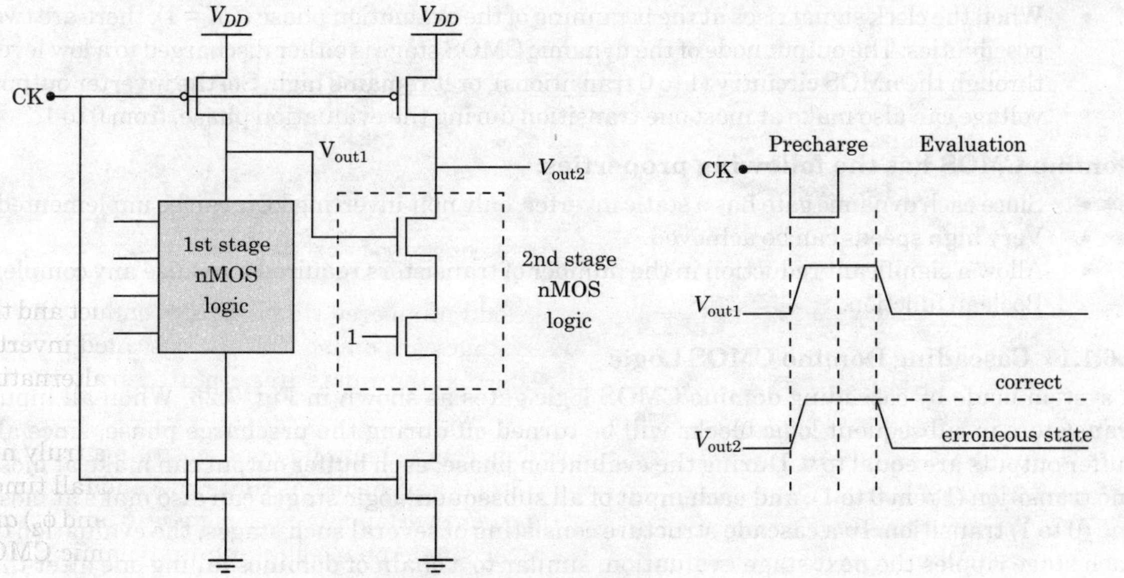

Fig. 7.24 Cascade of dynamic *n*-type network

7.6 HIGH-PERFORMANCE DYNAMIC CMOS CIRCUITS

High-Performance dynamic CMOS circuits are variants of the basic dynamic CMOS logic gate structure. The ultimate goal is to achieve reliable, high-speed, compact circuits using the least complicated clocking scheme possible.

7.6.1 Domino CMOS Logic **[UTU-2012]**

A domino logic module consists of an *n*-type dynamic logic block followed by a static inverter. The generalized circuit diagram of a domino CMOS logic gate shown in Fig. 7.25. The addition of the inverter allows us to operate a number of such structures in cascade, as explained in the following.

- When $CK = 0$ (pre-charge phase), the output node of the dynamic CMOS stage is pre charged to a high logic level, and the output of the CMOS inverter (buffer) becomes low.

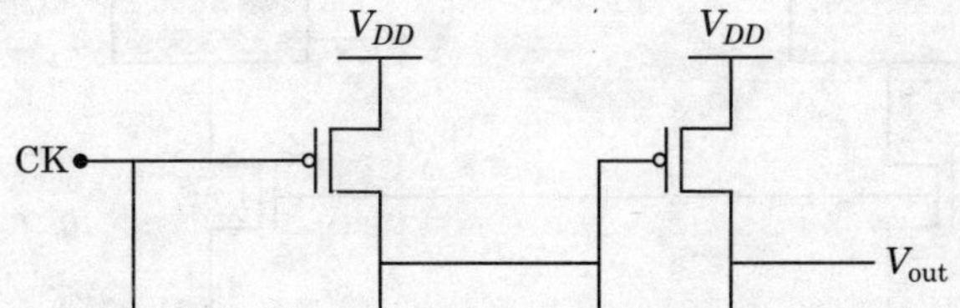

Fig. 7.25 Generalized circuit diagram of a domino CMOS logic gate.

- When the clock signal rises at the beginning of the evaluation phase ($CK = 1$), there are two possibilities: The output node of the dynamic CMOS stage is either discharged to a low level through the nMOS circuitry (1 to 0 transitions), or it remains high. So, the inverter output voltage can also make at most one transition during the evaluation phase, from 0 to 1.

Domino CMOS has the following properties:

- Since each dynamic gate has a static inverter, only non-inverting logic can be implemented.
- Very high speeds can be achieved.
- Allow a significant reduction in the number of transistors required to realize any complex Boolean function.

7.6.1.1 Cascading Domino CMOS Logic

A system made by cascading domino CMOS logic gates as shown in Fig. 7.26. When all input transistors in subsequent logic blocks will be turned off during the precharge phase, since all buffer outputs are equal to 0. During the evaluation phase, each buffer output can make at most one transition (From 0 to 1), and each input of all subsequent logic stages can also make at most one (0 to 1) transition. In a cascade structure consisting of several such stages, the evaluation of each stage ripples the next stage evaluation, similar to a chain of dominos falling one after the other. The structure is hence called *domino CMOS logic.*

Cascading domino CMOS logic is useful for reduce the problem in cascading conventional dynamic CMOS stages occurs when one or more inputs of a stage make a 1 to 0 transition during the evaluation phase.

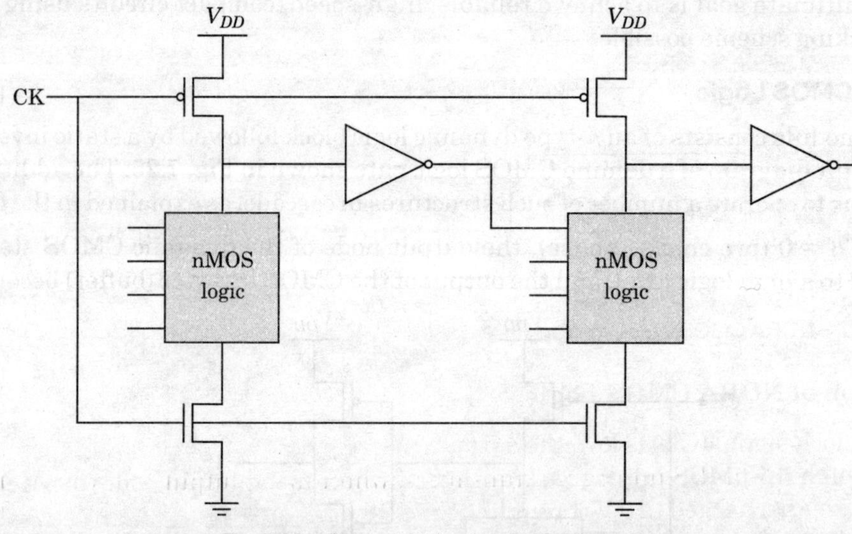

Fig. 7.26 Cascaded domino CMOS logic gates.

7.6.1.2 Limitations of Domino Logic Circuits

1. Number of inverting static logic stages in cascade must be even, so that the inputs of the next domino CMOS stage experience only 0 to 1 transition during evaluation.
2. Only non-inverting structures can be implemented using domino CMOS. If necessary inversion must be carried out using conventional CMOS logic.
3. Charge sharing between the dynamic stage output node and the inter mediate nodes of the nMOS logic block during the evaluation phase may cause erroneous output.

7.6.2 NORA CMOS Logic (NP-Domino Logic)

An alternative approach to cascading dynamic logic is provided by NP-CMOS. NP-CMOS uses two flavors (n-tree and p-tree) of domino logic and avoids an extra static inverter in the critical path that comes with domino logic. The nMOS transistors acting as pull-down networks, while the pMOS transistors are used to build a pull-up network as shown in Fig. 7.27. The precharge-and-evaluate timing of nMOS logic stages is accomplished by the clock signal CK, whereas the pMOS logic stages are controlled by the inverted clock signal, $\overline{\text{CK}}$.

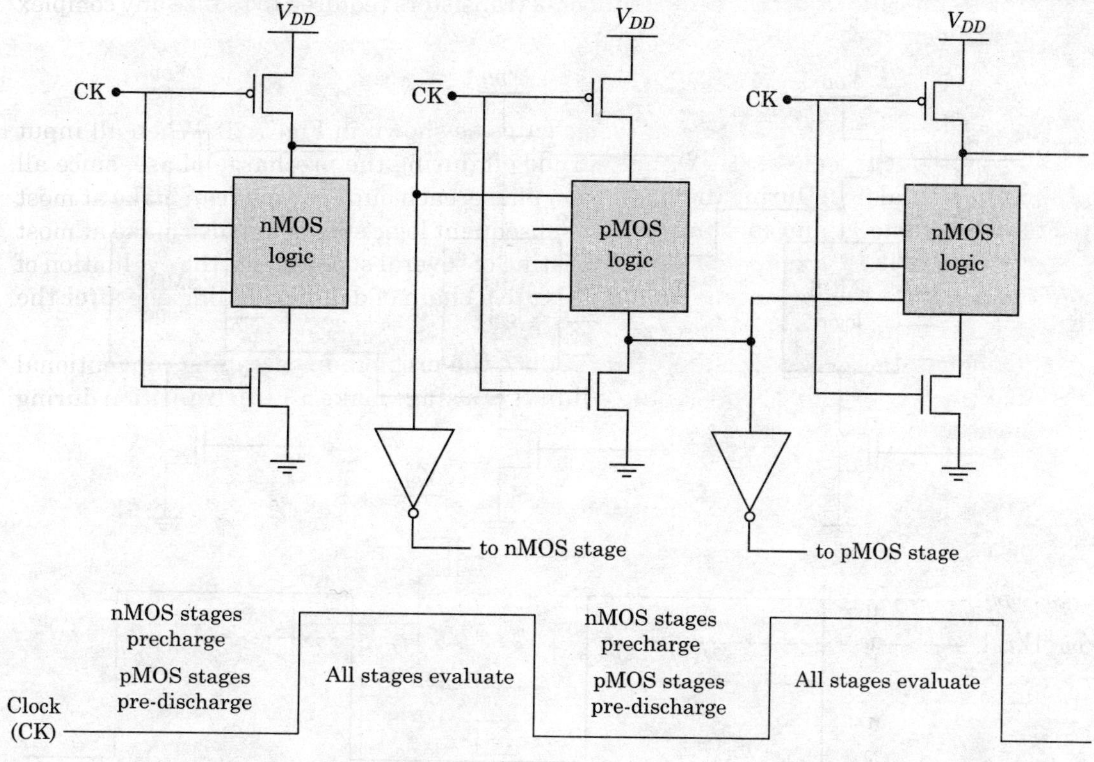

Fig. 7.27 NORA CMOS logic circuit and the scheduling of precharge/evaluation phases.

7.6.2.1 Operation of NORA CMOS Logic

1. When the clock signal (CK) is low, the output nodes of nMOS logic blocks are precharged to V_{DD} through the pMOS pre charge transistors, whereas the output nodes of pMOS logic blocks are pre-discharged to 0 V through the nMOS discharge transistors, driven by $\overline{\text{CK}}$.
2. When the clock signal (CK) makes a low-to-high transition, all cascaded nMOS and pMOS logic stages evaluate one after the other, much like the domino CMOS examined earlier.

7.6.2.2 Advantage of NORA CMOS logic

1. NORA CMOS logic is that a static CMOS inverter is not required at the output of every dynamic logic stage.
2. NORA CMOS logic is that it allows pipelined system architecture.

7.6.3 Zipper CMOS Circuits

The basic circuit architecture of Zipper CMOS is identical to NORA CMOS, with the exception of the clock signals. The generalized circuit diagram and the clock signals of the Zipper CMOS architecture are shown in Fig. 7.28. The Zipper CMOS clock scheme requires the generation of slightly different clock signals for the precharge (or discharge) transistors and for the pull-down (or pull-up) transistors. In particular, the clock signals which drive the pMOS precharge and nMOS discharge transistors allow these transistors to remain in weak conduction or near cut-off during the evaluation phase, thus compensating for the charge leakage and charge-sharing problems.

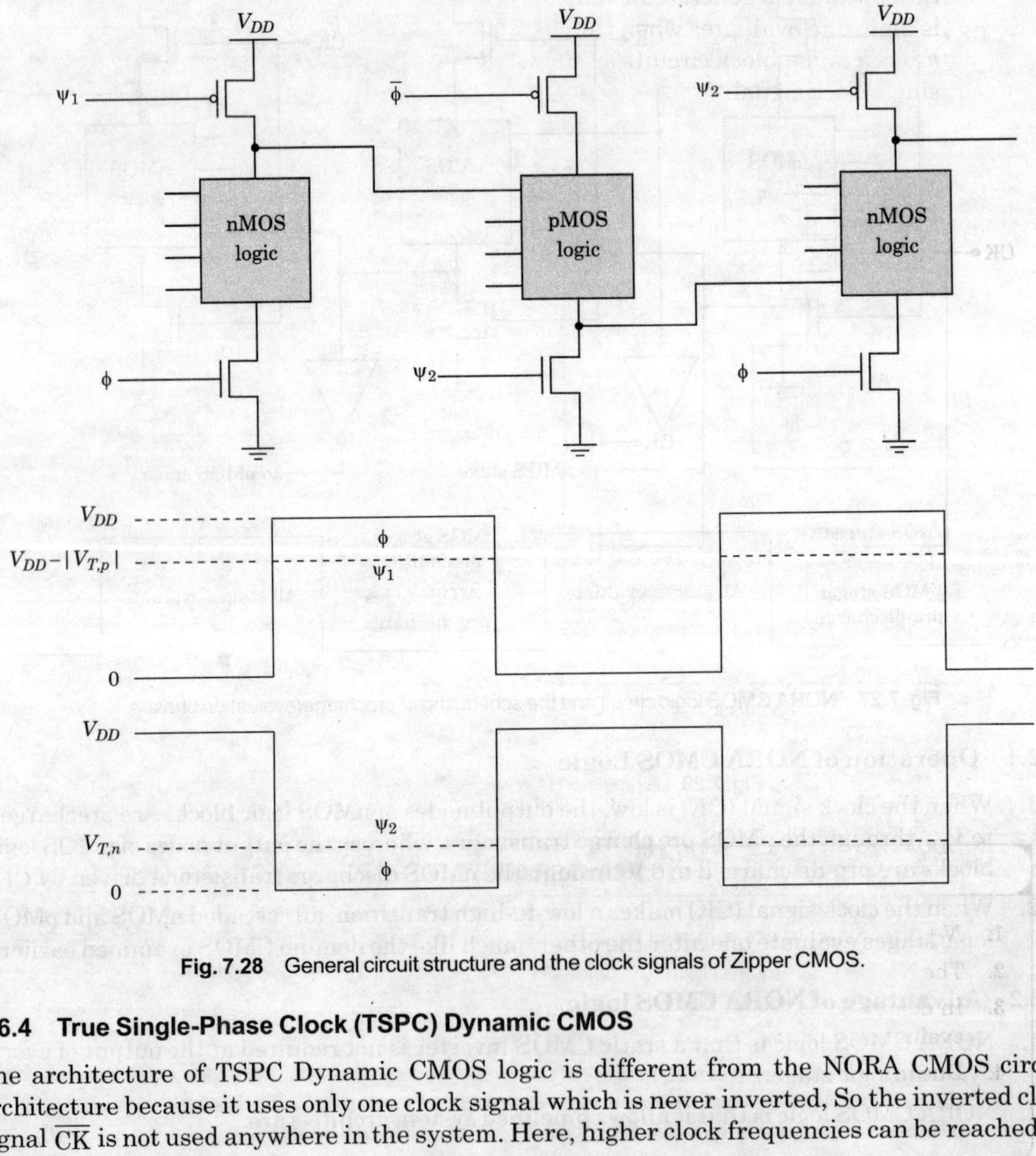

Fig. 7.28 General circuit structure and the clock signals of Zipper CMOS.

7.6.4 True Single-Phase Clock (TSPC) Dynamic CMOS

The architecture of TSPC Dynamic CMOS logic is different from the NORA CMOS circuit architecture because it uses only one clock signal which is never inverted, So the inverted clock signal $\overline{CK}$ is not used anywhere in the system. Here, higher clock frequencies can be reached for dynamic pipelined operation.

The circuit diagram of a pipelined true single-phase clock CMOS system is shown in Fig. 7.29. The circuit consists of alternating stages called *n*-blocks and p-blocks, driven by the same clock signal ϕ. An *n*-block is constructed by cascading a dynamic nMOS stage and a dynamic latch, while a *p*-block is constructed by cascading a dynamic pMOS stage and a dynamic latch.

7.6.4.1 Operation of TSPC Dynamic CMOS

1. When the clock signal is low, the output node of the n-block is being precharged to V_{DD} by the pMOS precharge transistor.
2. When the clock signal switches from low to high, the logic stage output is evaluated and the output latch generates a valid output level. The *p*-block pre-discharges when the clock is high, and evaluates when the clock is low. So that a cascade-connection of alternating *n*-block and p-block circuits as shown in Fig. 7.29 will allow pipelined operation using a single clock signal.

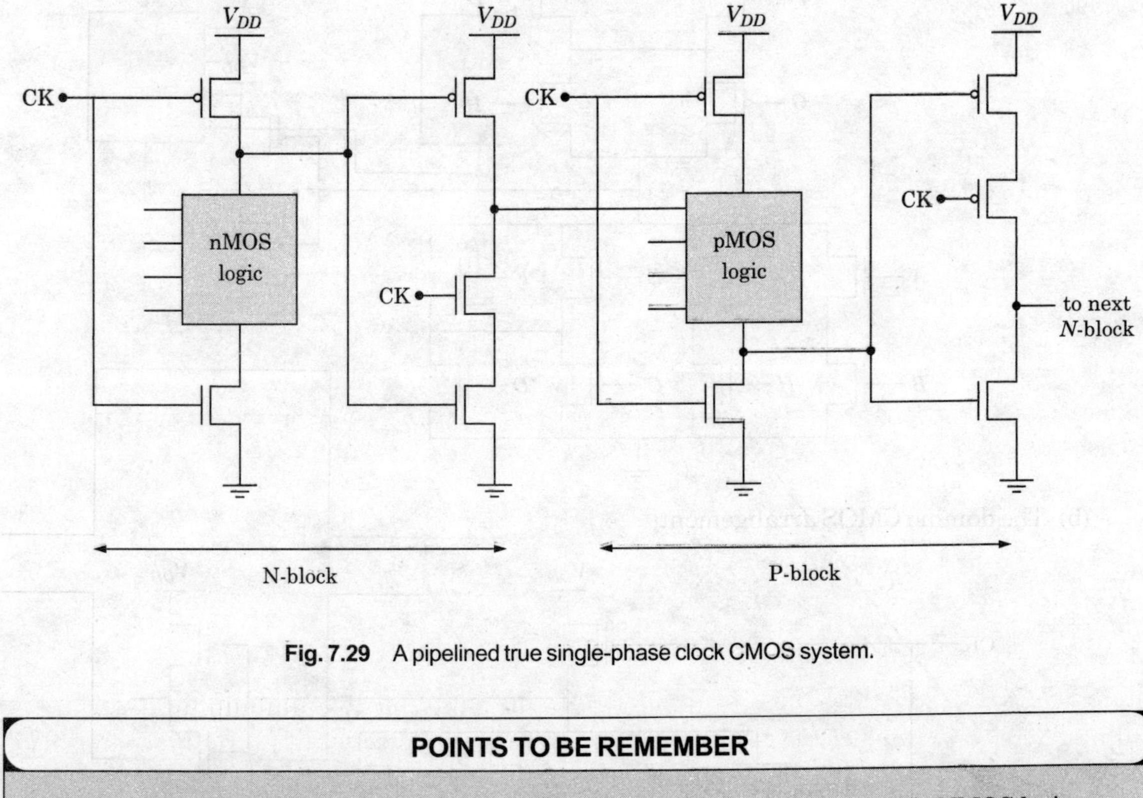

Fig. 7.29 A pipelined true single-phase clock CMOS system.

POINTS TO BE REMEMBER

1. $N + 1$ transistors are required to implement N-input logic gates in the preudo NMOS logic.
2. The NMOS precharge transistor drives the output low during precharge.
3. In domino logic, the butter (static inverter) output cannot make a high to low transition during evaluation mode regardless the applied input voltages on the dynamic CMOS stage.
4. Due to use of static inverters at the output stage, noise immunity increases because the fan-out of the gate is driven by static inverter with low impedance output. Buffer also reduces the capacitance of dynamic output node by separting internal and local capacitance.

SOLVED EXAMPLES

Example 7.1: Implemented the Boolean function, $Z = AB + (C + D)(F + E) + GH$ using standard CMOS and domino CMOS logic.

Solution: (a) The standard CMOS arrangement

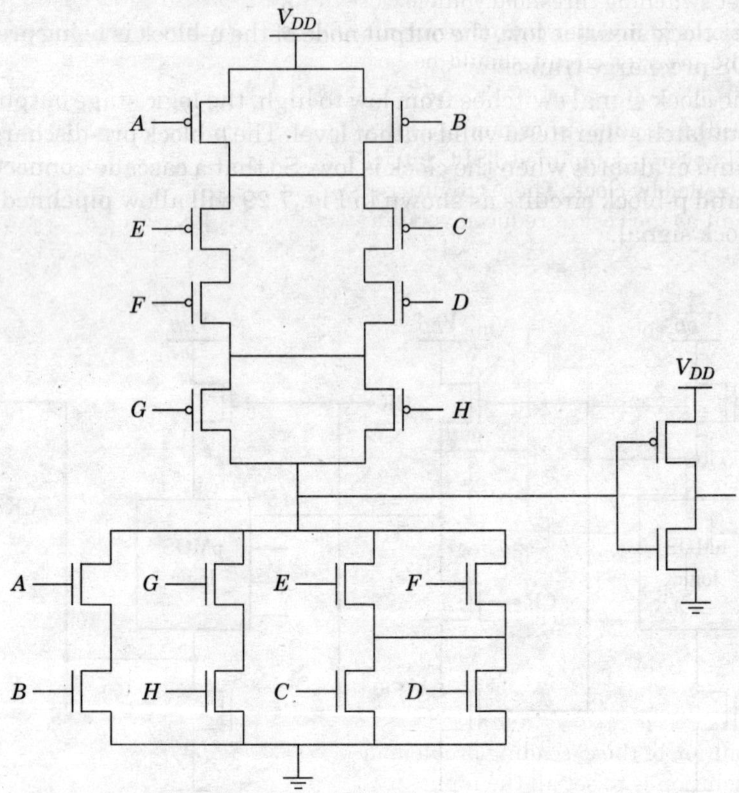

(b) The domino CMOS arrangement

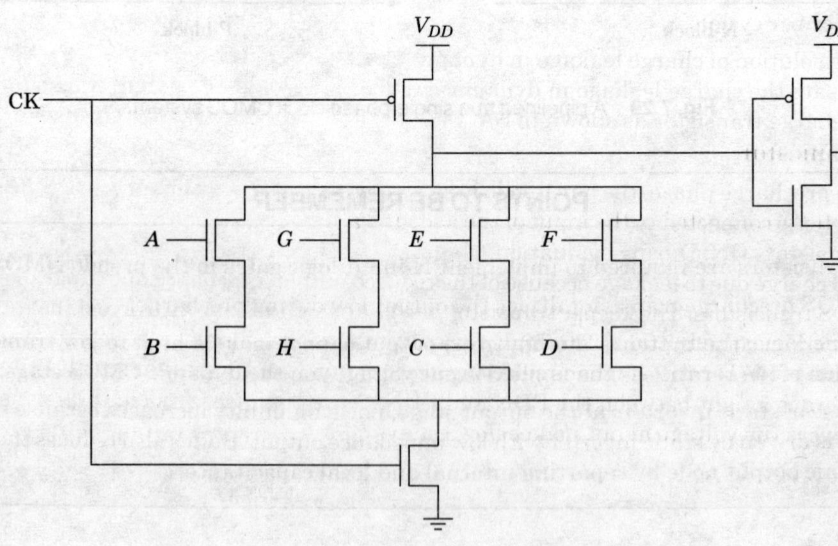

SHORT ANSWERS TYPE QUESTIONS

Q. 1. How as a VLSI designer can you optimize the performance of domino logic circuits?

Ans: Sizing of the transistors in the static inverter is one of the preferable solutions for optimization. Since the critical path during evaluation include the pull-down path of the dynamic gate and pull-up pMOS of the inverter circuit, to speed up the evaluation one should increase (*W/L*) ratio of the pMOS transistor to get switching threshold voltage close to V_{DD}. It is also necessary to reduce (*W/L*) ratio of nMOS transistor in inverter circuit which affects the precharge time. In short, we can say that beta ratio of the inverter circuit should be large. The larger value of beta will reduce the noise margin.

Q. 2. Explain the need of circuit as shown in SA. Fig. 7.1.

Ans: This circuit is known as simplified TSPC latch or split output. In these two circuits, only the first inverter is controlled by clock. There circuits reduce the number of transistor count by two and cut clock load in half at the cost of reduced voltage swings on internal nodes, thus lowering the noise margin.

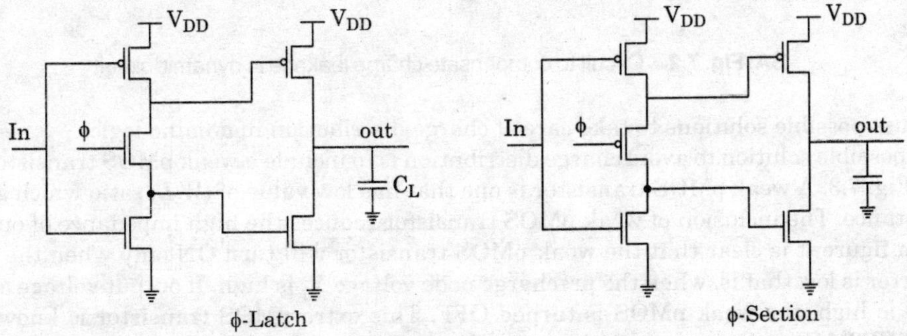

SA. Fig. 7.1

Q. 3. Discuss the solution of the cascading problem in the dynamic logic.

Ans: One possible solution is to set all the inputs to 0 during precharge mode so that all the transistors in the pull-down network are turned off after precharge. This will avoid any unwanted discharge of the storage capacitors during evaluation mode. Transistors are tuned ON only when required (at most once per cycle).

Q. 4. Discuss the solution of charge leakage in dynamic circuit.

Ans: To compensate the charge leakage in dynamic circuit, we can add one pMOS transistor parallel to pMOS precharge transistor as shown in SA. Fig. 7.2. The extra added pMOS transistor is known a **Keeper transistor.**

During the precharge phase, the out 1 node is at V_{DD}, which result in out = low = 0. Since the output of the inverter is connected to the input of the keeper transistors, the low inverter's output turns the keeper transistors ON. During evaluation, if all PDN transistors are OFF, the keeper compensates the drained charge due to leakage because of the reduced output impedance due to parallel connection of two pMOS transistors. The keeper transistor must be of high resistance to avoid the ratio problems and static power consumption. The high resistance in keeper circuit can be achieved by taking smaller value of (*W/L*) ratio. It means selecting keeper transistors of smaller size. If PDN transistors are ON, there is a fight between the PDN and PUN. Since circuit is ratioed, PDN wins eventually and this lowers the voltage at out node below the switching threshold voltage of the next gate.

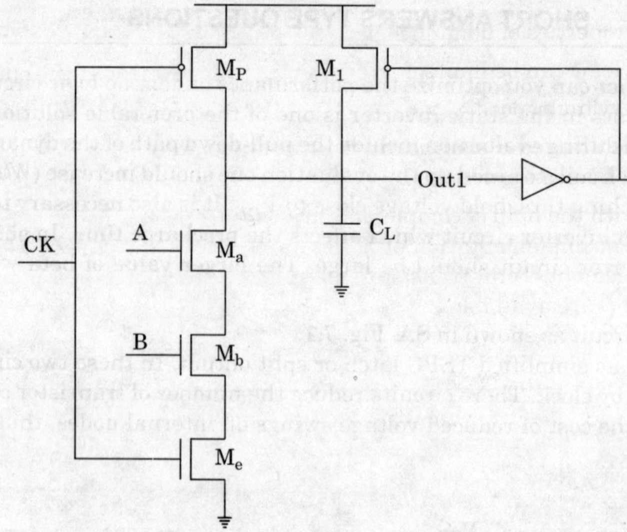

SA. Fig. 7.2 Circuit to compensate charge leakage in dynamic circuit

Q. 5. Discuss possible solutions to take care of charge distribution in domino logic.

Ans: One possible solution to avoid charge distribution is to include a weak pMOS transistor as shown in SA. Fig. 7.3. A weak pMOS transistor is one that has low value of (*W/L*) ratio which gives high ON resistance. The inclusion of weak pMOS transistor reduces the high impedance of output node Y_1. From figure it is clear that the weak pMOS transistor will turn ON only when the output of the inverter is low that is, when the precharge node voltage Y_1 is high. If output voltage at the inverter node is high, the weak pMOS is turned OFF. This extra pMOS transistor is known as **Keeper MOSFET** MP_2.

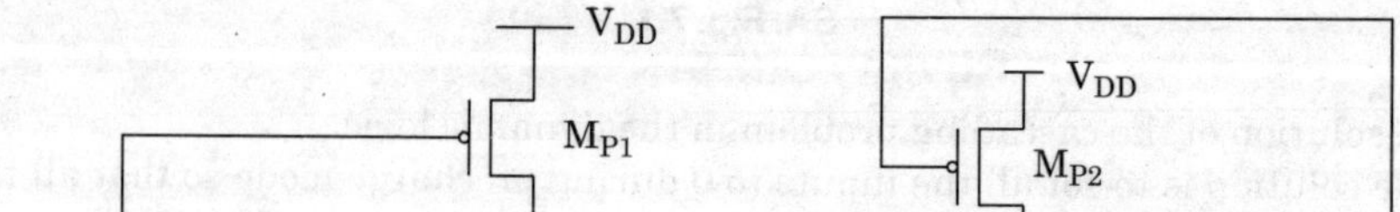

SA. Fig. 7.3 Weak pMOS pull-up transistor to take care of charging shearing

In other solution, we can use separate pMOS transistors to precharge all intermediate nodes in nMOS pull-down tree as shown in SA. Fig. 7.4. The precharging of high capacitance node within the circuit effectively eliminates the charge sharing problem during the evaluation phase. The inclusion of extra pMOS transistors will increase the delay, area and power consumption.

The third possible solution is to use a CMOS inverter of low threshold voltage so that the final stage output is not affected by lowering of node voltage due to charge sharing.

The following are the properties of domino logic:

1. Only non-inverting logic can be implemented due to static inverter at the output.
2. Domino logic has very high speed.
3. Domino CMOS logic allowed a significant reduction in the number of transistors required to realize any complex Boolean function.

Q. 6. Design a XOR circuit with the help of complementary pass transistor logic.

Ans: CPL provides an efficient implementation of XOR gate. SA. Fig. 7.4 shows the implementation of XOR gate and its complement. Total transistors used for the implementation are 8 assuming that inverse of inputs *A* and *B* are available.

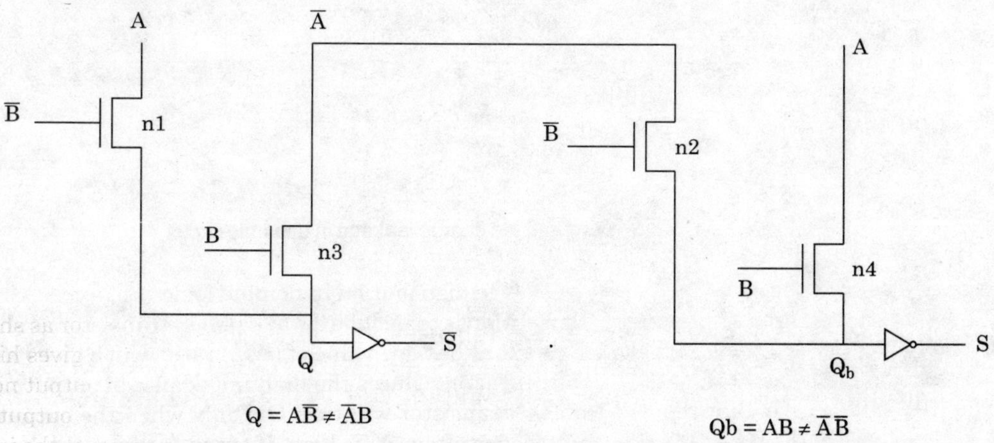

SA. Fig. 7.4 XOR/XNOR implementation using CPL

REVIEW QUESTIONS

1. Explain the precharge and evaluation phase of dynamic CMOS logic circuits
2. Discuss how the charge sharing problem can be overcome in dynamic CMOS logic.
3. Describe the Dynamic CMOS Gate in detail.
4. Explain the Domino CMOS logic with its all features.
5. Explain how Domino CMOS logic is cascaddble in VLSI circuit. **[UPTU 2007-08]**
6. Explain the basic principle of Transmission circuits.
7. Describe the term (a) Charge storage
   (b) Charge leakage
8. Discuss the feature of CMOS Transmission Gate logic and precharge evaluate logic.
9. Explain the working of the following circuits.
   (i) Domino CMOS logic
   (ii) NORA CMOS logic
   (iii) TSPC Dynamic logic
10. List of limitations of Domino CMOS logic gates.
11. Discuss the importance of dynamic logic circuits.
12. Write short note on the followings in connection with synchronous dynamic circuit techniques.
    (a) Dynamic pass transistor circuit.
    (b) Dynamic CMOS circuit techniques.

13. Explain the operation of logic "1" transfer for pass transistors.
14. Explain the behaviour of pass transistor in dynamic CMOS logic implementation. [UPTU 2011-12]
15. Explain how DOMINO CMOS logic overcomes the charge sharing problem. [UPTU 2011-12]
16. Enlist the advantages of dynamic logic circuit over static logic circuit. Explain DOMINO and NORA CMOS logic circuit with suitable example. [UPTU 2011-12]
17. Explain the term voltage Boot strapping in CMOS logic with suitable example. [UPTU 2011-12]

❑❑❑

8 Chapter

Sequential MOS Logic Circuits

Chapter Overview

- Introduction
- Mealy and Moore machines
- Behavior of Bi-stable Elements
- Non Voltage Static Latches
- Clock Latch and Flip-Flop Circuits
- Asynchronous and synchronous circuits
- Timing Metric for sequential Circuits
- Static Latches and Registers
- The SR Latch Circuit
- CMOS D-Latch and Edge-Triggered Flip-Flop

8.1 INTRODUCTION

All of the circuits so far considered in this book have been combinational. This means that their outputs are dependent only upon the inputs at that time. We now turn our attention to sequential logic circuits whose outputs are also dependent upon past inputs, and hence outputs. Put another way, the output of a sequential circuit may depend upon its previous outputs and so in effect has some form of 'memory'.

General forms of sequential circuit sequential circuits are essentially combinational circuits with feedback. A block diagram of a generalized sequential circuit is shown in Fig. 8.1. The generalized circuit contains a block of combinational logic, which has two sets of inputs and two sets of outputs. The inputs are:

1. *A*, the present (external) inputs to the circuit;
2. *y*, the inputs fed back from the outputs;
3. *Z*, the present (external) outputs from the combinational circuit;
4. *Y*, the outputs that are fed back into the combinational circuit.

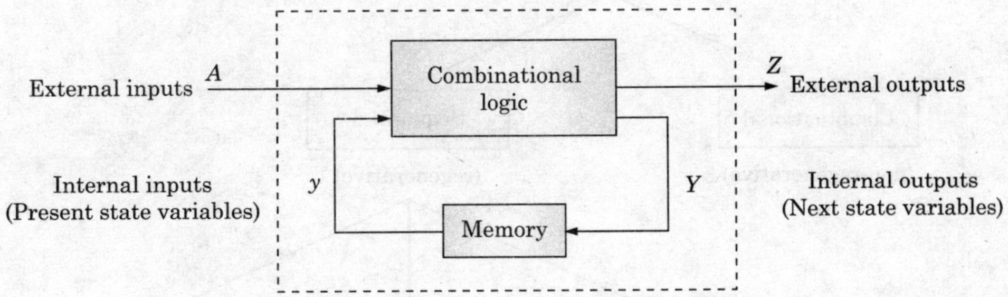

Fig. 8.1 The general form of a sequential logic circuit

Note that the outputs, Y, are fed back via the memory block to become the inputs, y, and that y are called the 'present state' variables because they determine the current state of the circuit, with Y the 'next state' variables as they will determine the next state the circuit will enter.

It is often useful to think in terms of two independent combinational circuits, one each for the two sets of outputs, Z (external) and Y (internal), as shown in Fig. 8.2. Both of these outputs will in general depend upon the external, A, and internal, y, (fed back) inputs.

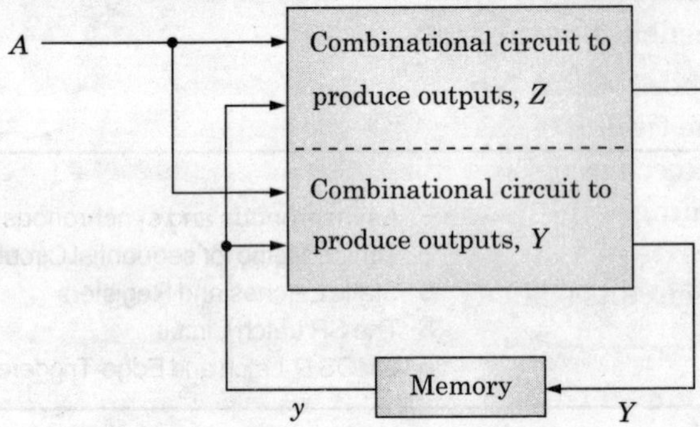

Fig. 8.2 A general sequential circuit emphasizing how the outputs from the combinational logic block is functions of both the external and internal inputs.

'States' and sequential circuits

An important concept to appreciate is that sequential circuits can be considered at any time to occupy a certain 'state'. These 'states' are dependent upon the internal feedback, and in the case of asynchronous sequential circuits, the external inputs as well. At this early stage we simply note that if the memory in a circuit has I digital lines leading to and from it then it can store 2^n different patterns and hence the circuit possesses 2^n internal states. (We call these internal states to distinguish them from the total states of the circuit, which are also dependent upon the external inputs).

This idea of the circuit possessing states is fundamental to sequential circuits since they are often designed and analyzed by the manner, or sequence, in which the available states are visited for given sequences of inputs. This is why in Fig. 8.2 the internal inputs, y, and outputs, Y, are labeled as the present and next state variables respectively (since they determine the current and next state of the circuit).The general classification of non-regenerative and regenerative logic circuits is shown in Fig. 8.3.

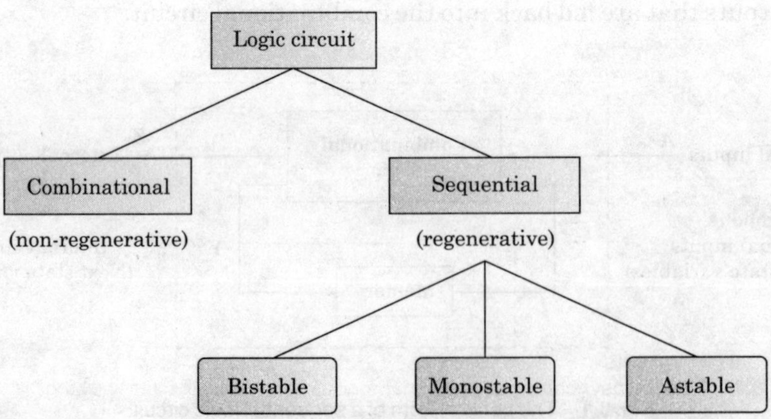

Fig. 8.3 Classification of logic circuits based on their temporal behavior.

8.2 ASYNCHRONOUS AND SYNCHRONOUS CIRCUITS

In this introductory text, we will define two broad classes of sequential circuits, namely: asynchronous and synchronous.

1. The timing of the operation of *asynchronous circuits,* as the name implies, is not controlled by any external timing mechanism. Rather, as soon as changes are made at the inputs of such a circuit they take effect at the outputs. The simplest form of memory in such circuits is just a wire forming the feedback connection.
2. *Synchronous circuits* are those which possess a clock of some sort which regulates the feedback process. Hence the timing of changes in the outputs, in response to changes at the inputs (which may have occurred sometime before), are controlled by the 'ticking' of a clock. Consequently, the timing of the operation of sequential circuits can be, and usually is, synchronized to other parts of a larger circuit. The memory in such circuits is itself made up of specialized logic circuits (called flip-flops) that essentially act as digital storage elements.

Fig. 8.4 shows the block diagrams for these two types of sequential circuits. For obvious reasons, synchronous sequential circuits are also referred to as clocked circuits, whilst asynchronous ones are known as unlocked or free running.

Although asynchronous circuits are more difficult to design than synchronous ones they do have certain benefits. These include the fact that because they are free running their speed of operation is limited solely by the characteristics of the components from which they are built and not by the speed at which they are clocked. Consequently, asynchronous circuits have the potential to work at higher speeds than synchronous ones. Also: some systems may require a circuit to respond immediately to changing inputs (i.e. the inputs cannot be synchronized to the rest of the circuit); in very large circuits the unavoidable delays as a signal traverses the whole circuit may mean exact synchronisation is not possible; and finally flip-flops which are essential digital circuit components are themselves asynchronous circuits.

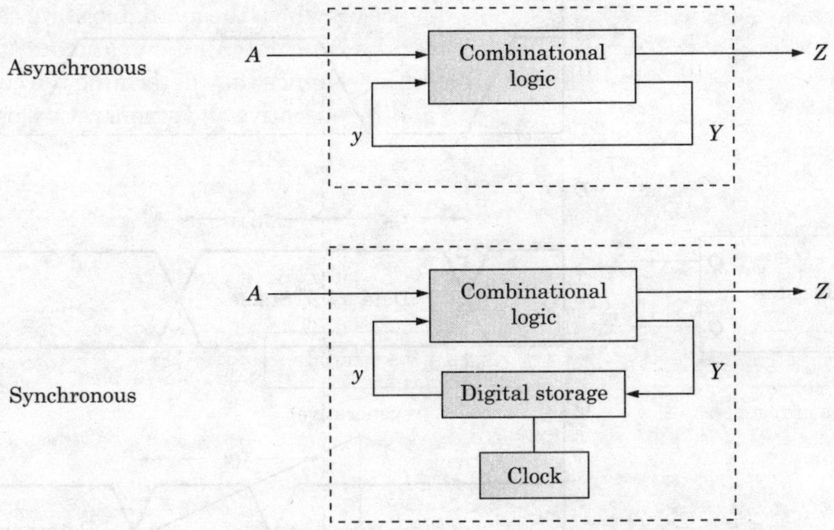

Fig. 8.4 General asynchronous and synchronous sequential circuits showing the difference between their respective 'memories'.

8.3 MEALY AND MOORE MACHINES

A basic model of a sequential circuit has inputs outputs and internal states. The sequential circuits are classified according to the type of output functions as:

- Mealy machines are a sequential circuit whose outputs are functions of both the present state and inputs. The state table of a mealy machine must include an output section that is a function of both the present state and inputs. When the outputs are directly taken from the flip-flops the state table can exclude the output section.
- Moore machines are a sequential circuit where outputs are functions of the present state only. The outputs of the sequential circuits are synchronized with the clock as they depend on the outputs of flip-flops only that are synchronized with the clock.

8.4 TIMING METRIC FOR SEQUENTIAL CIRCUITS

- For the timing metrics, we have consider a register as shown in Fig. 8.5. There are three important timing parameters associated with a register.
- The setup time (t_{su}) is the time that data input (D) must be valid before the clock transition. The hold time (t_{hold}) is the time that the data input must remain valid after the clock edge.
- Assuming that the setup and hold time are met, the data at the D input is copied to the Q output after a worst case propagation delay denoted by $\tau_{C\text{-}2}$.
- The clock period T, at which the sequential circuit operates, must thus accommodate the longest delay of any stage is network. Assume that $\tau_{p\,logic}$ denotes the worst-case propagation delay of logic τ_{cd}, the contamination delay. Then, the minimum clock period T required for proper operation of the sequential circuit is given by

$$T \geq \tau_{C\text{-}2} + \tau_{p\ logic} + \tau_{su} \quad \text{...(8.1)}$$

The hold time of the register impose an extra constraint for proper operation, namely

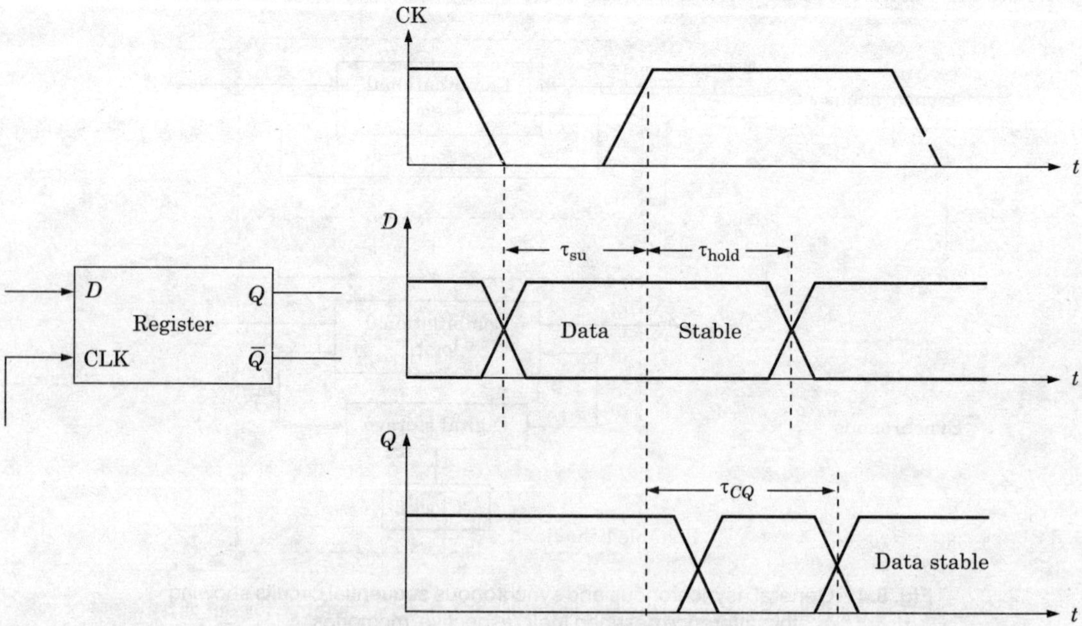

Fig. 8.5 Definition os setup time, hold time and propagation delay of a sychronous register.

$$\tau_{cd\ register} + \tau_{cd\ logic} \geq \tau_{hold} \quad ...(8.2)$$

where $\tau_{cd\ register}$ is the minimum propagation delay of the register.

- The constraint ensure that the input data of the sequential element is hold long enough after the clock edge and is not modified too soon by the new-wave of data coming in.
- According to Fig. 8.5, it is important to minimize the values of the timing parameters associated with the register, these directly affect the rate at which a sequential circuit can be clocked.
- In fact, modern high-performance system are characterized by a very low logic depth and the register propagation delay and set up times account for a significant portion of the clock period.
- The requirement of equation (8.2) is not difficult to meet, although it becomes an issue when there is title or no logic between register.

8.5 BEHAVIOR OF BI-STABLE ELEMENTS

- Static memories use positive feedback to create a bistable circuit- a circuit having two stable states that represent 0 and 1. Fig. 8.6(a) shows two identical cross coupled inverter circuits. Here, the output voltage of inverter (1) is equal to the input voltage of inverter (2) ($V_{o1} = V_{i2}$) and the output voltage of inverter (2) is equal to the input voltage of inverter (1) ($V_{o2} = V_{i1}$). The circuit serves as a memory, storing either a 1 or 0.
- For the analysis of static VTC of both inverters, First we draw the VTC of inverter (1) with respect to the $V_{o1} - V_{i1}$, axis pair Notice that the input and output voltages of inverter (2) correspond to the output and input voltages of inverter (1), respectively.
- Similarly, draw the VTC of inverter (2) using the same axis pair as shown in Fig. 8.6(b).
- From the Fig. 8.6(b), we can see that the two VTCs intersect at three points. Here, only two operating points are stable, as indicated in Fig. 8.6(b).

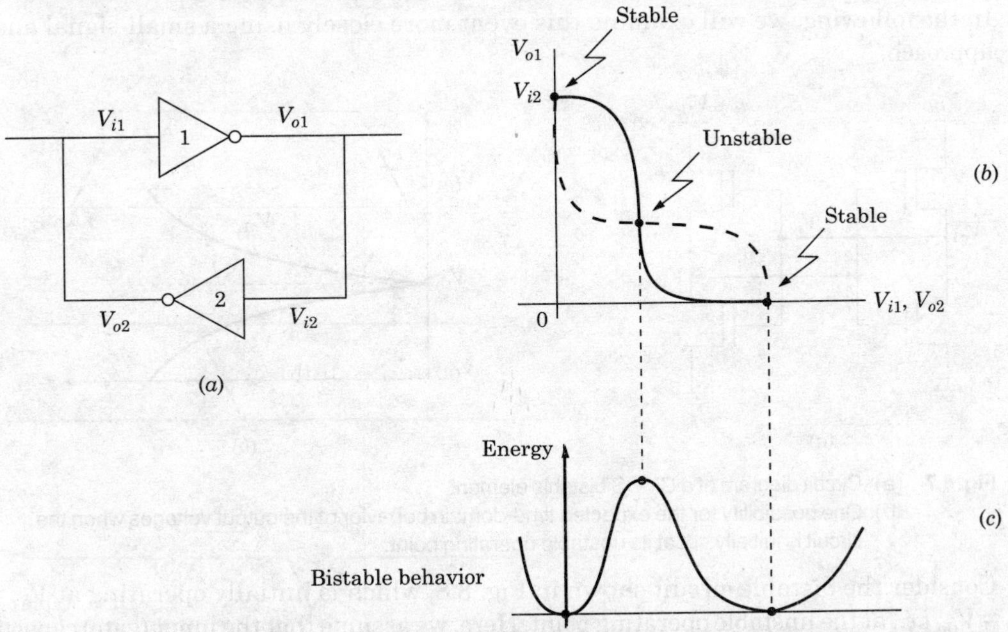

Fig. 8.6 Static behavior of the two-inverter basic bistable element: (a) Circuit schematic. (b) Intersecting VTC of the two inverters, showing the three possible operating points. (c) Qualitative view of the potential energy levels corresponding to the three operating points.

- The gain of each inverter circuit (the slope of the respective voltage transfer curves) is smaller than unity at the two stable operating points. A sufficiently large external voltage is required for the movement of one stable point to the other. Due to this the voltage gain of the inverter loop becomes larger than unity.
- At the third operating point the voltage gains of both inverters are larger than unity. Consequently, even if the circuit is biased at this point initially, a small voltage perturbation at the input of any of the inverters will be amplified, causing the operating point to move to one of the stable operating points. This leads to the conclusion that the third operating point is unstable. The circuit has two stable operating points; hence, it is called bistable.
- The bistable behavior of the cross-coupled inverter circuit can also be visualized qualitatively by examining the total potential energy level at each of the three possible operating points (Fig. 8.6(c)).
- It is seen that the potential energy is at its minimum at two of the three operating points, since the voltage gains of both inverters are equal to zero. The energy attains a maximum at the operating point at which the voltage gains of both inverters are maximum. Thus, the circuit has two stable operating points corresponding to the two energy minima, and one unstable operating point corresponding to the potential energy maximum.

8.5.1 CMOS Bi-stable Elements

- The circuit arrangement for a CMOS two-inverter bistable element is shown in Fig. 8.7(a). All four transistors are in saturation at the unstable operating point of this circuit, which gives maximum loop gain for the circuit.
- If the initial operating condition is set at this point, any small voltage perturbation will cause significant changes in the operating modes of the transistors. Thus, we expect the output voltages of the two inverters to diverge and eventually settle at V_{OH} and V_{OL}, respectively, as illustrated in Fig. 8.7(b).
- In the following, we will examine this event more closely using a small-signal analysis approach.

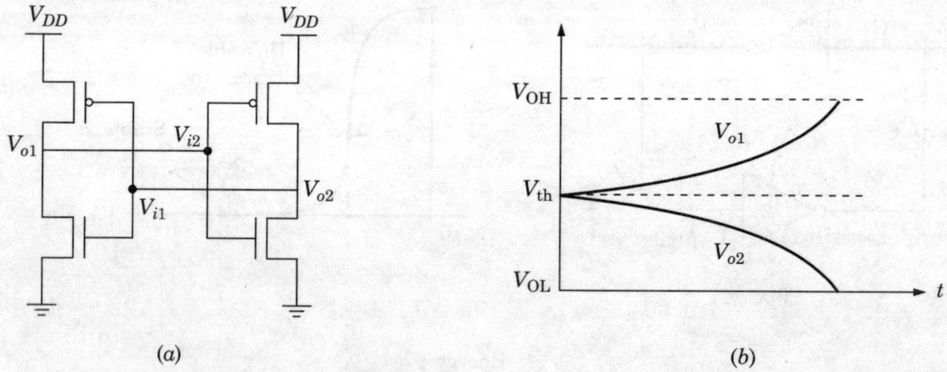

Fig. 8.7 (a) Circuit diagram of a CMOS bistable element.
(b) One possibility for the expected time-domain behavior of the output voltages when the circuit is initially set at its unstable operating point.

- Consider the bistable circuit shown in Fig. 8.8, which is initially operating at $V_{o1} = V_{o2} = V_{th}$, *i.e.*, at the unstable operating point. Here, we assume that the input (gate) capacitance C_g of each inverter is much larger than its output (drain) capacitance C_d ($C_g >> C_d$).

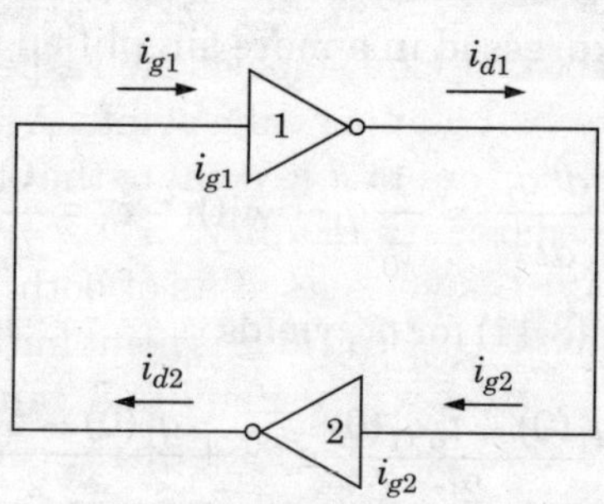

Fig. 8.8 Small-signal input and output currents of the inverters.

- The small-signal drain current supplied by each inverter (1 and 2) can be expressed, in terms of the small-signal gate voltage of that inverter, as follows.

$$i_{g1} = i_{d2} = g_m v_{g2}$$
$$i_{g2} = i_{d1} = g_m V_{g1} \quad \text{...(8.3)}$$

- Here, g_m represents the small-signal transconductance of the inverter. The gate voltages of both inverters can be expressed in terms of the gate charges, q_1, and q_2.

$$v_{g1} = \frac{q_1}{C_g}$$
$$v_{g2} = \frac{q_2}{C_g} \quad \text{...(8.4)}$$

- The small-signal gate current of each inverter can be written as a function of the time derivative of its small-signal gate voltage, as follows.

$$i_{g1} = C_g \frac{dv_{g1}}{dt}$$
$$i_{g2} = C_g \frac{dv_{g2}}{dt} \quad \text{...(8.5)}$$

From equations (8.3) and (8.5):

$$g_m\, v_{g2} = C_g \frac{dv_{g1}}{dt} \quad \text{...(8.6)}$$

$$g_m\, v_{g1} = C_g \frac{dv_{g2}}{dt} \quad \text{...(8.7)}$$

From equations (8.4), (8.6) and (8.7):

$$\frac{g_m}{C_g} q_2 = \frac{dq_1}{dt} \quad \text{...(8.8)}$$

$$\frac{g_m}{C_g} q_1 = \frac{dq_2}{dt} \quad \text{...(8.9)}$$

- Both differential equations given in (8.8) and (8.9) can now be combined to yield a second-order differential equation describing the time behavior of gate charge q_1

$$\frac{g_m}{C_g} q_1 = \frac{C_g}{g_m} \frac{d^2 q_1}{dt^2} \Rightarrow \frac{d^2 q_1}{dt^2} = \left(\frac{g_m}{C_g}\right)^2 q_1 \quad \text{...(8.10)}$$

This equation can also be expressed in a more simplified form by using ζ_0, the transit time constant.

$$\frac{d^2 q_1}{dt^2} = \frac{1}{\tau_0^2} q_1 \quad \text{with} \quad \tau_0 = \frac{C_g}{g_m} \qquad \text{...(8.11)}$$

The time-domain solution of (8.11) for q_1, yields

$$q_1(t) = \frac{q_1(0) - \tau_0 q_1(0)}{2} e^{-\frac{t}{\tau_0}} + \frac{q_1(0) - \tau_0 q_1(0)}{2} e^{+\frac{t}{\tau_0}} \qquad \text{...(8.12)}$$

Where the initial condition is given as:

$$q_1(0) = C_g \cdot v_{g1}(0) \qquad \text{...(8.13)}$$

Note that $v_{g1} = I_{02}$ and $v_{g2} = v_{o1}$. Replacing the gate charge of the both inverters with corresponding output-voltage variables, we obtain,

$$v_{o2}(t) = \frac{1}{2}(v_{o2}(0) - \tau_0 v'_{o2}(0))e^{-\frac{1}{\tau_0}} + \frac{1}{2}(v_{o2}(0) - \tau_0 v'_{o2}(0))e^{+\frac{1}{\tau_0}} \qquad \text{...(8.14)}$$

$$v_{o1}(t) = \frac{1}{2}(v_{o1}(0) - \tau_0 v'_{o1}(0))e^{-\frac{1}{\tau_0}} + \frac{1}{2}(v_{o1}(0) - \tau_0 v'_{o1}(0))e^{+\frac{1}{\tau_0}} \qquad \text{...(8.15)}$$

For large values of τ, the time-domain expressions (8.14) and (8.15) can be simplified as follows.

$$v_{o1}(\tau) \approx \frac{1}{2}(v_{o1}(0) + \tau_0 v'_{o1}(0))e^{+\frac{1}{\tau_0}}$$
$$v_{o2}(\tau) \approx \frac{1}{2}(v_{o2}(0) + \tau_0 v'_{o2}(0))e^{+\frac{1}{\tau_0}} \qquad \text{...(8.16)}$$

- Here, the magnitude of both output voltages increases exponentially with time. The magnitude is depends on the polarity of the initial small perturbations $dv_{01}(0)$ and $dv_{02}(0)$ and the output voltages of both inverters will diverge from their initial value of V_{th} to either V_{OL} or V_{OH}.
- Due to the charge-conservation principle, the polarity of the output-voltage perturbation dv_{01} must always be opposite to that of dv_{02}. Hence, the two output voltages always diverge into opposite directions, as expected.

$$\begin{aligned} v_{o1}&: \; V_{th} \to V_{OH} \;\text{ or }\; V_{OL} \\ v_{o2}&: \; V_{th} \to V_{OL} \;\text{ or }\; V_{OH} \end{aligned} \qquad \text{...(8.17)}$$

- The phase-plane representation of this event is shown in Fig. 8.9, which shows the operating point ($V_{01} = V_{th}$, $V_{02} = V_{th}$) is unstable. The two operating points ($V_{01} = V_{OL}$, $V_{02} = V_{OH}$) and ($V_{01} = V_{OH}$, $V_{02} = V_{OL}$) can be shown to be stable, using small-signal models at the corresponding operating points.
- The time-domain analysis presented here: When the bistable circuit is settling from its unstable operating point into one of its stable operating points, we can envision a signal traveling the loop consisting of the two cascaded inverters several times (Fig. 8.10). The time-domain behavior of the output voltage vol during this period is approximated as follows:

$$\frac{v_{o1}(t)}{v_{o1}(0)} = e^{+\frac{t}{\tau_0}} \qquad \text{...(8.18)}$$

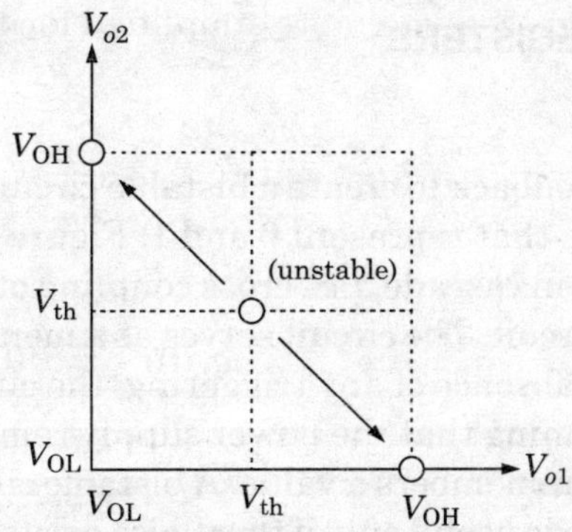

Fig. 8.9 Phase-plane representation of the bistable circuit behavior.

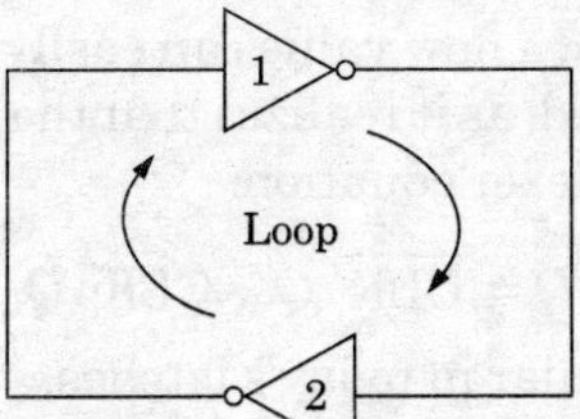

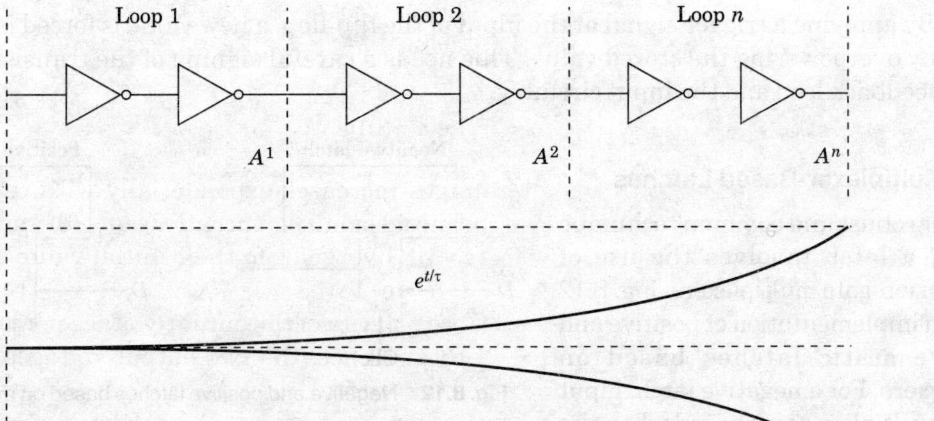

Fig. 8.10 Propagation of a transient signal in the two-inverter loop during settling.

- If during a time interval T, the signal travels the loop n times, then this is equivalent to the same signal propagating along a cascaded inverter chain consisting of $2n$ inverters. Expressing the loop gain (combined voltage gain of two cascaded inverters) with A, we obtain,

$$A^n = e^{+\frac{T}{\tau_0}} \quad \text{...(8.19)}$$

This expression describes the time-domain behavior of the diverging process until it reaches stable points, as depicted in Fig. 8.10.

8.6 STATIC LATCHES AND REGISTERS

8.6.1 Bistability Principle

Static memories use positive feedback to create a bistable circuit—a circuit having two stable states that represent 0 and 1. Figure 8.11 shows two inverters connected in cascade, *i.e.*, cross coupling of two inverters results in a bistable circuit. The circuit serves as a memory, storing either a 1 or a 0. In the absence of any triggering, the circuit remains in a single state (assuming that the power supply remains applied to the circuit) and thus remembers a value. A bistable circuit is also called *flip-flop*. A flip-flop is useful only if there also exists a means to bring it from-one state to the other one. In general, two different approaches may be used to accomplish the following:

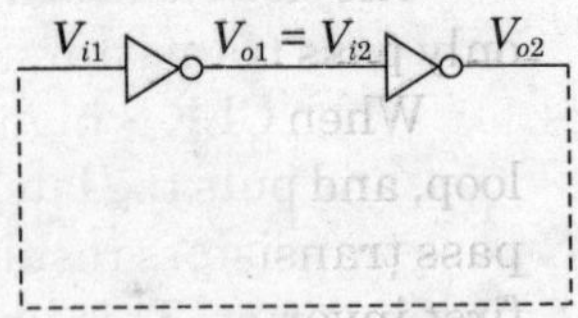

Fig. 8.11 Two cascaded inverters.

- **Cutting the feedback**
  Once the feedback loop is open, a new value can easily be written into Out (or Q). Such a latch is called multiplexer based, as it realizes that the logic expression for a synchronous latch is identical to the multiplexer equation:

  $$Q = \overline{\text{CLK}} \cdot Q + \text{CLK} \cdot \overline{Q}$$

  This approach is the most popular in today's latches.

- **Over powering the feedback loop**
  By applying a trigger signal at the input of the flip-flop, a new value is forced into the cell by overpowering the stored value. This needs a careful signing of the transistors in the feedback loop and the input circuitry.

8.6.2 Multiplexer-Based Latches

The most robust and common technique to build a latch involves the use of transmission-gate multiplexers. Fig. 8.12 shows an implementation of positive and negative static latches based on multiplexers. For a negative latch, input 0 of the multiplexer is selected when the clock is low, and the D input is passed to the output. When the clock signal is high, input 1 of the multiplexer, which connects to the output of the latch, is selected. The feedback ensures a stable output as long as the clock is high. Similarly in the positive latch, the D input is selected when the clock signal is high, and the output is held (using feedback) when the clock signal is low.

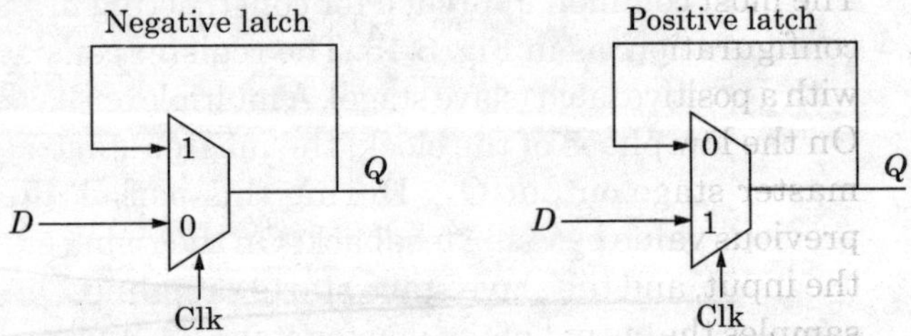

Fig. 8.12 Negative and positive latches based on multiplexers.

A transistor-level implementation of a positive latch based on multiplexers is shown in Fig. 8.13. When Clk is high, the bottom transmission gate is ON and the latch is transparent— i.e. the D input is copied to the output. During this phase, the feedback loop is open, since the top transmission gate is OFF. This particular latch

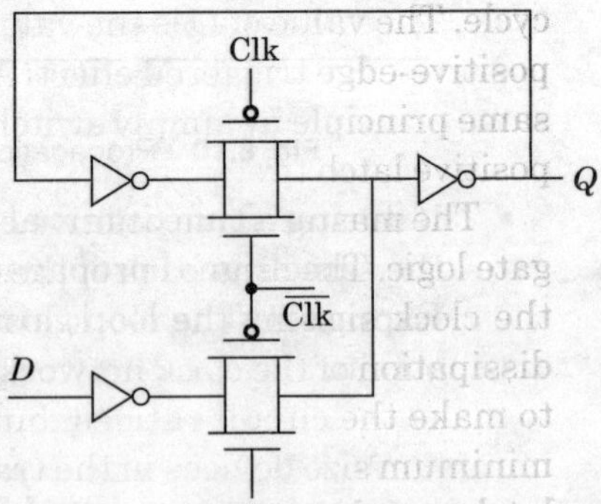

Fig. 8.13 Transistor-level implementation of a positive latch built by using transmission gates.

implementation is not very efficient from power perspective. It presents a load of four transistors to the Clk signal.

The clock load can be reduced to two transistors by implementing multiplexers that use NMOS-only pass transistors, as shown in Fig. 8.14.

When CLK is high, the latch samples the *D* input, while a low clock signal enables the feedback loop, and puts the latch in the hold mode. Even though the circuit is simple, the use of NMOS-only pass transistors results in the passing of a degraded high voltage of $V_{DD} - V_{Tn}$ to the input of the first inverter. This impacts both noise margin and the switching performance, especially in the case of low value of V_{DD} and high values of V_{Tn}. It also causes static power dissipation in the first inverter, because the maximum input voltage to the inverter equals $V_{DD} - V_{Tn}$, and the PMOS device of the inverter is never fully turned OFF.

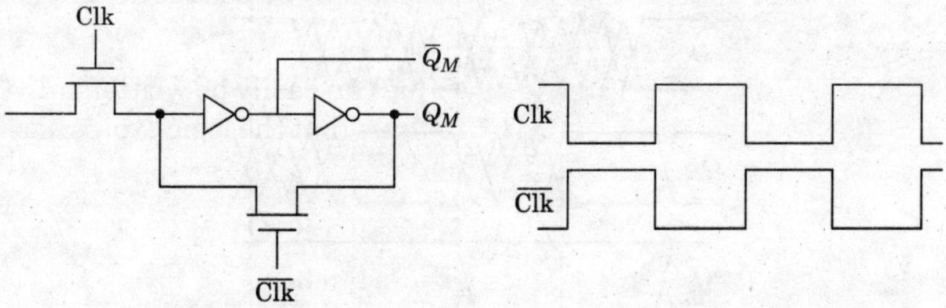

Fig. 8.14 Multiplexer-based NMOS latch by using NMOS-only pass transistors.

8.6.3 Master-Slave Edge-Triggered Register

The most common approach for constructing an edge-triggered register is to use a master-slave configuration, as in Fig. 8.15. The register consists of cascading a negative latch (master stage) with a positive latch (slave stage). A multiplexer-based latch is used in this particular implementation. On the low phase of the clock, the master stage is transparent, and the *D* input is passed to the master stage output Q_M. During this period, the slave stage is in the hold mode, keeping its previous value by using feedback. On the rising edge of the clock, the master stage stops sampling the input, and the slave stage starts sampling. During the high phase of the clock, the slave stage samples the output of the master stage (Q_M), while the master stage remains in a hold mode. Since Q_M is constant during the high phase of the clock, the output *Q* makes only one transition per cycle. The value of *Q* is the value of *D* input right before the rising edge of the clock, achieving the positive-edge triggered effect. A negative edge-triggered register can be constructed by using the same principle by simply switching the order of the positive and negative latches (i.e., placing the positive latch first).

The master-slave positive edge-triggered register can also be implemented with transmission gate logic. The drawback of the transmission-gate register is the high capacitive load presented to the clock signal. The clock load per register is important, since it directly impacts the power dissipation of the clock network. One approach to reduce the clock load at the cost of robustness is to make the circuit ratioed, but with increased design complexity. Using minimum or close-to-minimum size devices in the transmission gate is desirable to reduce the power dissipation in the latches and the clock distribution network.

Another problem with this scheme is reverse conduction—the second stage can affect the state of the first latch.

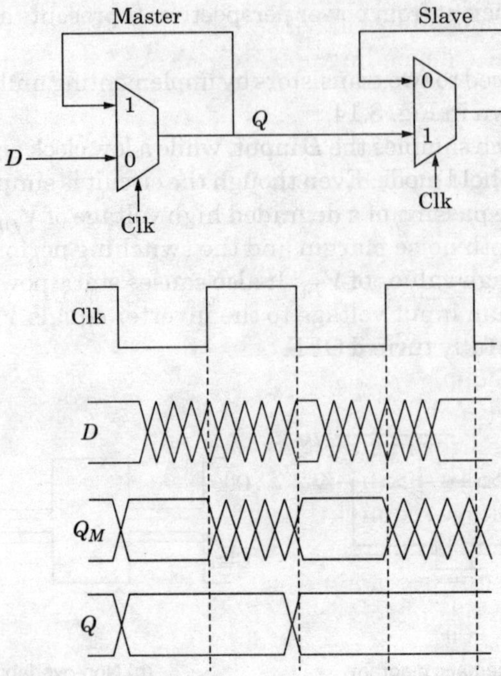

Fig. 8.15 Positive edge-triggered register based on a master-slave configuration.

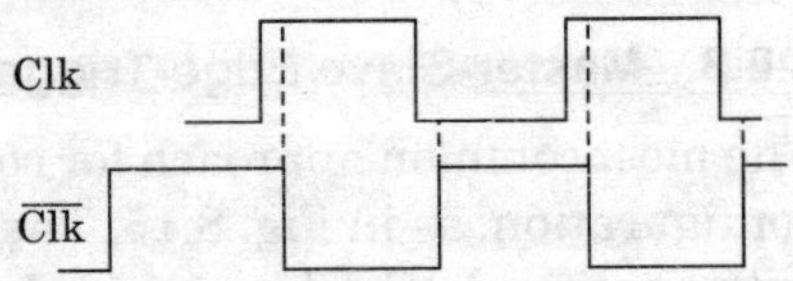

Fig. 8.16 Overlapping clock phases.

Non-ideal clock signals. So far, it is assumed that $\overline{\text{CLK}}$ is a perfect inversion of CLK, or in other words, that the delay of the generating inverter is zero. This is not true always, variations can exist in the wires used to route the two clock signals, or the load capacitances can vary based on data stored in the connecting latches. This effect is known as the clock skew and is a major problem, causing the two clock signals to overlap as in Fig. 8.16. Clock overlap can cause two types of failures:

1. When the clock goes high, the slave stage should stop sampling the master stage output and go into a hold mode. However, since $\overline{\text{CLK}}$ and CLK are both high for a short period of time (the overlap period), and there is a direct path from the D input to the Q output. As a result, the data at the output can change on the rising edge of the clock, which is undesired for a negative edge-triggered register. This is known as a race condition.
2. In a multiplexer-based register, clock overlap between $\overline{\text{CLK}}$ and CLK results in an undefined state between the input and the output.

The above mentioned problems can be avoided by using two non-overlapping clocks (Fig. 8.17) and by keeping the non-overlap time between the clock larger enough so that no overlap occurs even in the presence of clock-routing delays. During the non-overlap time, the flip-flop is in the high-impedance state. Leakage will destroy the state if this condition holds for

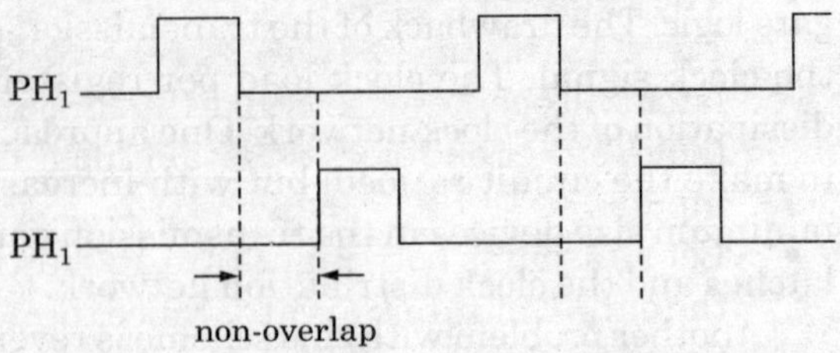

Fig. 8.17 Two-phase non-overlapping clocks.

too long. The register employs a combination of static and dynamic storage approaches, depending upon the state of the lock.

Generating non-overlapping clocks. Figure 8.18 shows one possible implementation of the clock generation circuitry for generating a two-phase non-overlapping clock.

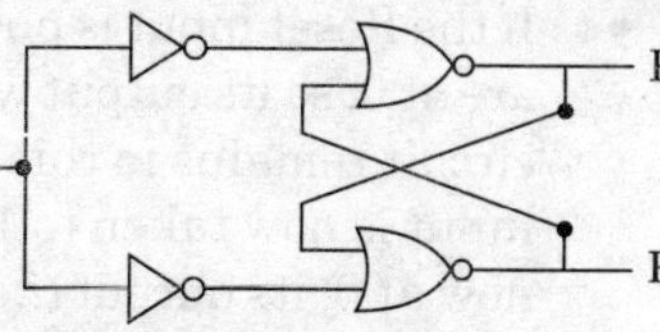

Fig. 8.18 Circuitry for generating a two-phase non-overlapping clock.

8.7 LOW VOLTAGE STATIC LATCHES

- The scaling of supply voltage is critical for low-power operation. Some latch structures do not function at reduced supply voltages.
- At very low power supply voltages, the input to the inverter cannot be raised above the switching threshold, resulting in incorrect evaluation. Even with the use of transmission gates, performance degrades significantly at reduced power supply.
- Scaling to low supply voltage requires the sue of reduced threshold voltages. This has the negative effect of exponentially increasing the sub-threshold leakage power. One solution of this problem involves the use of multiple threshold devices, where the low-threshold inverters are gated by using high-threshold devices to eliminate leakages.

8.8 THE SR LATCH CIRCUIT

(UTU-2012)

The bistable element consisting of two cross-coupled inverters has two stable operating modes, or states. The circuit preserves its state (either one of the two possible modes) as long as the power supply voltage is provided. Hence, the circuit can perform a simple memory function of holding its state.

8.8.1 The SR Latch Circuit using NOR-GATE

- A latch is a logic circuit with two outputs, which are the complement of each other. A basic latch can be constructed by cross coupling two *NOR* gates as shown in Fig. 8.19. The two inputs are labeled Set and Reset. The outputs Q and $\bar{Q}$ are always the complement of each other during normal operation.

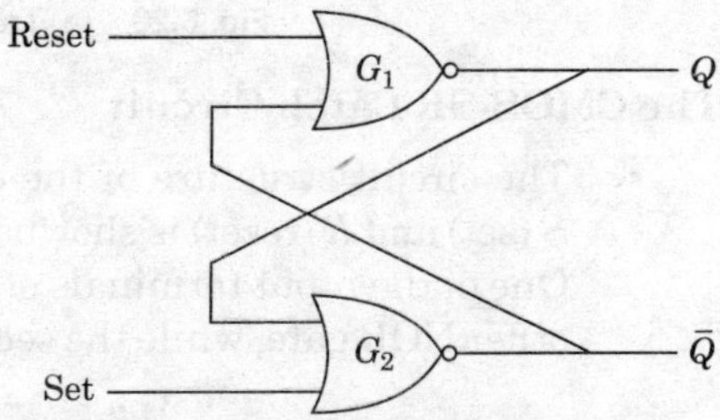

Fig. 8.19 Cross-coupled NOR flip-flop.

- The SR latch circuit has two complementary outputs, Q and $\bar{Q}$. By definition, the latch is said to be in its *set* state when Q is equal to logic "1" and $\bar{Q}$ is equal to logic "0." Or, the latch is in its *reset* state when the output Q is equal to logic "0" and $\bar{Q}$ is equal to "1."
- The gate-level schematics of the SR latch consisting of two NOR2 gates are shown in Fig. 8.19. It can easily be seen that when both input signals are equal to logic "0," the SR latch will operate exactly like the simple cross-coupled bistable element, *i.e.*, it will preserve (hold) either one of its two stable operating points (states) as determined by the previous inputs.
- Let us assume both inputs are at 0 initially, output Q is at 1, and output $\bar{Q}$ is at 0. Thus the output of gate G_1 is at 1. Since the output of gate G_1 is fed back to the input of gate G_2, the output of G_2 will be 0. The circuit is therefore stable with Q at 1 and $\bar{Q}$ at 0, as was originally assumed.

- If the Reset input is now taken to 1, the output of G_1 will change to 0. Both the inputs of G_2 are at 0, so its output will change to 1. The circuit is now stable with $Q = 0$ and $\bar{Q} = 1$. The circuit remains in this stable state even if the Reset input is changed back to 0. If the Set input is now taken to 1, the output of G_2 (i.e., $\bar{Q}$) will be at 0. Since both inputs of G_1 are now at 0, its output (*i.e.*, Q) will be at 1. The circuit remains stable with $\bar{Q} = 1$ and $Q = 0$ even when the Set input returns to 0.
- The input combination Set = 1 and Reset = 1 is not allowed because both Q and $\bar{Q}$ go to 0 in this case, which violates the condition that $\bar{Q}$ and Q should be complements of each other. Furthermore, when the Set and Reset inputs are returned to 0, an ambiguous situation arises. For example, if the propagation delay of G_1 is lower than that of G_2, then the output of G_1 (*i.e.*, Q) will change to 1 first. This in turn will make the output of G_2 (*i.e.*, $\bar{Q}$) = 0. On the other hand, if the output of G_2 changes to 1 first, then the output of G_1 will be forced to 0. In other words, it is impossible to predict the output. Therefore the Set = 1 and Reset = 1 combination is avoided during operation of this type of latch.
- The behavior of the latch circuit can be represented by the truth table of Fig. 8.20(*a*). The cross-coupled NOR latch is generally known as an SR (Set–Reset) latch. The logic symbol used to represent the SR latch is shown in Fig. 8.20(*b*).

S	R	Q_{n+1}	$\overline{Q_{n+1}}$	Operation
0	0	Q_n	$\overline{Q_n}$	hold
1	0	1	1	set
0	1	1	0	reset
1	1	d	d	not allowed

(*a*)

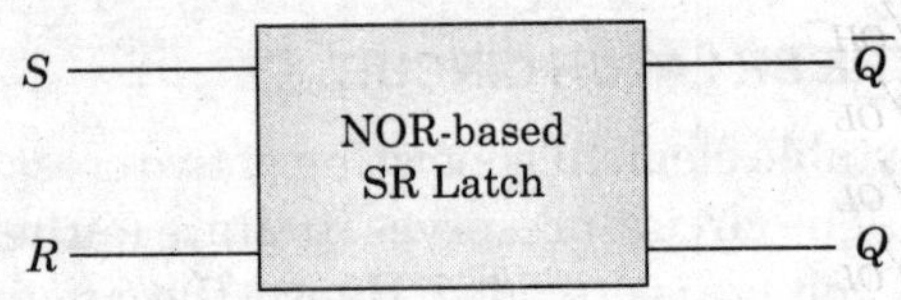

(*b*)

Fig. 8.20 (a) Truth table for NOR latch and (b) logic symbol.

The CMOS-SR Latch Circuit

- The circuit structure of the simple CMOS SR latch with two such triggering inputs, S (set) and R (reset) is shown in Fig. 8.21. The circuit consists of two CMOS NOR2 gates. One of the input terminals of each NOR gate is used to cross-couple to the output of the other NOR gate, while the second input enables triggering of the circuit.

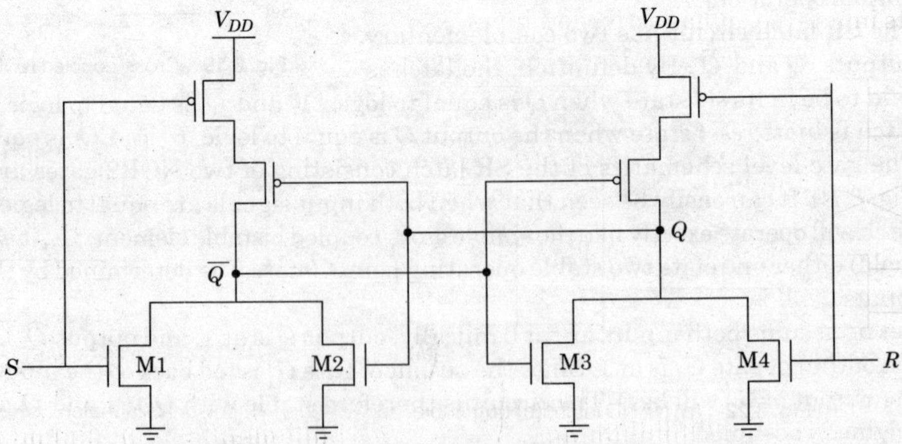

Fig. 8.21 CMOS SR latch circuit based on NOR2 gates.

- The operation of the CMOS SR latch circuit shown in Fig. 8.21 can be examined in more detail by considering the operating modes of the four nMOS transistors, M_1, M_2, M_3, and M_4.
- If the set input (S) is equal to V_{OH} and the reset input (R) is equal to V_{OL}, both of the parallel-connected transistors M_1 and M_2 will be on. Consequently, the voltage on node $\overline{Q}$ will assume a logic-low level of $V_{OL} = 0$. At the same time, both M_3 and M_4 are turned off, which results in a logic-high voltage V_{OH} at node Q. If the reset input (R) is equal to V_{OH} and the set input (S) is equal to V_{OL}, the situation will be reversed (M_1 and M_2 turned off and M_3 and M_4 turned on).
- When both of the input voltages are equal to V_{OL} on the other hand, there are two possibilities. Depending on the previous state of the SR latch, either M_2 or M_3 will be on, while both of the trigger transistors M_1 and M_4 are off. This will generate a logic-low level of $V_{OL} = 0$ at one of the output nodes, while the complementary output node is at V_{OH}. The static operation modes and voltage levels of the NOR-based CMOS SR latch circuit are summarized in the following Table 8.1.

Table 8.1 Operation modes of the transistors in the NOR-based CMOS SR latch circuit.

S	R	Q_{n+1}	$\overline{Q_{n+1}}$	*Operation*
V_{OH}	V_{OL}	V_{OH}	V_{OL}	M_1 and M_2 on, M_3 and M_4 off
V_{OL}	V_{OH}	V_{OL}	V_{OH}	M_1 and M_2 off, M_3 and M_4 on
V_{OL}	V_{OL}	V_{OH}	V_{OL}	M_1 and M_4 off, M_2 on or
V_{OL}	V_{OL}	V_{OL}	V_{OH}	M_1 and M_4 off, M_3 on

8.8.2 The SR Latch Circuit using NAND-GATE

- An SR latch can also be constructed by cross-coupling NAND gates as shown in Fig. 8.22(a). The circuit operates in the same manner as the NOR latch, but it does have a few subtle differences.
- Unlike the NOR latch, the NAND latch inputs are normally 1 and must be changed to 0 to change the output. An ambiguous output results when both the Set and the Reset inputs are at 0. Fig. 8.22 (b) shows the truth table of the NAND latch. The logic symbol for the NAND latch is shown in Fig. 8.22 (c); the circles denote that the latch responds to 0 on its inputs (i.e., it has active-low inputs).

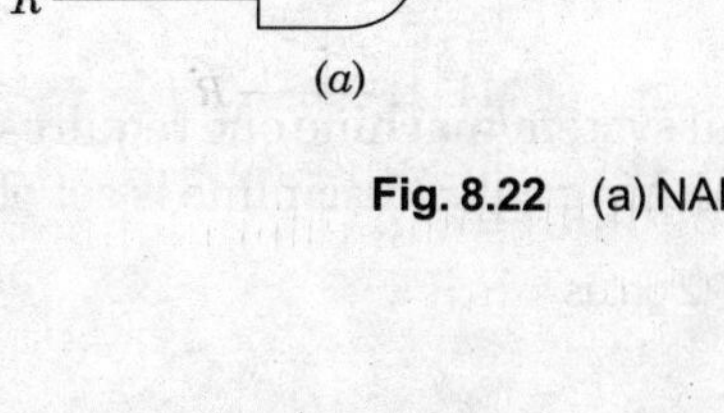

(a)

S	R	Q_{n+1}	$\overline{Q_{n+1}}$	Operation
0	0	1	1	not allowed
1	0	1	0	set
0	1	0	1	reset
1	1	Q_n	$\overline{Q_n}$	hold

(b)

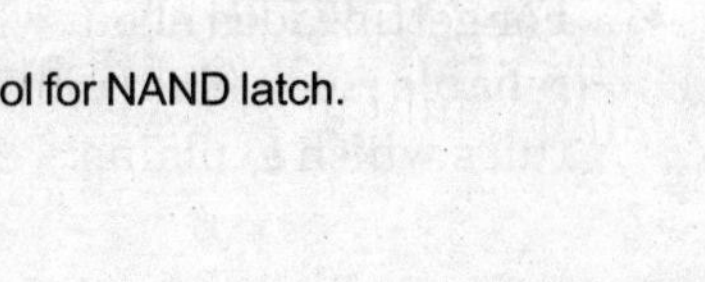

(c)

Fig. 8.22 (a) NAND latch, (b) truth table, and (c) logic symbol for NAND latch.

- The basic SR latch circuit using two NAND2 gates is shown in Fig. 8.23. Here, one input of each NAND gate is used to cross-couple to the output of the other NAND gate, while the second input enables external triggering.
- A close inspection of the NAND-based SR latch circuit reveals that in order to hold (preserve) a state, both of the external trigger inputs must be equal to logic "1". The state of the circuit can be changed only by pulling the set input to logic zero or by pulling the reset input to zero.
- We can observe that if S is equal to "0" and R is equal to "1," the output Q attains a logic "1" value and the complementary output Q becomes logic "0." Thus, in order to set the NAND SR latch, logic "0" must be applied to the set (S) input. Similarly, in order to reset the latch, logic "0" must be applied to the reset (R) input.
- The conclusion is that the NAND-based SR latch responds to active low input signals, as opposed to the NOR-based SR latch, which responds to active high inputs. Note that if both input signals are equal to logic "0," both output nodes assume a logic-high level, which is not allowed because it violates the complementarily of the two outputs.

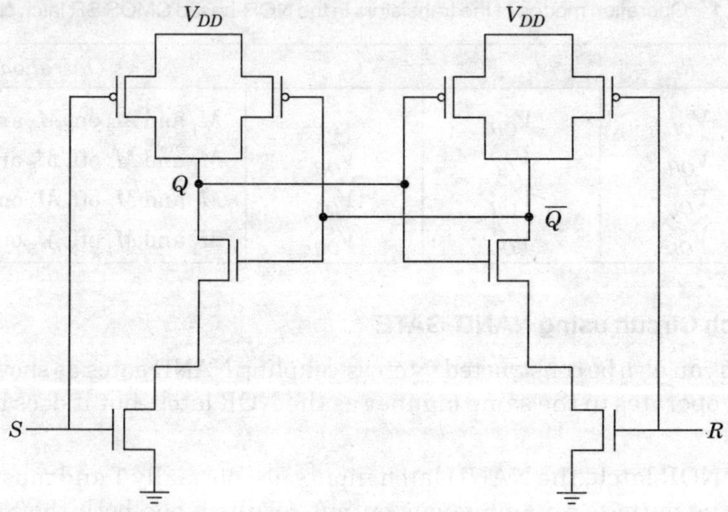

Fig. 8.23 CMOS SR latch circuit based on NAND2 gates.

8.9 SEQUENTIAL SYSTEMS AND CLOCKING RULES

- The sequential systems are built from combinational circuits/network and memory elements. The configuration of generic sequential system is shown in Fig. 8.24, which is also known as a finite-state machine.
- Memory elements hold the machine's state, the machine's input and outputs are also called primary inputs and primary outputs. If the primary outputs are a function of both the primary inputs and state, the machine is known as *mealy machine;* if the primary outputs depends only on the state, the machine is called a *Moore Machine.* A properly inter connected set of sequential systems is also a sequential system. It is often convenient to break a large system can be much easier to understand; it may also have a smaller layout and run faster.
- For getting idea about which time the circuit acts a sequential system/machine one requires reliable rules. The rules are governed by clocking rules. A clocking rules/discipline is set of rules which explains:

1. The construction of sequential system from gates and memory elements.
2. How to contain the behaviour of systems inputs overtime.

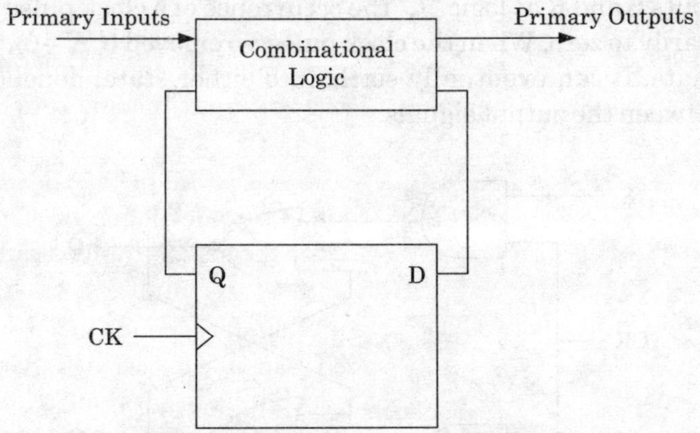

Fig. 8.24 A generic sequential system

- The constraints on system inputs are defines as signal types, which define both how signals behave overtime and what signals can be combined in logic gates or memory elements. By following there, we can ensure that the system will operate properly at some rate; we can then worry about optimizing the system to run as fast as possible while still functioning correctly. Different memory element types require different rules, so we will end up with a family of clocking disciplines. All disciplines have two common rules, however. The first is simple.

Clocking Rule 1: Combinational logic gates cannot be connected in a cycle. Gates connected in a cycle from a primitive memory element and case to be combinational — the gates outputs depend not only on the inputs to the cycle but the values running around the cycle. In fact, this rule is stronger than is absolutely necessary. It is possible to build a network of logic gates which has cycles but is still combinational — the values of its outputs depend only on the present input values, not past inputs values. However, careful analysis is required to ensure that a cycle network is combinational, whereas cycles can be detected easily. For most practical circuits, the cyclic rule is not restrictive.

Clocking Rule 2: All components must have bounded delay. This rule is easily satisfied by standard components, but does rule out synchronous for asynchronous signals.

8.10 CLOCK LATCH AND FLIP-FLOP CIRCUITS

8.10.1 Clocked SR Latch

- All the SR latch circuits will respond to the changes occurring in input signals at a circuit-delay dependent time point during their operation.
- To facilitate synchronous operation, the circuit response can be controlled simply by adding a gating clock signal to the circuit, so that the outputs will respond to the input levels only during the active period of a clock pulse. Here, we assumed that the clock pulse is a periodic square waveform, which is applied simultaneously to all clocked logic gates in the system.
- The gate-level schematic diagram of a clocked NOR-based SR latch is shown in Fig. 8.25.
- If the clock (CK) is equal to logic "0" then input signals have no influence upon the circuit response. The outputs of the two AND gates will remain at logic "0," which forces the SR latch to hold its current state regardless of the S and R input signals.

- If the clock (CK) goes to logic "1" then the logic levels applied to the S and R inputs are permitted to reach the SR latch and possibly change its state.
- With both inputs S and R at logic "1," the occurrence of a clock pulse causes both outputs to go momentarily to zero. When the clock pulse is removed ($CK = 0$), the state of the latch is indeterminate. It can eventually settle into either state, depending on slight delay differences between the output signals.

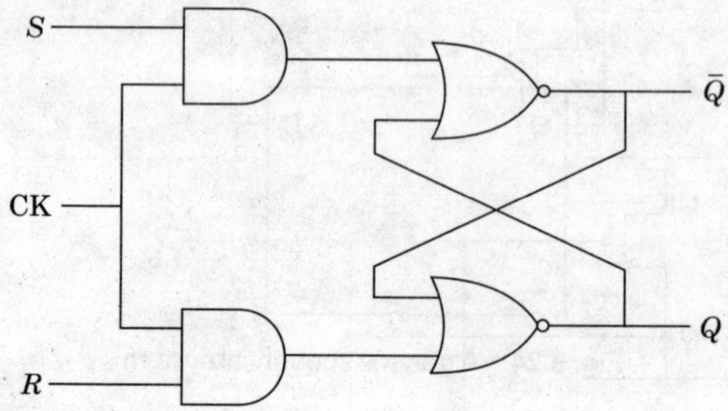

Fig. 8.25 Gate-level schematic of the clocked NOR-based SR latch.

Operation of Clocked SR Latch

- We can consider a sample sequence of CK, S, and R waveforms, and the corresponding output waveform Q are shown in Fig. 8.26 for illustrate the operation of the clocked SR latch.
- The circuit is strictly level-sensitive during active clock phases (any changes occurring in the S and R input voltages when the CK level is equal to "1" will be reflected onto the circuit outputs).
- Consequently, even a narrow spike or glitch occurring during an active clock phase can set or reset the latch, if the loop delay is shorter than the pulse width.

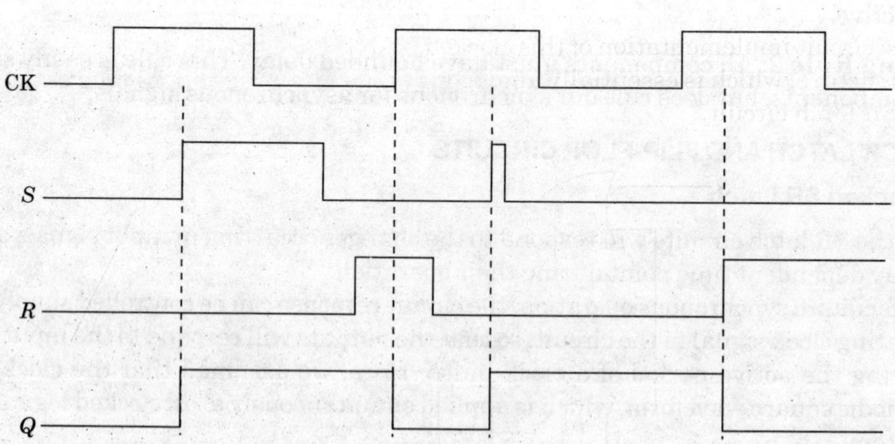

Fig. 8.26 Sample input and output waveforms illustrating the operation of the clocked NOR based SR latch circuit.

8.10.1.1 CLOCKED NOR-Based SR LATCH

A CMOS implementation of the clocked NOR-based SR latch circuit is shown in Fig. 8.27. The AOI-based implementation of the circuit results in a very small transistor count, compared with the alternative circuit realization consisting of two AND2 and two NOR2 gates.

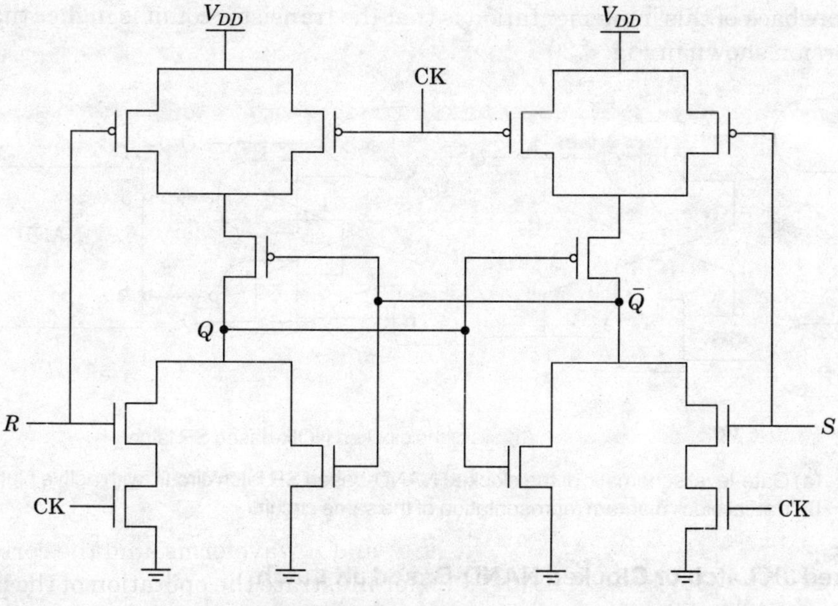

Fig. 8.27 AOI-based implementation of the clocked NOR-based SR latches circuit.

8.10.1.2 CLOCKED NAND-Based SR LATCH

- A CMOS implementation of the clocked NAND-based SR latch circuit is shown in Fig. 8.28. In this case, all the input signals S and R as well as the clock signal CK are active low. This means that changes in the input signal levels will be ignored when the clock is equal to logic " 1," and that inputs will influence the outputs only when the clock is active, *i.e.*, CK = "0".
- For the circuit implementation of this clocked NAND-based SR latch, we can use a simple OAI structure, which is essentially analogous to the AOI-based realization of the clocked NOR SR latch circuit.

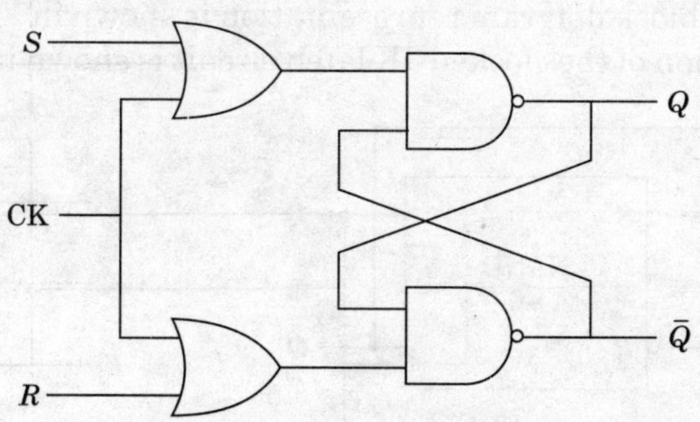

Fig. 8.28 Gate-level schematic of the clocked NAND-based SR latch circuit.

- A different implementation of the clocked NAND-based SR latch is shown in Fig. 8.29. Here, both input signals and the *CK* signal are active high (the latch output *Q* will be set when *CK* = "1" *S* = "1" and *R* = "0"). Similarly, the latch will be reset when *CK* = "1," *S* = "0," and *R* = "1." The latch preserves its state as long as the clock signal is inactive (when CK = "0").
- The drawback of this implementation is that the transistor count is higher than the active low version shown in Fig. 8.29

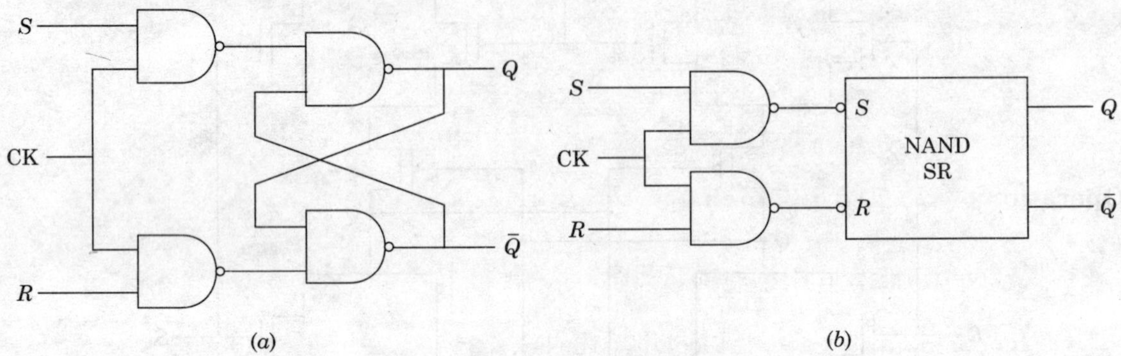

Fig. 8.29 (a) Gate-level schematic of the clocked NAND-based SR latch circuit, with active high inputs. (b) Partial block diagram representation of the same circuit.

8.10.2 Clocked JK Latch or Clocked NAND-Based JK Latch

- To prevent any possibility of a "race" condition occurring when both the *S* and *R* inputs are at logic 1 when the CLK input falls from logic 1 to logic 0, we must somehow prevent one of those inputs from having an effect on the master latch in the circuit. At the same time, we still want the flip-flop to be able to change state on each falling edge of the CLK input, if the input logic signals call for this. Therefore, the *S* or *R* input to be disabled depends on the current state of the slave latch outputs.
- Every simple latch circuit have a common problem of not-allowed input combination (their state becomes indeterminate when both inputs *S* and *R* are activated at the same time). This problem can be overcome by adding two feedback lines from the outputs to the inputs as shown in Fig. 8.30. The resulting circuit is called a JK latch.
- An NAND-based block diagram of the JK latch (JK flip-flop) with active high inputs and the corresponding block diagram representation is shown in Fig. 8.30. The NAND gate-level implementation of the clocked JK latch circuit is shown in Fig. 8.31.

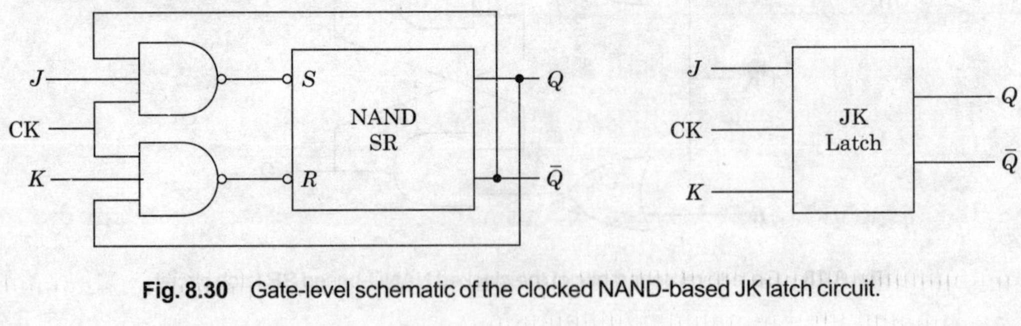

Fig. 8.30 Gate-level schematic of the clocked NAND-based JK latch circuit.

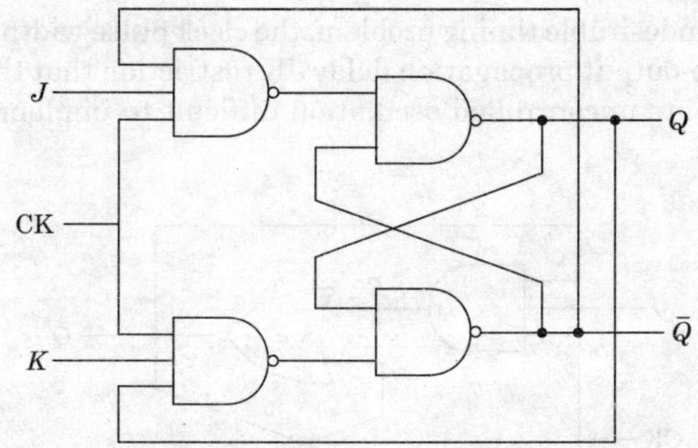

Fig. 8.31 All-NAND implementation of the clocked JK latch circuit.

Operation of Clocked JK Latch

- In this circuit, J and K inputs are corresponds to the set and reset inputs respectively.
- When the clock is active, the latch can be set with the input combination (J=1, K= 0) and it can be reset with the input combination (J = 0, K = 1).
- If both inputs are equal to logic "0", the latch preserves its current state.
- If both inputs are equal to logic "1" during the active clock phase, the latch simply switches its state due to feedback. So, we can see that the JK latch does not have a not-allowed input combination.
- As in the other clocked latch circuits, the JK latch will hold its current state, when the clock is inactive (CK = "0").The operation of the clocked JK latch is summarized in the truth table (Table 8.2).

Table 8.2 Detailed truth table of the JK latch circuit.

J	K	Qn	$\overline{Q_n}$	S	R	$Qn+1$	$\overline{Q_{n+1}}$	*Operation*
0	0	0	1	1	1	0	1	hold
		1	0	1	1	1	0	
0	1	0	1	1	1	0	1	reset
		1	0	1	0	0	1	
1	0	0	1	0	1	1	0	set
		1	0	1	1	1	0	
1	1	0	1	0	1	1	0	toggle
		1	0	1	0	0	1	

8.10.2.1 CLOCKED NOR-Based JK LATCH

- The NOR-based implementation of the clocked JK latch is shown in Fig. 8.32 (a). The CMOS realization of this circuit is shown in Fig. 8.32 (b). This circuit doesn't have the problem of undermined output, but there is a still problem arises.
- If both inputs are equal to logic " 1 " during the active phase of the clock pulse, the output of the circuit will oscillate (toggle) continuously until either the clock becomes inactive (goes to zero), or one of the input signals goes to zero.

- To prevent this undesirable timing problem, the clock pulse width must be made smaller than the input-to-output propagation delay JK restriction that the clock signal must go low before prevents uncontrolled oscillation difficult to implement for most practical applications.

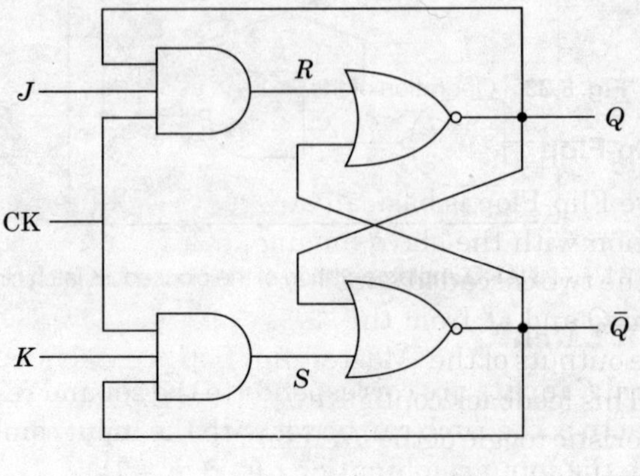

(a) Gate-level schematic of the clocked NOR-based JK latch circuit.

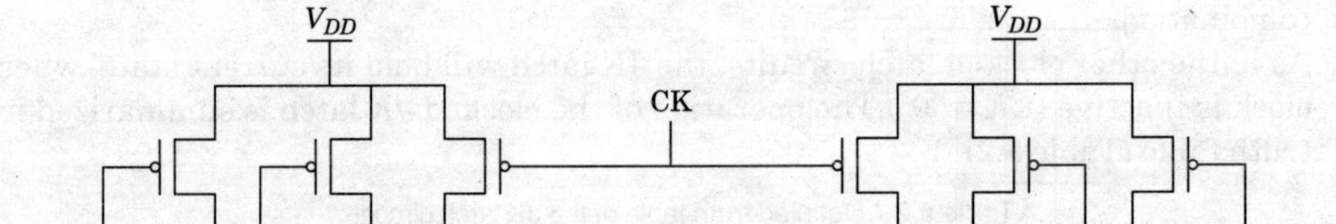

Fig. 8.32 (b) CMOS AOI realization of the JK latch.

- If the clock timing constraint described above is satisfied, the output of *JK* latch will toggle (change its state) only once for each clock pulse and both inputs are equal to logic "1" (as Fig. 8.33).A circuit which is operated exclusively in this mode is called a toggle switch.

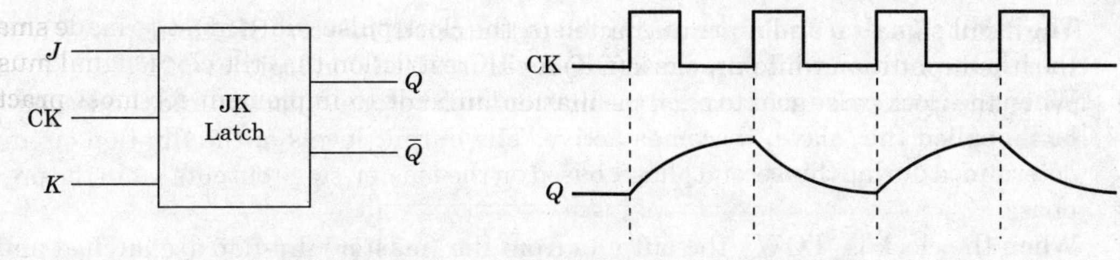

Fig. 8.33 Operation of the JK latch as a toggle switch.

8.10.3 Master-Slave Flip-Flop

- The Master-Slave Flip-Flop is basically two gated SR flip-flops connected together in a series configuration with the slave having an inverted clock pulse. The key operation principle is that the two cascaded stages are activated with opposite clock phases.
- The outputs from Q and Q' from the "Slave" flip-flop are fed back to the inputs of the "Master" with the outputs of the "Master" flip-flop being connected to the two inputs of the "Slave" flip-flop. This feedback configuration from the slave's output to the master's input gives the characteristic toggle of the *JK* flip-flop and JK latches are shown in Fig. 8.34 and Fig. 8.35 respectively.

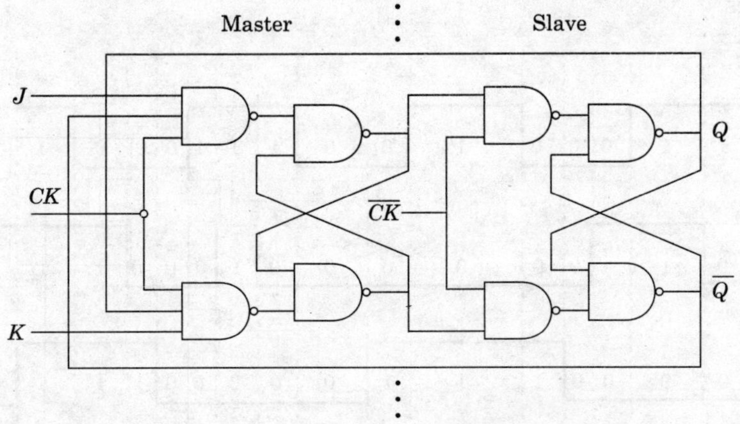

Fig. 8.34 Master-slave flip-flop using NAND-based JK flip-flop.

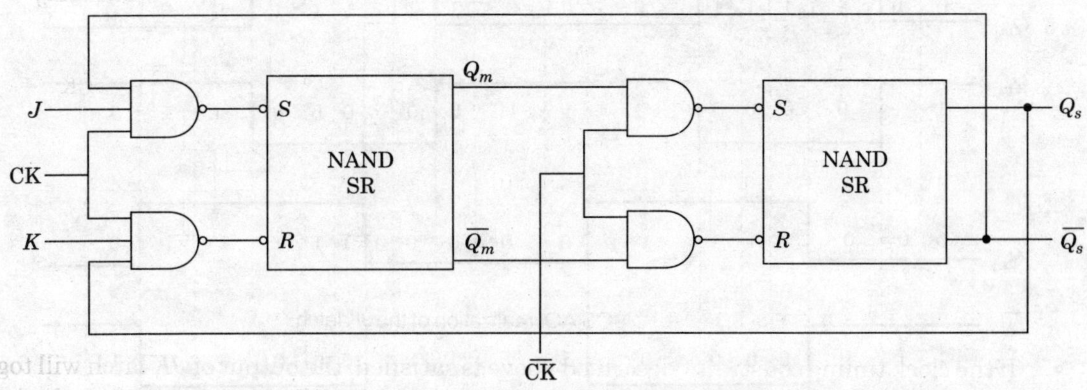

Fig. 8.35 Master-slave flip-flop consisting of NAND-based JK latches.

- The input signals J and K are connected to the gated "master" SR flip-flop which "locks" the input condition while the clock (CK) input is "HIGH" at logic level "1".
- When the clock pulse goes to zero, the master latch becomes inactive and the second-stage latch, called the "slave," becomes active. The output levels of the flip-flop circuit are determined during this second phase, based on the master-stage outputs set in the previous phase.
- When the clock is "LOW", the outputs from the "master" flip-flop are latched and any additional changes to its inputs are ignored. The gated "slave" flip-flop now responds to the state of its inputs passed over by the "master" section. Then on the "Low-to-High" transition of the clock pulse the inputs of the "master" flip-flop are fed through to the gated inputs of the "slave" flip-flop and on the "High-to-Low" transition the same inputs are reflected on the output of the "slave" making this type of flip-flop edge or pulse-triggered.
- Since the master and the slave stages are effectively decoupled from each other with the opposite clocking scheme, the circuit is never transparent, i.e., a change occurring in the primary inputs is never reflected directly to the outputs. This very important property clearly separates the master-slave flip-flop from all of the latch circuits examined earlier in this section. Fig. 8.36 shows a sample set of input and output waveforms associated with the JK master-slave flip-flop, which can help the reader to study the basic operation principles.

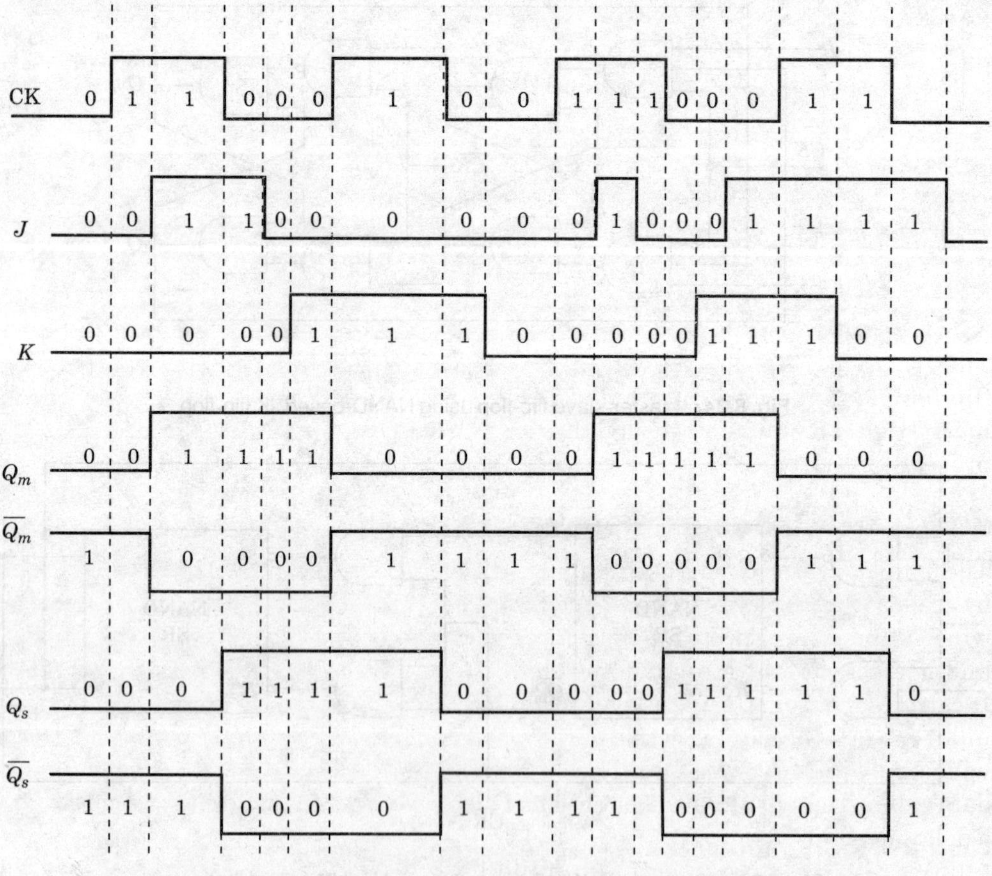

Fig. 8.36 Sample input and output waveforms of the master-slave flip-flop circuit.

- Because the master and the slave stages are decoupled from each other, the circuit allows for toggling when $J = K =$ "1". The circuit accepts input data when the clock signal is "HIGH", and passes the data to the output on the falling-edge of the clock signal. In other words, the Master-Slave JK Flip-flop is a "Synchronous" device as it only passes data with the timing of the clock signal.
- Figure 8.35 also shows that the master-slave flip-flop circuit examined here has the potential problem of "one's catching." When the clock pulse is high, a narrow spike or glitch in one of the inputs, for instance a glitch in the J line (or K line), may set (or reset) the master latch and thus cause an unwanted state transition, which will then be propagated into the slave stage during the following phase. This problem can be eliminated to a large extent by building an edge-triggered master-slave flip-flop, which will be examined in the following section.
- A NOR-based alternative realization for the master-slave flip-flop circuit is shown in Fig. 8.37.

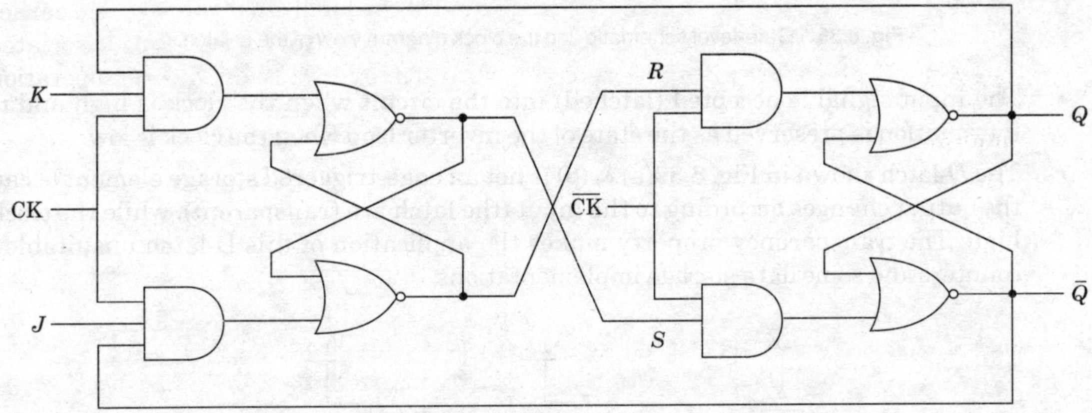

Fig. 8.37 NOR-based realization of the JK master-slave flip-flop.

- The JK flip-flop is a very versatile device, and is probably the most commonly used form of flip-flop in digital electronic and control circuits. The fact that it has two inputs often means that it is simpler to design the control logic required to ensure that it changes state properly in a circuit.

8.11 CMOS D-LATCH AND EDGE-TRIGGERED FLIP-FLOP

- The simple D-latch circuit is shown in Fig. 8.38. The gate-level D-latch is simply obtained by modifying the clocked NOR-based SR latch circuit. This circuit has a single input D, which is directly connected to the S input of the latch. The input variable D is also inverted and connected to the R input of the latch.
- It can be seen from the gate-level schematic that the output Q assumes the value of the input D when the clock is active ($CK =$ "1"). When the clock signal goes to zero, the output will simply preserve its state. Thus, the CK input acts as an enable signal which allows data to be accepted into the D-latch. The D-latch is used for temporary storage of data or as a delay element.

- In the following, we will discuss about the CMOS implementation of *D* latch. The circuit arrangement consisting a basic two-inverter loop and two CMOS transmission gate (*TG*) switches are shown in Fig. 8.38.

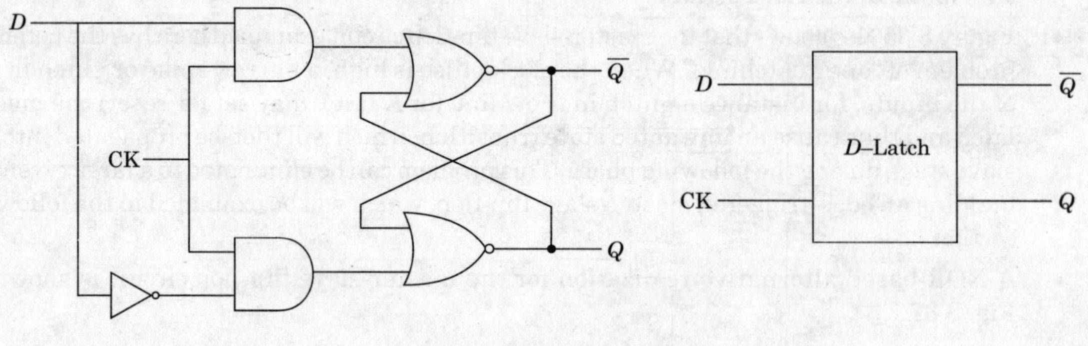

Fig. 8.38 Gate-level schematic and the block diagram view of the *D*-latch.

- The input signal is accepted (latched) into the circuit when the clock is high and this information is preserved as the state of the inverter loop when the clock is low.
- The *D*-latch shown in Fig. 8.39(a) & (b) is not an edge-triggered storage element because the output changes according to the input (the latch are transparent), while the clock is high. The transparency property makes the application of this D-latch unsuitable for counters and some data storage implementations.

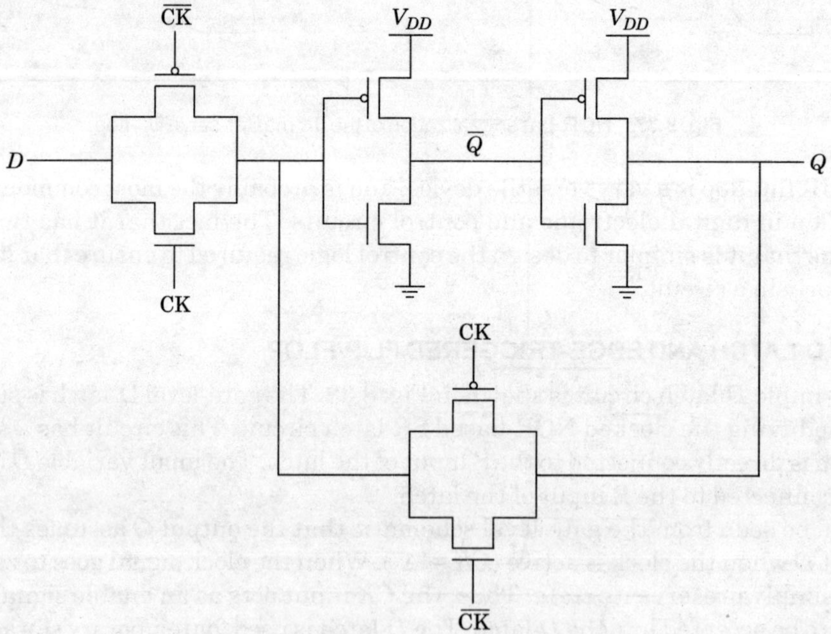

Fig. 8.39 (a) CMOS implementation of the D-latch.

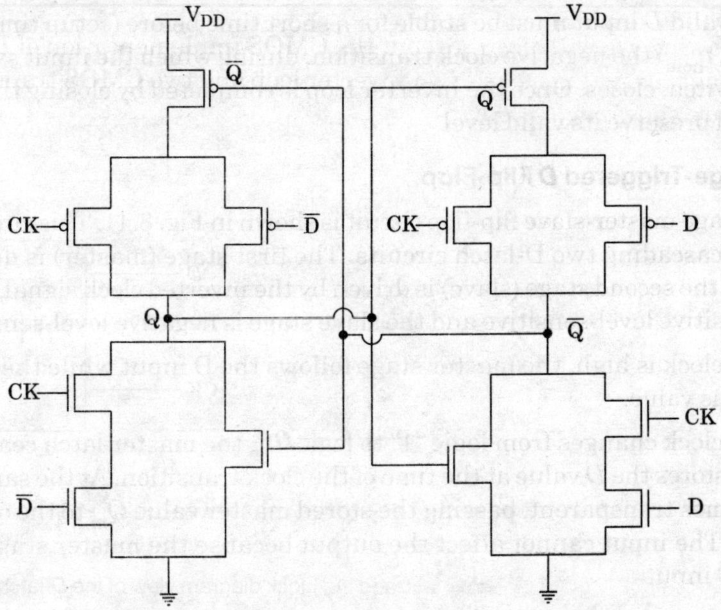

Fig. 8.39 (b) CMOS implementation of D flip-flop

- The operation of the CMOS *D*-latch circuit can be better visualized by replacing the CMOS transmission gates with simple switches as shown in Fig. 8.40. A timing diagram accompanying this figure shows the time intervals during which the input and the output signals should be valid (un-shaded).

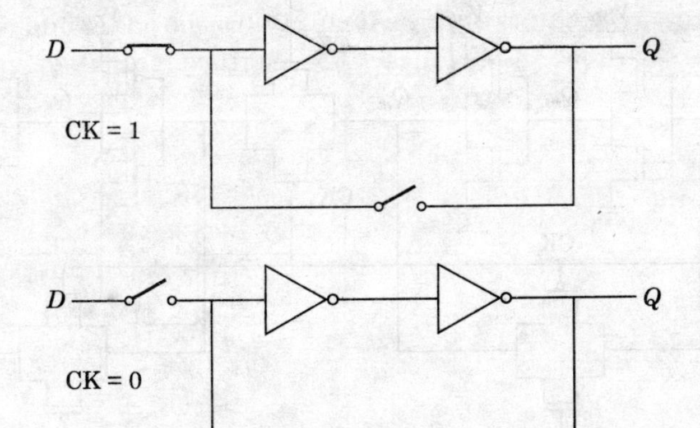

Fig. 8.40 Simplified schematic view and the corresponding timing diagram of the CMOS *D*-latch circuit, showing the setup time and the hold time.

- Here, the valid D input must be stable for a short time before (setup time, t_{setup}) and after (hold time, t_{hold}) the negative clock transition, during which the input switches opens and the loop switch, closes. Once the inverter loop is completed by closing the loop switch, the output will preserve its valid level.

8.11.1 CMOS Edge-Triggered *D* Flip-Flop

- The two-stage master-slave flip-flop circuit is shown in Fig. 8.41. This circuit is constructed by simply cascading two D-latch circuits. The first stage (master) is driven by the clock signal and the second stage (slave) is driven by the inverted clock signal. Thus, the master stage is positive level-sensitive and the slave stage is negative level-sensitive.
- When the clock is high, the master stage follows the D input while the slave stage holds the previous value.
- When the clock changes from logic "1" to logic "0", the master latch ceases to sample the input and stores the D value at the time of the clock transition. At the same time, the slave latch becomes transparent, passing the stored master value Q_m to the output of the slave stage, Q_8. The input cannot affect the output because the master stage is disconnected from the D input.
- When the clock changes again from logic 0″ to 1″, the slave latch locks in the master latch output and the master stage starts sampling the input again. Thus, this circuit is a negative edge-triggered D flip-flop by virtue of the fact that it samples the input at the falling edge of the clock pulse.

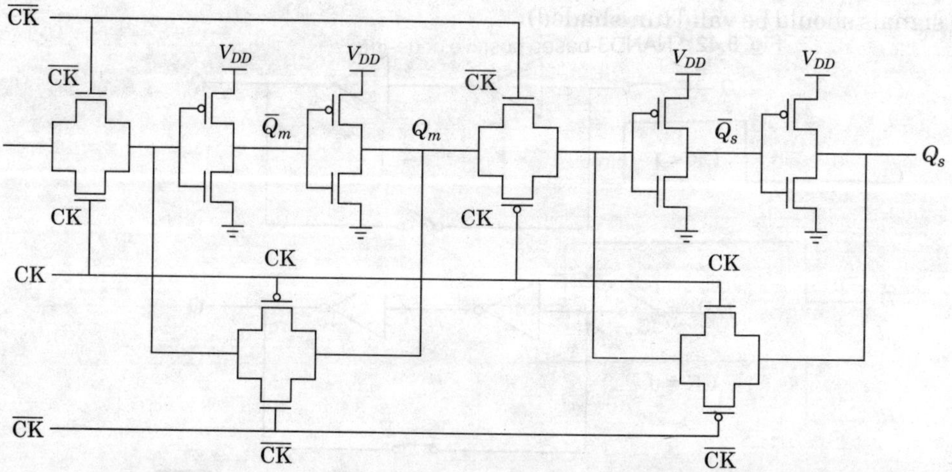

Fig. 8.41 CMOS negative (falling) edge-triggered master-slave *D* flip-flop (DFF).

8.11.1.1 CMOS Edge-Triggered D Flip-Flop using NAND3 gates

- The edge-triggered D flip-flop using six NAND3 gates is shown in Fig. 8.42. This D flip-flop is positive edge-triggered as illustrated in the waveform chart in Fig. 8.43.
- Initially, all the signal values except for S are 0, *i.e.*, $(S, R, CK, D) = (1, 0, 0, 0)$, and $Q = 0$.
- In the second phase, both D and R switch to 1, *i.e.*, $(S, R, CK, D) = (1, 0, 1, 1)$, but no change in Q occurs and the Q value remains at 0.

- In the third phase, if CK goes to high, *i.e.*, $(S, R, CK, D) = (1, 1, 1, 1)$, the output of gate 2 switches to 0, which in turn sets the output of the last stage SR latch to 1.
- Thus, the output of this D flip-flop switches to 1 at the positive-going edge of the clock signal, CK.

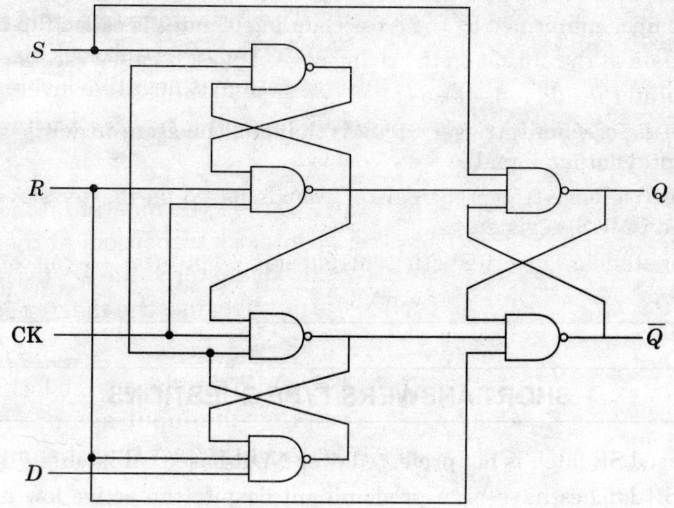

Fig. 8.42 NAND3-based positive edge-triggered *D* flip-flop circuit.

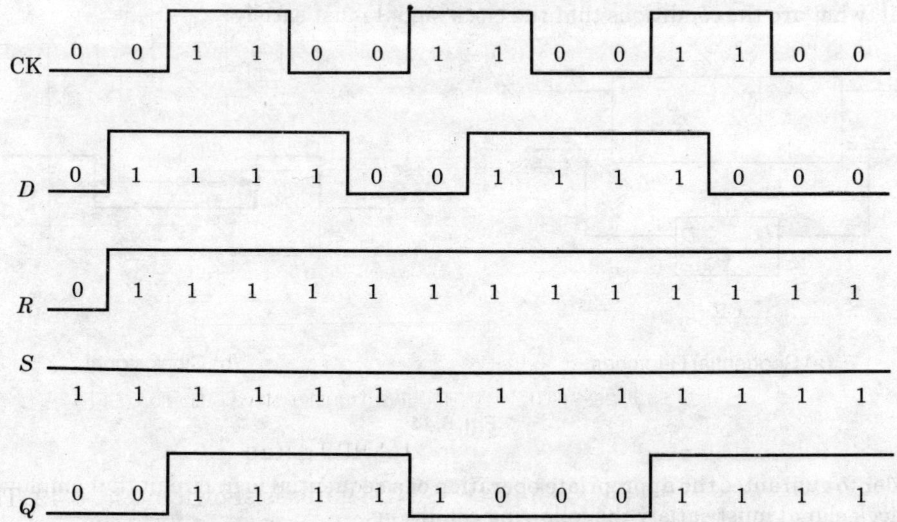

Fig. 8.43 Timing diagram of the positive edge-triggered *D* flip-flop.

POINTS TO BE REMEMBERS

1. The time that the data input must be valid before clock transistion is known as setup time (SUT).
2. A latch is a level sensitive device.
3. An edge triggered storage element is called register.
4. Any bistable component formed by the cross coupling of gates is called flip-flop.
5. The change of data at the output on the rising edge of clock is called race condition.
6. Monostable vibrator is otherwise also called one shot vibrator.
7. One of the main use of schmitt trigger circuit is changing a noising or slowly varying input signal into a clean digital output signal.
8. A circuit of positive edge-triggered register that is based on master-slave concept is called C^2MOS (clocked CMOS) register.
9. By replacing the standard inverter with a current starved inverter, we can convert standard ring oscillator in V_{CO}.

SHORT ANSWERS TYPE QUESTIONS

Q.1. Why NAND-based SR latch is not preferred over NOR-based SR latch?

Ans. NAND-based SR latches have been predominant despite the active low inputs because NAND gates are cheaper to produce than the NOR gates in CMOS technology, NAND gates are still performed.

Q.2. A clock signal as shown in Fig. 8.44(b) is applied on the sequential logic circuit employing D-latches as shown in Fig. 8.44(a). In order to guarantee the appropriate operation of a sequential logic circuit, what are the conditions that the clock signal must satisfy?

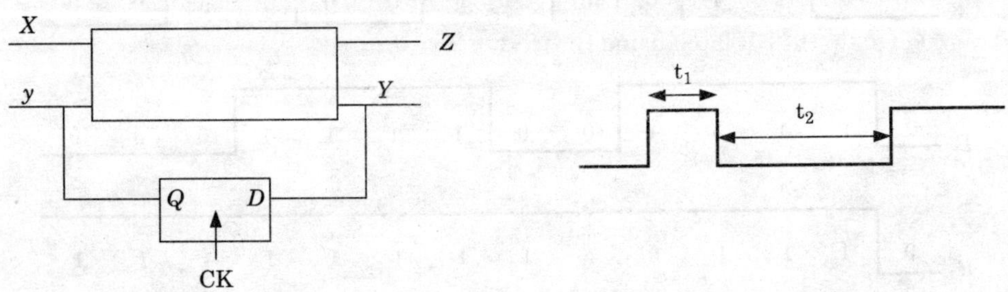

(a) Sequential D-latches (b) Clock signal

Fig. 8.44

Ans. In order to guarantee the appropriate operation of a sequential logic circuit that employs D-latches, the clock signal must satisfy the following conditions:

1. $t_1 > \Delta_{latch}$, where Δ_{latch} is the delay associated with D-latch. This condition ensure that the next variable state Y has enough time to travel from D to Q so that it can be latched correctly.
2. $t_1 < \Delta_{c/L(min)}$, where $\Delta_{c/L(min)}$ is the shortest delay of the combinational logic circuit. This condition prevents the potential state transistion hazard problem in an asynchronous sequential logic circuit. When this condition is satisfied, the changes caused by the new Y still being latched will not come back to the combinational circuit input in the same clock cycle.

3. $t_2 > \Delta_{c/L(max)}$, where $\Delta_{c/L(max)}$ is the longest delay of the combinational circuit. This requirement allows the slowest signal to settle down before the next transition. This ensures the latching of the correct new state variables.
4. Input X should not be changed until the latches settle down at their outputs (Q). This can be ensured by changing X during time internal t_2. In this case, time t_2 has to be lengthened according to the time of changing X.

REVIEW QUESTIONS

1. Design a JK Flip-flop logic static CMOS minimize size devices with $V_{DD} = 5$ V.
2. Design a D Flip-flop using complementary CMOS and TSPC logic design style.
3. Explain latch up in the CMOS inverter.
4. Explain the difference between combinational circuits and sequential circuits. **[UPTU 2004-05]**
5. Design CMOS implementation of SR Flip-flop.
6. Explain the difference between mealy and moore machine.
7. Explain CMOS transistor based NAND SR latch circuit and explain its working.
8. Explain the working of Master-slave Edge Triggered Register.
9. Discuss about the working of Multiplexer Based Latches.
10. Why we use master-slave FF? Explain with proper diagram and draw the input and output waveform.
11. Explain negative edge triggered master-slave Flip-flop with proper diagram.
12. Explain the differences between Asynchronous and Synchronous sequential circuits.
13. Write short note on SRAM. **[UPTU 2004-05]**
14. Design a pass transistor network that implements the function

$$F = ABC + A\overline{B}\overline{C} + \overline{A}\overline{B}C + \overline{A}B\overline{C}$$

15. Draw and explain the SR latch using CMOS devices and based on NAND logic. Also draw the stick diagram. **(UTU-2012)**

❑❑❑

Chapter 9

Semiconductor Memories

Chapter Overview

- Introduction
- Memory Classification
- Dynamic Random access Memory (Dram)
- Read Only Memory
- Erasable programmable read only memory (EPROM)
- Electrically Erasable Programmable Read Only Memory (E^2PROM)
- Designing Issues/factors for Semiconductor Memories
- Memory Organization
- Static Random Access Memory (SRAM)
- Flash Memory

9.1 INTRODUCTION

Modern digital systems require the capability storing and retrieving large number of information at high speeds. *Memories* are circuits or systems that store digital information in large quantity.

The amount of memory required in a particular system depends on the type of the application but, in general, the number of transistors for the information storage function is much large than the number of transistors used for logic operations and other purposes. Information can be stored in semiconductors based on either positive feedback or capacitive storage.

9.2 DESIGNING ISSUES/FACTORS FOR SEMICONDUCTOR MEMORIES

- The number of stored data bits per unit area, is one the key design criteria of the memory array.
- Other important issues is the memory access time. The time required to store/or retrieve a particular data bit in the memory array is known as *memory access time.*
- An important performance key for memory array is speed. The memory speed is determine by the memory access time.
- Another important significant factor is the static and dynamic power consumption for design for memory array.

9.3 MEMORY CLASSIFICATION

The semiconductor memory is generally classified according to the types of data storage and data access. Every type of memory has a different cell design, but the basic structure, organization and access mechanisms are largely the same. The overview of semiconductor memory types is shown in Fig. 9.1.

- Read/write memory must permit the modification of data bits stored in memory. This memory is known as *Random Access Memory (RAM).* The stored data is volatile. RAM's are classified into two main categories, which are based on the operation type of individual data storage cells. These two categories are DRAM and SRAM.

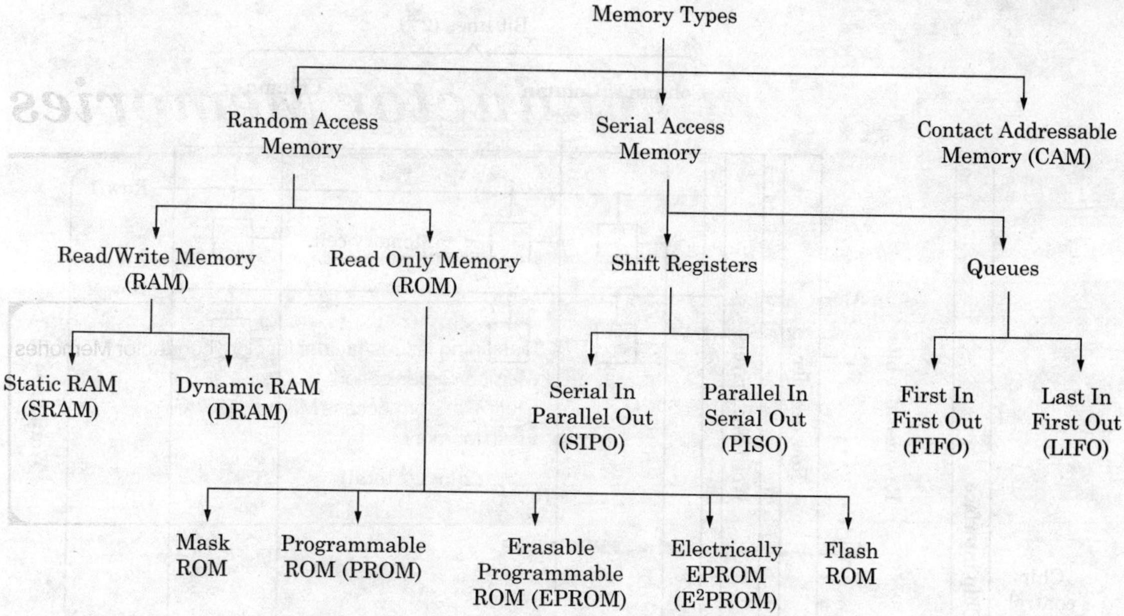

Fig. 9.1 Overview of semiconductor memory types

The DRAM cell consists of a capacitor to store binary information and a transistor to access the capacitor. Cell Information (voltage) is degraded mostly due to Junction Leakage current at storage node.

The SRAM cell consists of a latch. Due to this the cell data is kept as long as the power is turned on and refresh operation is not required. SRAM have high power and low power consumption.

- Read only memory (ROM) allows only retrieval of previously stored data and does not permit modifications of the stored information contents during format operation. ROM are nonvolatile memories. ROM's are classified into various types, depending on the data storage or data erasing characteristic.
  - ⇒ In the mask ROM, the data are written during chip fabrication by using a photo mask.
  - ⇒ In PROM, data are written electrically after the chip is fabricated.
  - ⇒ In EPROM and E^2PROM data can be rewritten, but the number of subsequent rewrite operations is limited to 10^4–10^5.
  - ⇒ Flash memory is similar to E^2PROM, where data in the block can be erase by using a high electrical voltage.

9.4 MEMORY ORGANIZATION

- A conceptual memory organization is shown in Fig. 9.2. This memory organization is a random-access architecture.
- The data storage structure or core is made by simple cell circuits, which are arranged to share the connections in horizontal rows and vertical columns. Each cell is capable of storing one bit of binary information. The horizontal rows (2^N), driven only from the outside the storage array, are known as *wordlines*.

The vertical lines (2^M) along which data flow into and out of cells, are known as *bit lines*. Thus, the total number of memory cells in this array is $2^N \times 2^M$.

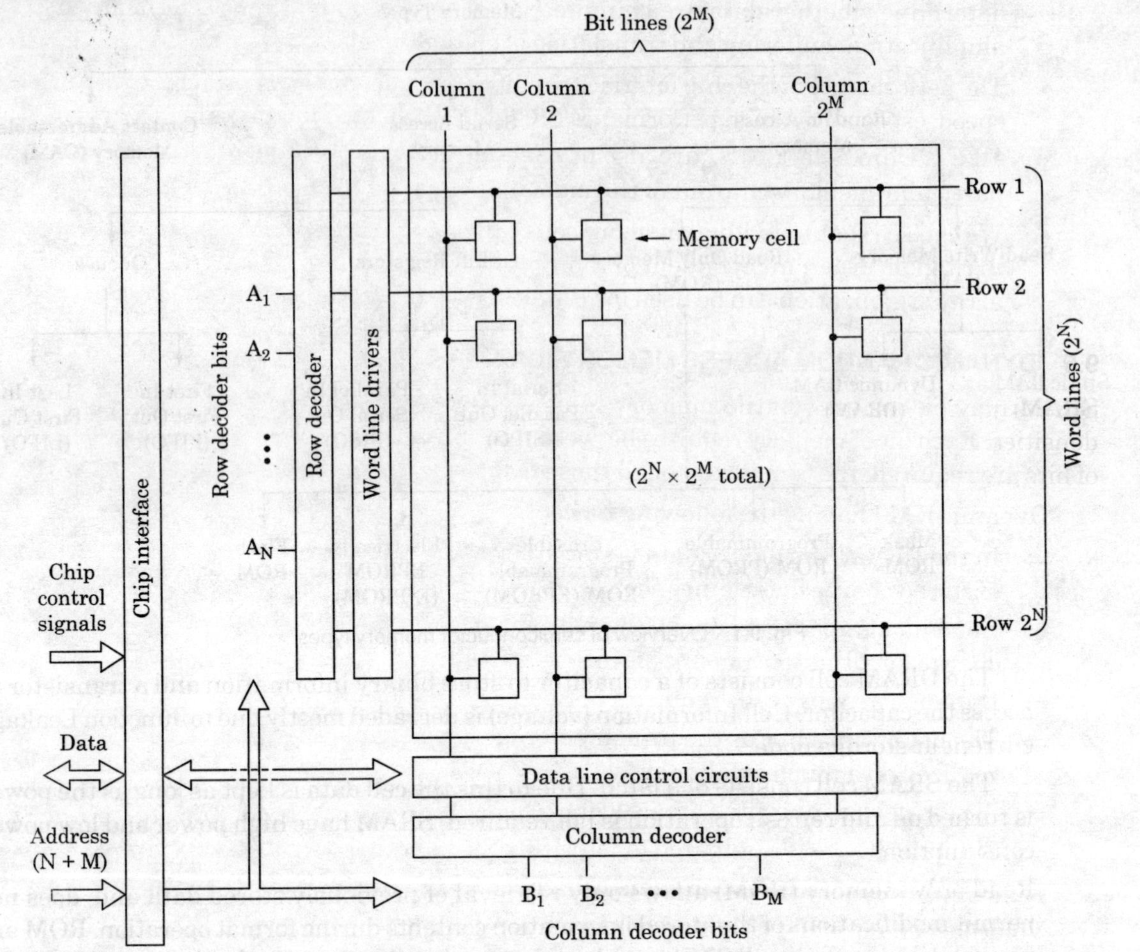

Fig. 9.2 Conceptual random–access memory array Organisation.

- A cell is accessed for reading or writing by selecting its row and column. Each cell can store 0 or 1. Memories may simultaneously select 4, 8, 16, 32 or 64 columns in one row depending on the application. The row and column (or group of columns) to be selected are determined by decoding binary address information. These addresses for reading or writing are provided by the memory controller on the processor directly.
- The row and column selection operations are accomplished by row and column decoders, respectively. The row decoder circuit selects one out of 2^N word lines across to an N-bit address, while the column decoder circuits selects and out of 2^M bit lines according to an M-bit column address.
- When an information read or write operation is performed on the selected signal or multiple bits on a particular row, then the column address serves the double duties of selecting the particular columns and routing the corresponding data content is selecting row to the output through the memory chip interface. The columns are known as *data out buffers*.
- The data out buffers are required to drive relatively large current load on the board, so large transistors are used.
- These circuits (e.g. decoders, sense amplifiers, column precharge, data buffer, etc., are designed so that they may be shared among many memory cells. Read/write (R/W) circuits

determine whether data are being retrieved or stored and they perform any necessary amplification, buffering and translation of voltage levels.

- The performance of the chip interface circuit determines a major portion of the total memory speed, especially in high performance SRAMs. Other chip control signals (e.g. Write Enable $(\overline{WE})$, Chip Select $(\overline{CS})$ are also provided to activate the read or write operation of the particular memory chip out of the memory cluster on the board.
- We can see that individual memory cells can be accessed for data read or write operations in random order, independent of their physical locations in the memory array. Thus, the array organization can be used for Random Access Memory (RAM) structure.

9.5 DYNAMIC RANDOM ACCESS MEMORY (DRAM)

SRAM may be easiest volatile memory to use, but it is not the least expensive in significant densities. Each bit of memory requires between four and six transistors. When millions or billions of bits are required, the complexity of all those transistor becomes substantial.

Dynamic RAM having the following concepts:

⇒ In the DRAM, the binary data is stored charge in capacitor where the presence of absence of stored charge determines the value of the stored bit.

⇒ But data in capacitor cannot be stored for a long time because a capacitor holds an electrical charge for a limited amount of time as the charge gradually drains away.

⇒ Dynamic memory cells require a periodic refreshing of the stored data.

⇒ The use of capacitor as the primary storage device generally enables the DRAM cell to be realized on a much smaller area.

⇒ Access devices or switch are used for read or write operations.

⇒ No static power is dissipated for storing charge on the capacitance.

⇒ Dynamic RAM arrays can achieve higher integration densities than SRAM arrays.

Volatile memory must be both quick to access and not be subject to write cycle limitations– both of which are restrictions of nonvolatile memory technologies. When a capacitor is designed to have its charge quickly and easily manipulated, the down side of rapid discharge emerges.

A very efficient volatile storage element can be created with a capacitor and a single transistor is shown in Fig. 9.3, but that capacitor loses its contents soon after being charged. This is where the term *dynamic* comes from in DRAM—the memory cell is indeed dynamic under steady stage conditions.

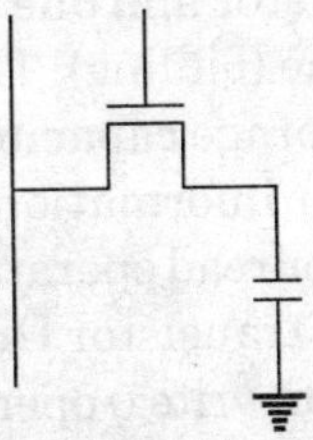

Fig. 9.3 DRAM bit structure

The solution to this problem of solid-state amnesia is to periodically refresh, or update each DRAM bit before it completely losses its charge. As with SRAM, the pass transistor enables both reading and writing the state of the storage element.

However, a single capacitor takes the place of multitransistor latch. This significant reduction in bit complexity enables much higher densities and lower per-bit costs when memory is implemented in DRAM rather than SRAM. This is why main memory in most computers is implemented using DRAM.

The different configuration of DRAM cell is shown in Fig. 9.4.

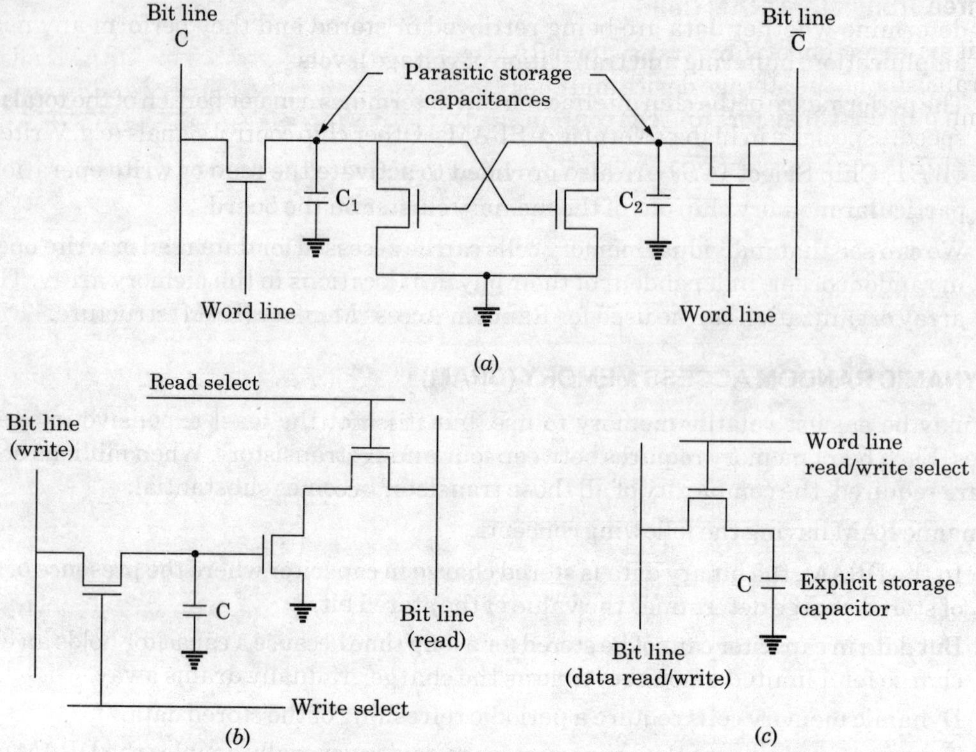

Fig. 9.4 Various configurations of DRAM cell (a) DRAM cell using 4 transistor, (b) DRAM cell using 3 transistor, (c) DRAM cell using 1 transistor

9.5.1 One Transistor DRAM Cell

The circuit arrangement for one transistor DRAM cell is shown in Fig. 9.5. Here we use one transistor and one explicit capacitor. This cell has one read/write control in (word line) and one I/O line (bit line). The word line is controlled by the row address decoder. The capacitor C_1 and C_2 are storage capacitor and parasitic column capacitor respectively. In this type of DRAM cell, the data or information is stored between capacitor C_2 and C_1.

For read operation, an elaborate read refresh circuit has to be built because data read operation in one transistor DRAM cell is destructive.

For write 0 operation, the bit line (*D*) is pulled to logic 0 and the word line is pulled high by the row address decoder. The storage capacitor discharges through the access transistor, resulting in a stored 0 (zero) bit.

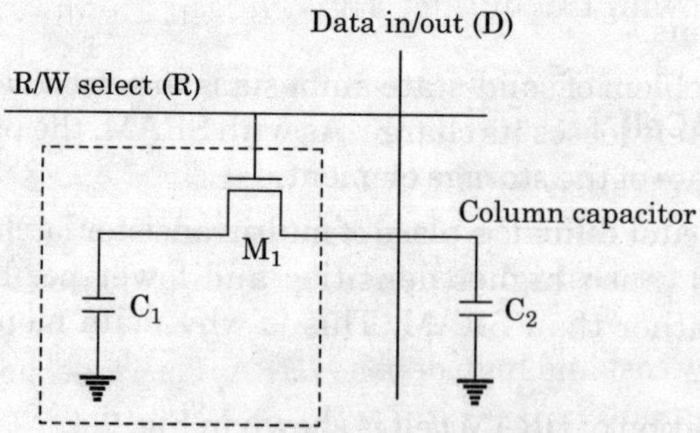

Fig. 9.5 One transistor DRAM cell.

9.5.2 Three Transistor DRAM Cell

The circuit arrangement of three transistor DRAM cell is shown in Fig. 9.6. This DRAM cell utilize a single transistor as the storage device and one transistor each for read and write access switches. The column pull up transistors and the column read/write circuitry is also shown in Fig. 9.6.

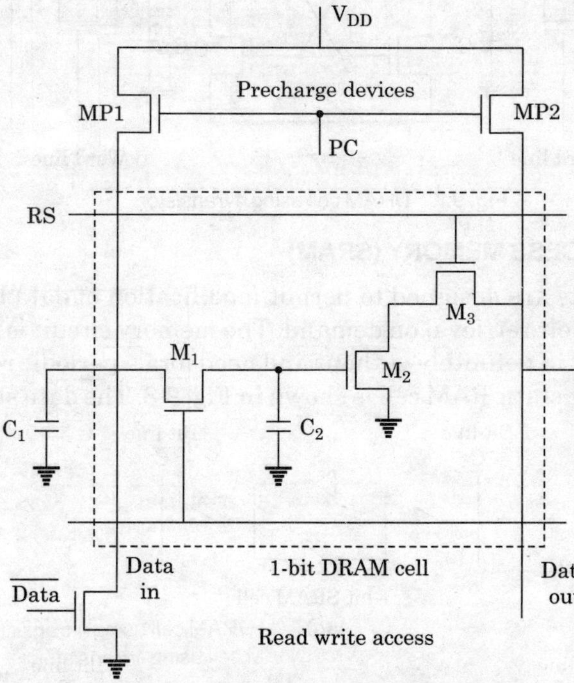

Fig. 9.6 DRAM cell using three transistor.

Here, the binary data is stored in the form of charge in capacitor C_1. The cell has two control and two I/O lines, which separate read and write select lines make relatively fast. The storage transistor m_2 is turned-off or on depending on the charge stored in C_1 and the pass transistors m_3 and m_1 act as access switches for data write and read operations.

9.5.2.1 Properties of 3-transistor DRAM Cell

1. No constraint exist on device ratio. The choice of device size is totally based on performance and reliability considerations.
2. The data value stored in the cell is not affected by read operation.
3. The value stored on storage node X when writing 1 equals $V_{WWL} - V_{Th}$. This threshold loss reduces the current flowing through M_2 during a read operation and increases the read access time.

9.5.3 Four Transistor DRAM Cell

The circuit arrangement of four transistor DRAM cell is shown in Fig 9.7. There are two storage nodes *i.e.*, the parasitic oxide and diffusion capacitances of the nodes. Since no current path is provided to the storage nodes for restoring the charge being last on leakage, the cell must be refreshed periodically.

Due to the advantage of low cost and high density, DRAM is widely used for the main memory in personal and main frame computers.

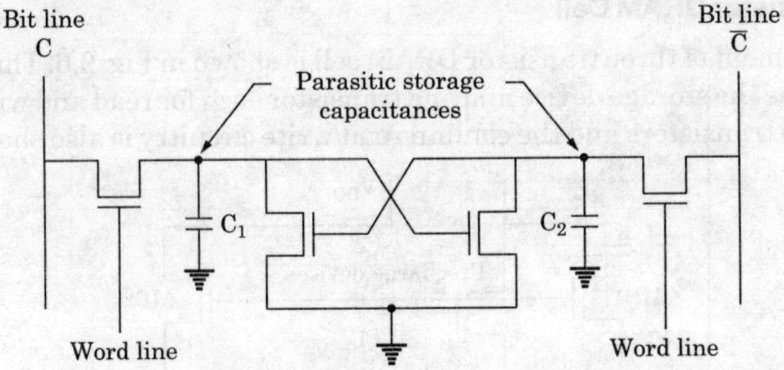

Fig. 9.7 DRAM cell using 4 transistor.

9.6 STATIC RANDOM ACCESS MEMORY (SRAM) (UTU-2012)

Read write memory circuits are designed to permit modification of data bits to be store in the memory array as well as their retrieval on demand. The memory circuit is said to be static if the stored data can be retained in definitely without and need fora a periodic refresh operation. The various configuration of the static RAM cell is shown in Fig. 9.8. The data storage cell or the 1-bit

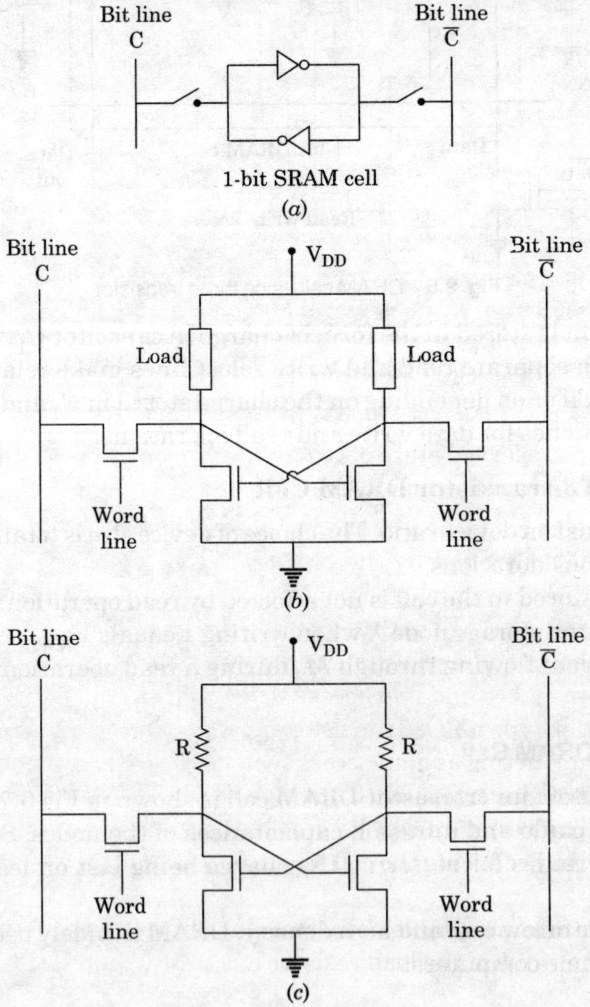

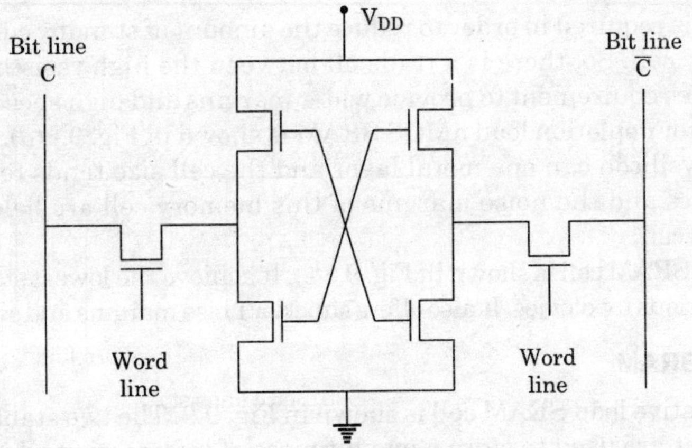

(d)

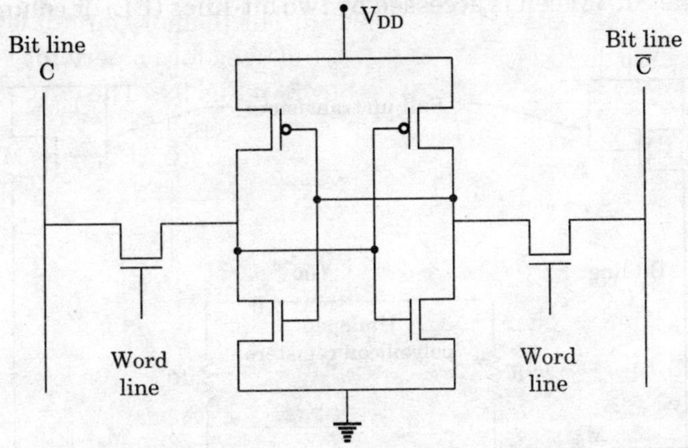

(e)

Fig. 9.8 Various configuration of static RAM cell.

memory cell in the static RAM arrays consists of a simple latch circuit with two stable operating points. Depanding on the reserved state of two inverter latch circuit. that data being held in the memory cell will be incorporated either as a logic 0 or as a logic 1.

- To access the data contained in the memory cell via the bit line at least one switch is needed which is controlled by the corresponding word line (Fig. 9.8(*a*)). Usually, the complementary access switches consisting of nMOS pass transistors are implemented to connect the 1-bit SRAM cell to the complimentary bit lines (low).
- Figure 9.8(*b*) shows a generic structure of the MOS static RAM cell. It is made by two cross-coupled inverters and two access transistors. The load devices may be polysilicon resistors, depletion type nMOS transistor or pMOS transistor depending out type of memory cell. The pass gates acting as data access switches are enhancement type nMOS transistors.
- The resistive load inverters with undoped polysilicon resistors in latch is shown in Figure 9.8(*c*). The resistors can be stacked on top of the cell using double polysilicon technology, thereby reducing the cell size to four transistor as opposed to six transistor cell topology.

In order to attain acceptable noise margins and output pull-up times for the resistance load inverter the value of the load resistor has to be kept relatively low. But a high valued load resistor is required in order to reduce the amount of standby current being drawn by each memory cell. So, there is a trade off between the high resistance required for low power and the requirement to provide wider margins and high speed.

- A six transistor depletion load nMOS SRAM is shown in Fig. 9.8(*d*). This is implemented with one polysilicon can one metal layer and the cell size tends to be small. The static characteristics and the noise margins of this memory cell are below than those of the resistive load cell.
- A full CMOS, SRAM cell is shown in Fig. 9.8(*e*). It achieve the lowest static power dissipation among the various topologies. It also offers superior noise margins and switching speed as well.

9.6.1 Operation of SRAM (UTU-2012)

A four transistor resistive load SRAM cell is shown in Fig. 9.9. The two stable operation points of this basic latch circuit are used to store a one-bit piece of information which is a cross coupled inverter. To perform read and write operations two nMOS pass transistors driven by row select signal (RS) is used. The SRAM cell is accessed by two bit-lines (BL) or columns.

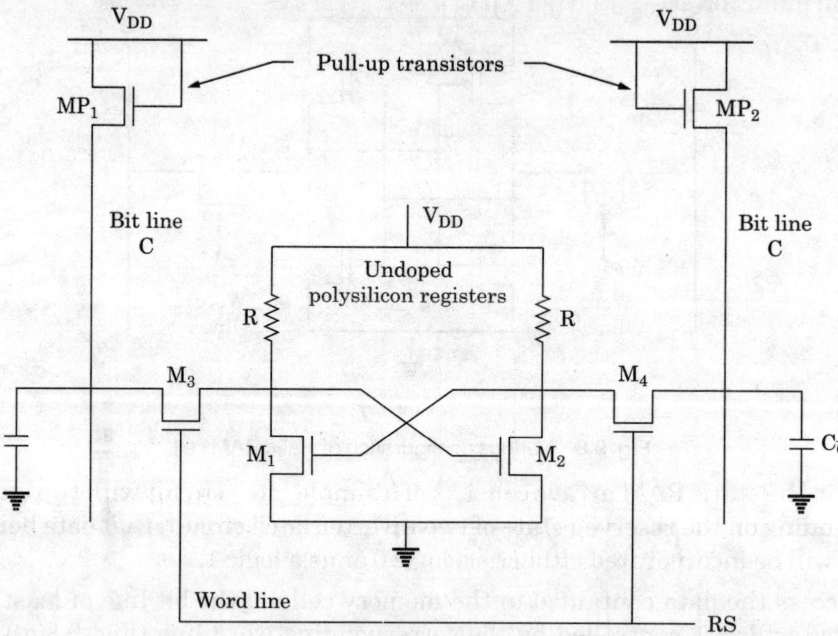

Fig. 9.9 Basic structure of resistive load SRAM cell.

The operations of this cell is discussed below:

⇒ When the word line (RS) is not selected, then the pass transistors M_3 and M_4 are turned off. The simple latch circuit consisting of two cross connected inverters pressures one of its two stable operating points and so data is held.

⇒ If all word lines in the SRAM array are inactive, the relatively large column capacitances are charged up by the column pull-up transistors MP_1 and MP_2. These both transistors operate in saturation.

⇒ When the memory cell is selected by raising its word line voltage to logic 1, the pass transistors m_3 and m_4 are turned on, then the memory cell is selected the following operations are performed.

1. **Read "1" operation:** The voltage of column certains its precharge level while the voltage of column $\bar{C}$ is pulled down by m_2 and m_4. The data read circuitry detects the small voltage difference and amplifies it as a logic "1" data output.
2. **Write "0" operation:** The data write circuitry forces the voltage level of column C to a logic low. The driver transistor m_2 turns-off. The voltage V_2 attains a logic high level and V_1 goes low. So a logic "0" is stored.
3. **Write "1" operation:** The voltage level of column $\bar{C}$ is forced to logic low by the data write circuitry. The driver transistor m_1 turns off through transistor m_4.
4. **Read "0" operation:** The voltage of column $\bar{C}$ retains its precharge level while the volltage of column C is pulled down by m_1 and m_3. The data read circuitry detects the small voltage difference and amplifiers it as a logic "0" data output.

The circuit arrangement of the CMOS cell is shown in Fig. 9.10. Here CMOS inverters are used in place the resistive load nMOS inverters. pMOS column pull up transistors are also used on the complementary bit lines. The circuit operation is same as that of the resistive load SRAM. The pMOS column pull-up devices used in CMOS RAM cell allow the column voltages to reach full V_{DD} level.

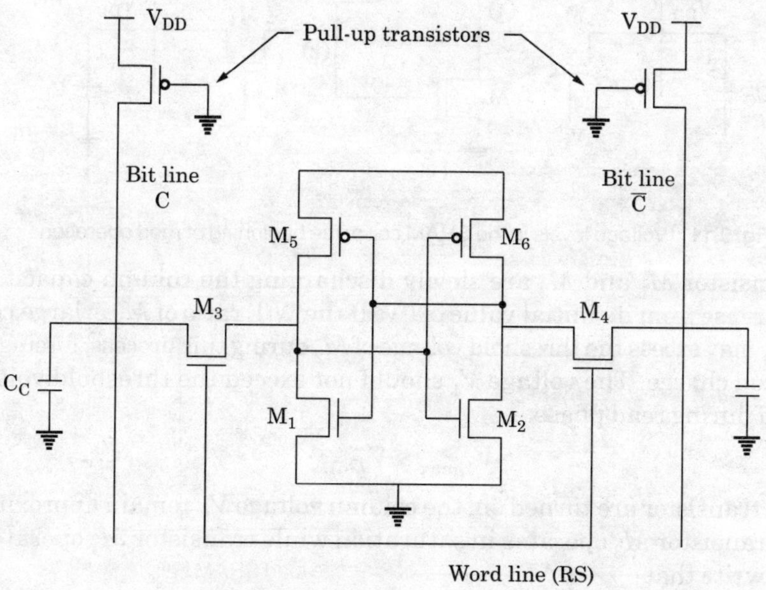

Fig. 9.10 CMOS SRAM cell.

9.6.2 CMOS SRAM Cell Design Strategy

For circuit designing W/L ratios of the transistors is to be determines. The basic requirements that dictate the W/L ratio are:

(a) The data read operation should not destroy the stored information in the SRAM cell.
(b) The cell should allow modification of the stored information during the data write phase.

For Read Operation **(UTU-2012)**

If we assume that a logic 0 is stored in the cell. The voltage levels is shown in Fig. 9.11. Then transistor M_1 and M_6 operate in the linear mode and transistor M_2 and M_5 are turned off. The interval node voltage are $V_1 = 0$ and $V_2 = V_{DD}$ before the cell access transistors M_3 and M_4 are turned on. At this stage the voltage level of column $\bar{C}$ will not show any significant variation since no current will flow through M_4. But M_1 and M_3 will conduct a non-zero current and the voltage level of column C will begin to drop slightly. The column capacitance C_C is very large, so the amount of decrease in the column voltage is limited to a few hundred millivolts during the read phase. The data read circuitry detects this small voltage drop and amplifiers it as a stored "0".

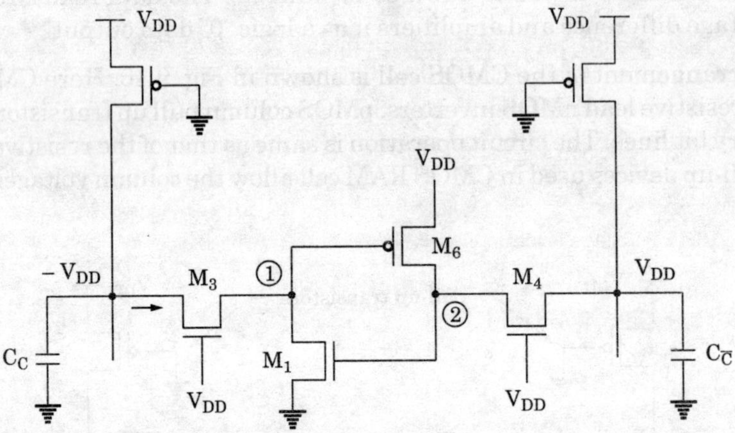

Fig. 9.11 Voltage levels in the SRAM cell at the beginning of read operation.

When the transistor M_1 and M_3 are slowly discharging the column capacitance, the node voltage V_1 will increase from its initial value of 0V. If the W/L ratio of M_3 is large compared to M_1 the node voltage V_1 may excess the threshold voltage of M_2 during this process, forcing an unintended change of the stored charge. The voltage V_1 should not exceed the threshold voltage of M_2 as it remains turned off during read phase.

$$V_{1,\max} \le V_{T,\, M_2} \quad \text{...(9.1)}$$

After the pass transistor are turned on, the column voltage V_C remain approximately equal to V_{DD}. Due to this transistor M_3 operates in saturation while transistor M_1 operates in the linear region. So we can write that

$$\frac{K_{n,M_3}}{2}(V_{DD} - V_1 - V_{T},n)^2 = \frac{K_{n,M_1}}{2}\left[2(V_{DD} - V_{T,n})V_1 - V_1^2\right]$$

$$\Rightarrow \quad \frac{K_{n,M_3}}{K_{n,M_1}} = \frac{(W/L)_{M_3}}{(W/L)_{M_1}} < \left[\frac{2(V_{DD} - 1.5\,V_{T,n})}{(V_{DD} - 2V_{T,n})}\right] \quad \text{...(9.2)}$$

Thus, transistor M_2 will remain in cut-off during the read "0" operation if the above write condition is satisfied.

For Write Operation **(UTU-2012)**

We assume that the logic 1 is stored and a write '0' operation is to be performed. The voltage level at the beginning of the data write operation is shown in Fig. 9.12.

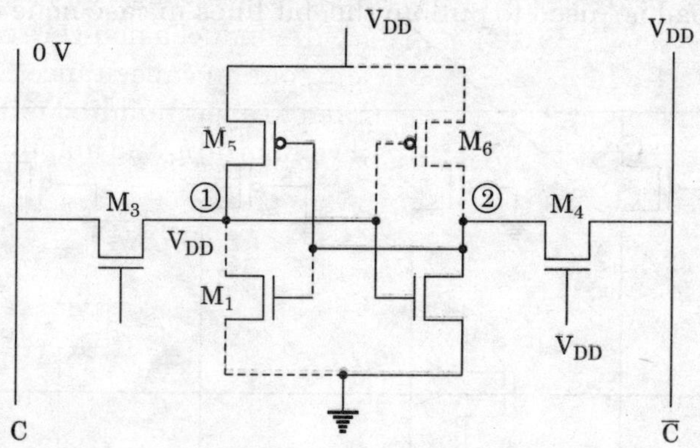

Fig. 9.12 Voltage levels in the SRAM cell at the beginning of the write operation.

The internal node voltage are $V_1 = V_{DD}$ and $V_2 = 0$ before the cell access transistor M_3 and M_4 are turned on. The transistors M_1 and M_6 are turned off and transistor M_2 and M_5 operate in the linear mode.

By the row selection circuitry, the pass transistors M_3 and M_4 are turned on. The node voltage V_2 is expected to remain below the threshold voltage of M_1. Since M_2 and M_4 are designed according to the condition in equation 9.2. Due to this, the voltage level at node (2) would not be sufficient to turn on M_1.

To change the stored information *i.e.*, to force V_2 to V_{DD} and V_1 to 0V the node voltage V_1 must be reduced below the threshold voltage of $M_2 \cdot M_2$ turn off first. When $V_1 = V_{T,n}$ the transistor M_3 operates in the linear region while M_5 operates in saturation. So that

$$\frac{K_{p,M_5}}{2}(0 - V_{DD} - V_{T,p})^2 = \frac{K_{n,M_3}}{2}[2(V_{DD} - V_{T,n})\,V_{T,n} - V_{T,n}^2] \quad ...(9.3)$$

or

$$\frac{K_{p,M_5}}{K_{n,n_3}} = \frac{(W/L)_{M_5}}{(W/L)_{M_3}} < \frac{2\mu_n(V_{DD} - 1.5\,V_{T,n})\,V_{T,n}}{\mu_p\,(V_{DD} + V_{T,p})} \quad ...(9.4)$$

Thus, the transistor M_2 will be forced into cut-off mode during the write "0" operation if condition in equation (9.4) is satisfied. This generates that M_1 subsequently turns on, charging the stored information.

9.7 READ ONLY MEMORY

The read only memory is an non-volatile memory. The contents of a ROM cell are permanently fixed. The ROM cell or array can also be seen as a simple combinational Boolean network which produces a specified output value for each input combination, *i.e.*, for each address. Thus, storing binary information at a particular address location can be achieved by the presence or absence of a data path from the selected row (word line) to the selected column (bit line), which is equivalent to the presence or absence of a device at the particular location.

Here, we discussed about the MOS ROM cell. A MOS ROM cell one can achieve perfect isolation, lower power consumption and higher density at the cost of reduced speed. This cell requires the bit line to be resistively clamped to supply voltage *i.e.*, the default value at the output must equal 1. The absence of a transistor between *WL* and *BL* means that a 1 is stored. The 0-cell is realized by providing an MOS device between bit line and ground. A 4 × 4 MOS ROM array is shown in Fig. 9.13. Applying a high voltage on the word line turns on the device which in turn pulls down the bit line. A pMOS load is used to pull up the bit lines in case none of the attached NMOS devices are enabled.

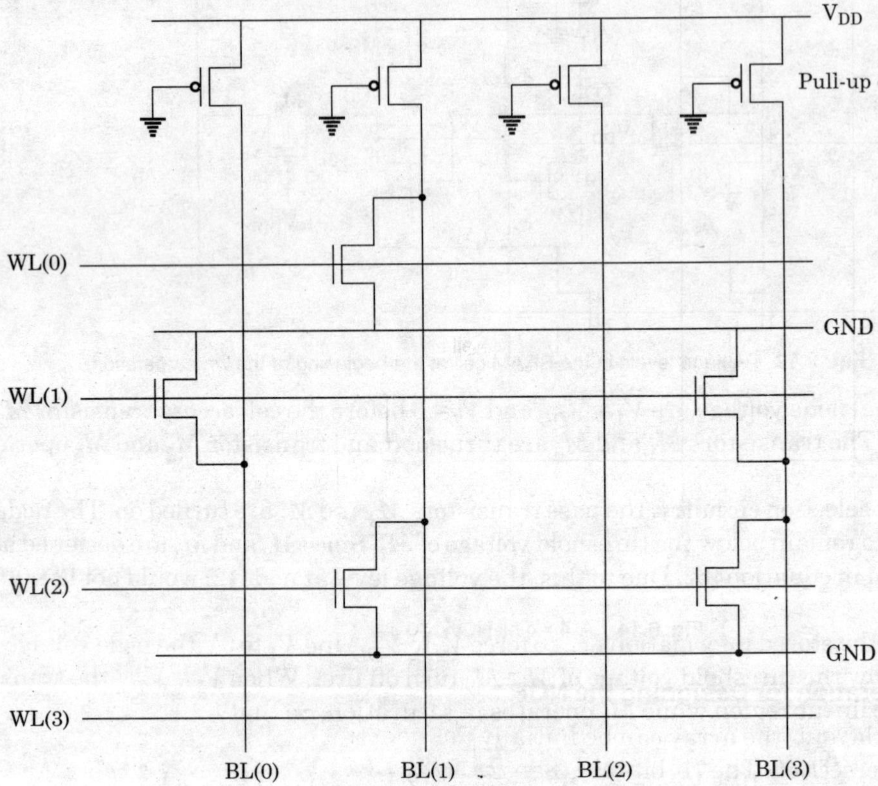

Fig. 9.13 A 4 × 4 MOS ROM cell.

9.7.1 ROM Using MOS NOR

The 4 bits × 4 bit NOR based ROM array is shown in Fig. 9.14. Here each column consists of a pseudo-nMOS NOR gate driven by some of the row signal (word lines). Their are some points may be considered.

1. Only one word line is activated at time by raising its voltage to V_{DD} while all other rows are held at a low voltage level.
2. If an active transistor exists at the crosspoint of a column and the selected row, the column voltage is pulled down to the logic low level by the transistor.
3. If no active transistor exist at the cross point, the column voltage is pulled high by the pMOS load device.
4. A logic "1" bit is stored as the absence of an active transistor and a logic 0 bit is stored as the presence of an active transistor at the cross point.
5. To reduce static power consumption, the pMOS load transistors in the ROM array can also be driven by a periodic precharge signal resulting in a dynamic ROM.

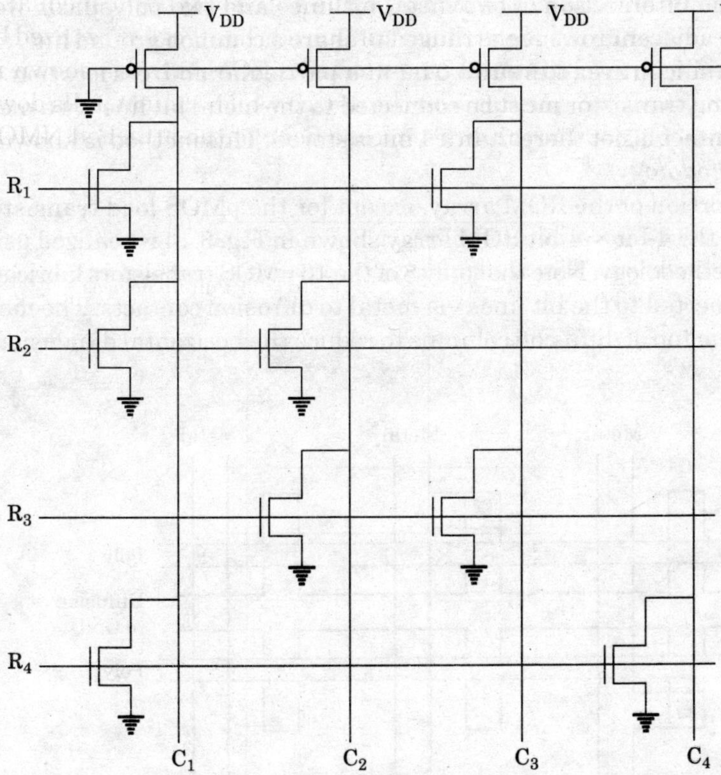

R1	R2	R3	R4	C1	C2	C3	C4
1	0	0	0	0	1	0	1
0	1	0	0	0	0	1	1
0	0	1	0	1	0	0	1
0	0	0	1	0	1	1	0

Fig. 9.14 A 4 × 4 bit NOR based ROM array.

9.7.2 Consideration for Designing the NOR based ROM Array

In actual ROM layout, the array can be initially manufactured with nMOS transistors at every row-column intersection. The "1" bits are than realized by omitting the drain or source connection or the gate electrode of the corresponding nMOS transistors in the final metallization step. The layout of NOR ROM array in shown in Fig. 9.15.

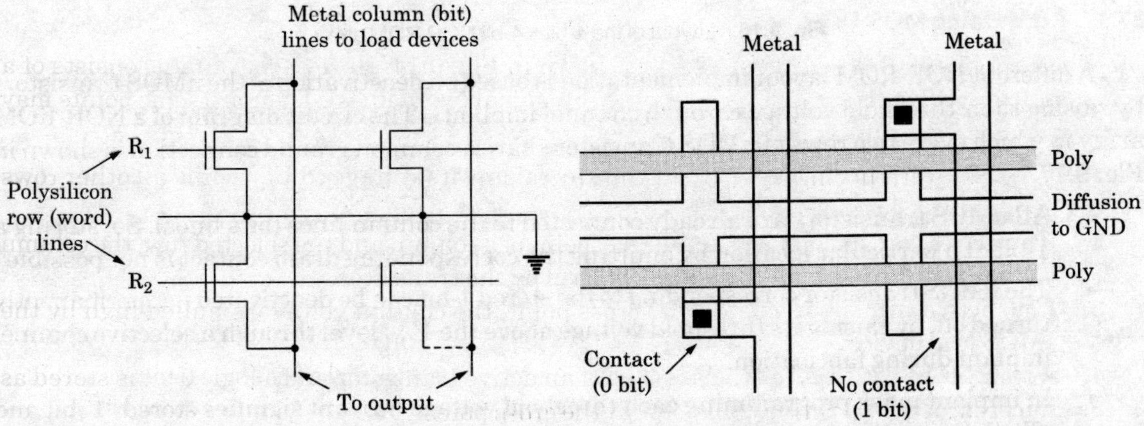

Fig. 9.15 Layout of a NOR ROM array using four transistors.

The nMOS transistor is at the intersection of two metal bit lines and two polysilicon word lines. The transistors in every two adjacent rows are arranged to share a common ground line, also routed in n-type diffusion to save silicon area to store a 0 bit at a particular address location the drain diffusion of the corresponding transistor must be connected to the metal bit line via a metal to diffusion contact. When this contact is not there than a 1 bit is stored. This method is known as *contact* mask *programming methodology.*

Figure 9.16 shows a larger portion of the ROM array, except for the pMOS load transistors connected to metal column. Here, the 4-bit × 4-bit ROM array shown in Fig. 9.14 is realized using the contact-mask programming methodology. Note that only 8 of the 16 nMOS transistors fabricated in this structure are actually connected to the bit lines via metal to diffusion contacts. The metal column lines are laid out directly on top of diffusion columns to reduce the horizontal dimension of the ROM array.

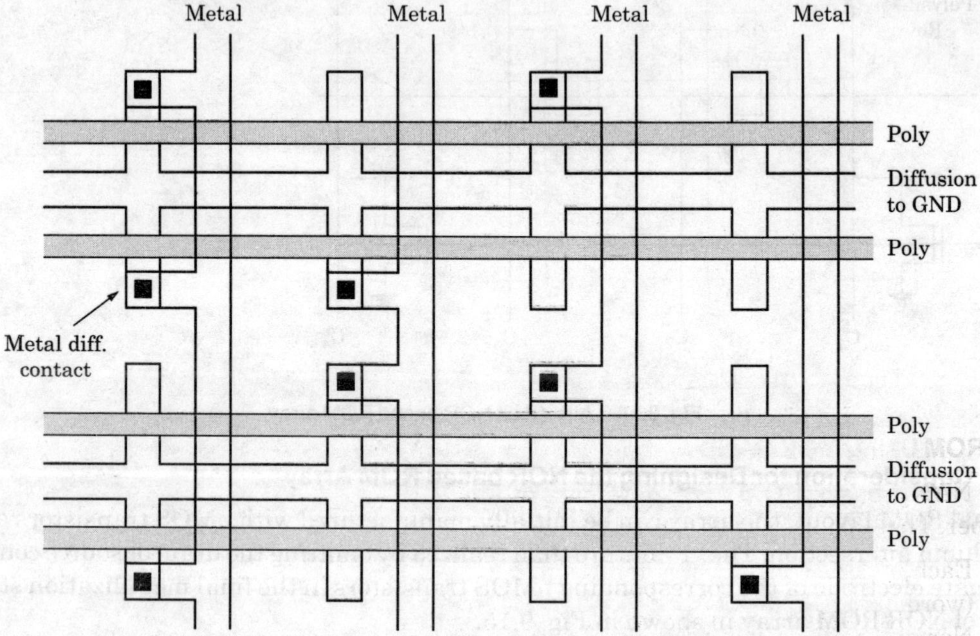

Fig. 9.16 Layout of the 4-bit × 4-bit NOR ROM array.

A different NOR-ROM layout implementation is based on deactivation of the nMOS transistors by raising their threshold voltage through channel implants. The circuit diagram of a NOR-ROM array in which every two rows of nMOS transistor share a common ground connection is shown in Fig. 9.17.

- All nMOS transistors are already connected to the column lines (bits lines). So, storing a 1-bit at a particular location by omitting the corresponding drain contact is not possible.
- The nMOS transistor corresponding to the stored 1-bit can be deactivated *i.e.*, permanently turned off, by raising its threshold voltage above the V_{OH} level through a selective channel implant during fabrication.
- In implant mask programming each threshold voltage implant signifies stored "1" bit and all other transistor correspond to stored '0' bits.

- Since each diffusion to metal contact in this structure is shared by two adjacent transistors the implant mask ROM layout can yield a higher core density *i.e.*, a smaller silicon area per stored bit, compared to the contact mask ROM layout.

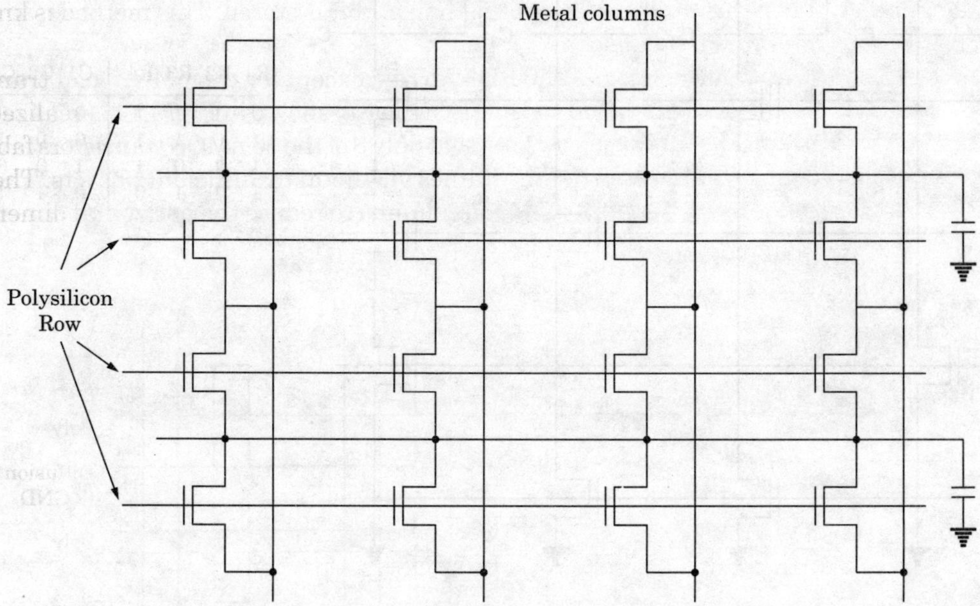

Fig. 9.17 The nMOS transistors in the implant-mask programmable NOR-ROM array.

9.7.3 ROM Using MOS NAND

A 4 × 4 NAND based ROM array is shown in Fig. 9.18. There are various important points to remember. These points are discussed below:

- Each bit line consists of a depletion load NAND gate driven by some of the row signals (word line).
- Normally, all work lines are held at the logic high voltage level except for the selected line which is pulled down to logic low level.
- If a transistor exist at the cross point of a column and the selected row, that transistor is turned off and the column voltage is pulled high by load device.
- If no transistor exist at that particular cross point, the column voltage is pulled low by the other nMOS transistors in the multi-inpoint NAND structure.
- A logic '1' bit is stored by the presence of a transistor that can be deactivated while a logic '0' bit is stored by a shorted or normally on transistor at the cross point.
- A NAND based ROM array can be fabricated initially with a transistor connect present at every row- column intersection.
- A '0' bit is then stored by lowering the threshold voltage of the corresponding nMOS transistor at the cross point through a channel implant. In this case, the transistor remains on regardless of the gate voltage.
- Because of the above mentioned reason the depletion-type nMOS load transistor are used instead of pMOS loads.

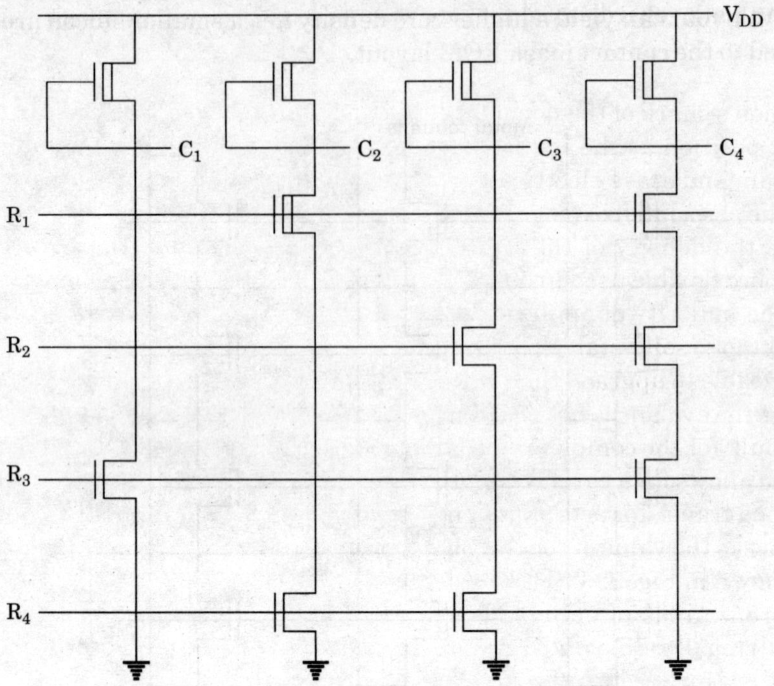

R1	R2	R3	R4	C1	C2	C3	C4
0	1	1	1	0	1	0	1
1	0	1	1	0	0	1	1
1	1	0	1	1	0	0	1
1	1	1	0	0	1	1	0

Fig. 9.18 A 4 bit × 4-bit NAND based ROM array.

The implant mask implementation of a sample 4 bit × 4 bit NAND ROM array layout is shown in Fig. 9.19. In the layout the vertical columns of *n*-type diffusion intersect at regular intervals with horizontal rows of polysilicon which results in an nMOS transistor at each intersection point.

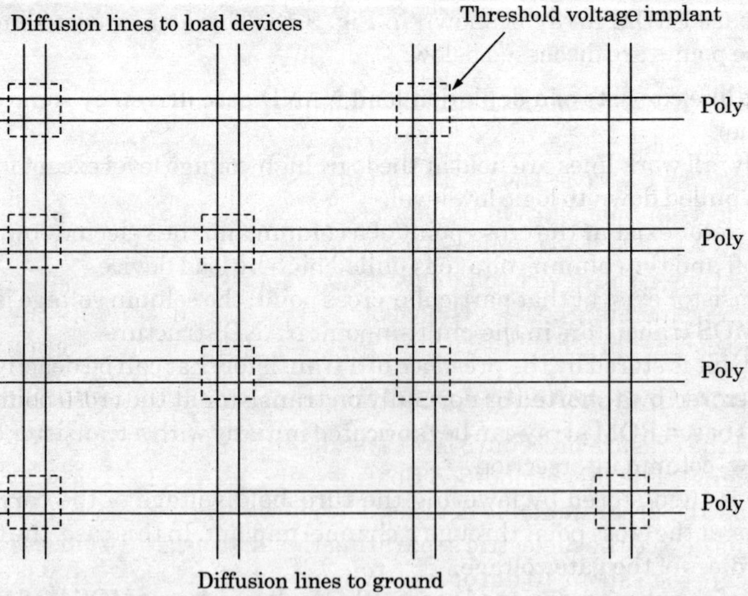

Fig. 9.19 Implant-Mask layout of the NAND ROM array. The threshold voltages of '0'-bit transistors are lowered below 0V through implant.

The transistors with threshold voltage implants operate as normally on depletion device thus providing a continuous current path regardless of the gate voltage level.

9.8 FLASH MEMORY

Flash memory captured the lion's shark of the nonvolatile memory market from EPROM's in the 1990s and holds a dominant position as the industry leader to this day. Flash is an enhanced EPROM that can both program and erase electrically without time consuming exposure to UV light and it has no need for the associated expensive ceramic and quartz packaging.

Flash memory combines the density of the EPROM with the versatility of the E^2PROM structure. Flash memory is more flexible used in terms of electronic ensures more than makes up a small cost differential in the majority of applications. It is found in everything from cellular phones to automobiles to desktop to solid-state disk drives. It has enabled a whole class of hexible computing platforms, that are able to upgrade their software easily.

Flask E^2PROM devices use the avalanche hot-electron injection approach to program the devices. The erasure is performed in bulk for the complete chip or for a subsection of memory. Erasing the complete memory core at once allows for a careful monitoring of the device characteristics during erasure and the unprogrammed transistor acts as enhancement devices. The control hardware on the memory chip regularly checks the value of the threshold during erasure.

The ETOX flash cell is shown in Fig. 9.20. This flash memory has a very thin tunneling gate oxide of about 10 nm. Different areas of gate oxide are used for programming and erasure. Programming is done by applying 0 (zero) volt or grounded on drain and a high voltage of about 12 V on gate terminals, while erasure occurs with the gate grounded and source of 12 V.

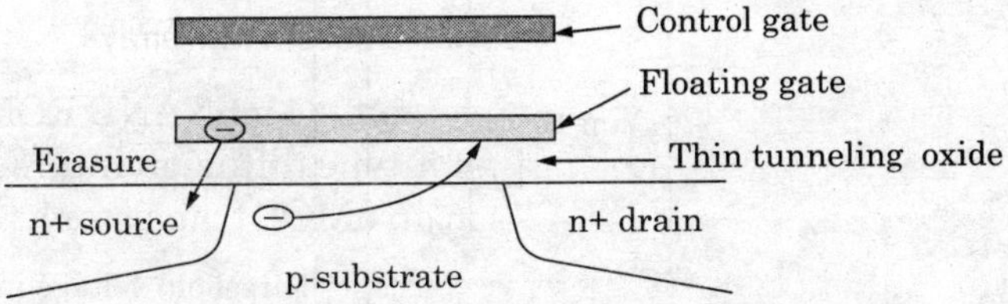

Fig. 9.20 ETOX device as used in flash memory.

9.9 ERASABLE PROGRAMMABLE READ ONLY MEMORY (EPROM)

When strong ultraviolet light falls on an EPROM, it is erased. A transparent window in provided in the package for the UV rays. The UV radiation rendes the oxide slightly conductive by the direct generation of electron-hole pairs in the material. The erase process is slow and depends on the intensity of the UV source. The number of erase/program cycles is generally limited to a maximum of one thousand times. The device reliability is also questionable. The device thresholds may vary with repeated programming cycles. Thus most of the EPROM chips contain on chip circuitry to control the value of the threshold to within a specified range during programming.

9.10 ELECTRICALLY ERASABLE PROGRAMMABLE READ ONLY MEMORY (E^2PROM)

In an E^2PROM the erasure procedure is off system. In E^2PROM the erasure is done by providing an electrical signal. For this a modified floating-gate device FLOTOX (floating gate tunneling oxide) is used as the programmable devices. The cross section of a FLOATOX device is shown in Fig. 9.21. In this device the dielectric separating the floating gate from the channel and drain is reduced in thickness to about 10 nm or less. When a voltage of about 10 V is applied over the thin insulator, electrons travel to and from the floating gate through a mechanism known as Fowler-Nordheim tunneling.

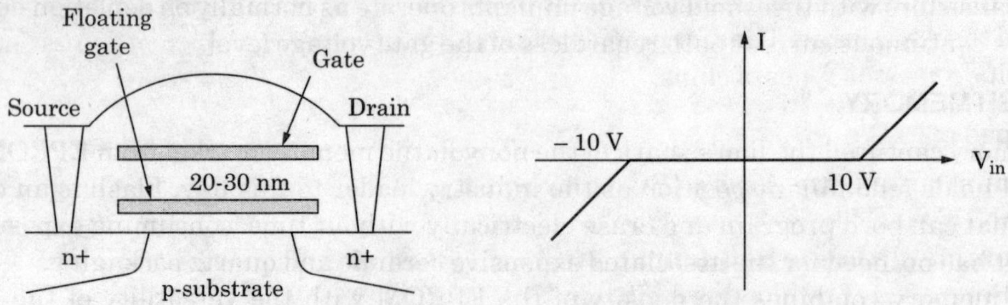

Fig. 9.21 FLOTOX Transistor.

In this programming approach erasing is simply achieved by reversing the voltage applied during the writing process. Injecting electrons on to the floating gate raises the threshold while the reverse operation lowers the threshold.

But this creates a threshold control problem. An extra transistor connected series with the floating gate transistor is added to the E^2PROM cell to solve this problem. This extra transistor acts as the access device during the read operation while the FLOTAX transistor forms the storage function as shown in Fig. 9.22.

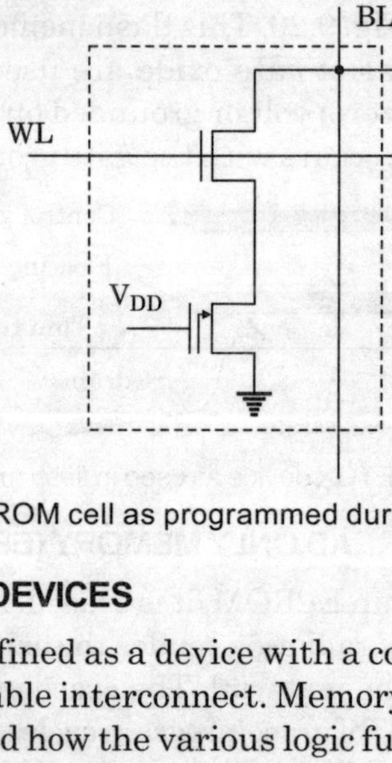

Fig. 9.22 E^2PROM cell as programmed during read operation.

9.11 PROGRAMMABLE LOGIC DEVICES

- Programmable logic is defined as a device with a configurable logic and flip-flops linked together with programmable interconnect. Memory cells control and define the function that the logic performs and how the various logic functions are interconnected.
- In the programmable design mode, IC's known as programmable logic devices (PLD's) are used in the design. There are several types of PLD's each with their own pattern of gates and interconnection paths pre-fabricated on the chip.
- These ICs programmed by the user with special programming equipment (known as PLD programmers). That is, they are field programmable.
- Some PLDs allow erasing the program to reprogram then and some do not. In the early development stages of the digital system, field-programmable devices are typically used to allow the flexibility of design alterations.
- Once the design is tested and the circuits performance is deemed acceptable, mask programmable logic IC's can be used to implement the system.

- A PLD or programmable logic device is composed of two types of gate arrays the OR array and the AND array. Thus for providing sum of products algorithms representations PLD's include various types of chips.

9.11.1 Types of PLD's

1. **Programmable Gate Array (PGA):** Provides a single level of logic, e.g., array of multi-input AND gates.
2. **Programmable Logic Sequencer (PLS):** Provides two levels o logic, usually an AND/OR array. Programmer by blowing fusible links at specified points.
3. **Programmable Logic Array (PLA):** PLA is the most flexible device. It is having both the AND and OR gate arrays programmable.

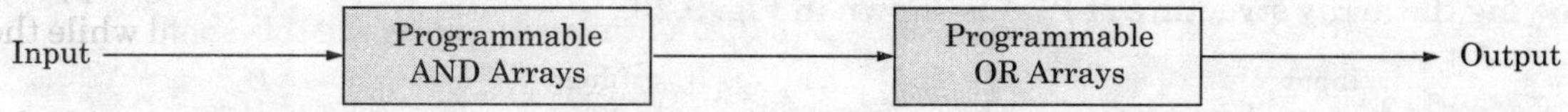

4. **Programmable Array Logic (PAL):** PLA consists programmable AND array with a fixed OR array and bi-directional input/outputs pins. In a PAL device only the AND arrays are programmable.

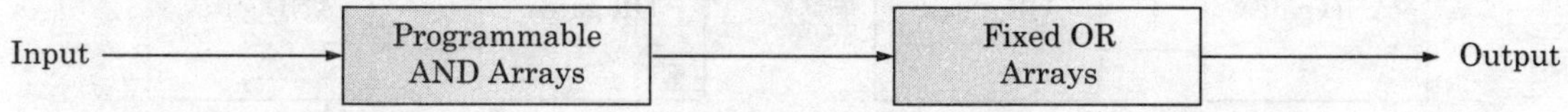

5. **Programmable Read Only Memory (PROM):** PROM consists of fixed AND array and programmable OR array with programmable interconnect.

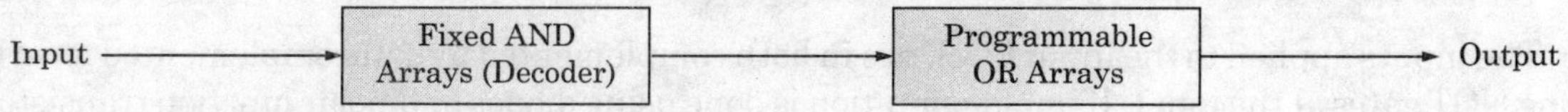

- In a PLD, the software is used to decide how to connect the groups of transistors or the groups of gates together in a particular way.
- PLD's are programmed in one of three ways,
  - (i) Connections between gates are linked but physically breaking some lines.
  - (ii) Alternatively, connections are made by melting some lines.
  - (iii) The third method connects the gate by using the software-controlled switches within the PLD's.

9.11.2 Advantages of PLD's

1. The key value proposition for these products, include faster time to mark due to a shorter design cycle.
2. PLD's are cheaper for low volume production.
3. PLD's can be reprogrammed even after a piece of equipment is shipped a customer. A number of equipment manufacturers now about the ability add new features or upgrade products that already are in the field. Due to this, they simply upload a new programming fiel to the PLD, via the Inter creating new hardware logic in the system.
4. Many types of PLDs available according applications.
5. PLDs ofer much more flexibility during the design cycle because design iterations simply a matter of changing the programming file and the results of design chip can be seen immediately in working parts.

9.11.3 Disadvantages of PLD's

1. Higher manufacturing cost per unit.
2. Poor performance than ASIC.
3. Due to the need of programmable, PLD's must include extra circuitry in add to what is necessary to perform a specific function. This results in increased die sizes.

9.12 PROGRAMMABLE LOGIC ARRAYS (PLA's) (UTU-2012)

A programmable logic array (PLA) is a combinational module that provides a NOT-AND-OR or NOT-OR-AND implementation of switching modules. The modules are usually Instated and the enable input E is used to select the module for designing purpose. For $E = 1$, a high or a low binary value can be obtained and for $E = 0$, the outputs are in high impedance state. The block diagram showing the array structure of PLA is shown in Fig. 9.23.

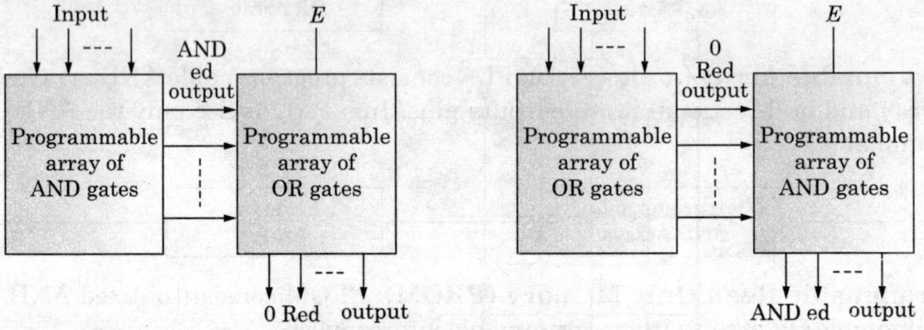

Fig. 9.23 (*a*) PLA block where output is in SOP form, (*b*) PLA block showing outputs in POS form.

The inputs applied to the input block are in both complemented and uncomplemented form by using NOT gates at the input. If implementation is done using diodes or bipolar junction transistors in the bit position then AND-OR array is used and if MOS implementation is done then OR-AND array is used. The AND gates are arranged in a regular structure known as AND array and the OR gates are also arranged in a regular structure known as OR array. This is shown in Fig. 9.24.

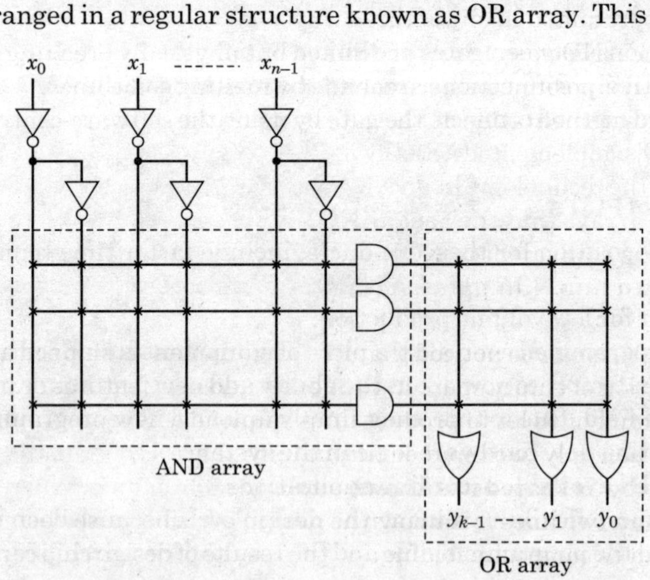

Fig. 9.24 Regular arrangement of AND array and OR array.

The crosspoints shown in the figure are the points that are to be included in the product term in the AND array. Each crosspoint represents the input (in complemented or uncomplemented form). If a certain input is used in the product term then a dot is shown. Same is the case with OR array. If a dot is placed then the resultant ANDed terms are ORed. For example if the following function is to be implemented on PLA

$$y_1 = x_0 + x_1$$

$$y_2 = x_0x'_1 + x'_0\,x_2 + x_1$$

From the function it is clear that 5 × 6 AND matrix and 5 × 2 OR array is required. The size is usually determined by the number of inputs and the number of product and ORed terms as shown in Fig. 9.25.

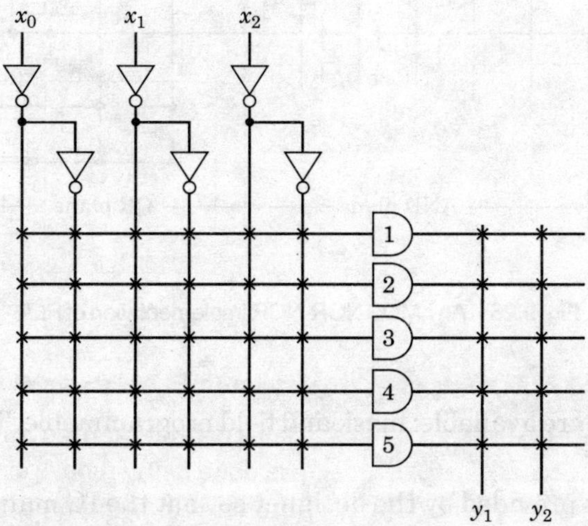

Fig. 9.25 Ok AND and OR matrix to implement y_1 and y_2.

The output of the AND gate 1 is x_0 and that of AND gate 2 is x_1. In the OR array, y_1 is given by the sum of all the dot positions where each dot correspond to the ANDed result. The following examples show the implementation of PLA.

When the device is implemented actually in PLA, the crosspoints are either diodes or transistors and this depends on the technology. In an NMOS implementation, the transistors are so connected in the AND and OR arrays that they act as NOT gates. So we have NOR-NOR implementation. The AND-OR array is equivalent to NOR-NOR array. We have already seen how AND and OR gates can be converted into NOR gates. A NMOS NOR-NOR implementation of PLA is shown in Fig. 9.26.

From the Fig. 9.26, it can be noticed that PLA utilizes less space as compared to ROM. A more better way of utilization of empty space in PLA can be done by allowing two or more input lines to share the same column, or to let to product lines share the same row or by both. This sharing is known as folding. When only two input or output lines share a column, the folding is called simple column fold and if a row is shared it is known as simple row folding. When single column or a row is shared by more than two lines it is known as multiple folding and when both columns and rows are folded it is known as composite folding.

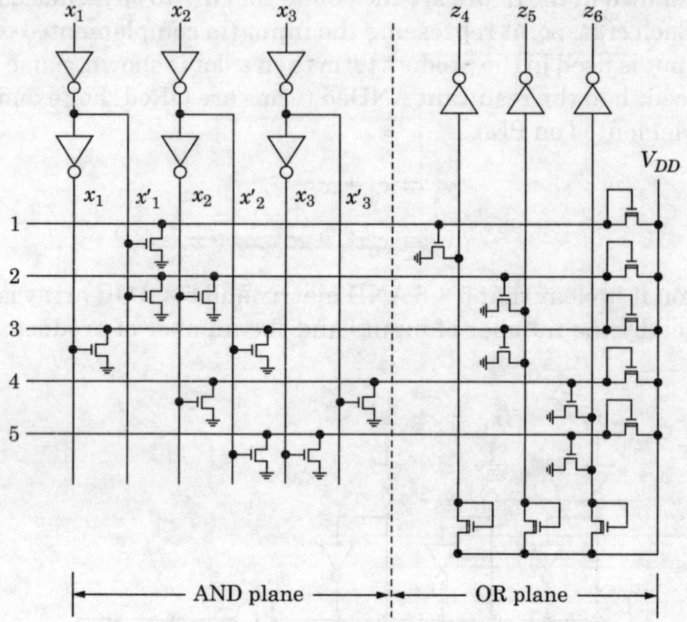

Fig. 9.26 An nMOS NOR-NOR implementation of PLA

9.12.1 Types of PLAs

- The types of PLAs are available: mask and field programmable. They are identical to ROM types.
- A PLA program is provided by the designer so that the IC manufacturer can fabricate a mask-programmed PLA, which can never be altered once programmed. Special devices are used to program the field-programmable PLA (FPLA). For this type of PLA, switching devices are fabricated at each row-column intersection and the connections are established by blowing the fusible links, as in ROM programming.
- Fig. 9.27 shows the general model for a $n \times k \times m$ PLA (e.g., with n inputs, m outputs and k product terms). The complexity of circuit implementable by a PLA device thus is constrained by the values of n, m, and k, which are in turn functions of the technology. When the compleity of the circuit exceeds these limits, multiple module implementations (similar to the ones using ROMs) are used.
- Fig. 9.27 shows the implementation of the serial two's complement controller using a PLA.
- Fig. 9.28 shows the details of the Signetics PLC105, a field-programmable logic sequencer. This is essentially a PLA along with some flip-flops on the chip to enable the implementation of Mealy-type sequential circuits.
- This CMOS chip contains an AND-OR gate array with user programmable connections that control the inputs of the on chip state and output registers.
- The state register is made up of six SR flip-flops, and the output register is made up of eight SR flip-flops.

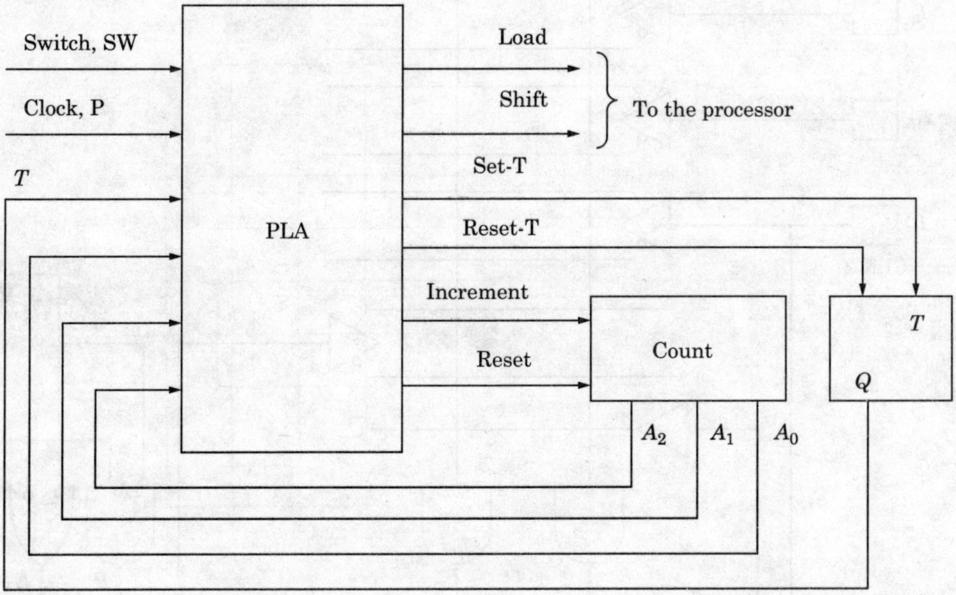

(a)

Stat	SW	$T(t)$	A_2	A_1	A_0	$T(t+1)$	Load	Shift	Reset-T	Set-T	Rest count	Increment count
S_1	×	0	×	×	×	0	0	0	1	0	0	0
	1	0	×	×	×	1	1	0	0	1	1	0
S_1	×	1	0	0	0	0	1	0	0	1	0	1
	×	1	0	0	0	1	1	0	0	1	0	1
	×	1	0	1	1	0	1	0	0	1	0	1
	×	1	0	1	1	1	1	0	0	1	0	1
	×	1	1	0	0	0	1	0	0	1	0	1
	×	1	1	0	0	1	0	0	1	0	0	1

(b)

Fig. 9.27 PLA implementation of the serial twos complementer controller
(a) circuit (b) operation

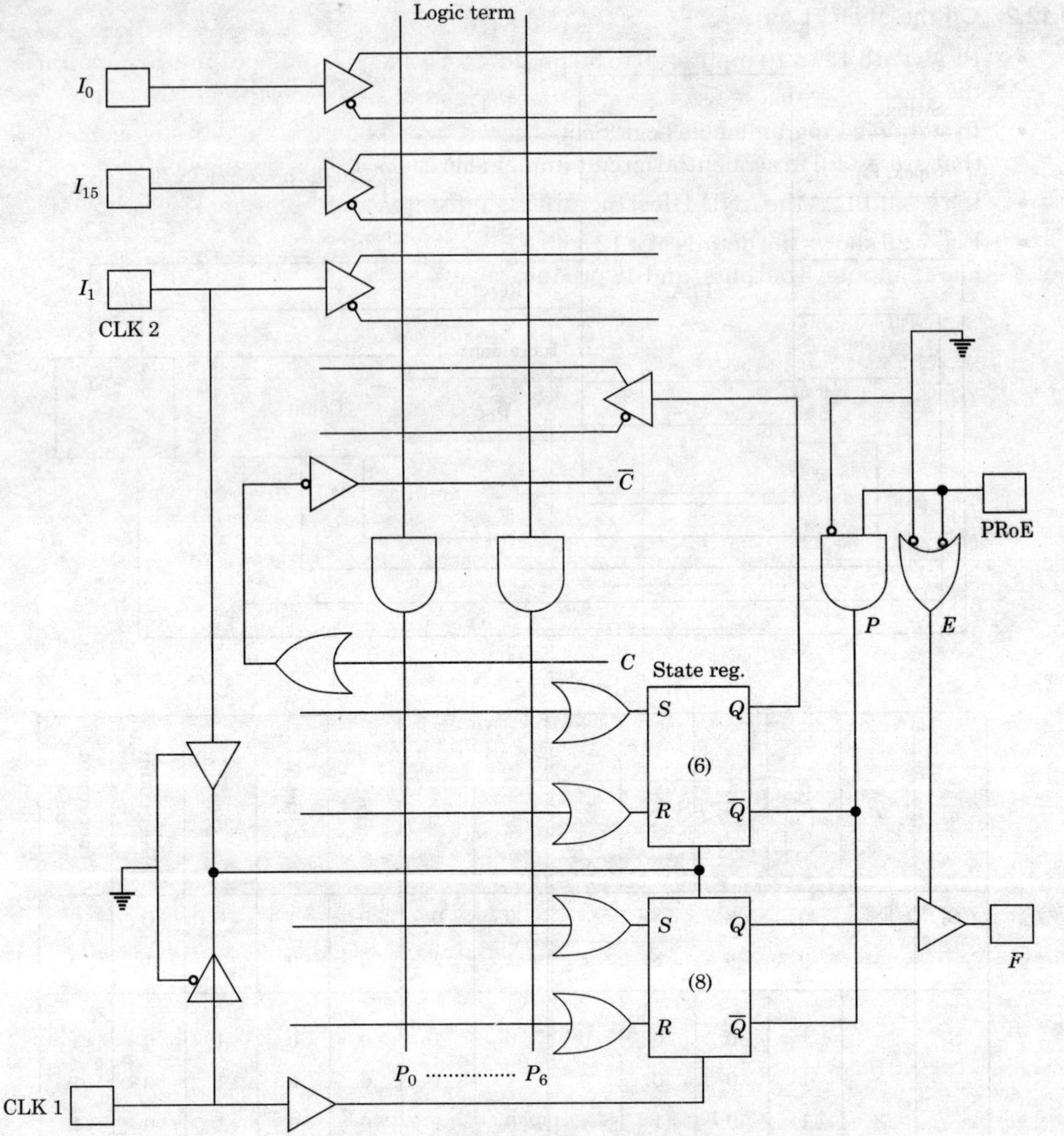

Fig. 9.28 Field programmable logic array

- These flip-flops are positive edge-triggered with a direct set (preset) input. During the power-on, all the flip-flops are set to 1.
- The AND array combines the 16 eternal inputs ($I_0 - I_{15}$) and the seven internal inputs ($P_0 - P_6$), which are fed back from the state registers to form up to 48 AND terms. These AND terms are then merged by the OR array to realize the output and the next-state circuits.
- An internal input variable C is available that enables the generation of either true or complement AND terms, if needed. As a user programmable option, the PRESET input can also be used as an Output Enable.

9.12.2 Off-the-shelf PLAs

- PLAs with 12 to 20 inputs, 20 to 50 product terms, and 6 to 12 outputs are available off-the-shelf.
- In addition, Programmable Logic Sequencers (PLSs) that are PLAs with storage capabilities that are useful in sequential circuit implementations are also available.
- In recent PLAs the AND-OR structure is augmented with additional logic capabilities.
- Fig. 7.29 shows the details of a 12 × 48 × 8 bipolar FPLA (Signetics PLS161). This device has 12 inputs, 8 outputs, and 48 product terms.

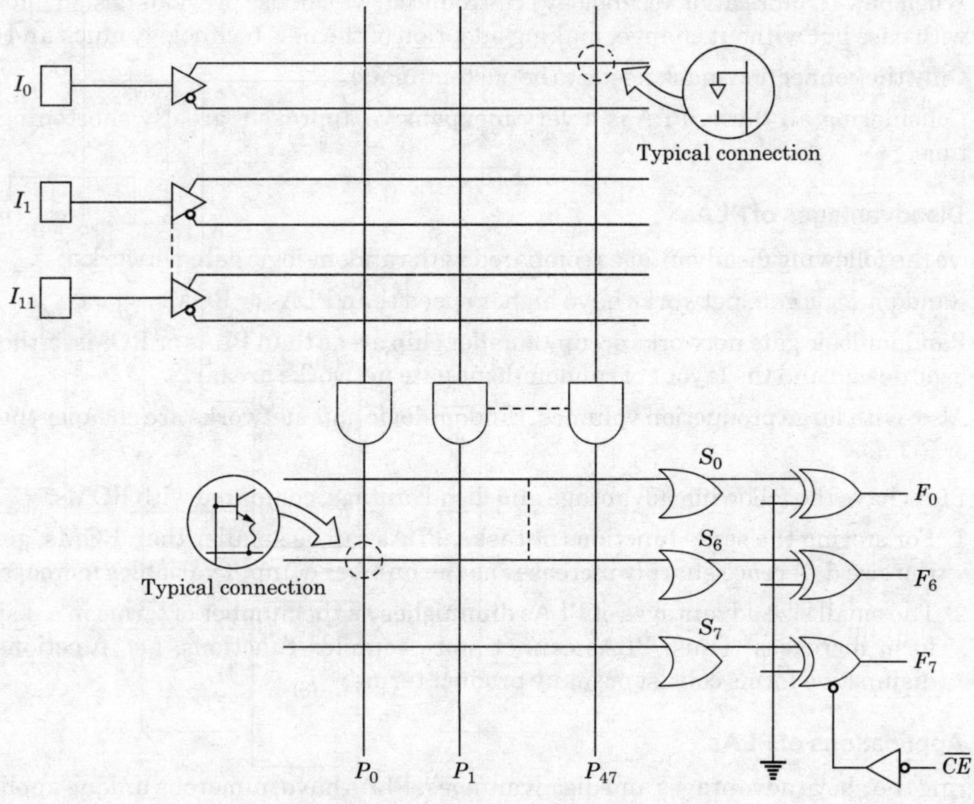

Fig. 9.29 Mask programmable logic array

- The outputs can be programmed as active-high or active-low. There is an enable input (CE).
- In the unprogrammed state, this device is characterized by the following :

1. All Ni-Cr links intact.

- The typical propagation delay of this device is 35 ns.
- Several CAD tools are available to enable designing with PLAs. These tools generate PLA programs, optimize PLA layout for custom fabrication, and simulate PLA designs.

9.12.3 Advantages of PLAs

PLAs, like ROMs, which are more general, have the following advantages over random-logic gate networks, where random-logic gate networks are those that are compactly laid out on an 1C chip:

1. There is no need for the time-consuming logic design of random-logic gate networks and even.
2. Design checking is easy, and design change is also easy.
3. Layout is far simpler than that for random-logic gate networks, and thus is far less time-consuming.
4. When new IC fabrication technology is introduced, we can use previous design information with ease but without change, making adoption of the new technology quick and easy.
5. Only the connection mask needs to be custom-made.
6. Considering all these, PLA is a very inexpensive approach, greatly shortening design time.

9.12.4 Disadvantages of PLAs

PLAs have the following disadvantages compared with random-logic gate networks:

1. Random-logic gate networks have higher speed than PLAs or ROMs.
2. Random-logic gate networks occupy smaller chip areas than PLAs or ROMs, although the logic design and the layout of random-logic gate networks are far.
3. Also, with large production volumes, random-logic gate networks are cheaper than PLAs or ROMs.

- PLAs have the following advantage and disadvantage, compared with ROMs:
    1. For storing the same functions or tasks, PLAs can be smaller than ROMs; generally, the size difference sharply increases as the number of input variables increases.
    2. The small size advantages of PLAs diminishes as the number of terms in a disjunctive form increases. Thus, PLAs cannot store complex functions, *i.e.*, functions whose disjunctive forms consist or many product terms.

9.12.5 Applications of PLAs

Considering the above advantages and disadvantages, PLAs have numerous unique applications.

1. A microprocessor chip uses many PLAs because of easy of design change and check.
2. In particular, PLAs are used in its-control logic, which is complex and requires many changes, even during its design. Also, PLAs are used for code conversions, microprogram address conversions, decision tables, bus priority resolvers, and memory overlay.
3. When a new product is to be manufactured in small volume or test-marketed, PLAs is a choice. When the new product is well received in the market and does not need further changes, PLAs can be replaced by random-logic gate networks for low cost for high volume production and high speed.
4. Also, a full-custom design approach is very time-consuming, probably taking months or years, but if PLAs are used in the control logic, a number of different custom-design chips with high performance can be made quickly by changing only one connection mask for the PLAs, although these chips cannot have drastically different performance and functions.

9.13 PROGRAMMABLE ARRAY LOGIC (PAL)

1. A PAL is a programmable logic device that consists of a programmable AND matrix whose outputs drive fixed OR gates. This was one of the earliest forms of programmable logic.
2. PALs can typically implement small functions easily (up to a hundred gates) and run very fast, but they are inefficient for large functions.
3. The PAL represents another alternative to PROMS.
4. PAL is in the form of standard two level logic circuit for evaluating an SOP logical expression.
5. The input AND gate connections are programmable in PAL, whereas the output OR gate connections arc programmable in PROM This difference usually in reduction of gate count in PAL compared to PROM performing same function.
6. The advantage of PALs is the elimination of fuses in the OR array and special electronic circuits to blow these fuses.
7. Since these special electronic circuits and programmable OR array occupy a very large area, the area is significantly reduced in PAL.
8. Since single-output, two-level networks (i.e., many AND gates in the first level and one OR gate as the network output) are needed most often in design practice, many single-output two-level networks, which are mutually unconnected, are placed in some PAL packages.
9. An elementary non-programmed PAL structure is shown below. The device has two input lines, two programmable 4-input AND gates and one fixed 2-input OR gate:

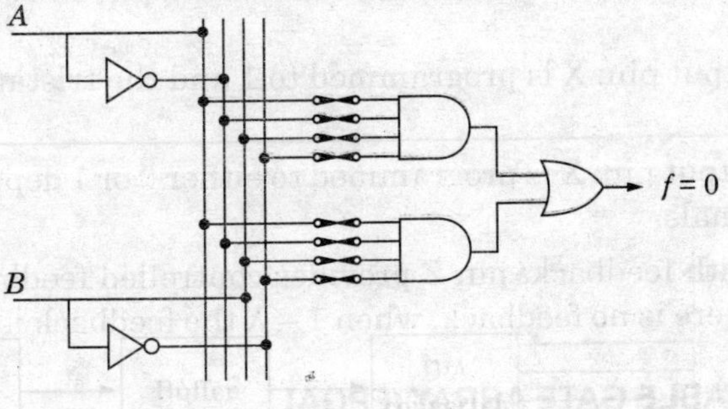

Fig. 9.30

A programmed PAL that implements the XOR function $f = AB + AB$ is shown below:

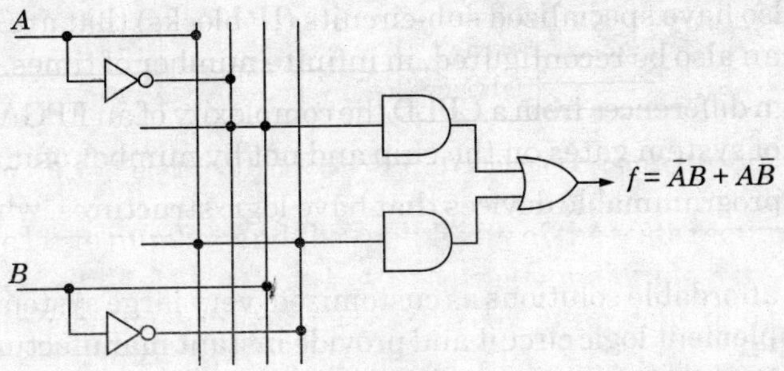

Fig. 9.31

- Additional flexibility can be provided by placing tri-state buffers between the outputs of the OR gates and the input output pins.

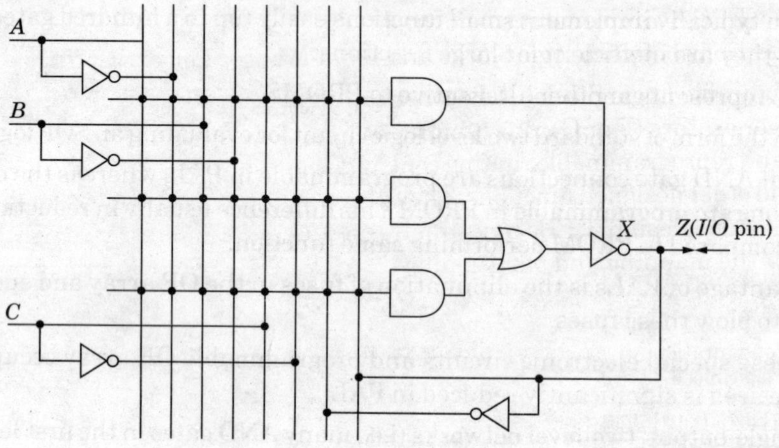

Fig. 9.32

- **The pin Z can be used in four modes:**
  1. Dedicated input pin: X is programmed to 0 and the tri-state buffer is permanently disabled.
  2. Dedicated output pin: X is programmed to 1 and the tristate buffer is permanently enabled.
  3. Controlled output pin: X is programmed to either 0 or 1 depending upon the present state of its signals.
  4. Output pin with feedback: pin *Z* provides controlled feedback to the AND matrix. When 0 = *X* there is no feedback, when 1 = *X* the feedback is enable.

9.14 FIELD PROGRAMMABLE GATE ARRAY (FPGA)

- A FPGA is larger than CPLD. An FPGA has a different architecture (design) than CPLD, which enables it to do more complicated functions. FPGAs often contain memory in which data can be stored.
- Some FPGAs also have specialized sub-circuits (IP blocks) that are optimized for specific functions and can also be reconfigured, in infinite number of times.
- Because of design differences from a CPLD the complexity of an FPGA is typically measured by the number of system gates on the chip and not by number of macro cells.
- FPGA are field programmable devices that have logic structures, which can be configured by enc user.
- FPGAs provide affordable solutions as customized, very large system integrated chip with the ability to implement logic circuit and provide instant manufacturing turnaround with very low cost prototypes.

- FPGAs can be programmed in minutes to offer the same functionality and performance as ASIC/ASSPs thus providing a significant time to market advantage.
- The basic FPGA architecture shown in Fig. 9.33 is built a logic blocks and the interconnections between logic blocks and input & output blocks (IOBs)
- The Configurable Logic Block (CLB) constitutes the basic FPGA cell. It includes two 16-bit function generators (F or G), one 8-bit function generator (H), two registers (flip-flops or latches), and reprogrammable routing controls (multiplexers). CLBs are used to implement macros and other designed functions. They provide the physical support for an implemented and downloaded design. CLBs have inputs on each side, and this versatility makes them flexible for the mapping and partitioning of logic.

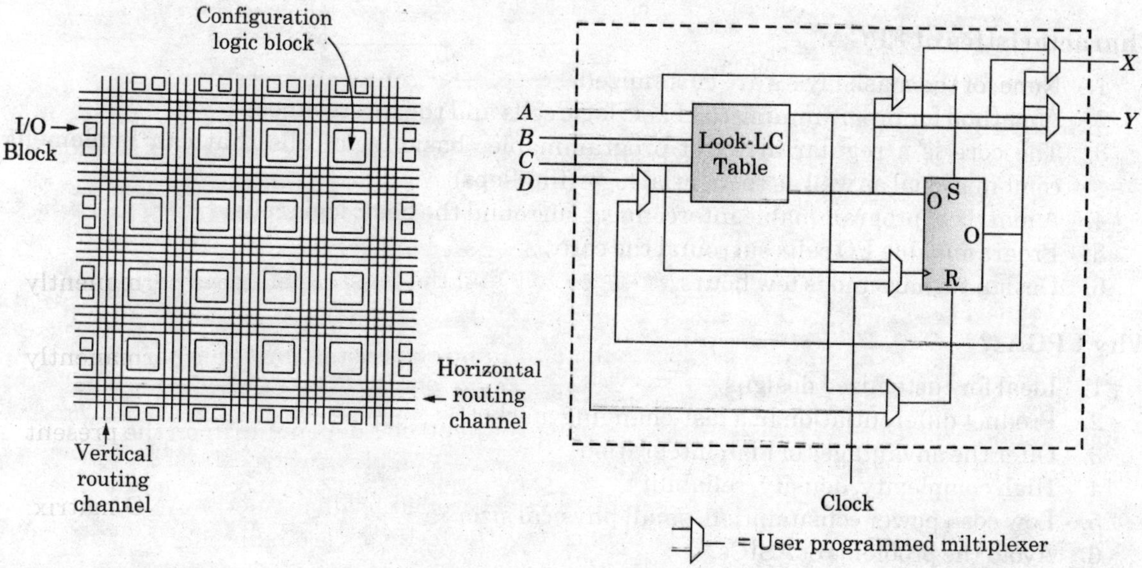

Fig. 9.33 Configurable logic block

- Logic blocks have look up table where the sequential circuit is implemented. The interconnection or routing structure consist of wire segments and programmable switches that connect the different logic blocks.
- The IBOs provide an interface between the package pins and the internal user logic while FPGAs today primarily consist of look up table, interconnections and IOBs with the growing momentum of the system on chip concept.
- A-RAM-based FPGA is an FPGA whose configuration data is programmed into random access memory. These devices are reprogrammable.
- FPGA applications include fast counters, fast pipelined designs; register intensive designs and battery powered multi-level logic.

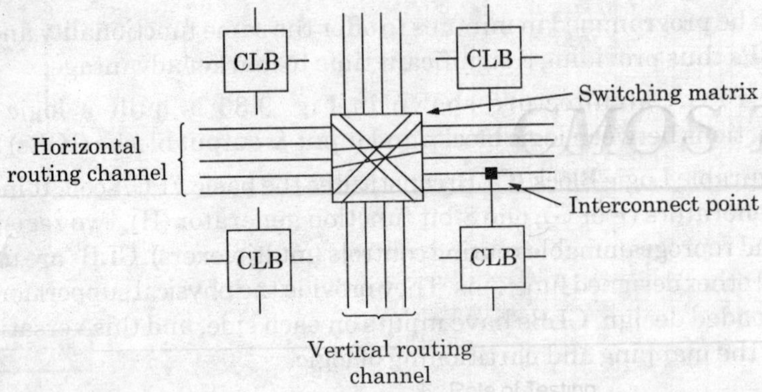

Fig. 9.34 Routing structure of FPGA

Characteristics of FPGA:

1. None, of the mask layers are customized.
2. A method for programming the basic logic cells and the interconnect.
3. The core is a regular array of programmable basic logic cells that can implement combinational as well as sequential logic (flip-flops).
4. A matrix of programmable interconnect surround the basic logic cells.
5. Programmable I/O cells surround the core.
6. Design turnaround is few hours.

Why FPGAs?

1. Ideal for customized designs.
2. Product differentiation in a fast-changing market.
3. Offer the advantages of high integration.
4. High complexity, density, reliability.
5. Low cost, power consumption, small physical size.
6. Avoid the problems of ASICs.
7. High NRE cost, long delay in design and testing.
8. Increasingly demanding electrical issues.
9. Fast time-to-market, fast response to market changes.

9.14.1 Basic Types of FPGAs

- In the case of mask-programmable gate arrays, designers have to wait for delivery of finished gate arrays from semiconductor manufacturers because the semiconductor manufacturers must prepare custom masks (although the number of custom masks for gate arrays is fewer than the case of the standard-cell library).
- With FPGAs, designers can realize their design on FPGA chips by themselves in minutes. Thus, FPGAs are becoming popular.
- Several different types of structures for FPGAs are available commercially. All of them have a basic structure that consists of many logic blocks or logic cells, accompanied by a large number of pre-laid lines for connecting these logic blocks. So, some manufacturers call FPGAs Logic Block Arrays (LBAs).
- One has a structure similar to a gate array with routing channels where each logic cell in a gate array is replaced with a logic block, as shown in Fig. 9.35(*a*).

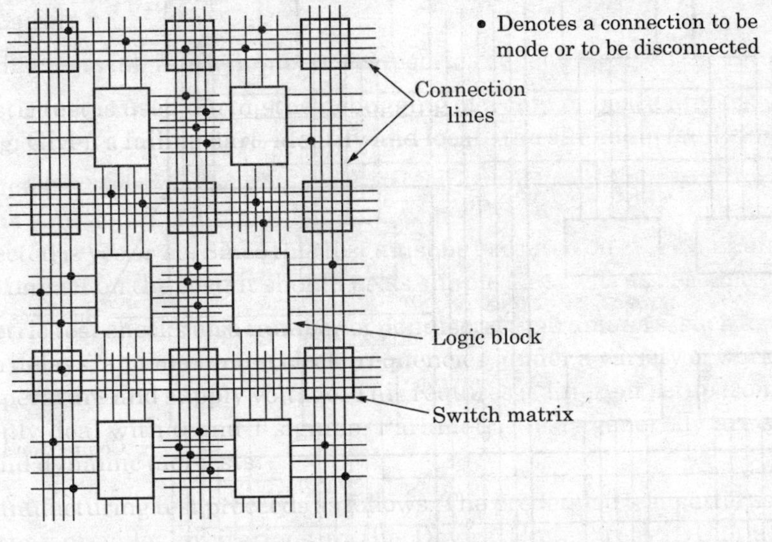

Fig. 9.35 (a) FPGA type of gate array with routing channels

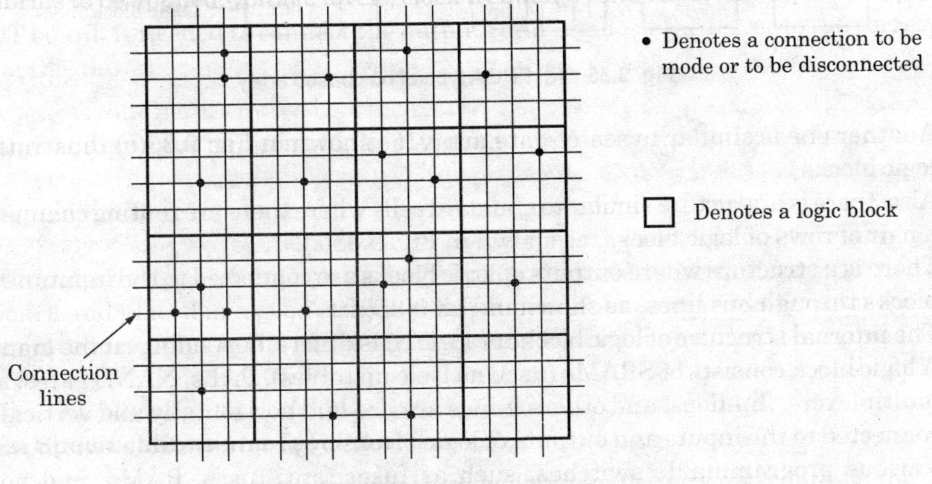

Fig. 9.35 (b) FPGA type of sea-of-gate

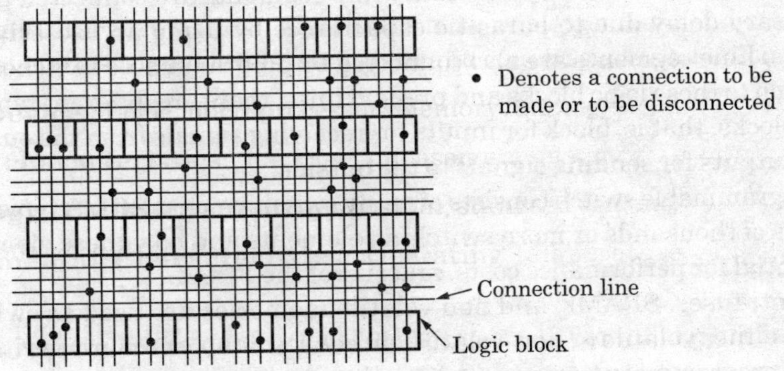

Fig. 9.35 (c) FPGA type of row-based array

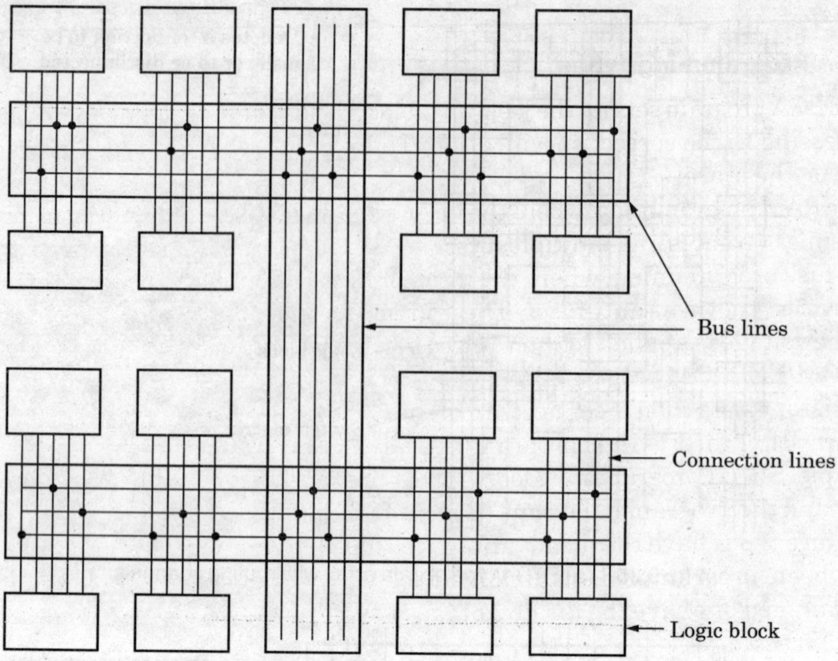

Fig. 9.35 (d) FPGA type of bus based array

- Another one is similar to sea-of-gate array, as shown in Fig. 9.35(*b*) illustrated with 16 logic blocks.
- Also, there is a structure similar to standard cells where there are routing channels between a pair of rows of logic blocks, as shown in Fig. 9.35(*c*).
- There is a structure where outputs of logic blocks are connected to the inputs of other logic blocks through bus lines, as shown in Fig. 9.35(*d*).
- The internal structure of logic blocks or logic cells differs, depending on the manufacturer. A logic block consists of SRAMs (used as look-up tables), PALs, NAND gates, along with multiplexers, flip-flops, and others. Lines are pre-laid horizontally and vertically and are connected to the inputs and outputs of logic blocks by programmable switches.
- Various programmable switches, such as fuses, anti-fuses, RAMs, and non-volatile memories, are provided by different manufacturers. Each line actually consists of many short line segments and only necessary line segments are connected in order not to add unnecessary delay due to parasitic capacitance by using an excessive number of line segments. Line segments are also connected by programmable switches.
- In addition to these logic blocks and pre-laid lines, there are different types of input/output control blocks, that is, block for inputs for receiving signals from the outside of the FPGA and for outputs for sending signals to the outside.
- Each programmable switch consists of many switching elements. A typical FPGA contains hundreds of thousands or more switching elements and how these elements are realized are essential for performance costs, and size of the FPGA.
- Fuses, anti-fuses, SRAMs, and non volatile memories are used as switching elements. They are either volatile or non-volatile, either rewritable or non-rewritable, and all have different programming times (or writing times).
- Thus, depending on which of these FPGAs have significantly different applications.

9.14.2 FPGA Programming Types

- The FPGA program sets up the look-up table (if it has one), switching routes and multiplexer addressing to the correct specification. There are three main types of programming for FPGAs, these are:
- Static RAM programming, where the interconnected routes are set up using pass transistors, transmission gates and multiplexers.
- Anti-fuse programming, where the interconnection routes are programmed permanently by devices, known as anti-fuses. These are programmed into a low-impedance state when a large voltage is applied to them, else they are in a high-impedance state.
- EPROM/EEPROM programming, where transistors are programmed to exist or not. If a transistor exists then it connects in an input to a logic block.
- The way that the FPGA is programmed has a great effect on the usage of the device. For example, SRAM programming looses the contents of the program when the power is taken away and the program must thus be reloaded each time the device is powered-up.
- Anti-fuse types have the disadvantage in that they cannot be re-programmed once they have been programmed. EPROM devices have the advantage that they can be re-programmed but have the disadvantage in that they require a special programmer to re-program them.

9.14.2.1 Static RAM Programming

1. SRAM programming uses one or two-way switches and multiplexers to define paths. 1 in the case of the switches with a single NMOS transistor then if the RAM cell Associated with that switch is programmed with a 1 then a path will be created for the signal (in one direction).
3. For two-way switches then a transmission gate can be made from a NMOS transistor in parallel with a PMOS transistor.
4. Figure 9.36 shows three connections between two logic blocks.
5. In this example the top route and the bottom route exists because the SRAM has logic 1 for these transistors, while the middle route does not exist because the switching transistors gate is set to logic 0.
6. The diagram also shows a transmission gate, which, in this case, passes signals in both directions because its gate is at logic 1.
7. Note that the look-up table within the logic block (if one exists) is also stored in the SRAM.

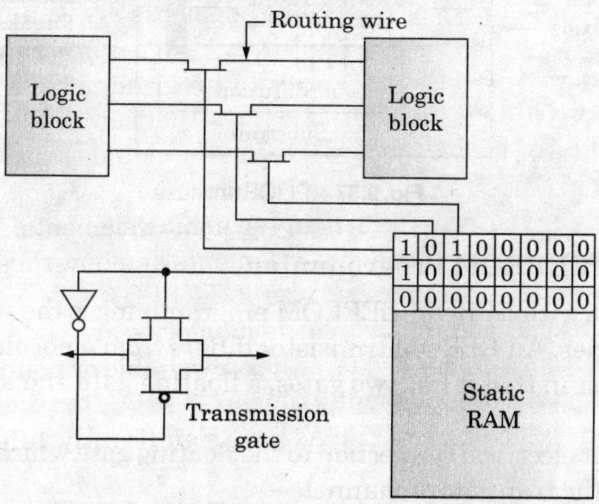

Fig. 9.36 Pass transistors and transmission gates in SRAM programming

Advantage of SRAM programming:

1. Advantage of SRAM programming is that there is no need to take the FGPA out of the circuit to re-program it.
2. All that is required is to change the program stored in the non-volatile memory.
3. SRAM devices also require a relatively large area of silicon because at least five transistors are required for each memory cell of the memory.

Disadvantage of SRAM programming:

Disadvantage of SRAM programming is that it looses the program stored in the memory when the power is taken away. Thus, some form of non-volatile memory is required to store the program and it also requires some form of program loader.

9.14.2.2 Anti-fuse Programming

1. An anti-fuse is a high impedance contact until it has a relatively high voltage applied to it which turns it into a low-impedance state.
2. Once fused it cannot be un-fused, thus anti-fuse FPGAs cannot be reprogrammed.
3. The PLICE anti-fuse is used by Actel and consists of a layer of positively doped silicon (n+ diffusion), a layer of dielectric (Oxygen-Nitrogen-Oxygen) and a layer of polysilicon, as illustrated in Fig. 9.37.
4. A relatively high voltage of 18V is applied across the device.
5. The current that flows (approximately 5mA) causes the dielectric to melt and form a conductive link between the polysilicon and the diffusion.
6. Two layers of metal connect to the polysilicon and the diffusion, thus causing a low-impedance connection between the two layers of metal.
7. A typical resistance for a fused contact is between 300 and 500Ω.
8. Note that the reason they are named anti-fuses is that metal fuses normally open circuit when they are fused, which is opposite to the effect of the anti-fuse.

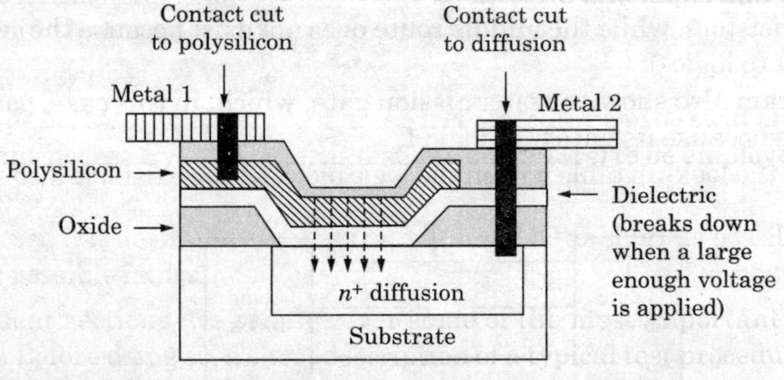

Fig. 9.37 PLICE anti-fuse

9.14.2.3 EPROM and EEPROM Programming

1. The method used with EPROM/EEPROM programming is the same as the method used in EPROM memories. An EPROM transistor differs from a normal.
2. NMOS transistor in that it has two gates, a floating gate and a select gate, as illustrated in Fig. 9.38.
3. There is no direct electrical connection to the floating gate which is positioned between the select gate and the transistors channel.

4. When no charge exists on the floating gate then the transistor operates in the normal way, which is the un-programmed state.
5. To program the device a relatively large current flows between the drain and source that causes an electrical charge to be trapped under the floating gate. This charge causes the transistor to be permanently off.
6. An EPROM transistor is re-programmed by removing the trapped charge from the floating gate. This is achieved by exposing the gate to ultra-violet light, which excites the trapped electrons.
7. When they gain enough energy they pass through the gate oxide into the substrate.
8. In the SRAM and anti-fuse types of FPGA, switched connections are made between two metal lines.
9. With EPROM programming the transistors are used as 'pull-down' interfaces to the logic block inputs, as illustrated in Fig. 9.38.
10. As the Fig. 9.38 shows the signal line connects to the select gate of the EPROM transistor. If the transistor has not been programmed in the off state then this acts as an input to the logic block.

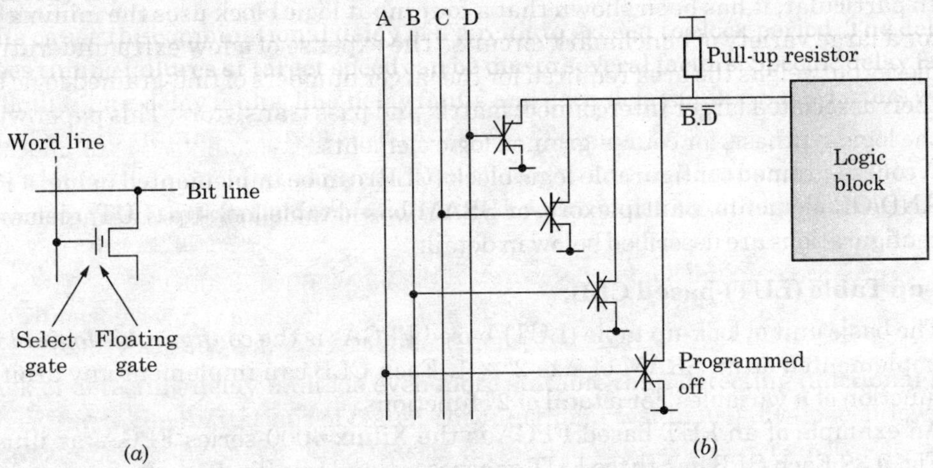

Fig. 9.38 EPROM transistors and example usage of EPROM transistors

11. If it is programmed in the off state then the connected transistor has no effect on the inputs. In the example in Fig. 9.38 the input logic function is *BD*. Because the transistors connecting the A and C lines have been programmed in the off state.
12. Typically many EPROM transistors connect to the same input lines but have different signal lines.
13. The input function to the bit line is a wired-AND function because of the pull-up resistor on the input line. Thus the transistors act both as inputs and implement an AND function.
14. The pull-up resistor unfortunately leads to an increased amount of power dissipation.
15. An improved device is an EEPROM, which is re-programmed electrically and thus, can be re-programmed in-circuit.

9.14.3 FPGA Structures **(UTU-2012)**

1. An FPGA consists of reconfigurable logic elements, flip-flops, and a reprogrammable interconnects structure. The logic elements are typically arranged in a matrix. The interconnect is arranged as a mesh of variable-length metal wires and pass transistors to interconnect the logic elements.

2. The logic elements are programmed by downloading binary control information from an external ROM, a build-in EPROM, or a host processor. After download, the control information is stored on the device and used to determine the function of the logic elements and the state of the pass transistors. Unlike a PLA, the FPGA can be used for multi-level logic functions.
3. The *granularity* of an FPGA refers to the complexity of the individual logic elements. A *fine-grain* logic block appear to the user to be much like a standard mask-programmable gate array. Each logic block consists of only a few transistors, and is limited to implementing only simple functions of a few variables.
4. A *course-grain* logic block (such as those from Xilinx, Actel, .Quicklogic, and Altera) provides more general functions of a larger number of variables. Each Xilinx 4000-series logic block, for example, can implement any Boolean function of five variables, or two Boolean functions of four variables.
5. It has been found that the course-grain logic blocks generally provide better performance than the fine-grain logic blocks, as the course-grained devices require less space for interconnect and routing by combining multiple logic functions into one logic block.
6. In particular, it has been shown that a four input logic block uses the minimal chip area for a large variety of benchmark circuits. The expense of a few extra underutilized logic blocks outweighs the area required for the larger numbers of fine-grained logic blocks and their associated larger interconnect matrix and pass transistors. This paper will focus on the logic synthesis for course-grained logic elements.
7. A course-grained configurable logic block (CLB) can be implemented using: a PLA-based AND/OR elements, multiplexors, or SRAM-based table look-up (LUT) elements. These configurations are described below in detail.

1. Look-up Table (LUT)-based CLB:

1. The basic unit of look-up table (LUT)-based FPGAs is the *configurable logic block* (CLB), implemented as an SRAM of size $2^n \times 1$. Each CLB can implement any arbitrary logic function of n variables, for a total of 2^n functions.
2. An example of an LUT-based FPGA is the Xilinx 4000-series FPGA, as illustrated in Fig. 9.39 Each CLB has three LUT generators, and two flip-flops.

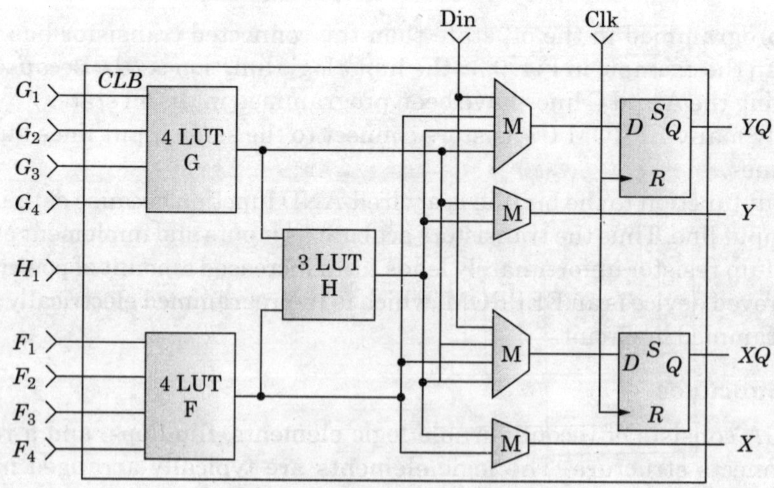

Fig. 9.39 Xilinx 4000-scrics CLB

3. The first two LUTs implement any function of four variables, while the third LUT implements any function of three variables.
4. Separately, each CLB can implement two functions of four variables. Combined, each CLB can implement any one function of five variables, or some restricted functions of nine variables (such as AND, OR, XOR).

2. PLA-based CLB:

1. PLA-based FPGA devices evolved from the traditional FLDs. Each basic logic block is an AND-OR block consisting of wide fan-in AND gates feeding a few-input OR gate.
2. The advantage of this structure is that many logic functions can be implemented using only a few levels of logic, due of the large number of literals that can be used at each block. It is, however, difficult to make efficient use of all inputs to all gates.
3. Even so, the amount of wasted area is minimized by the high packing density of the wired-AND gates.
4. To further improve the density, another type of logic block, called the *logic expander*, has been introduced. It is a wide-input NAND gate whose output could be connected to the input of the ANDOR block.
5. While its delay is similar, the NAND block uses less area than the AND-OR block, and thus increases the effective number of product terms available to a logic block.

3. Multiplexer-based CLB:

1. Multiplexer-based FPGAs utilize a multiplexer to implement different logic function by connecting each input to a constant or a signal.
2. The ACT-1 logic block, for example, has three multiplexers and one logic gate. Each block has eight inputs and one output, implementing:

$$f = \overline{(S_3 + S_4)}\,(S_1 w + S_1 x) + (S_3 + S_4)\,(S_2 y + S_2 x)$$

3. Multiplexer-based FPGAs can provide a large degree of functionality for a relatively small number of transistors. Multiplexer-based CLBs, however, place high demands on routing resources due to the large number of inputs.

4. Interconnect:

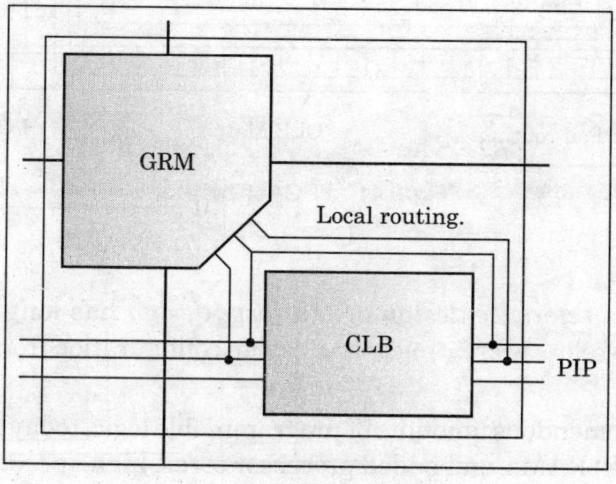

Fig. 9.40 Xilinx routing matrix

1. In all structures, a reprogrammable routing matrix interconnects the configurable logic blocks. A portion of the routing matrix for the Xilinx 4000-series FPGA for example, is illustrated in Fig. 9.40. Local interconnects are used to join adjacent CLBs. Global routing modules are used to route signals across the chip.
2. The routing and placement issues for the FPGAs are some what different from those of custom logic. For a large fan-out node, for example, an optimal placement for the elements for the fan-out would be along a single row or column, where the routing-could be done using a long line.
3. For custom logic, the optimal placement would be as a cluster, where the optimization attempted to minimize the distance between nodes.
4. For the FPGA, the routing delay is more influenced by the number of pass transistors for which the signal must cross rather than by the length of the signal line.
5. The power of the FPGA comes from the flexibility of the interconnect. A block diagram of a typical third generation.
6. FPGA device is shown in Fig. 9.41. The CLB matrix and the mesh of the interconnect occupy most of the chip real area. Macro blocks, when present, implement functions such as high-density memory or microprocessing cores. The I/O blocks surround the chip and provide connectivity to external devices.

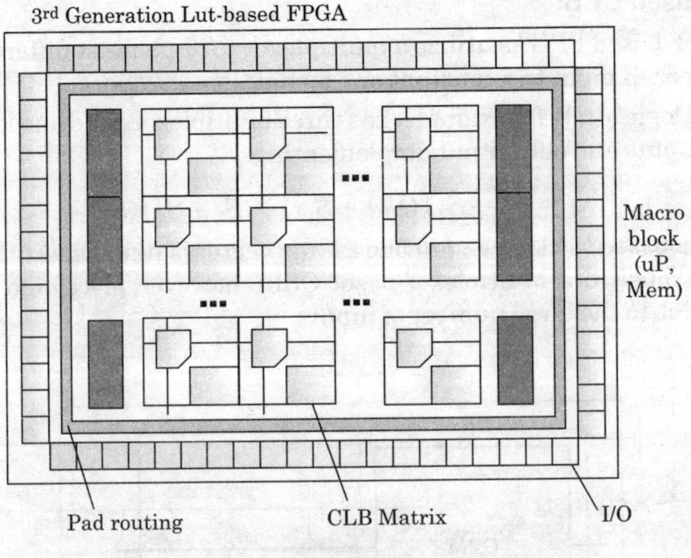

Fig. 9.41 FPGA chip layout

5. Platform FPGAs:

1. The concept of a *reference design* or *platform design* has long been used at the circuit board level. This refers to creating a base design configuration from which multiple products can be derived.
2. In addition to tremendous amounts of programmable logic, today's high-end FPGAs feature embedded (block) RAMs, embedded processor cores, high-speed I/O blocks, and so forth.

3. Furthermore, designers have access to a wide range of IP. The end result is-the concept of the *platform FPCA*.
4. A company may use a platform FPGA design as a basis for multiple products inside that company, or it may supply an initial design to multiple other companies for them to customize and differentiate.

9.14.4 Advantages of FPGA (UTU-2012)

1. Very fast custom logic
2. Massively parallel operation
3. Faster than microcontrollers and microprocessors
4. Much faster than DSP engines
5. More flexible than dedicated chipsets
6. Allows unlimited product differentiation
7. More affordable and less risky than ASICs
8. No NRE, minimum order size, or inventory risk
9. Reprogrammable at any time
10. In design, in manufacturing, after installation

9.14.5 Applications of FPGAs (UTU-2012)

- FPGAs have gained rapid acceptance and growth over the past decade because they can be applied to a very wide range of applications. A list of typical applications includes: random logic, integrating multiple SPLDs, device controllers, communication encoding and filtering, small to medium sized systems with SRAM blocks, and many more.
- Other interesting applications of FPGAs are prototyping of designs later to be implemented in gate arrays, and also emulation of entire large hardware systems.
- Another promising area for FPGA application, which is only beginning to be developed, is the usage of FPGAs as custom computing machines. This involves using the programmable parts to "execute" software, rather than compiling the software for execution on a regular CPU.
- When designs are mapped into CPLDs, pieces of the design often map naturally to the SPLD-like blocks. However, designs mapped into an FPGA are broken up into logic block-sized pieces and distributed through an area of the FPGA. Depending on the FPGAs interconnect structure, there may be various delays associated with the interconnections between these logic blocks. Thus, FPGA performance often depends more upon how CAD tools map circuits into the chip than is the case for CPLDs.

POINTS TO BE REMEMBER

1. Before any "read" and "write" operation, bit-lines must be precharged to a specific voltage and during read and write operation, the precharge circuit must be turned OFF.
2. Look up table is the most commonly used logic blocks in FPGA.
3. Making connections between cells and blocks is called routing.
4. FSM can be designed as Mealy machine or Moor machine.

SOLVED EXAMPLES

Example 9.1: Design and implement the full adder and full subtractor circuit using PAL. **(UTU-2012)**

Solution: The function of full adder is given by

$$\text{Sum} = \bar{A}\bar{B}\bar{C} + \bar{A}B\bar{C} + A\bar{B}\bar{C} + ABC$$

$$\text{Carry} = AB + BC + A$$

and the function of full subtractor is given by

$$\text{Difference} = \bar{A}\bar{B}C + \bar{A}B\bar{C} + ABC + A\bar{B}\bar{C}$$

$$\text{Borrow} = \bar{A}B + BC + C\bar{A}$$

where A, B and C are the inputs.

The implementation of the full adder and full substractor using PAL given by the following circuit diagram.

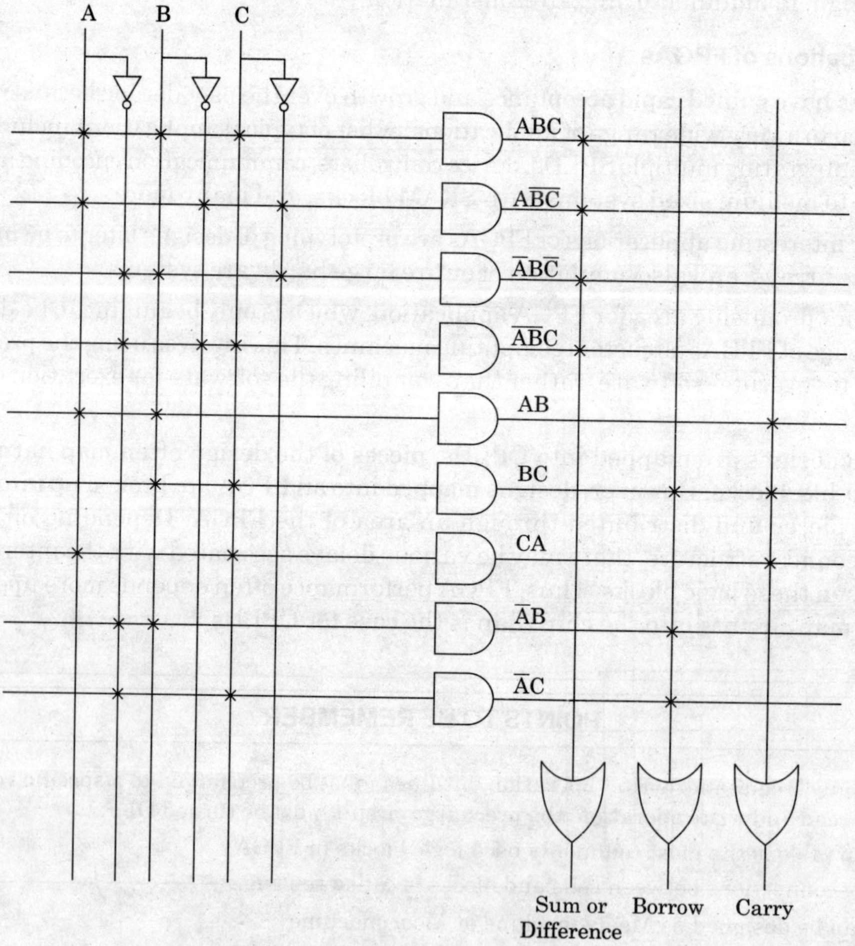

Example 9.2: Design and implement the excess-3 to BCD code converter using PAL. **(UTU-2012)**

Solution: Truth table for excess-3 to BCD code converfer

Inputs (Excess-3 code)					**Outputs (BCD-code)**			
W	**X**	**X**	**Z**		**A**	**B**	**C**	**D**
0	0	1	1		0	0	0	0
0	1	0	0		0	0	0	1
0	1	0	1		0	0	1	0
0	1	1	0		0	0	1	1
0	1	1	1		0	1	0	0
1	0	0	0		0	1	0	1
1	0	0	1		0	1	1	0
1	0	1	0		0	1	1	1
1	0	1	1		1	0	0	0
1	1	0	0		0	0	0	1

$$A = W\bar{X}YZ + WX\bar{Y}\bar{Z}$$

$$B = \bar{W}XYZ + W\bar{X}\bar{Y}\bar{Z} + W\bar{X}\bar{Y}Z + W\bar{X}Y\bar{Z}$$

$$C = \bar{W}X\bar{Y}Z + \bar{W}XY\bar{Z} + W\bar{X}\bar{Y}Z + W\bar{X}Y\bar{Z}$$

$$D = \bar{W}X\bar{Y}\bar{Z} + \bar{W}XY\bar{Z} + W\bar{X}\bar{Y}\bar{Z} + W\bar{X}Y\bar{Z} + WX\bar{Y}\bar{Z}$$

The circuit diagram using PAL for this converter is shown in figure.

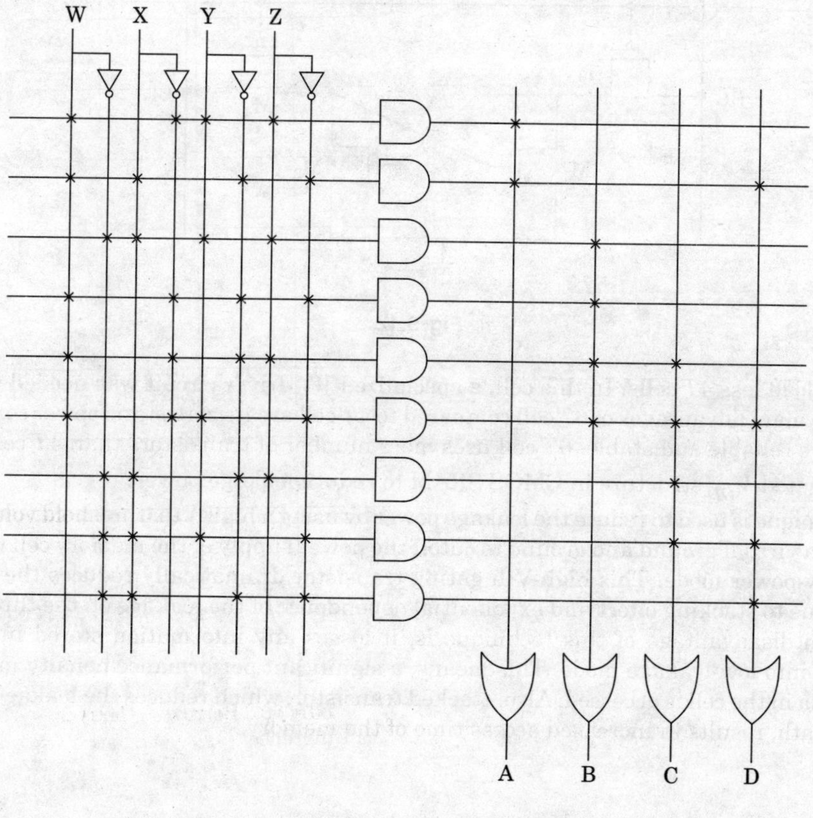

SHORT ANSWERS TYPE QUESTIONS

Q.1. What are the main differences between 1-*T* DRAM cell and 3*T* DRAM cell?

Ans. 1. 1-*T* DRAM cell requires the presence of a sense amplifier for each bit line to be functional.

2. The read out of 1-*T* DRAM is destructive, whereas in 3*T* DRAM cell read out is non-destructive.

3. Unlike 3*T* cell that relies on charge storage on a gate capacitance, the 1-*T* cell requires an extra storage capacitor. Due to this reason, 3-*T* DRAM cell is an attractive choice for embedded memory applications.

Q.2. What are the advances of using address multiplexing scheme in DRAM cell?

Ans. The advantages of using multiplexing scheme are:

(a) To reduce the chip package size by reducing pin count and

(b) Reduction of packaging cost.

Q.3. What is main significance to precharge the bit-lines up to $D_{DD/2}$ instead of V_{DD}?

Ans. This is an efficient technique to reduce the operating current of memory array. During the sensing operation, one of the but-line switches from $V_{DD/2}$ to V_{DD} and other switches from $V_{DD/2}$ to zero. This results in a power saving of almost equal to half compared to full V_{DD} precharge case. This will also reduce the peak current which results in reduction of noise. This technique also reduces precharge time and the cycle time.

Q.4. Design a SRAM cell using 4-transistors. What are the advantages and disadvantages of using 6-transistor SRAM compared to 4-transistors SRAM?

Ans. The SRAM cell using four transistors are implemented as shown in Fig. 9.39.

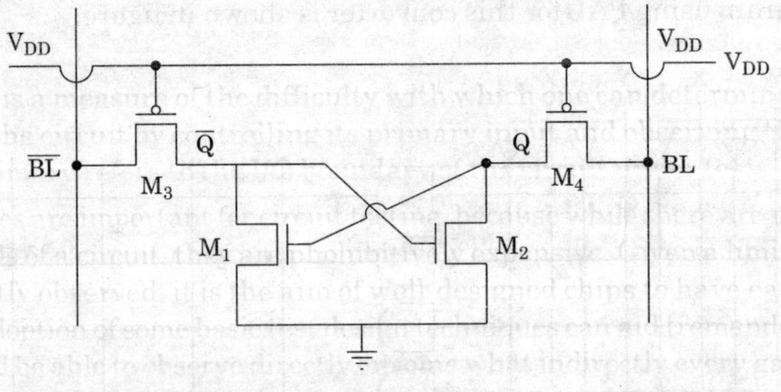

Fig. 9.42

This is a load less 4*T* cells. In this cell, a specialized W_L driver circuit was needed to hold the cell data. The main advantages of 6*T* cell compared to 4*T* cell are its zero static power consumption and its is more reliable and stable. 6*T* cell uses more number of transistors than 4*T* cell.

Q.5. What is gated-V_{DD} structure in CMOS SRAM to reduce leakage power?

Ans. This technique is used to reduce the leakage power by using a high-Vth (threshold voltage) transistor between a virtual ground and ground to cutoff the power supply of the memory cell when the cell is set to low-power mode. This high-Vth gating transistor dramatically reduces the leakage of the circuit, due to stacking effect and exponential dependence of the leakage on the threshold voltage. The main disadvantage of this technique is, it losses any information stored in the cell when switched into low-leakage mode. This means, a significant performance penalty may be incurred when data in the cell is accessed. Also, stacked transistor, which reduces the leakage current in the critical path, results in increased access time of the memory.

Q.6. What are the two commonly quoted parameters of a memory element?

Ans. (a) Setup time
(b) Hold time

Q.7. What are the four types of devices that are classified as PLD's?

Ans. (a) PROM
(b) PAL
(c) PLA
(d) GAL

REVIEW QUESTIONS

1. Explain read/write operation of SRAM memory cell. How 1-bit cell is used in bigger memory systems? **[UPTU 2006]**
2. Explain basic organization of nMOS NAND ROM and its layout. Compare different types of ROM structures. **[UPTU 2006]**
3. Explain the difference between SRAM and DRAM.
4. Explain the difference between different types of ROM.
5. Draw and compare three transistors and one transistor dynamic RAM. Explain row and column decoder design for RAM. **[UPTU 2007]**
6. What is the flash E^2PROM? Discuss briefly tunneling phenomena. Give a comparison of various types of ROM. How power dissipation is reduced in memories. **[UPTU 2007]**
7. Classify semiconductor memories.
8. What is the role of parasitic capacitance in DRAM, explain?
9. Why address decoders are used in memory architecture?
10. Draw the circuit diagram of SRAM and explain read and write operation.
11. What do you mean by field programmable?
12. What are the different types of FPGA programming types.
13. Explain the difference between PAL, PLA, FPGA.
14. What is PLA? What are different types of array structure of PLA? Explain. **[UPTU 2003-04]**
15. What do you understand by PLD? **[UPTU 2004-05]**
16. Explain the different type of array structure of PLA. **[UPTU 2005-06]**
17. Explain the working of 6T SRAM cell. **[UPTU 2009-10]**
18. Discuss with a neat diagram the operation of CMOS SRAM cell. **[UPTU 2011-12]**
19. Write short note on DRAM cell. **[UPTU 2011-12]**
20. With CMOS transistor circuit design, explain the READ/WRITE operation of static RAM cell. **[UTU-2012]**
21. Discuss the different applications and advantages of FPGA. What are different components of a simple Xilin FPGA? **[UTU-2012]**

❑❑❑

10

Chapter

CMOS Testing

Chapter Overview

- Introduction
- VLSI Testin
- Defects, Errors and Faults
- Introduction to Fault Coverage
- Design Strategiesfor Test
- Role of Testing
- Need of VLSI Testing
- Manufacturing Test Principles
- Controllability and Observability
- Built-In Self Test (BIST) Techniques

10.1 INTRODUCTION

Once a logic function has been designed, it must be tested and the faulty chips should be identified from good chips. As the number of transistors integrated into a single chip increase, the task of chip testing to ensure correct functionality becomes increasingly more difficult. For its checking, we must develop test methodology. VLSI testing deals with techniques that are used to determine if a chip behaves properly after the fabrication sequence is completed. A fault may be caused by the mechanisms ranging from – crystalline dislocations to lithographic errors to bad electing to vias.

In this chapter, we discuss types of faults, the corresponding fault models, design of testable circuits and self testing circuits. The term testability will be defined in terms of absorbability and controllability, which are also used in control system theory.

10.2 ROLE OF TESTING

The role of testing is detect whether something went wrong and the role of diagnosis it to determine exactly what went wrong and where the process needs to be altered. Therefore, correctness and effectiveness of testing is most important for quality products (perfect products).

The benefits of testing are *quality* and *economy*. These two attributes are not independent and neither can be defined without the other. Quality means satisfying the user's needs at a minimum cost. A good test process can weed out all bad products before they reach the user.

However, if too many bad items are being produced then the cost of those bad items will have to be recovered from the price charged for the few good items that are produced. It will be impossible for an engineer to design a quality product without profound understanding of the physical principles underlying the proces of manufacturing and test.

10.3 VLSI TESTING **[V. Imp.]**

- Although integrated circuit are batch processed during wafer fabrication, each and every circuit chip must be individually tested at least twice during the manufacturing cycle: once at the electrical sort step prior to scribing and die separation, and a second time after completion of the assembly operation.

- This second test cycle, called the final test step, includes a detailed testing of all circuit parameters according to the minimum and maximum specifications given in the product data sheet. Unless specified otherwise, both the electrical sort and the final test operations are done at room temperature.
- A large number of tests are often necessary to characterize an analog integrated circuit. Consequently, testing is one of the most important, expensive, and time-consuming parts of the overall IC manufacturing process. The types of testing that can be performed on integrated circuits can be divided into three categories:

1. **DC testing:** This measures the static parameters of the circuit, such as operating voltage and current levels, input bias currents, offsets, and so on. It is performed by forcing preprogrammed current or voltage levels to various circuit terminals and then sensing the resulting voltage or current levels.
2. **AC testing:** This evaluates the circuit performance under operating bias, with sinusoidal ac signals applied to it.
3. **Dynamic testing:** This includes testing the circuit in an environment or operating condition which simulates its actual application and includes pulse, amplitude, and time measurements as well as complex waveforms.

- In normal production testing of analog integrated circuits, extensive dc testing with some ac testing is utilized. Dynamic testing is usually quite complex and time consuming. Therefore, it is used mainly for very specialized circuits which combine analog, and digital functions on the same chip, and whose end applications are very well defined.

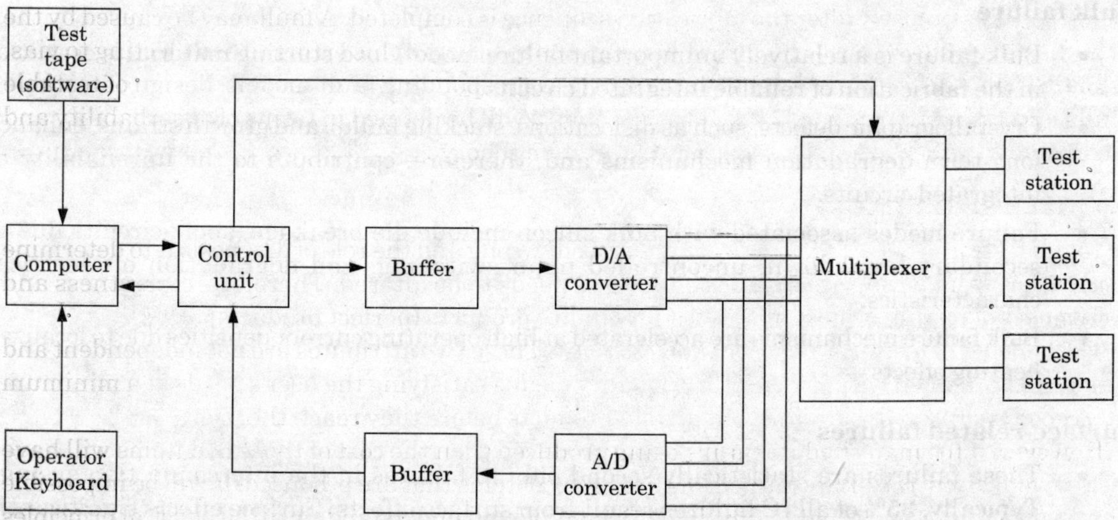

Fig. 10.1 Functional block diagram of computer controlled test system

- Because of the large number and the complexity of the tests required, computer controlled and fully automated test systems have become a major element in IC testing. Fig. 10.1 shows a generalized block diagram of such a computer-controlled test system.
- The test program can be loaded into the computer by punched cards, paper tape, or magnetic tape. Instructions from the computer are then sent to an interface or control unit which controls the various elements of the system.

- Stimulus instructions for the integrated circuits are buffed, converted, into analog voltages, and delivered to the pins of multiplexed test stations or wafer probes, which are time shared under computer control.
- Analog to Digital (A/D) converters convert the output functions of the integrated circuits into digital form.
- This information is then buffered and returned to the computer for processing.
- The computer makes a "go" or "no go" decision based on the test result and automatically sorts the device into the appropriate container.
- Normally, such a tester would also have a data-logging capability such that preliminary modes of failure can be recorded for failure analysis and yield improvement purposes.

Reliability considerations

- One of the most significant attributes of monolithic integrated circuits is high reliability. Integrated circuits are far more reliable than their discrete component counterparts, and their reliability is improving rapidly with increased knowledge of the processing techniques and an understanding of the possible failure mechanisms.

Failure modes and mechanisms

- The most commonly encountered failure mechanisms in integrated circuits can be attributed to one of the following four sources: (1) bulk failures, (2) surfacerelated failures, (3) failures of metallization or interconnections, and (4) package-related failures.

Bulk failure

- Bulk failure is a relatively unimportant failure mode. Good starting material is essential in the fabrication of reliable integrated circuits.
- Crystallographic defects, such as dislocations, stacking faults, and growth strains, enhance long-term degradation mechanisms and, therefore, contribute to the unreliability of integrated circuits.
- Failure modes associated with bulk silicon include die breakage, short-circuits due to secondary breakdown, uncontrolled pnpn switching, and degradation of electrical characteristics.
- Bulk failure mechanisms are accelerated at high operating current densities due to localized heating effects.

Surface-related failures

- These failures are statistically second only to failures in the interconnection system. Typically, 35% of all IC failures result from surface effects. Surface effects significantly influence p-n junction characteristics and tend to control transistor gains, junction breakdown voltages, and leakage currents.
- Charge migration along the silicon surface, especially in the vicinity of a p-n junction, is a major mechanisms of surface instability. This instability is often caused by ionic contaminants in the oxide on the surface or near the silicon-oxide interface.
- The charge buildup due to ionic contamination may be high enough to cause the formation of an inversion layer along the surface, where the resistivity type of the underlying silicon may be reversed.

- Failures due to surface effects are accelerated by increasing the temperature and reverse biasing the p-n junctions. Both conditions tend to increase the ionic charge mobility and enable ionic contaminants to induce inversion layers near the junction.

Metallization and interconnection failures

- The most common failure modes in integrated circuits are open or short circuits in the circuit metallization and in bonding. These conditions contribute to more failures than all other failure types combined (typically between 50 and 60%).
- Under high current densities (≥5 X 10 A/cm^2), electromigration effects become a dominant failure mode. Electromigration is a mass transport effect which causes the atoms of the interconnection metal to migrate gradually toward the more positive end of the conductor. Tins mass transport phenomenon takes place along the grain boundaries of the metal interconnections and results in the formation of voids in the interconnection pattern which may eventually lead to an open circuit. Electromigration effects are enhanced at elevated temperatures.
- An additional failure mechanism associated with the metal interconnections is formation of micro-cracks in the aluminum interconnections as a metal trace passes over a step on the SiO_2 layer.

Package-related failures

- One of the serious reliability problems in integrated circuits is associated with the bonding of the wire leads between the package and the chip. A serious failure mechanism associated with gold-wire bonds on aluminum is known as purple plagud and is due to the formation of gold-rich intermetallic compounds such as Au_2Al, Au_4AL and Au_5Al_2.
- These compounds create porous regions in the bond which are mechanically weak are electrically nonconductive.
- As these intermetallic compounds are formed, the differences in their structure thermal expansion can stress the porous interface layer to the point of rupture. However "plague" formations arc a serious problem only at elevated temperatures (typically 200°C) and are more likely to occur in step stressing than in actual use.
- Therefore, gold-wire-aluminum metallization and bonding systems are still considered be the most reliable under normal operating conditions.

10.4 NEED OF VLSI TESTING (UTU-2012)

1. While designers tend to spend numerous hours on the analysis, optimization, and layout of their circuits, one issue is often overlooked when a component returns from the manufacturing plant, how does one know if it actually works? Does it meet the functionality and performance specifications ?
2. The customer expects a delivered component to perform as described in the specification sheets. Once a part is shipped or deployed in a system, it is expensive to discover that it does not work. The later a fault is detected, the higher the cost of correction.
3. For instance, replacing a component in a sold television set means replacement ' : complete board as well as the cost of labor. Shipping a nonworking or partially functional device should be avoided if at all possible.

4. A correct design does not guarantee that the manufactured component will be operational. A number of manufacturing defects can occur during fabrication, either due to faults in the base material (e.g., impurities in the silicon crystal) or as a result of variations in (he process, such as misalignment.
5. Other faults might be introduced during the stress tests that are performed alter the manufacturing. These tests expose a part to cycles of temperature and mechanical stress is ensure its operation over a wide range of working conditions.
6. Typical faults include short-circuits between wires or layers and broken interconnections. This translates into network nodes that are either shorted to each other or to the supply rails, or that may be floating.
7. Making sure a delivered part is operating correct!)' under all possible input conditions not as simple as it would seem at a first glance. When analyzing the circuit behavior during the design phase, the designer has unlimited access to all the nodes in the network. He is free to apply input patterns and observe the resulting response node he desires. This is not the case once the part is manufactured. The only access < has to the circuit is through the input-output pins.
8. A complex component such as a microprocessor is composed of millions of transistors and contains an uncountable number of possible states. It is a very lengthy process to bring such a component into a particular state and to observe the resulting circuit response through the limited bandwidth offered by the input-output pads if at all possible. Hardware testing equipment tends to be very expensive and every second a part spends m the tester adds to its price.
9. It is therefore advisable to consider the testing early in the design process. Some modifications in a circuit can help make it easier to validate the absence of faults. This approach to design has been dubbed Design for testability (DFT).
10. While often despised by circuit designers who prefer to concentrate on the exciting aspects of design, such as transistor optimization. DFT is an integral and important part of the design process and should be considered as early as possible in the design flow.
11. A DFT strategy contains two components:
    - Provide the necessary circuitry so that the test procedure can be swift and comprehensive.
    - Provide the necessary test patterns (excitation vectors) to be employed during the test procedure.
12. For reasons of cost, it is desirable that the test sequence be as short as possible while covering the majority of possible faults.
13. In the subsequent sections, we briefly cover some of the most important issues in each of these domains. Before doing so, a short description of a typical test procedure can help to put things in perspective.
14. Chip testing is usually multi-purpose and attempts to detect faults in fabrication, design and failures due to stress operating conditions, namely the reliability problems. Testing of a chip can occur:
    - at the water level
    - at the packaged-chip level
    - at the board level
    - in field

    By detecting a malfunctioning chip at an earlier chip, the manufacturing cost may be kept low.

10.4.1 Test Procedure

(UTU-2012)

1. Manufacturing tests fall into a number of categories depending upon the intended goal.
2. The diagnostic test is used during the debugging of a chip or board and tries to accomplish the following: Given a failing part, identify and locate the offending fault.
3. The functional test (also called go/no go test) determines whether or not a manufactured component is functional. This problem is simpler than the diagnostic test since the only answer expected is yes or no. Since this test must be executed on every manufactured die and has a direct impact on the cost, it should be as simple and swift as possible.
4. The parametric test checks on a number of nondiscrete parameters, such as noise margins, propagation delays, and maximum clock frequencies, under a variety of working conditions, such as temperature and supply voltage. This requires a different setup from the functional tests that only deal with 0 and 1 signals. Parametric tests generally are subdivided into static (dc) and dynamic (ac) tests.
5. A typical manufacturing test proceeds as follows. The predefined test patterns are loaded into the tester that provides excitations to the Device Under Test (DUT) and collects the corresponding responses. The test patterns arc defined in a test program that describes the waveforms to be applied, voltage levels, clock frequency, and expected response. A probe card, or DUT board, is needed to connect the outputs and inputs of the tester to the corresponding pins on the die or package.
6. A new part is automatically fed into the tester. The tester executes the test program, applies the sequence of input patterns to the DUT, and compares the obtained response with the expected one. If differences are observed, the part is marked as faulty (e.g., with an ink spot). In case of a probe card, the probes are automatically moved to the next die on the wafer. During the scribing process that divides the wafer into the individual dies spotted parts will be automatically discarded. For a packaged part, the tested component is removed from the test board and placed in a good or faulty bin, depending upon the out-come-of the test. The whole procedure takes in the range of a few seconds per part making it possible for a single tester to handle thousands of parts in an hour.

10.4.2 Test Specifications and Test Plan

The device specification documents initiates the development activity and contains the following information:

- **Functional Characteristics:** Algorithms to be implemented, I/O signal characteristics (timing waveforms, signal levels, etc.), data and control signal behaviour, clock rate.
- **Type of Device:** Logic, microprocessor, memory, analog, etc.
- **Physical characteristics:** Package, pin assignments, etc.
- **Technology:** CMOS (or gate array), custom, standard cell, etc.
- **Environmental Characteristics:** Operating temperature range, supply voltage humidity, etc.
- **Reliability:** Acceptance quality level (defective parts per million), failure rate per 1000 hours, noise characteristics, etc.

Test specifications, if not given explicitly, are derived from the above data based on these specifications, a test plan is generated. In the test plan the type of test equipment and the type of tests are specified. Selection of a tester depends on such parameters are throughput, clock rate, timing-accuracy, test sequence length, tester availability and cost. The types of test may include parametric, functional, burn-in, margin, speed sorting, etc. The fault coverage requirement should also be specified. There are two types of tests; functionality tests and manufacturing tests.

10.4.3 Functionality Tests

The functionality tests verify that the chip performs its intended function, etc., it performs a digital filtering, acts, as a microprocessor or communicates using a particular protocol. These tests assert that all the gates in the chip achieve a desired function. These tests are usually used early in the design cycle to verify the functionality of the circuit.

For some cases, functionality test involves proving that the circuit is – functionally equivalent to some specification. The specification might be a verbal description, plain-language textual specification, description in some high-level computer language such as *C*, Pascal or in a hardware description language such as VHDL, ELLA or verilog or simply a table of inputs and required outputs.

Functional equivalence involves running a simulator at some level on the two descriptions of the chip, one at the gate level and one at a functional level, and ensuring for all inputs applied, the outputs are equivalent at some convenient check-points in time. The functional equivalence may be carrier out at various levels of the design hierarchy. If the description is in a behavioural language, the behaviour at a system level may be verifiable.

There is no good theory on how to ensure that good functional tests be written. The best way is to simulate the chip or system as closely a possible to the way it will be used in the real world.

10.4.4 Manufacturing Tests

Manufacturing tests verify that every gate and register in the chip functions correctly. These are used after the chip is manufactured to verify that the silicon is intact. The functionality tests verify the function of chip as a whole whereas manufacturing tests are used to verify that every gate-operates as expected. This is necessary due to a number of manufacturing defects that might occur during chip fabrication or during accelerate life testing. Typical defects include the following.

- Layer to layer shorts
- Discontinuous wires
- Thinoxide shorts to substrate or well

Apart from the verification of internal gates, I/O integrity is also tested through completing the following tests:

(i) I/O level test
(ii) Speed test
(iii) I_{DD} test

The I/O (input/output) level tests checks for the noise margins for TTL, ECL, CMOS I/O pads. These three tests check the leakage if the circuit is composed of complementary logic. Wafer test may be done at high speed or low speed (1 MHz) due to possible power and around bounce effects.

In general, manufacturing test generation assumes that the chip functions correctly and ways of exercising all gate inputs and of monitoring all gate outputs are required.

10.5 DEFECTS, ERRORS AND FAULTS

In correctnesses in electronic systems are described in several ways:

Defect: A defect in an electronic system is the unintended difference between the implemented hardware and its intended design.

Some typical defects in VLSI chips are given below:

1. **Process Defects:** Missing contact windows, parasitic transistors, oxide breakdown etc.
2. **Material Defects:** Bulk defects (cracks, crystal imperfections), surface impurities, etc.
3. **Age Defects:** Dielectric breakdown, electromigration, etc.
4. **Package Defects:** Contact degradation, seal leaks, etc.

Defects occur either during manufacture or during the use of devices. Repeated occurrence of the same defect indicates the need for improvements in the manufacturing process or the design of the device. Procedures for diagnosing defects and finding their causes are known as *failure mode analysis*.

Error: A wrong output signal produced by a defective system is called *an error*. An error is an "effect" whose cause is some "defect".

Fault: A representation of a "defect" at the abstracted function level is called *a fault*.

The difference between a defect and a fault is rather subtle. They are the imperfections in the hardware and function, respectively.

10.6 MANUFACTURING TEST PRINCIPLES

A critical factor in a all LSI and VLSI designs is the need to incorporated methods of testing the circuits. A combinational circuit with n-inputs shown in Fig. 10.2(*a*). To test this circuit exhaustively, a sequence 2^n inputs (to test vectors) must be applied and observed to fully exercise the circuit. The combinational circuit is converted into a sequential circuit with the addition of m-storage elements as shown in Fig. 10.2(*b*). The state of the circuit is determined by the inputs and the previous state. A minimum of 2^{n+m} test vectors must he applied to exhaustively test the circuit or $2^{(n+m)}$ inputs are required to exhaustively test circuit for $n = 25$, $m = 50$, 1 μs/test, the test time is over 1 billion years.

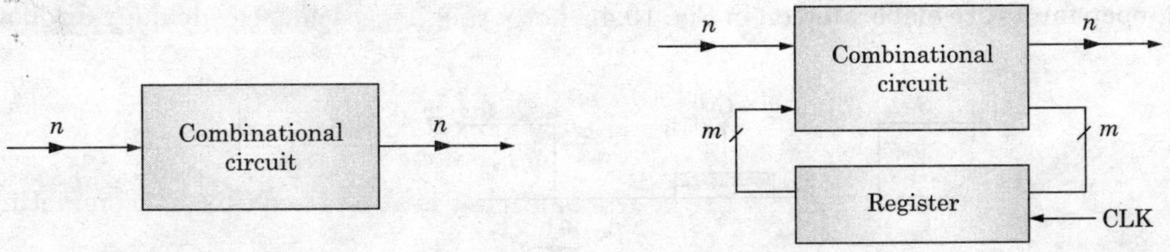

(a) 2^n inputs required to exhaustively test circuit

(b) The combinational explosion in test vectors

Fig. 10.2 Testing the circuits

10.6.1 Fault Models

10.6.1.1 Stuck-at Faults

(UTU-2012)

In order to deal with the existence of good and bad parts, it is necessary to propose a 'fault model', i.e., a model for how faults occur and their impact on circuits. The most popular model is called the 'stuck-at' model. This fault is modeled by assigning a fixed (0 or 1) value to a signal line in the circuit. A signal line is an input or an output of a logic gate or a flip-flop. These are logic-based faults: stuck-at-0 (s-a-0) and stuck-at-1 (s-a-1). These faults occur when a node is a accidentally connected to the power supply (s-a-1) or ground (s-a-0). The effect of a stuck at fault varies with the location.

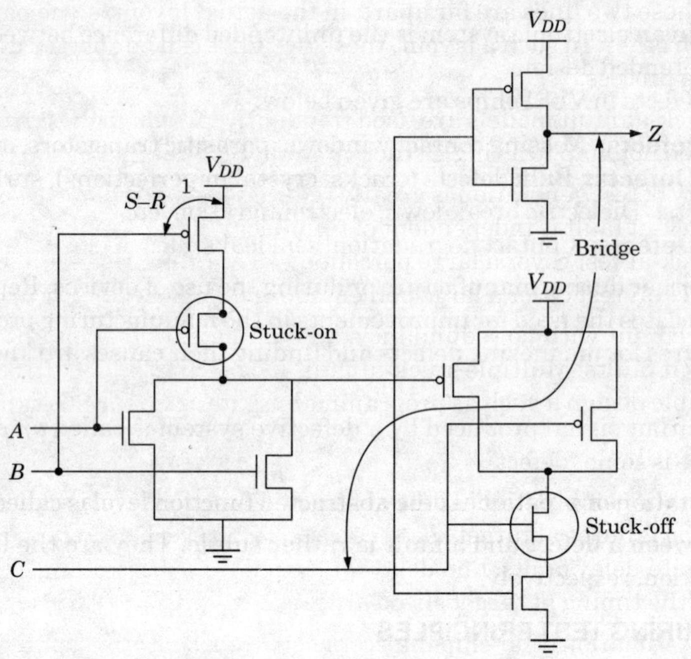

Fig. 10.3 A CMOS circuit consisting of NOR2, NAND2, and inverter gate.

For example, we can consider a CMOS circuit consisting of NOR2, NAND2 and inverter gates is shown in Fig. 10.3. In this circuit, the input line *B* can be stuck-at-1 (s-a-1) since some part of the input line is shorted to the power rail. The pMOS transistor of the first stage NOR2 gate is stuck-on due to a process problem that causes a short between its source and drain-terminals. The top nMOS transistor in the NAND2 gate, is stuck-open due to either an incomplete contact of the source or drain node or due to a large separation of drain or source diffusion from the gate, which causes permanent turn-off of the transistor regardless of the input *C* value. The stuck-on and stuck-open faults are elaborated on in Fig. 10.4.

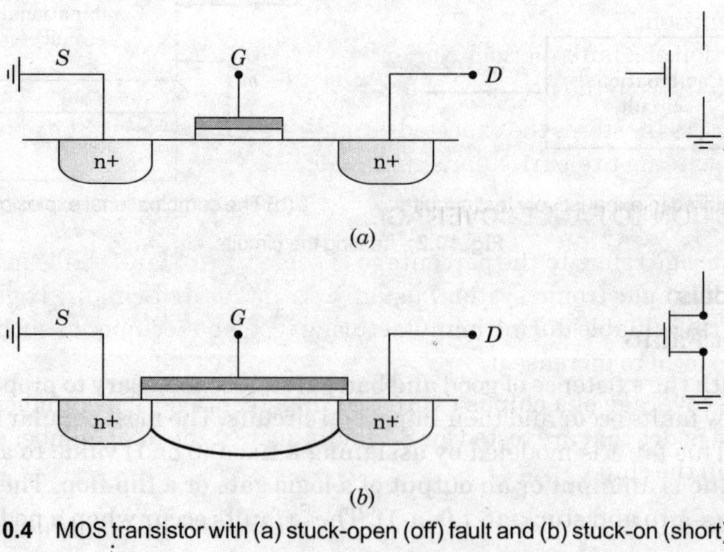

Fig. 10.4 MOS transistor with (a) stuck-open (off) fault and (b) stuck-on (short) fault.

The brideing fault between the output line of the inverter and the input line C can be due to a fabrication defect which causes a short between any two parts of the two lines. Although in the circuit diagram, these two lines are far apart, in the actual layout, some parts of these two lines can be close to each other. In such a layout, these two lines can be shorted due to underteching in the line patterning process.

The single stuck-at fault models are used frequently, although design under test (DUT) can have defects that do not map to a single stuck-at fault. Some of the reasons are:

- Complexity of test generation is greatly reduced.
- Single stuck-at fault is independent of technology, design style.
- Single stuck-at test cover a large percentage of multiple stuck-at faults.
- Single stuck-at test cover a large percentage of unmodeled physical defects.

In a two-level circuit with no redundancy, any compete test set for all single stuck-at faults can over all stuck-at faults. Multiple stuck-at fault models find applications for fuse or anti-fuse based programmable design a such as programmable gate arrays, FPGA's and RAMs.

10.6.1.2 Delay Fault

These faults cause the combinational delay of a circuit to exceed to clock period. The delay fault which causes timing failures at target speed can be due to several factors. Specific delay faults are transition faults, gate-delay faults, line-delay faults and path-delay faults. Some factors which are useful for failures the timing at target speed are given below.

- Improper estimation of on-chip interconnect delays and other timing considerations,
- Excessive variations in the fabrication process which cause significant variations in circuit delays and clock skews,
- Opens in metal lines connecting parallel transistors which make the effective transistor size much smaller.
- Aging effects such as hot-carrier induced delay increase.

The task of detecting delay faults is even more suitable than detecting functional faults in steady stage. Due to the limitations of tester, the functional test in usually done at lower speed than the target speed. Special clocking is used to apply delay tests on a slow tester. The fault models mentioned above are used in fault simulation aimed at;

- Test generation,
- Construction of a fault dictionaries, or
- Circuit analysis in the presence of faults.

Each fault dictionary stores the expected output response of every faulty circuit to a particular test vector corresponding to a particular simulated fault.

10.7 INTRODUCTION TO FAULT COVERAGE

- Fault coverage refers to the percentage of some type of fault that can be detected during the test of an electronic system, usually an integrated circuit. High (null coverage is particularly valuable during manufacturing test, and techniques such as Design for Test (DFT) are used to increase it.
- The fault coverage of a chip is expressed in percentage. This number is the ratio of the number of nodes that are tested for SA faults against the total number of faults that can be detected in the chip.

- A high value of fault coverage like 97% is a good sign for the semiconductor vendor. This is because when the device is in the wafer and also after the packaging, the health of device can be determined to a greater accuracy.
- Most of the foundries have it as a policy to go for high fault coverage, close to 95%. This number is used as an indicator of the quality of the parts delivered from a foundry.

10.7.1 Fault Modeling

1. The basic assumptions regarding the nature of physical and logical fault a combinational logic circuit as referred to as fault model. The most widely used fault model for digital logic circuits is that of a single node being permanently stuck at a logic value. This type of fault model is referred to as single stuck-at-fault mode.
2. Fault modeling has been in the practice, for over three decades, to resort to the use of stuck-at models to imitate the effects of defects. This model was more realistic when (small-scale integration) (SSI) was predominant.
3. However, the stuck-at model, for practical reasons, is still widely used by commercial tools. Basically put, this model assumes that an input or output of a logic gate (e.g.. an inverter, an AND gate, an OR gate, etc.) is stuck to a logic value 0 or 1 and is insensitive to signal changes from the signal that drives it.
4. With this faulting mechanism the process, in rather general terms, proceeds as follows. Computer models of digital circuits are created, and faults are injected into the model. The fault-free circuit and the faulted circuit are simulated. If there is a difference in response at an observable 1/0 pin, the fault is classified as detected. After many faults are evaluated in this manner, fault coverage is computed as;

   Fault coverage = No. faults detected / No. faults modeled
5. Given a fault coverage number, there are two questions that occur: How accurate is it, and for a given fault coverage, how many defective chips are likely to become tester escapes? Accuracy of fault coverage will depend on the faults selected and the accuracy of the fault model relative to real defect mechanisms.
6. Fault selection requires a statistically meaningful random sample, although it is often the practice to fault simulate a universal sample of faults, meaning faults applied to all logic elements in a circuit.
7. The fault model, like any model, is an imperfect replica. It is rather simplistic when compared to the various, complex kinds of defects that can occur in a circuit; therefore, predictions of test effectiveness based on the stuck-at model are prone to error and imprecision. The number of tester escapes will depend on the thoroughness of the test that is the fault coverage, the accuracy of that fault coverage, and the process yield.

10.7.2 Fault Simulation

1. Fault simulation is the process of measuring the quality of a test. Test stimuli that will eventually be applied to the product on a tester are themselves first evaluated by applying them to circuit models that have been slightly altered to imitate the effects of faults.
2. If the response at the circuit outputs, as determined by simulation, differs from the response of the circuit model without the fault, then the fault is detectable by those stimuli. After the

process is performed for a sufficient number of modeled faults, an estimate T, called the fault coverage, or test coverage, is computed. The equation is

T = (# faults detected)/(# faults simulated)

3. The variable T reflects the quality or effectiveness of the test stimuli. Fault simulation is performed on a structural model, meaning that the model describes the system in terms of realizable physical components.
4. The term can, however, refer to any level except behavioural, depending upon whether the designer was creating a circuit using geometrical shapes or functional building blocks.
5. The fault simulator is a structural level simulator in which some part of the structural model has been altered to represent behaviour of a fault. The fault simulator is instrumented to keep track of all differences in response between the unfaulted and the faulted circuit.
6. Fault simulation is most often performed using gate-level models, because of their granularity, although fault simulation can also be performed using functional or circuit level models.
7. The stuck-at fault model, in conjunction with logic gates, makes it quite easy to automatically inject faults into the circuit model by means of a computer program. Fault simulation serves several purposes besides evaluating stimuli:
    - It confirms detection of a fault for which an ATPG generates a test.
    - It computes fault coverage for specific test vectors.
    - It provides diagnostic capability.
    - It identifies areas of a circuit where fault coverage is inadequate.

Confirm detection

- When creating a test, an automatic test pattern generator (ATPG) makes simplifying assumptions. By restricting its attention to logic behaviour and ignoring element delay times, the ATPG runs the risk of creating test vectors that are susceptible to races and hazards.
- A simulator, taking into account element delays and using hazard and race detection techniques, may detect anomolous behavior caused by the pattern and conclude that the fault cannot be detected with certainty.

Compute fault coverage

- The ability to identify all faults detected by each vector can reduce the number of iterations through an ATPG. As will be seen in the next chapter, an ATPG targets specific faults.
- If a fault simulator identifies faults that were detected incidentally by a vector created to detect a particular fault, there is no need to create test vectors to detect those other faults.
- In addition, the fault simulator can identify vectors that detect no faults, potentially reducing the size of a test program.

Example

- Suppose the pattern A, B, C, D, E, F = (0, 1, 1, 1, 0, 0) is created to test for the output of gale H SA1 in the circuit of Fig. 10.5 Simulating the fault-free circuit produces an output of 0.

- Simulating the same circuit with a SA1 on the output of// produces a 1 on the circuit output; hence the fault is detected. But, when the effects of a SA1 on the upper input to gate G are simulated using the same pattern, we find that this fault also causes the circuit to respond with a 1 and therefore is detected by the pattern.
- Several other faults arc detected by the pattern. We leave it as an exercise for the reader to find them.

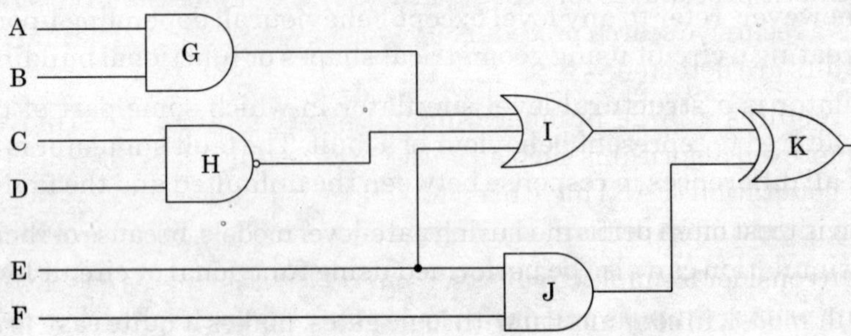

Fig. 10.5 Circuit with fault

10.8 CONTROLLABILITY AND OBSERVABILITY (UTU-2012)

Controllability, observability and predictability are the three most important factors that determine the complexity of deriving a test for a circuit.

10.8.1 Observability (UTU-2012)

The observability is a measure of the difficulty with which one can determine the signal value at any logic node in the circuit by controlling its primary input and observing the primary outputs. Here, the term primary refers to the I/O boundary of the circuit under test.

These measures are important for circuit testing, because while there are methods of observing the internal signals of a circuit, they are phohibitively expensive. Given a limited number of nodes that may be directly observed, it is the aim of well-designed chips to have easily observable gate outputs and the adoption of some basic test design techniques can aid tremendously in this respect. Ideally, one should be able to observe directly or some what indirectly every gate output within IC.

10.8.2 Controllability (UTU-2012)

Controllability for a digital circuit is defined as the difficulty of setting a particular logic signal to a 0 or a 1. (OR)

The controllability of a circuit is a measure of the difficulty with which the controller (test engineer) can establish a specific signal value at each node by setting values at the circuit input terminals.

This measure is of importance when assessing the degree of difficulty of testing a particular signal within a circuit. An easily controllable node would be directly settable via an input pad. A node with little controllability might require many hundreds or thousands of cycles to get it to the right state. Often one finds it impossible to generate a test sequence (bench) to set a number of poorly controllable nodes into the right state. It should be the aim of well-designed circuit to have all nodes easily controllable.

- The controllability and observability measures are useful because they approximately quantify how hard it is to set and observe internal signals of a circuit. Testability analysis usually has two significant attribute:
  (i) It involves circuit topological analysis, but no test vectors. It is a static type of analysis.
  (ii) It has linear complexity, because otherwise testability analysis is pointless and one might as well use automatic test pattern generation (ATPG) or fault simulation.

There are deterministic procedures for test generation for combinational circuit such as the D-algorithm which uses a recursive search procedure advancing one gate at a time and backtracking, necessary, untill all faults are detected.

The *D*-algorithm requires a large number (amount) of computer time. To overcome such shortcomings, many improved algorithms such as Path-Oriented DEcision Making (PODEM) and FAN-out oriented test generation (FAN) have been introduced. Sequential circuit test generation is several orders of magnitude more difficult than these algorithms.

For example, we can consider a simple circuit consisting of four logic gates as shown in Fig. 10.6. This circuit having four primary inputs and one primary output. To detect any detect any defect on line 8, the primary inputs *A* and *B* must be set to logic 1. However, such a setting forces line 7 to logic 1. So, any stuck-at-1 (s-a-1) fault on line 7 cannot be tested at the primary output although in the absence of such a fault, the logic value on line 7 can be fully controllable through primary inputs *B*, *C* and *D*. Hence, this circuit is not fully testable.

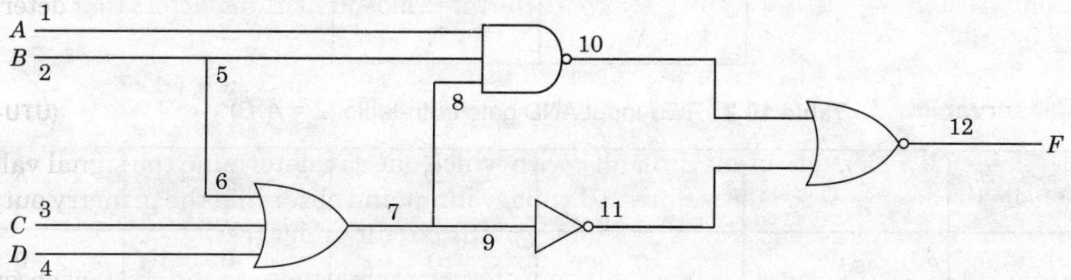

Fig. 10.6 A simple circuit consisting of four gates.

The main cause of this difficulty in this circuit is the fact that input *B* fans out to lines 5 and 6 and then after the OR3 gate, both line signals are combined in the AND 3 gate. Such a fan-out is known as reconvergent fan-out.

If a large number of input vectors are required to set a particular node value to 1 or 0 and to propagate an error at the node to an output (fault-effect propagation), then the testability is low.

The circuits with the feedbacks, decoders and clock generators having poor controllability.

The circuits with poor observability includes sequential circuits with long feedback, loops and circuits with reconvergent fan-outs, redundant nodes, and embedded memories such as RAM, ROM and PLA.

10.8.3 Predictability

- Predictability is the ability to obtain know output values in response to given input. Some factors affecting predictability are the initial state of a circuit, races, hazards and free-running oscillators.
- Most DFT, techniques deal with the ways for improving controllability, observability, predictability and design efforts specially employed to ensure that a device is testable.

10.8.4 Automatic Test Pattern Generation (ATPG)

A physical fault can be transformed into a logical fault model that allows one to develop sets of test vectors. Many techniques have been developed for testing CMOS VLSI chips that use common circuit design styles.

Most Automatic Test Pattern Generation (ATPG) approaches have been based on simulation. A five-valued logic form is commonly used to implement test generation algorithms. This consists of the states 1, 0, D, $\bar{D}$ and X, where 0 and 1 represent logical zero and logical one respectively. D represents a logic 1 in a good machine and a logic 0 in faulty machine, $\bar{D}$ represents a logic 0 in a good machine and a logic 1 in faulty machine. X represents the don't care states.

The truth table for NOT, AND and OR gates for these states are shown in Table 10.1, 10.2 and 10.3.

Table 10.1 Inverter truth table ($Z = \bar{A}$)

A	$Z = \bar{A}$
0	1
1	0
D	$\bar{D}$
$\bar{D}$	D
X	X

Table 10.2 Two-Input AND gate truth-table ($Z = A \cdot B$)

A \ B	0	1	D	$\bar{D}$	X
0	0	0	0	0	0
1	0	1	D	$\bar{D}$	X
X	0	X	X	X	X
D	0	D	D	0	X
$\bar{D}$	0	$\bar{D}$	0	$\bar{D}$	X

Table 10.3 Two-Input OR gate truth-table ($Z = A + B$)

A \ B	0	1	X	D	$\bar{D}$
0	0	1	X	D	D
1	1	1	1	1	1
X	X	1	X	X	X
D	D	1	X	D	1
$\bar{D}$	$\bar{D}$	1	X	1	D

10.9 DESIGN STRATEGIES FOR TEST

10.9.1 Testability

1. **Definition:** Testability is the property of a circuit that makes it easy (and sometimes possible!) to test.
2. The perception of testability is highly dependent on one's involvement with testing. For example, a design engineer might consider the complexity of test vector generation as a measure of circuit testability.
3. To a test engineer, testability may mean compatibility of design with the test equipment. Quality engineers often relate testability to fault coverage.
4. The testability of a circuit is an abstract concept that deals with a variety of the costs associated with testing.
5. By increasing the testability of a circuit it is implied that some function of these costs is being reduced, though not necessarily each individual cost. For example, scan designs may lower the cost of test generation but increase the number of I/O pins, area, and test time.
6. Testability is a design characteristic that influences various costs associated with testing usually it allows for the status (normal inoperable, degraded) of a device to be determined and the isolation of faults within the device to be performed quickly, to reduce both the test time and cost, and the cost-effective development of the tests to determine this status.
7. Design for testability techniques are design efforts specifically employed to ensure that a device is testable.
8. Two important attributes related to testability are controllability and observability.
    - Testability is property of a circuit, board, or systems that test are no longer considered separate issues
    - Testability involves a design approach aimed at easing the detection of faults on the chip, on the board and in the system.
9. Testability analysis usually has two significant attributes:
    - It involves circuit topological analysis, but no test vectors. It is a static type of analysis.
    - It has linear complexity, because otherwise testability analysis is pointless and one might as well use Automatic Test-pattern Generation (ATPG) or fault simulation.

10.9.2 Design for Testability

There are two key concepts to designing circuits that are testable. They are:

1. Controllability
2. Observability

Controllability is the ability to set (to 1) and reset (to 0) every node internal to the circuit. Observability is the ability to observe either directly or indirectly the state of any node in the circuit. Quite simply, these concepts ensure that the designer considers the provision of means of setting or resetting key nodes in the system and of observing the response at key points.

Three main approaches to design for testability are:

(i) Ad-hoc testing
(ii) Scan-based approaches
(iii) Self-teat and built-in testing

10.9.3 Ad hoc testing

[UPTU Exam: 2011-12]

Ad hoc techniques are collections of ideas aimed at reducing the combinational explosion of testing. Ad hoc testing increases the observability and controllability of a design. Common techniques involve:

- Partitioning large sequential circuits
- Adding test points
- Adding multiplexers
- Providing for easy state reset

10.9.3.1 Partition-and-Mux Technique

- We know that the sequence of many serial gates, functional blocks, or large circuits are difficult to test, such circuits can be partitioned and multiplexors (muxes) can be inserted such that some of the primary inputs can be fed to partitioned parts through multiplexers with accessible control signals.
- The number of accessible nodes can be increased and the number of test patterns can be reduced in this design technique.
- A case in point would be the 32-bit counter. Dividing this counter into two 16-bit parts would reduce the testing time in principle by a factor of 2^{15}.
- However, circuit partitioning and addition of Multi plexers may increase the chip area and circuit delay. This practice is not unique and is similar to the divide-and-conquer approach to large, complex problems. Fig. 10.7 illustrates this method.

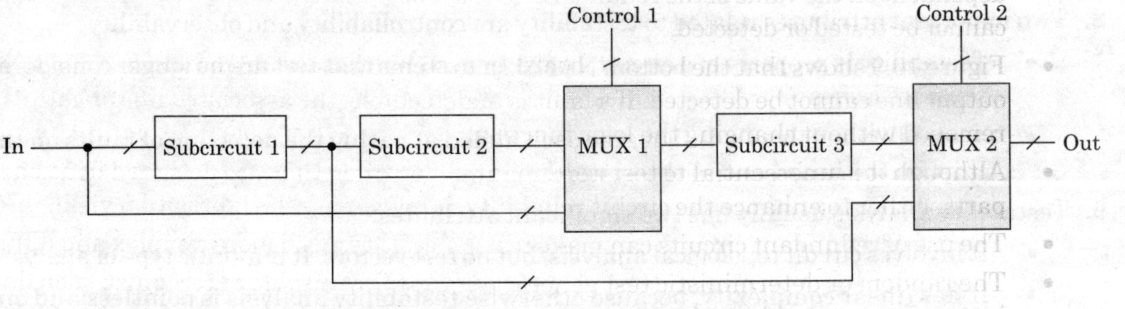

Fig. 10.7 Partition-and-Mux method for large circuits.

10.9.3.2 Initialize Sequential Circuit

- When the sequential circuit is powered up, its initial state can be a random, unknown state. In this case, it is not possible to start the test sequence correctly. The state of a sequential circuit can be brought to a known state through *initialization*.
- In many designs, the initialization can be easily done by connecting asynchronous preset or clear-input signals from primary or controllable inputs to flip-flops or latches.

10.9.3.3 Disable Internal Oscillators and Clocks

- Internal oscillators and clocks should be disabled to avoid the synchronization problems during testing.
- For example, rather than connecting the circuit directly to the on-chip oscillator, the clock signal can be ORed with a disabling signal followed by an insertion of a testing signal as shown in Fig. 10.8.

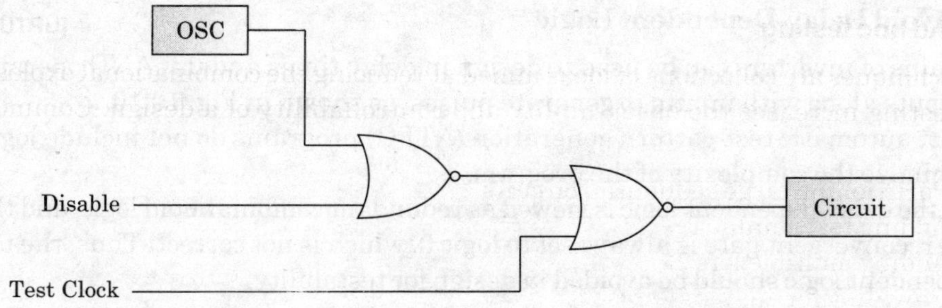

Fig. 10.8 Avoid synchronization problems-via disabling of the oscillator.

10.9.3.4 Avoid Asynchronous Logic and Redundant Logic

- The enhancement of testability requires serious tradeoffs. The speed of an asynchronous logic circuit can be faster than that of the synchronous logic circuit counterpart.
- The design and test of an asynchronous logic circuit are more difficult than for a synchronous logic circuit, and its state transition times are difficult to predict.
- The operation of an asynchronous logic circuit is sensitive to input test patterns, often causing race problems and hazards of having momentary signal values opposite to the expected values. Sometimes, designed-in logic redundancy is used to mask a static hazard condition for reliability.
- The redundant node cannot be observed since the primary output value cannot be made dependent on the value of the redundant node. Hence, certain faults on the redundant node cannot be tested or detected.
- Figure 10.9 shows that the bottom NAND2 gate is redundant and the stuck-at- fault on its output line cannot be detected. If a fault is undetectable, the associated line or gate can be removed without changing the logic function.
- Although it is unessential to test redundant nodes when they are designed in as backup parts, either to enhance the circuit reliability or to increase the fabrication yield.
- The use of redundant circuits can make test generation much more complex and difficult.
- The random or deterministic test generators would not be able to recognize such design intent. Some redundancy in circuits may be unintentional due to lack of design efficiency.

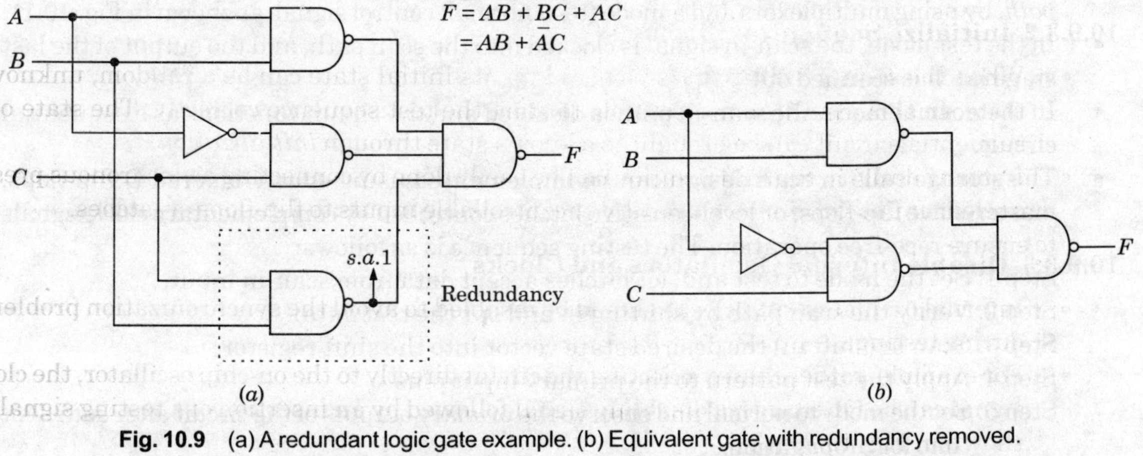

Fig. 10.9 (a) A redundant logic gate example. (b) Equivalent gate with redundancy removed.

10.9.3.5 Avoid Delay-Dependent Logic

- Chains of inverters can be used to design in delay times and use AND operation of their outputs along with inputs to generate pulses, as shown in Fig. 10.10.
- Most automatic test pattern generation (ATPG) programs do not include logic delays to minimize the complexity of the program.
- So, the delay-dependent logic is viewed as redundant combinational logic, and the output of the reconvergent gate is always set to logic 0, which is not correct. Thus, the use of delay-dependent logic should be avoided in design for testability.

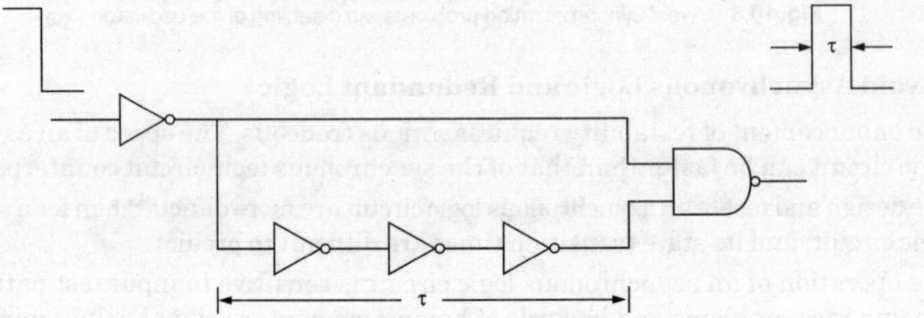

Fig. 10.10 A pulse-generation circuit using a delay chain of three inverters.

10.9.4 Scan-Based Techniques

- The controllability and observability can be enhanced by providing more accessible logic nodes with use of additional primary input lines and multiplexors. However, the use of additional I/O pins can be costly not only for chip fabrication but also for packaging.
- A popular alternative is to use scan registers with both shift and parallel load capabilities. The scan design technique is a structured approach to design sequential circuits for testability. The storage cells in registers are used as observation points, control points, or both.
- By using the scan design techniques, the testing of a sequential circuit is reduced to the problem of testing a combinational circuit.

In the scan-based design, the storage elements are replaced by the combination of multiplexers.

- The storage elements are connected to form a long serial shift register, the so-called *scan path,* by using multiplexors and a mode (test/ normal) control signal, as shown in Fig. 10.11.
- In the test mode, the scan-in signal is clocked into the scan path, and the output of the last stage latch is scanned out.
- In the normal mode, the scan-in path is disabled and the circuit functions as a sequential circuit.
- The storage cells in scan design can be implemented using edge-triggered D flip-flops, master-slave flip-flops, or level-sensitive latches controlled by complementary clock signals to ensure race-free operation. The testing sequence is as follows:
  Step 1: Set the mode to test and, let latches accept data from scan-in input,
  Step 2: Verify the scan path by shifting in and out the test data.
  Step 3: Scan in (shift in) the desired state vector into the shift register.
  Step 4: Apply the test pattern to the primary input pins.
  Step 5: Set the mode to normal and observe the primary outputs of the circuit after sufficient time for propagation.,

Step 6: Assert the circuit clock, for one machine cycle to capture the outputs of the combinational logic into the registers.

Step 7: Return to test mode; scan out the contents of the registers, and at the same time scan in the next pattern.

Step 8: Repeat steps 3-7 until all test patterns are applied.

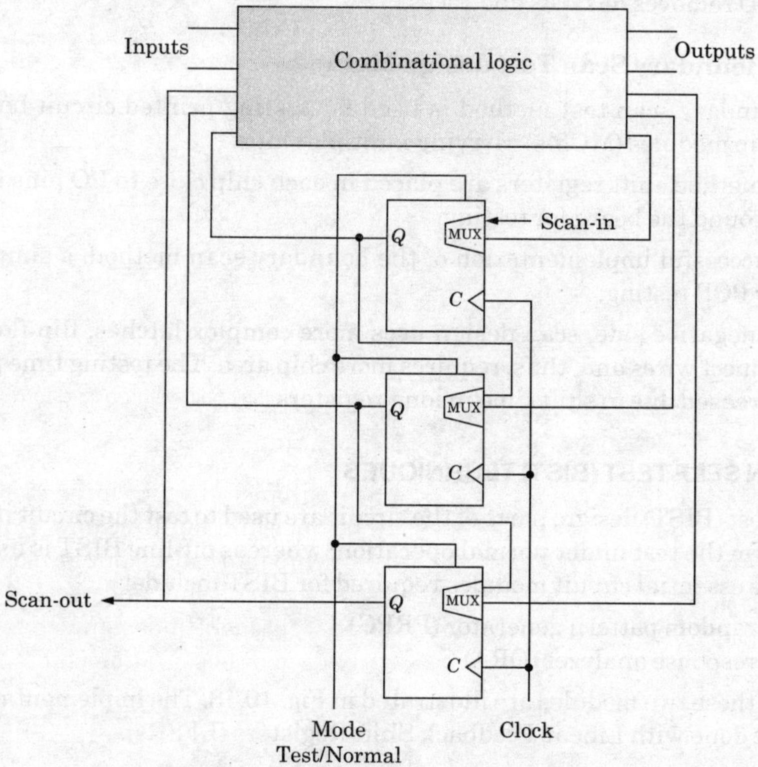

Fig. 10.11 General structure of scan-based design.

A scan-based design of an edge-triggered D flip-flop is shown in Fig. 10.12.

- Optimizing a single clock signal for both the normal operation and shift operation is difficult in large high-speed circuits.
- To overcome this difficulty, two separate clocks are used for each operation (normal & shift). Since, the shift operation does not have to be performed at the target speed, its clock is much less constrained.

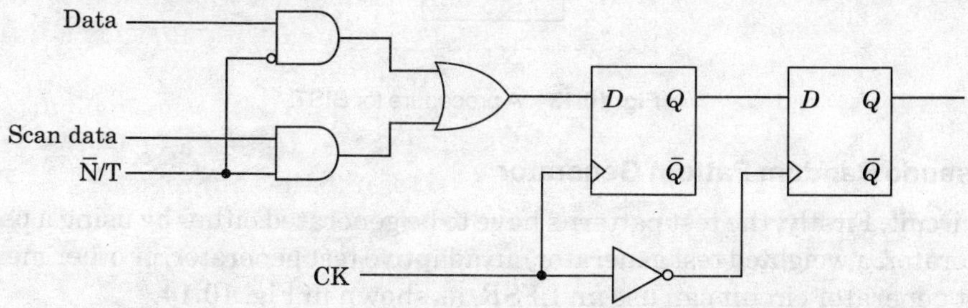

Fig. 10.12 Scan-based design of an edge-triggered D flip-flop.

10.9.4.1 The Level Sensitive Scan design

- The level sensitive scan design (LSSD) is an important approach among scan-based designs. This incorporates both the level sensitivity and the scan path approach using shift registers. The level sensitivity is to ensure that the sequential circuit response is independent of the transient characteristics of the circuit, such as the component and wire delays.
- So, LSSD removes hazards and races.

10.9.4.2 The Boundary Scan Test design Method

- The boundary scan test method is used for testing printed circuit boards (PCBs) and multichip modules (MCMs) carrying multiple chips.
- In this method shift registers are placed in each chip close to I/O pins in order to form a chain around the board for testing.
- With successful implementation of the boundary scan method, a simpler tester can be used for PCB testing.
- On the negative side, scan design uses more complex latches, flip-flops, I/O pins, and interconnect wires and, thus, requires more chip area. The testing time per test pattern is also increased due to shift time in long registers.

10.10 BUILT-IN SELF TEST (BIST) TECHNIQUES [V. Imp.]

In built-in self test (BIST) design, parts of the circuit are used to test the circuit itself. Online BIST is used to perform the test under normal operation, whereas off-line BIST is used to perform the test off-line. The essential circuit modules required for BIST include:

- Pseudo random pattern generator (PRPG)
- Output response analyzer (ORA)

The roles of these two modules are illustrated in Fig. 10.13. The implementation of both PRPG and ORA can be done with Linear Feedback Shift Registers (LFSRs).

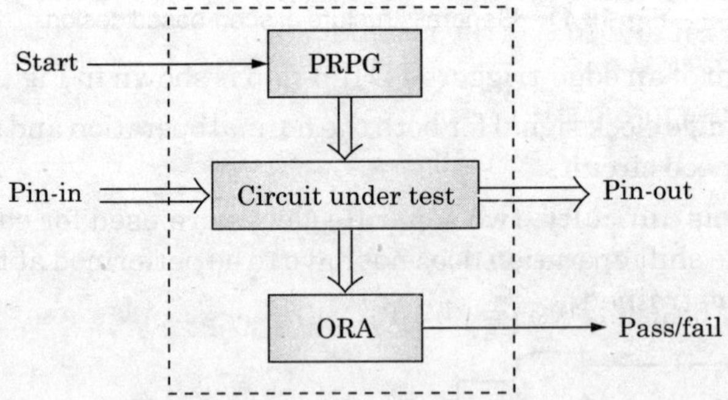

Fig. 10.13 A procedure for BIST.

10.10.1 Pseudo Random Pattern Generator

To test the circuit, Firstly, the test patterns have to be generated either by using a pseudo random pattern generator, a weighted test generator, an adaptive test generator, or other means. A pseudo random test generator circuit can use an LFSR, as shown in Fig. 10.14.

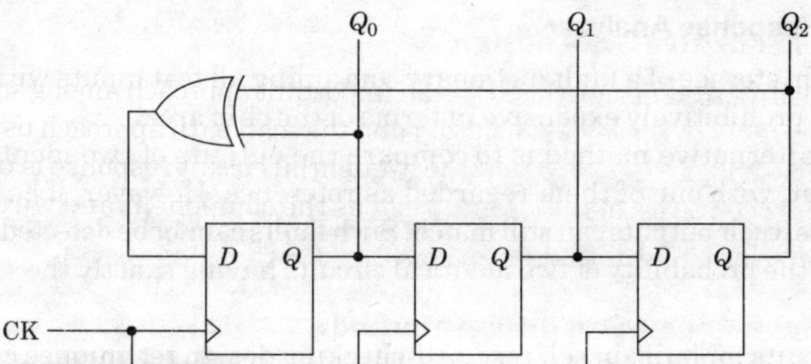

Fig. 10.14 A pseudo-random sequence generator using LFSR.

10.10.2 Linear Feedback Shift Register as ORA **(UTU-2012)**

- Data compressions schemes are used to compare the compacted test responses instead of the entire raw test data for reduce the chip area penalty.
- The popular data compression scheme is the signature analysis, which is based on the concept of cyclic redundancy checking.
- The signature analysis uses the polynomial division. The polynomial division divides the polynomial representation of the test output data by a characteristic polynomial and then finds the remainder as the signature.
- The signature is then compared with the expected signature to determine whether the device under test is faulty. It is possible that the output of a faulty circuit can match the output of the fault-free circuit; thus, the fault can go undetected in the signature analysis. Such a phenomenon is called *aliasing*.

In its simplest form, the signature generator consists of a single-input linear feedback shift register (LFSR), as shown in Fig. 10.15, in which all the latches are edge-triggered.

- In this case, the signature is the content of this register after the last input bit has been sampled. The input sequence {an) is represented by polynomial $G(x)$ and the output sequence by $Q(\mathbf{x})$. It can be shown that $G(x) = Q(x)\ P(x) + R(x)$, where $P(x)$ is the characteristic polynomial of LFSR and $R(x)$ is the remainder, the degree of which is lower than that of $P(x)$.
- For the simple case in Fig. 10.15, the characteristic polynomial is

$$P(x) = 1 + X^2 + X^4 + X^5$$

For the 8-bit input sequence {1 1 1 1 0 1 0 1}, the corresponding input polynomial is

$$G(x) = X^7 + X^6 + X^5 + X^4 + X^2 + 1$$

and the remainder term becomes $R(x) = X^4 + X^2$, which corresponds to the register contents of {0 0 1 0 11}.

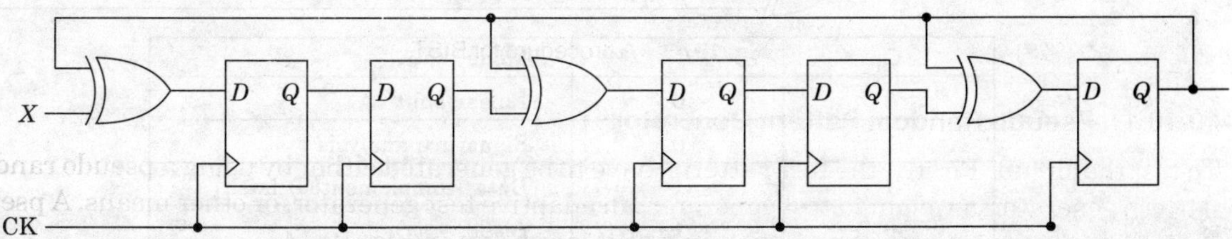

Fig. 10.15 Polynomial division using LFSR for signature analysis.

10.10.3 Output Response Analyzer (UTU-2012)

- The on-chip storage of a fault dictionary containing all test inputs with the corresponding outputs is prohibitively expensive in terms of the chip area.
- A simple alternative method is to compare the outputs of two identical circuits for the same input, with one of them regarded as reference. However, if both circuits have the same faults, their outputs can still match. Such faults cannot be detected with this technique, although, the probability of two identical circuits having exactly the same faults would be very low.

The above circuits for built-in self test, self-checking design techniques can be used to detect faults autonomously during on-line operation. Usually a checker circuit is inserted such that the checker generates and sends out a signal when on-line faults occur. The distribution of checkers throughout a very large digital circuit or system can provide prompt detection of the fault location by tracing the checker which sent the fault signal.

The use of self-checking circuits simplifies the development of software diagnostic programs. However, some additional hardware is required, and the checker itself needs to have self-checking capability. When self-checking capability of the checker itself is required, a single-output checker is not sufficient since that output may have a stuck-at fault, thus preventing the detection of actual faults in the circuit under test. A checker with a *pair* of outputs can be used instead to overcome this problem.

10.10.4 Built-In Logic Block Observer (UTU-2012)

- The built-in logic block observer (BILBO) register is a form of ORA which can be used in each cluster of partitioned registers. A basic BILBO circuit is shown in Fig. 10.16. This circuit is allows four different modes controlled by C_0 and C_1 signals.
- The BILBO operation allows monitoring of circuit operation through exclusive- ORing into LFSR at multiple points, which corresponds to the signature analyzer with multiple inputs.

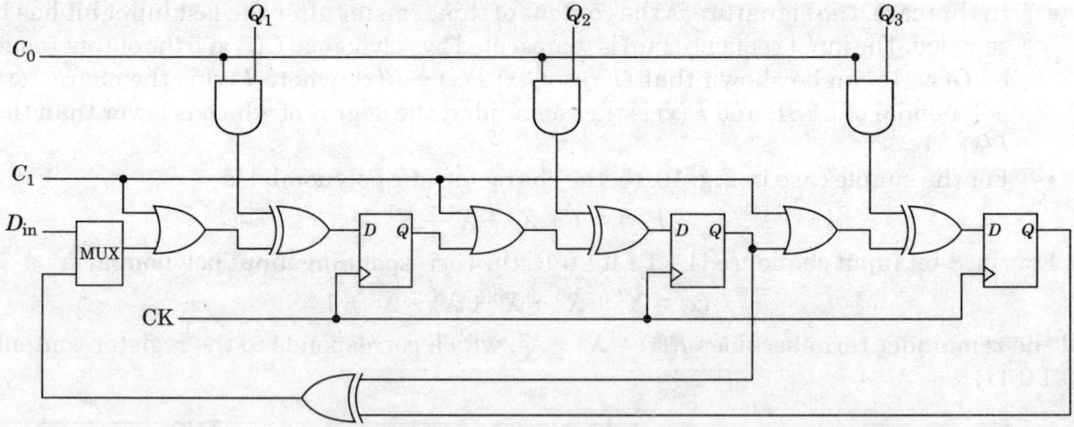

C_0	C_1	Mode
0	0	Linear shift
1	0	Signature analysis
1	1	Data (complemented) latch
0	1	reset

Fig. 10.16 3-bit built-in logic observer (BILBO).

10.10.5 Current Monitoring I_{DDQ} Test

- The I_{DDQ} test is an often-used technique for testing fabrication defects. When the signal inputs are stable (not switching), the quiescent leakage current I_{DDQ} can be measured.
- For example, we consider a circuit as shown in Fig. 10.17. Every chip design is found to have a range of 'normal' levels. I_{DDQ} testing is based on the assumption that an abnormal reading of the leakage current indicates a problem on the chip. I_{DDQ} testing is usually performed at the beginning of the testing cycle. If a chip fails, it is reflected and no further tests are required.

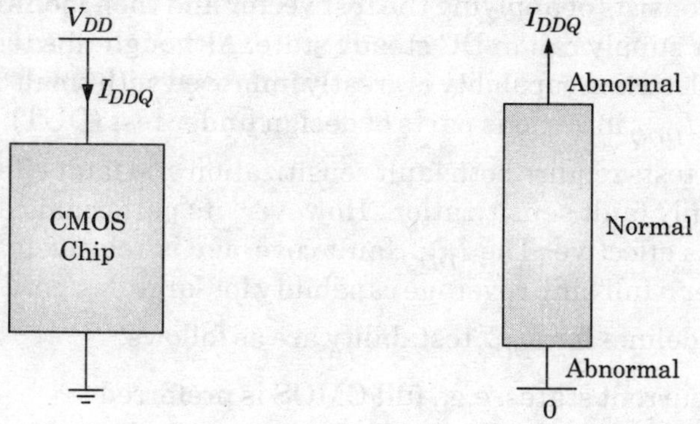

Fig. 10.17 Basic I_{DDQ} test.

- The components of a basic I_{DDQ} measurement system is shown in Fig. 10.18. The test chip is modeled as being in parallel with the testing capacitance C_{test}. A power supply with a value V_{DD} is connected to the chip by a switch that is momentarily closed at time $t = 0$. The current I_{DDQ} is monitored by a buffer and gives the output voltage.

$$I_{DD} = C[\Delta V_0/\Delta t]$$

Where the voltage falls by an amount ΔV_0 in a time Δt. The total capacitance C in the equation is the sum

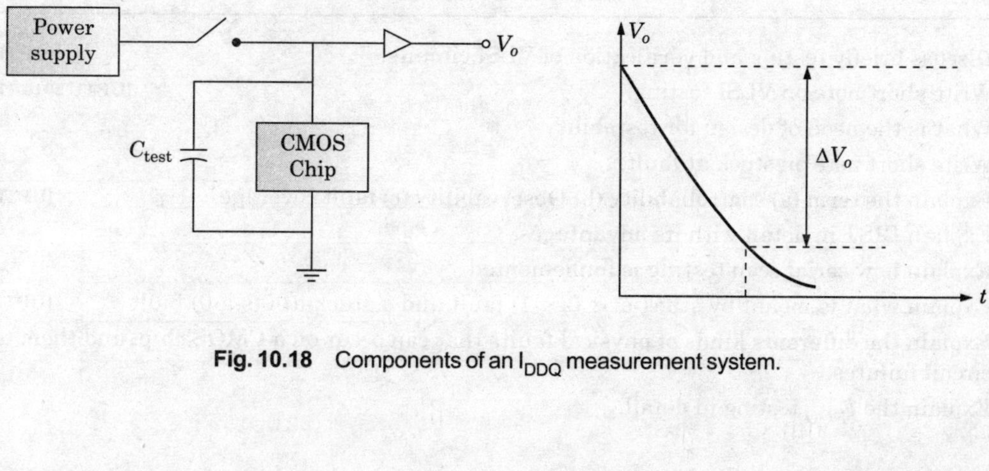

Fig. 10.18 Components of an I_{DDQ} measurement system.

An other example, if the drain node of the pMOS transistor in a CMOS inverter is shorted to the power supply rail due to a bridging fault, its IDDQ current can be very high even when the input is high. It can also detect other fabrication defects not easily detected by other test methods, including:

⇒ Gate oxide short

⇒ Channel punch-through

⇒ p-n diode leakage

⇒ Transmission-gate defect

- The I_{DDQ} test consists of applying the test vector and then monitoring the current drawn from the power supply rail in DC steady state. Although this test requires more testing time, the fault detection capability is greatly improved with small circuit overhead required to monitor the I_{DDQ} in various parts of design under test (DUT).
- While stuck-at tests require both fault sensitization and fault effect propagation, the I_{DDQ} test requires only fault sensitization. However, its performance in open drain and open gate test is less effective. The I_{DDQ} fault coverage is relatively easy to obtain and may potentially offer a full-chip coverage capability for large designs.
- The design guidelines for I_{DDQ} testability are as follows:

  ⇒ Low static current states, e.g., full CMOS is preferred

  ⇒ No active pull-ups or pull-downs

  ⇒ No internal drive conflicts, e.g., drivers share a bus

  ⇒ No floating nodes in the circuit

  ⇒ No degraded voltages, e.g., must have $V_{OH} = V_{DD}$ and $V_{OL} = 0$

10.10.6 Advantages of BIST

1. Lower cost due to elimination of external tester.
2. In system, at-system, high-quality testing.
3. Faster fault detection, ease of diagnosis.
4. Overcomes pin limitations and related interfacing problems.
5. Reduces maintenance and repair costs at system level.

REVIEW QUESTIONS

1. Discuss briefly testing and verification of VLSI circuits. **[UPTU 2006-07]**
2. Write short note on VLSI testing. **[UPTU 2010-11, 2003-04]**
3. What is the need of design for testability.
4. Write short note on stuck at fault.
5. Explain the term (a) controllability (b) Observability (c) fault coverage. **[UPTU 2007-08]**
6. Explain BIST in detail with its advantages.
7. Explain how serial scan testing is implemented.
8. Explain what is meant by stuck-at-1 (s-a-1) fault and a stuck-at-0(s-a-0) fault. **[UPTU 2007-08]**
9. Explain the differents kinds of physical faults that can occur on a CMOS chip, and them to typical circuit failures.
10. Explain the I_{DQQ} testing in detail.

11. Explain gate array and sea of gates design.

12. What are stuck-at-zero and stuck-at-one faults? Explain briefly test and testability of VLSI circuits. **[UPTU 2010-11, 2007-08]**

13. Write short notes on scan-based techniques.

14. Explain ad-hoc testable design techniques.

15. Describe the commonly use VLSI testing procedure. **[UPTU 2005-06]**

16. Briefly explain the issues involved in Built-in Self Test (BIST) techniques. **[UPTU 2011-12, 2007-08, 2009-10]**

17. Define the terms controllability and observability. Discuss in brief Ad-hoc testable design techniques. **[UPTU 2011-12]**

18. Explain stuck at models with components level circuit in detail. **[UTU-2012]**

19. What is the need of VLSI testing? Describe the VLSI testing procedure. **[UTU-2012]**

20. Define and explain the following. **[UTU-2012]**

(*a*) LFSR
(*b*) BILBO
(*c*) Analyzer
(*d*) Controllability and Observability

❑❑❑

Appendix I

International System of Units

Quantity	Unity	Symbol	Units
Length	metter	m	
Mass	kilogram	kg	
Time	second	s	
Temperature	kelvin	k	
Current	ampere	A	
Frequency	hertz	Hz	1/s
Force	newton	N	kg-m/s^2
Pressure	pascal	Pa	N/m^2
Energy	joule	J	N-m
Power	watt	W	J/s
Electric charge	coulomb	C	A-s
Potential	volt	V	J/C
Conductance	siemens	S	A/V
Resistance	ohm	Ω	V/A
Capacitance	farad	F	C/V
Magnetic flux	weber	Wb	V-s
Magnetic induction	tesla	T	Wb/m^2
Inductance	henry	H	Wb/A

II

Appendix

Physical Constants

Quantity	Symbol	Value
Angstrom unit	Å	1 Å = 10^{-1} nm = 10^{-4} mm = 10^{-8} cm
Avogadro constant	N_{AVO}	6.0204×10^{23} mol^{-1}
Bohr radius	a_B	0.52919 Å
Boltzmann constant	k	1.38066×10^{-23} J/K or 8.62×10^{-5} eV/K
Elementary charge	q	1.60218×10^{-19} C
Electron rest mass	m_0	0.91095×10^{-30} kg
Electron volt	eV	1 eV = 1.60218×10^{-19} J = 23.053 kcal/mol
Gasa constant	R	1.98719 cal-mol^{-1} K^{-1}
Permeability in vacuum	μ_0	1.25663×10^{-8}H/cm ($4\pi \times 10^{-9}$)
Permittivity in vacuum	ε_0	8.85418×10^{-14} F/cm ($1/\mu_0 c^2$)
Planck constant	h	6.62617×10^{-34} J-s/4.135×10^{-15} eVs
Reduced Planck constant	$\hbar$	1.05458×10^{-34} J-skg
Proton rest mass	M_P	1.67264×10^{-27} kg
Speed of light in vacuum	C	2.99792×10^{10} cm/s
Standard atmospheric pressure		1.01325×10^5 N/m^2
Thermal voltage at 300 K	KT/q	0.0259 V
Wavelength of 1 eV quantum	λ	1.23977 V
Dielectric constant of SiO_2	ε_{ox}	$3.97\varepsilon_0$ F/cm

III

Silicon, Gallium Arsenide and Germanium Properties (T = 300 K)

Appendix

Property	Si	GaAs	Ge
Atoms (cm^{-3})	5.0×10^{22}	4.42×10^{22}	4.42×10^{22}
Atomic weight	28.09	144.63	72.60
Crystal structure	Diamond	Zincblende	Diamond
Density (g/cm–3)	2.33	5.32	5.33
Lattcie constant (Å)	5.43	5.65	5.65
Melting point (°C)	1415	1238	937
Dielectric constant	11.7	13.1	16.0
Bandgap energy (eV)	1.12	1.42	0.66
Electron affinity, X (volts)	4.01	4.07	4.13
Effective density of states in conduction band, N_C (cm^{-3})	2.8×10^{19}	4.7×10^{17}	1.04×10^{19}
Effective density of states in valence band, N_C (cm^{-3})	1.04×10^{19}	7.0×10^{18}	6.0×10^{18}
Lintrinsic carrier concentration (cm^{-3})	1.5×10^{10}	1.8×10^{18}	2.4×10^{13}
Mobility (cm^2/V-s)			
Electron, μ_n	1350	8500	3900
Hole, μ_p	480	400	1900
Effective mass (m^*/m_o)			
Electrons	$m_l^* = 0.98$ $m_t^* = 0.19$	0.067	1.64 0.082
Holes	$m_{lh}^* = 0.16$ $m_{hh}^* = 0.49$	0.082 0.45	0.044 0.28
Effective mass (density of states)			
Electrons (m_n^*/m_o)	1.08	0.067	0.55
Holes (m_p^*/m_o)	0.56	0.48	0.37

Electrical Properties of SiO_2

Property	Value
Dielectric constant	3.9
Dielectric strength	$\sim 10^7$ V/cm*
Bandgap	~ 9 eV
Resistivity	$\sim 10^{12} - 10^{16}\ \Omega - cm^+$
Electron mobility	20 – 40 cm^2/V.S
Hole mobility	$\sim 2 \times 10^{-5}$ cm^2/V.S

* Function of growth conditions and thickness
+ Function of applied field

Conversion factors

	Prefixes	
1 Å (angstrom) = 10^{-8} cm	10^{-15}	Femto – = f
1 μm (micron) = 10^{-4} cm	10^{-12}	pico – = p
1 mil = 10^{-3} in = 25.4 mm	10^{-9}	nano – = n
2.54 cm = 1 m	10^{-6}	micro – = m
1 eV = 1.6×10^{-19} J	10^{-3}	milli – = m
1 J = 10^7 erg	10^{+3}	Kilo – = K
	10^{+6}	mega – = M
	10^{+9}	giga – = G
	10^{+12}	tetra = T

B. TECH

VLSI CIRCUIT DESIGN

(Sem. VI) Examination, 2012

TEC–602

Time : 3 Hours *Maximum Marks : 100*

Note: Attempt ALL the questions. Assume suitable data if necessary.

Q.1. Attempt any FOUR parts of the following. **5 × 4 = 20**

(a) Consider a *N*-channel MOSFET with, $N_A = 3 \times 10^{16}/\text{cm}^3$ and $t_{ox} = 450°\text{A}$. Let $L = 1.25\ \mu\text{m}$ assume that $r_j = 0.5\ \mu\text{m}$ calculate the threshold voltage shift due to short channel effect.

(b) Explain the term "self aligned gate". What is the requirement of self aligned gate? How it is achieved.

(c) A MOS transistor has threshold voltage of 0.85 V. The body effect coefficient $\gamma = 0.64$. Calculate the threshold voltage for $V_{SB} = 4$ V and $2\varphi F = -0.58$ V.

(d) What do you understand by strong inversion in MOSFET?

(e) Explain the electrical properties of the MOSFET, CMOS and BCMOS.

(f) What is pass transiter? Explain with suitable example.

Q.2. Attempt any FOUR parts of the following. **5 × 4 = 20**

(a) Explain the CMOS static transfer characteristics.

(b) Derive the total power consumption in CMOS inverter.

(c) Calculate the noise margin of CMOS inverter with $V_{DD} = 5$ V, $V_{\text{Ton}} = 1$ V, $V_{\text{Top}} = -1$ V, $K_n = 210\ \mu\text{A/V}^2$, $K_p = 210\ \mu\text{A/V}^2$.

(d) Calculate the average propagation delay of a symmetric CMOS inverter with $C = 0.1$ pF. Given that $K_n = 210\ \mu\text{A/V}^2$ and $W/L = 4$.

(e) Design X-OR and X-NOR logic gates using CMOS transmission logic gates.

(f) What is stick diagram? Draw the circuit diagram and coloured stick diagram of 2-input CMOS NAND gate.

Q.3. Attempt any TWO parts of the following. **10 × 2 = 20**

(a) Draw and explain the SR latch using CMOS devise and based on NAND logic. Also draw the coloured stick diagram.

(b) Write a short note on skew-tolerant Domino Circuits and OTB Domino.

(c) With CMOS transistor circuit design, explain the READ/WRITE operation of static RAM cell.

Q.4. Attempt any TWO parts of the following. **10 × 2 = 20**

(a) What is PLA? What are the different type array structures of PLA? Explain.

(i) Full subtractor and full adder

(ii) Excess-3 to BCD code convertor.

(b) Design and implement the following multiple output combinational circuits using PAL.

(c) Discuss the different applications and advantages of FPGA. What ate different components of a simple Xilinx FPGA?

Q.5. Attempt any TWO parts of the following. **10 × 2 = 20**

(a) Define and explain the following

(i) LFSR

(ii) BILBO

(iii) signature analyzer

(iv) controllability and observability

(b) What is the need of VLSI testing? Describe the VLSI testing procedure.

(c) Explain stuck at models with component level circuit in detail.

Answers/Solution Key

Q.1. (a) Refer Example 1.15

(b) Refer Heading 1.2

(c) Refer Example 1.22

(d) Refer Heading 1.6

(e) Refer Heading 1.10

(f) Refer Heading 6.6

Q.2. (a) Refer Heading 4.8

(b) Refer Heading 5.3

(c) Refer Example 4.3

(d) Refer Example 4.2

(e) Refer Example 6.6

(f) Refer Heading 3.4 and 4.4.2.1

Q.3. (a) Refer Heading 8.8

(b) Refer Heading 7.6.1

(c) Refer Heading 9.6, 9.6.1, 9.6.2

Q.4. (a) Refer Heading 9.12 and 9.12.1

(b) Refer Example 9.1 and 9.2

(b) Refer Heading 9.14.4 and 9.14.5

Q.5. (a) (i) Refer Heading 10.10.2

(ii) Refer Heading 10.10.4

(iii) Refer Heading 10.10.3

(iv) Refer Heading 10.8

(b) Refer Heading 10.4 and 10.4.1

(c) Refer Heading 10.6.1

Index

Index

P

R

S

T

Z

❑❑❑

ATTENTION READERS

Please help us to improve ... request you for your frank assessment regarding some of the aspects of the book given as under:

VLSI CIRCUIT DESIGN

Deepak [illegible]

Please fill up the given space in neat capitals.

(i) What topics of your syllabus that are not covered in the book?

(ii) What are the chapters which, in the book, are organised/presented ...?

(iii) Have you come across misprints/mistakes in the book? Please specify the details and the page numbers.

(iv) What other books on the same subject do you consider better than the present book? Please specify in terms of titles and authors.

(v) Further suggestions and comments for the improvement of the book.

PS: If need be please attach a separate sheet to write your views.

Win Prizes!

ATTENTION READERS

We request you for your frank assessment regarding some of the aspects of the book given as under:

VLSI CIRCUIT DESIGN (UTU)

By

Deepak Garg

Please fill up the given space in neat capital letters.

(i) What topics of your syllabus that are important from you examination point of view are not covered in the book ?

(ii) What are the chapters wherein the treatment of the subject matter is not systematic or organised or updated ?

(iii) Have you come across misprints/mistakes/factual inaccuracies in the book? Please specify the chapters and the page numbers.

(iv) What other books on the same subject have you found/heard better that the present book? Please specify in terms of price and quality.

(v) Further suggestions and comments for the improvement of the book:

(PS: If need be please attach a separate sheet to write your views.)

Other Details :

(i) Who recommended you the book ? (Please tick in the box near the option relevant to you.)

☐ Teachers ☐ Friends ☐ Bookseller

(ii) Name of the recommending teacher, his designation and address.

....

....

....

....

(iii) Name and address of the bookseller you purchased the book form :

....

....

....

....

(iv) Name and address of your institution (Please mention the University or Board, as the case may be)

....

....

....

....

(v) Your name and complete postal address :

....

....

....

....

(vi) Course you are enrolled in :

....

....

....

....

The best assessment will be awarded each month. The award will be in the form of our publications as decided by the Editorial Board.

Please mail the filled up coupon at your earliest to:

S.K. Kataria & Sons®

4885/109, Prakash Mahal, Dr. Subhash Bhargav Lane,
Opposite Delhi Medical Association, Daryaganj, New Delhi-110002
Phone: +91-11-23243489, +91-1123269324; Telefax: +91-11-23243489
e-mail: katariabooks@yahoo.com; katariabook@gmail.com
Website: www.skkatariaandsons.com